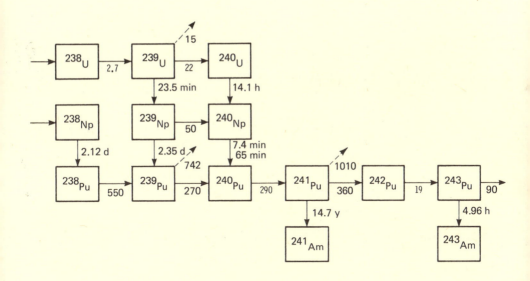

NUCLEAR
ENERGY
TECHNOLOGY

NUCLEAR ENERGY TECHNOLOGY

Theory and Practice of Commercial Nuclear Power

RONALD ALLEN KNIEF
Three Mile Island
Middletown, Pennsylvania

◉ HEMISPHERE PUBLISHING CORPORATION   *Washington*   *New York*   *London*

McGRAW–HILL BOOK COMPANY   *New York*   *St. Louis*   *San Francisco*
Auckland   *Bogotá*   *Hamburg*   *Johannesburg*   *London*   *Madrid*   *Mexico*
Montreal   *New Delhi*   *Panama*   *Paris*   *São Paulo*   *Singapore*
Sydney   *Tokyo*   *Toronto*

NUCLEAR ENERGY TECHNOLOGY:
Theory and Practice
of Commercial Nuclear Power

1 2 3 4 5 6 7 8 9 0   B C B C   8 9 8 7 6 5 4 3 2 1

This book was set in Press Roman
by Hemisphere Publishing Corporation.
The editors were Christine Flint and Elizabeth Dugger;
the designer was Sharon Martin DePass;
the production supervisor was Miriam Gonzalez;
and the typesetters were Sandra F. Watts and Wayne Hutchins.
BookCrafters, Inc., was printer and binder.

Library of Congress Cataloging in Publication Data

Knief, Ronald Allen, date.
Nuclear energy technology.

(McGraw-Hill series in nuclear engineering)
Bibliography: p.
Includes indexes.
1. Nuclear engineering. 2. Atomic power.
I. Title.
TK9145.K58   1981   621.48   80-25862
ISBN 0-07-035086-8

To my grandparents

CONTENTS

FOREWORD

This timely volume on nuclear energy technology is an outgrowth of Professor Ron Knief's popular course at The University of New Mexico. The course has been extremely well received as a clear and concise treatment of relevant subject matter, including such current topics as the accident at Three Mile Island.

I have followed Ron Knief's career with some interest since his graduate student days when he was assigned to the Safeguards R&D Group at Los Alamos in the summer of 1969. Upon completion of his Ph.D., he chose to take a position with Combustion Engineering in order to gain direct practical experience in the commercial nuclear industry. His eventual return to the Southwest as a faculty member of The University of New Mexico marked the beginning of his professional commitment and significant contributions to nuclear education and training that have continued ever since.

Ron Knief is well equipped to write a book of this scope and orientation. He has acquired firsthand knowledge of most of the major areas of commercial nuclear power. His industrial experience in reactor design, coupled with consulting activities in nuclear criticality safety and nuclear material safeguards, has led to professional interactions with a number of reactor sites and most of the nation's fuel cycle facilities. He is familiar with both sides of the regulatory process, having served as a consulting fuel-facility inspector for the U.S. Nuclear Regulatory Commission, as a licensed operator and chief reactor supervisor at The University of New Mexico's AGN-201M reactor facility, and presently as manager of training activities at General Public Utilities' Three Mile Island facilities.

The success of the subject matter of this book in addressing a diverse audience can be attributed in large measure to Ron Knief's extensive experience and effectiveness in teaching undergraduate, graduate, and continuing education courses. He also continues to be very active in public education and information programs on nuclear energy and energy-related issues.

Of particular importance in the wake of Three Mile Island and its worldwide impact are the timeliness and international orientation of this book with its emphasis on operational experience and current practice in today's nuclear industry and ongoing fuel cycle operations under existing political and technical constraints.

G. Robert Keepin
Los Alamos, New Mexico

PREFACE

This book is designed as an introductory text in nuclear engineering. It may be used for an overview course with a general audience of junior or senior science and engineering students or for more comprehensive coverage with senior and graduate nuclear engineering students. The book may also be used for continuing education of practicing engineers (as has worked particularly well throughout its evolution).[†]

Despite the underlying prerequisite of about two years of physical science or engineering (including some familiarity with differential equations), the verbal descriptions are generally complete enough to stand alone. A number of individuals with mathematical skills limited to basic algebra have demonstrated, in fact, that they can read, understand, and enjoy the vast majority of the manuscript by merely skipping over the equations.

The book is the culmination of over five years of effort, which began when I was an assistant professor in my first semester at The University of New Mexico. Based in part on my industrial experience with Combustion Engineering, Sandia Laboratories asked me to prepare and teach a special introductory course on commercial nuclear energy. Their goal was to provide education for "competent scientists and engineers who had little working knowledge of fuel cycle and reactor applications" (since Sandia's primary role in the nuclear weapons area was at that time being augmented by increased amounts of commercially oriented activities). It was recognized that an appropriate course could ease job and career transitions.

I began a course entitled "Nuclear Engineering Orientation" [NEO] at Sandia in February 1975. Although it was taught on Saturday mornings in a 3-hour block, 75 persons (including technicians, staff members, and managers with associate degrees through Ph.D.'s) enrolled. Minimum attendance was about 60. Text material consisted of copies of handwritten lecture outlines, as no single book or small group

---

[†]The instructor's manual includes suggestions for implementation of such multiple uses.

of books appeared adequate to describe both the unique aspects of nuclear theory and the current practices.

The 16-week NEO course was videotaped. The tapes and lecture notes were then used on a proctored basis at Sandia from 1975 through 1977 and at Los Alamos Scientific Laboratory in 1976. They were also used on a self-study basis by a number of students at The University of New Mexico, as well as by staff members at Sandia, Los Alamos, and the Lovelace Inhalation Toxicology Research Institute.

The NEO material was updated selectively over time. However, when a complete revision was deemed appropriate in spring 1978, the "Nuclear Energy Technology" [NET] course was born. This time the text material was written in rough prose form. Videotaping again allowed proctored use by Sandia and self-study by The University of New Mexico students and nearby professionals. At their request, the Nuclear Regulatory Commission [NRC] was also provided with the course materials for use by staff members (many in the Office of Safeguards) needing an overview of commercial nuclear power.

The very warm response to NET by participants of both the in-class and self-study activities suggested that formal publication was viable. With the encouragement and support of Hemisphere and McGraw-Hill, I prepared the final version of the manuscript during the summer and fall of 1979. It was a thorough revision of the earlier NET course material, expanded for use as a stand-alone textbook. Testing in two fall courses—one for senior and first-year graduate nuclear engineering students, the other for junior and senior (nonnuclear) engineering students—demonstrated that the material was easy to use and flexible enough to meet the needs of different audiences. During the spring semester of 1980, Chaps. 16-18 and 20 also served as the starting point for a new course on the methods of reactor safety and safeguards.

The book is organized into six parts, each with certain goals in mind. The first part provides a general overview of the commercial nuclear fuel cycle and power reactors so that the reader will have some perspective before attacking the second, or theory, part of the book. This second part emphasizes the theory that is important to understanding the unique aspects of nuclear energy, including use of several simple calculational methods. The following two parts then return to the nuclear fuel cycle and power reactors, respectively, for an in-depth look at their structure and functions. The fifth part considers the increasingly important subject of reactor safety (including the accident at Three Mile Island, federal regulations, and safeguards). The main body of the book ends with a discussion of nuclear fusion and its prospects as a commercial energy source in the next century. Appendixes provide useful data as well as a discourse on the role of nuclear power in averting a serious energy crisis.

The book is designed to be used in an iterative manner with the recognition that theory makes more sense when related to practice, and vice versa. Thus, theoretical concepts are reintroduced frequently as they relate to specific applications.

The numerical problems at the end of each chapter are designed to provide additional practical insights into topics covered in that chapter. They allow the user to test calculational skills, while demonstrating principles that may not be spelled out specifically in the text. The selected bibliographies provide an extensive reading list for the interested reader.

I have included instructional objectives at the head of each chapter to provide minimum-performance standards that the reader can use to gauge his or her own comprehension. The objectives also serve to highlight those concepts that I consider to be most important and that are most useful later in the book. The exercises—questions and numerical problems—are also tied closely to the objectives.

It is appropriate to acknowledge a number of persons for their varying contributions to my ultimate emergence as a published author. Were it not for the influence of several dedicated teachers (whom I have also considered to be good friends)—my grandmother Xarifa Ross Ryder, my uncle Glenn Pinkham, junior high math teacher Richard Dick, undergraduate advisor Charles Ricker, and graduate advisors Bernie Wehring and Marv Wyman—I might never have entered the professorial ranks.

The basic premise of the book evolved from the successful NEO and NET course offerings at Sandia Labs. Identification of educational objectives and preparation of text material proceeded smoothly with the help of the course supervisors—Peter McGrath, Dick Coates, Bob Jefferson, Gene Ives, and Jon Reuscher—and the education and training specialists—Don Hosterman and Kathy Pitts.

The publishers' final reviewers, Jack Courtney and Charles Bonilla, provided insightful comments and suggestions, which added greatly to the quality of the final manuscript. Other valuable guidance came from those who reviewed one or a few chapters in their areas of expertise: Gary Cooper and Dave Woodall at The University of New Mexico; Augie Binder, Dick Lynch, Frank Martin, Doug McGovern, Leo Scully, and Jim Todd at Sandia; Bob Keepin at Los Alamos Scientific Laboratory; Bob Long at GPU; and Bob Erickson at the NRC. Of no less importance are the 200 or so students at The University of New Mexico, Sandia, and Los Alamos, who proof-tested the material at various stages of its development (and who will always be able to say, "Let me tell you how hard he worked us while he was writing this book...."). Of special note are the contributions of former students and friends Mark Hoover, Ken Boldt, Ted Luera, Jim Morel, Donnie Cutchins, and Larry Sanchez. Other important, even if unidentified, individuals include those who responded to often frantic phone calls requesting an explanation, data, a report, a figure, or a photograph for use in the book.

I would like to thank my parents, sister, brother, other family members, and friends for their long-term encouragement and willingness to put up with my eccentricities, especially during the "final push." I am grateful to all who helped maintain my morale and sense of equilibrium.

Very significant acknowledgment belongs to Roberta Benecke who typed the final manuscript and translated what I wrote into what I meant to say. Christine Flint and other Hemisphere and McGraw-Hill folks did a super job in the final production of this book.

Ronald Allen Knief

I

OVERVIEW

Goals

1. To introduce the basic concepts of both the nuclear fuel cycle and the world's five major nuclear power reactor systems
2. To provide a context for a better understanding of the theoretical concepts presented in Part II

Chapter in Part I

Introduction

1

INTRODUCTION

Objectives

After studying this chapter, the reader should be able to:

1. Explain the two major advantages and the two major disadvantages of nuclear fission as an energy source.
2. Describe the basic components of the commercial nuclear fuel cycle currently employed in the United States.
3. Identify the steps in a closed fuel cycle not found in an open one.
4. Explain the five characteristics upon which power reactors may be classified.
5. Identify the world's five major reactor types and describe the steam cycle and general fuel characteristics for each.
6. Perform basic calculations of energy equivalents and material flows.

The current basis for commercial application of nuclear energy is the *fission process*. Figure 1-1 shows a neutron striking an atom of uranium-235 [^{235}U] to produce a fission, or splitting of that atom. From the standpoint of energy production, the reaction has the major advantage that each such splitting provides nearly one hundred million [100,000,000] times as much energy as the "burning" of one carbon atom in a fossil fuel. The production of more neutrons from fission allows the process to participate in a *chain reaction* for continuous energy production in a device called a *reactor*. A material that can produce a self-sustaining chain reaction by itself is said to be *fissile*. *Fissionable* and *fertile* materials can contribute to a chain reaction without being able to sustain one by themselves. When the reaction is exactly balanced in a steady-state condition, the system is said to be *critical*.

One major disadvantage of using the process as an energy source is the generation of *radiation* at the time of fission. Another problem is the presence of the fission

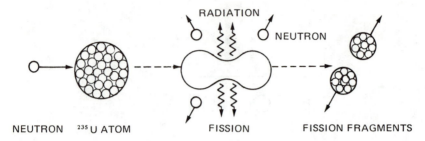

FIGURE 1-1
Fission of uranium-235 by a neutron.

fragments, which are *radioactive* and will themselves give off radiation for varying periods of time after the fission events.

These characteristics each have major impacts on the design and operation of nuclear fission systems. The six chapters in the second part of this book treat the basic theories and principles that contribute to the ultimate utilization of fission energy. The remaining parts then build on this framework to provide descriptions of the nuclear fuel cycle, design and operation of nuclear fission reactors, and administrative aspects of nuclear energy. The final part of the book considers *nuclear fusion*, which has long-term prospects as a commercial energy source.

Since theory and practice interact thoroughly, an overview of the current development of commercial nuclear power can aid the understanding of the basic underlying principles. Thus, the remainder of this chapter provides a brief overview of the nuclear fuel cycle and current reactor designs. The reader should note that only basic understanding, and not thorough knowledge, is expected at this stage, since each and every definition and concept is clarified and treated in greater detail in later chapters.

NUCLEAR FUEL CYCLES

The production of energy from any of the current fuel materials is based on a *fuel cycle*. Typical cycles, such as those for the fossil fuels, consist of at least the following components:

- exploration to identify the compositions and amounts of a resource available at various locations
- mining or drilling to bring the resource to the earth's surface in a usable form
- processing or refining to convert raw materials into a final product
- consumption of the fuel for energy production
- disposal of wastes generated in all portions of the cycle
- transportation of materials between the various steps of the cycle

The nuclear fuel cycle is substantially more complicated for the following reasons:

1. ^{235}U, which is the only practical naturally occurring fissile material, is less than 1 percent abundant in all known uranium deposits (the remaining uranium is fissionable ^{238}U).

2. Two other fissile materials, ^{233}U and ^{239}Pu [plutonium-239], are produced by neutron bombardment of ^{232}Th [thorium-232] and ^{238}U, respectively. (For this reason, the latter two materials are said to be fertile.)
3. All fuel cycle materials contain small to large amounts of radioactive constituents.
4. A neutron chain reaction [criticality] could occur outside a reactor under appropriate conditions.
5. The same chain reaction that can be used for commercial power generation also has potential application to a nuclear explosive device.

Each of these five concerns is considered in the following paragraphs and later in the book as related to the structure of the nuclear fuel cycle.

Uranium Fuel Cycle

A schematic representation of the uranium fuel cycle is shown in Fig. 1-2. This cycle is employed for the light-water reactor [LWR] systems that currently dominate worldwide nuclear power. The solid arrows in Fig. 1-2 connect components of the presently "open" cycle that exists in the United States. The dashed arrows show pathways that would complete, or "close," the uranium fuel cycle.

One inherent step of the fuel cycle that is not named explicitly on Fig. 1-2 is *recycle* of plutonium and uranium. *Transportation* between the various steps is indicated by the arrows. Waste disposal from operations other than reprocessing is not shown explicitly.

Nuclear safety, which is charged with protecting operating personnel and the public from potentially hazardous materials in the fuel cycle, must be superimposed on appropriate portions of the cycle. Also superimposed are *material safeguards* to preclude use of fuel cycle materials for nuclear explosives.

The steps preceding reactor use, which generally have little radioactivity, are often considered to form the *front end* of the fuel cycle. Those that follow reactor use are characterized by high radiation levels and are said to be part of the *back end* of the cycle.

Exploration

The exploration process typically begins with geologic evaluation to identify potential uranium deposits. Areas that have characteristics similar to those of known content usually receive first consideration. The actual presence of uranium may be verified by chemical and/or radiological testing.

Drilling into the deposit accompanied by detailed analysis of the samples provides information on uranium ore composition and location. Only after completion of a very detailed mapping of the ore body will mining operations begin.

Mining

Uranium mining in the United States is divided approximately equally between surface [open pit] and underground operations. Most uranium mining in the United States is concentrated in the West, with about one-half of current production coming from New Mexico. Major worldwide resources are located in Australia, Canada, South Africa, and the U.S.S.R.

Typical U.S. uranium ore deposits assay at 0.20-0.25 wt %* [weight percent]

---

*Weight percent is also sometimes abbreviated as w/o.

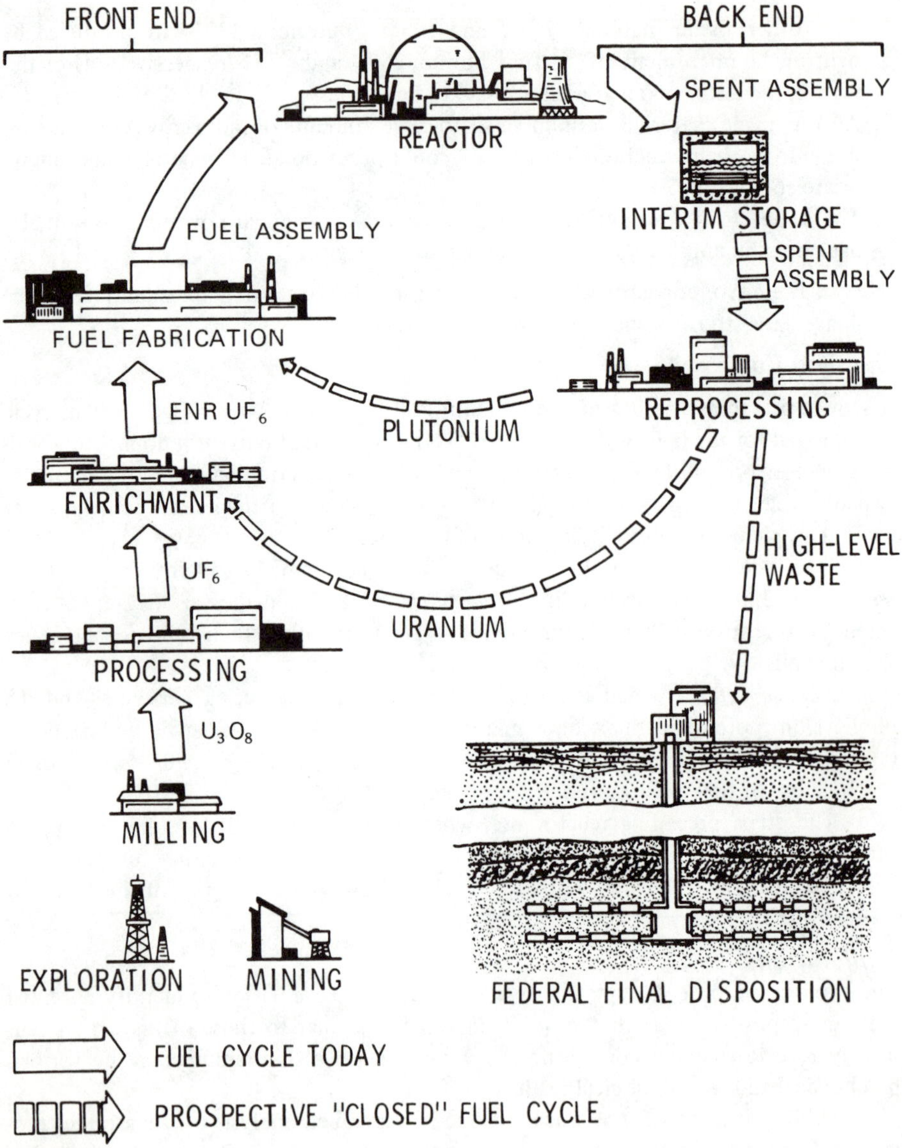

FIGURE 1-2
Uranium fuel cycle for light-water reactors in the United States. (Adapted courtesy of U.S. Department of Energy.)

uranium metal. Despite the low fractional content, uranium ore is 30–50 times more efficient than coal on the basis of energy per ton mined. Since many environmental impacts are proportional to the amount of ore removed, clear advantages for nuclear energy may accrue here.

Milling

One type of milling operation removes uranium from the ore by employing the following steps:

- crushing and grinding of ore to optimum size
- leaching in acid to dissolve the metals away from predominantly nonmetal ore content
- ion-exchange or solvent-extraction operations to separate uranium from other metals
- production of U_3O_8, usually in the form of *yellow cake*, so named because of its color

The major problems associated with milling operations are related to chemical effluent releases and some radioactivity in the ore residues [*tailings*].

Conversion and Enrichment

Natural uranium is composed of two isotopes—fissile ^{235}U (0.711 wt %) and fissionable ^{238}U (99.3 wt %)—which cannot be separated by chemical means. Since many reactor concepts require that the ^{235}U fraction of the total uranium content be higher than this, *enrichment*—separation of the isotopes by physical means—has been implemented.

The conversion step begins by purifying the U_3O_8 [yellow cake]. Then, through chemical reaction with fluorine, uranium hexafluoride [UF_6] is produced.

UF_6—a gas at temperatures above 56°C [134°F] at atmospheric pressure—is readily employed in one of several enrichment schemes. The gaseous diffusion method that has been the world's "workhorse" is based on forcing UF_6 against a porous barrier. The lighter $^{235}UF_6$ molecules penetrate the barrier more readily than do the heavier $^{238}UF_6$ molecules. (According to the kinetic theory of gases, each molecule has the same average kinetic energy, so that greater speed, and thus barrier penetration probability, belongs to the lighter molecule.) By cascading the barrier stages, any desired enrichment can be obtained. At the present time, *slightly enriched uranium* at 2-4 wt % ^{235}U is produced for LWR use. The uranium left behind in the process is called the depleted stream [*enrichment tails*] and is typically 0.2-0.35 wt % ^{235}U.

The gaseous diffusion process is the largest energy-consuming step of the LWR fuel cycle, using about 95 percent of the total energy. However, this is no more than 4 percent of the energy that may be extracted from the nuclear fuel. Development of gas-centrifuge, laser, or other isotope-separation technologies can be expected to reduce these energy requirements at some future date.

Fabrication

The fabrication step of the cycle produces fuel in the final form that is used for power production in the reactor. LWR fabrication begins by converting the slightly enriched uranium hexafluoride to uranium dioxide [UO_2]—a black ceramic composition. The UO_2 powder is then formed into cylindrical *pellets* roughly the size of a thimble.

The pellets are loaded into long *cladding tubes* to form individual *fuel pins*. The final *fuel assembly* consists of an array of fuel pins plus some other hardware. Fuel assemblies for the light-water and other reactor systems are described in more detail later in this chapter.

Reactor Use

The completed fuel assemblies are loaded into the *reactor core*, where the fission chain reaction is initiated to generate heat energy. As fissions occur, the ^{235}U atoms are consumed. An amount of ^{239}Pu is produced as ^{238}U absorbs some of the extra neutrons. The buildup of fission fragments and their radioactive products tends to produce a "poisoning" effect by absorbing neutrons that could otherwise participate in the chain reaction. Since the loss of ^{235}U and the poison effect are dominant, the fuel must eventually be replaced as it becomes unable to sustain a chain reaction.

Current practice is to replace one-quarter to one-third of the fuel assemblies in the reactor core on a roughly annual cycle. By using careful *fuel management*, fuel assemblies are shuffled to maximize the energy extracted from each during its 3–4 years in the reactor.

Interim Spent Fuel Storage

Since the fuel assemblies are very highly radioactive when they are discharged from the reactor, they are allowed to "cool" for a period of time in a water basin. The fuel cycle that exists in the United States today calls for spent fuel to be stored at the reactor site or in a special off-site facility. Current federal policy has not established the actual mechanisms or locations of any later, more permanent steps in the nuclear fuel cycle (Fig. 1-2).

Reprocessing

If spent fuel processing [*reprocessing*] were to be implemented, the residual uranium and the plutonium could be extracted for further use in the fuel cycle. The fission-product and other wastes produced would be handled in the waste disposal step.

In the initial steps of the reprocessing operations, the fuel assemblies are mechanically disassemblied (i.e., chopped into small pieces) and dissolved in acid. The uranium and plutonium are separated from the wastes, then separated from each other. The large amounts of highly radioactive byproducts contained in the spent fuel necessitate very stringent environmental controls for the processing steps and the storage of wastes.

Recycle

The residual uranium and the plutonium extracted from the spent fuel by the reprocessing operation may be reintroduced into the fuel cycle. Use of these recycled materials can reduce uranium resource requirements by up to 25 percent.

The residual uranium is returned to the fuel cycle for reenrichment, as shown in Fig. 1-2. The plutonium is transported to the fabrication operation where it is mixed with natural or depleted uranium to produce a *mixed oxide* [$PuO_2 + UO_2$] with a fissile content [effective enrichment] roughly comparable to that of slightly enriched uranium.

Waste Disposal

All steps of the fuel cycle (including the waste disposal step itself) produce some amounts of radioactive waste. Near-surface burial of the "low-level" wastes from the front end of the fuel cycle is generally appropriate.

Reactor use and reprocessing produce "high-level" wastes. Current federal policy calls for storage of spent fuel assemblies that may be reprocessed or disposed of directly at some indeterminant future time.

Waste from reprocessed fuel would likely be stored as a liquid for approximately 5 years, solidified and stored on-site for an additional period of approximately 5 years, and then transferred to a federal repository for final disposal. Final disposal is likely to be in a stable geologic formation, in the seabed, or in outer space.

Transportation

Since the various fuel cycle operations take place at a number of different locations, transportation is a very important component. The design and operation of effective transportation systems should minimize the risks of:

- release of dangerous chemical or radioactive materials to the environment
- accidental nuclear chain reaction outside of a reactor core
- damage to expensive components
- theft of valuable and potentially dangerous materials

Based on the nature of these risks, specially designed containers and/or vehicles may be used between various steps of the fuel cycle.

Nuclear Safety

Nuclear safety in fuel cycle facilities is usually divided into categories of *radiation safety* and *nuclear criticality safety*. The former includes shielding and containment of radiation sources plus effluent control to minimize exposures to operating personnel and the general public.

Reactors are designed to handle the effects of a fission chain reaction while fuel cycle facilities generally are not so designed. Thus, nuclear criticality safety is charged with prevention of such chain reactions in all environments outside of reactor cores. Since accidental criticality is not credible for natural uranium, these safety concerns begin at the enrichment step (Fig. 1-2).

Material Safeguards

All fissile materials have potential use for nuclear explosives and must, therefore, be safeguarded against theft or diversion. Physical-security and material-accountancy systems are designed to minimize the *terrorist threat* for theft by a *subnational* group. International safeguards based on inventory verification have been developed to deter *proliferation*, i.e., diversion by a *nation* for the purpose of acquiring nuclear weapons capability.

Safeguard measures should be commensurate with the risks perceived for given materials. The slightly enriched uranium in the LWR fuel cycle, for example, could only be used for a nuclear explosive if it were enriched further. The extreme complexity of the enrichment technology would seem to make implementation of the required clandestine operations highly unlikely.

Since spent fuel contains fissile plutonium that can be separated chemically, it is a somewhat more attractive target. Only a national effort, however, would seem to be able to handle the complexity and hazard (as well as detectability) of reprocessing operations.

By contrast, recycle with the presence of separated plutonium would appear to offer the best theft target for the terrorist or other subnational groups. Material safeguard measures, therefore, should be most stringent for this portion of the fuel cycle.

Other Fuel Cycles

Reactor concepts other than the LWR require fuel cycles that have many similarities to the cycle just considered. Figure 1-3 is a schematic diagram of a generic fuel cycle that encompasses most options.

The greatest differences occur for systems that use thorium. Since the main constituent is ^{232}Th, present for production of fissile ^{233}U, the conversion and enrichment steps are not required for thorium. The reprocessing step, of course, must be capable of separating ^{233}U from ^{232}Th and the wastes. As described later in this chapter, there is a unique fuel assembly design for the high-temperature gas-cooled reactor [HTGR], which allows ^{233}U and ^{235}U to be separated and recycled directly without requiring enrichment technology.

Uranium enrichment requirements vary from use of unenriched, natural uranium

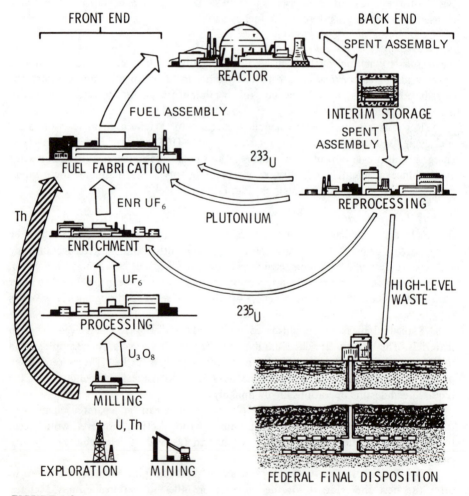

FIGURE 1-3
Generic nuclear fuel cycle material flow paths.

in the pressurized-heavy-water reactor [PHWR] to 93 wt % ^{235}U in the HTGR. The liquid-metal fast-breeder reactor [LMFBR] uses depleted uranium—i.e., enrichment tails—plus plutonium as its fuel material.

Other possible nuclear fuel cycles include the "symbiotic" or "cross-progeny" cycles based on interchange of various fuel materials among two or more different reactor types. One such possibility is the exchange of plutonium between LWR and LMFBR systems.

Material safeguards are required for all separated plutonium and for separated uranium whenever it has greater than 20 wt % enrichment in ^{233}U and/or ^{235}U. The spent fuel usually may be protected at a somewhat lower level because reprocessing would be required to obtain the fissile content.

NUCLEAR POWER REACTORS

All nuclear reactors are designed and operated to achieve a self-sustained neutron chain reaction in some combination of fissile, fissionable, and other materials. The power reactors use the fission process for the primary purpose of producing usable energy in the form of electricity.

Common characteristics of power reactors, which are used for classification purposes, include:

1. Coolant—primary heat extraction medium
2. Steam cycle—the total number of separate coolant "loops," including secondary heat transfer systems (if any)
3. Moderator—material (if any) used specifically to "slow down" the neutrons produced by fission
4. Neutron energy—general energy range for the neutrons that produce most of the fissions
5. Fuel production—system is referred to as a *breeder* if it produces more fuel (e.g., fissile ^{239}Pu from fertile ^{238}U) than it consumes; it is a *converter* otherwise

The first two features relate to the current practice of converting fission energy into electrical energy by employing a steam cycle.

Neutrons from fission are emitted at high energies. However, neutrons at very low energies have a higher likelihood of producing additional fissions. Thus, many systems employ a moderator material to "slow down" the fission neutrons. Neutrons with very low energies (roughly in equilibrium with the thermal motion of surrounding materials) are called *thermal* neutrons, with the slowing down process sometimes called *thermalization*. Neutrons at or near fission energies are *fast* neutrons.

Any reactor that contains fertile materials will produce some amount of new fuel. The major distinction between breeder and converter reactors is that the former is designed to produce more fuel than is used to sustain the fission chain reaction. By contrast, the converter replaces only a fraction of its fissile content.

Five basic nuclear power reactor designs are currently employed in the world. These are identified in Table 1-1 and classified on the basis of the reactor characteristics noted above. (A substantially expanded version of the table is contained in App. IV.) The remainder of this chapter considers each of the reference designs in some detail.

TABLE 1-1
Basic Features of Five Major Reactor Types

Feature	Boiling-water reactor [BWR]	Pressurized-water reactor [PWR]	Pressurized-heavy-water reactor [PHWR]	High-temperature gas-cooled reactor [HTGR]	Liquid-metal fast-breeder reactor [LMFBR]
Steam cycle/ coolant(s)					
Number of loops	1	2	2	2	3
Primary coolant	Water	Water	Heavy water	Helium	Liquid sodium
Secondary coolant(s)	–	Water	Water	Water	Liquid sodium/ water
Moderator	Water	Water	Heavy water	Graphite	–
Neutron energy	Thermal	Thermal	Thermal	Thermal	Fast
Fuel production	Converter	Converter	Converter	Converter	Breeder

Steam Cycles

Most of the world's electric power is generated via a steam cycle. Water in a *boiler* is heated to produce steam by burning fossil fuel. The steam then turns a *turbine-generator* set to produce electricity.

Nuclear steam cycles have many of the same features as the fossil-fuel case. One major conceptual difference is that the fission-energy heat source is a fuel *core* located physically inside the boiler, or *reactor pressure vessel*. In one reactor design, steam is produced directly in the core; in the others, heat is transferred from the core to generate steam in a secondary system.

The boiling-water reactor [BWR] steam cycle employs a single loop as shown in Fig. 1-4. In this direct-cycle system, the water coolant flows through the fuel core and acquires an amount of energy sufficient to produce boiling, and thus steam, within the reactor vessel. In this and most other conventional steam cycles, *condenser cooling* water is used to condense steam in the turbine and enhance net conversion efficiency.

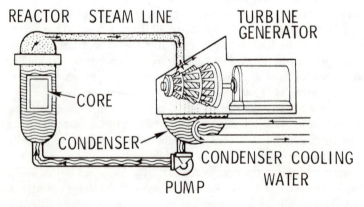

FIGURE 1-4
Direct, single-loop steam cycle. (Adapted courtesy of U.S. Department of Energy.)

The indirect-cycle reactors maintain high-pressure conditions to prevent boiling in the vessel. Instead, the heat acquired from the core by the coolant is carried to a *heat exchanger.* Three of the reactor concepts—pressurized-water reactor [PWR], pressurized-heavy-water reactor [PHWR], and high-temperature gas-cooled reactor [HTGR] —employ a two-loop steam cycle, as shown in Fig. 1-5. As the name suggests, the *steam generator* is a heat exchanger that produces steam for the turbine-generator. It may be noted that steam generators in these systems play the same role of heat source as do the fossil-fuel boiler and BWR vessel.

As the name implies, the pressurized-water reactor relies on high pressure to maintain water in a liquid form within its primary loop. Despite the steam cycle difference between the PWR and BWR, they have many design similarities resulting from the use of ordinary water as their coolant. The two are, therefore, grouped together as the *light-water reactors* [LWR] (which were mentioned in connection with the nuclear fuel cycle discussions earlier in this chapter).

The pressurized-heavy-water reactor [PHWR] uses heavy water as coolant in a cycle that is otherwise similar to that of the PWR. The high-temperature gas-cooled reactor [HTGR] employs helium gas as its primary coolant.

The liquid-metal fast-breeder reactor [LMFBR] is based on a three-loop steam cycle, as shown in Fig. 1-6. Primary and secondary liquid-sodium loops are connected by an intermediate heat exchanger. The secondary sodium loop transfers energy to the steam generator.

The steam cycles are just one aspect of integrated reactor system concepts. The remaining sections of this chapter provide some insight into the selection of coolants for the reference reactors. More detailed information on the subject is postponed until Chaps. 11–15.

Moderators

With the exception of the LMFBR, the remaining four reactors use moderator material to reduce fission-neutron energies to the thermal range. Light elements are found to be the most effective moderators, as described more fully in Chap. 4.

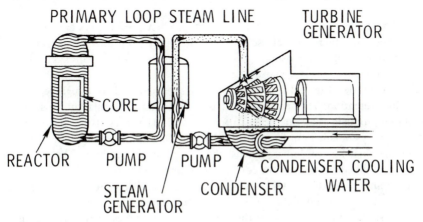

FIGURE 1-5
Two-loop steam cycle. (Adapted courtesy of U.S. Department of Energy.)

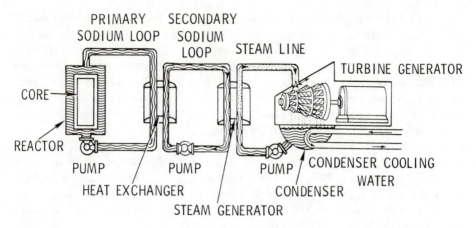

FIGURE 1-6
Three-loop steam cycle. (Adapted courtesy of U.S. Department of Energy.)

In the two LWR types, the coolant water also serves as the moderator. The PHWR uses separate supplies of heavy water for the coolant and the moderator. Carbon in the form of graphite serves as the moderator for the HTGR.

The LMFBR concept is based on a chain reaction with fast neutrons. Thus, the relatively heavy liquid-sodium coolant was purposely selected to minimize moderation effects.

Reactor Fuel

The designs for the fuel used in the reference reactors are as varied as the steam cycles and moderators. In one system the fuel and moderator form an integral unit, while in the others these constituents are separated. The features of *fuel assemblies*—the units ultimately loaded into the reactor vessel—for each system are summarized in Table 1-2. (Appendix IV contains more detailed information, including representative dimensions.)

Since the fission process creates radioactive products, reactor systems must be designed to minimize the risk of release of these potentially hazardous materials to the general environment. The philosophy of *multiple barrier containment* has evolved from this requirement.

As a first barrier, the fuel is formed into particles designed for a high degree of fission product retention. The second barrier is typically an encapsulation capable of holding those products that do escape from the fuel. The integrity of the reactor vessel and primary-coolant loop form a third barrier. One or more containment structures form the final line of defense against the release of radioactivity. The last two barriers that are outside of the reactor core are considered in Chap. 17.

Since uranium dioxide [UO_2] and uranium carbide [UC] are relatively dense ceramic materials with good ability to retain fission products, they are favored first-barrier compositions. Encapsulation of the particles in metal tubes or in other coatings provides the second barrier. These principles have major impact on the fuel-assembly designs for the reference reactors.

TABLE 1-2
Typical Characteristics of the Fuel Cores of Five Major Reactor Types[†]

Component	Boiling-water reactor [BWR]	Pressurized-water reactor [PWR]	Pressurized-heavy-water reactor [PHWR]	High-temperature gas-cooled reactor [HTGR] [‡]	Liquid-metal fast-breeder reactor [LMFBR]
Fuel particle(s)					
Geometry	Short, cylindrical pellet	Short, cylindrical pellet	Short, cylindrical pellet	Multiply coated microspheres	Short, cylindrical pellet
Chemical form	UO_2	UO_2	UO_2	UC/ThC	Mixed oxides UO_2 and PuO_2
Fissile	2–4 wt % ^{235}U	2–4 wt % ^{235}U	Natural uranium	93 wt% ^{235}U microsphere	10–20 wt% Pu
Fertile	^{238}U	^{238}U	^{238}U	Th microsphere	^{238}U in depleted U
Fuel pins	Pellet stacks in long Zr-alloy cladding tubes	Pellet stacks in long Zr-alloy cladding tubes	Pellet stacks in short Zr-alloy cladding tubes	Microsphere mixture in short graphite fuel stick	Pellet stacks in medium-length stainless steel cladding tubes
Fuel assembly	8 × 8 square array of fuel pins	16 × 16 or 17 × 17 square array of fuel pins	37-pin cylindrical arrangement	Hexagonal graphite block with stacked fuel sticks	Hexagonal array of 271 fuel pins
Reactor core [§]					
Axis	Vertical	Vertical	Horizontal	Vertical	Vertical
Number of fuel assemblies along axis	1	1	12	8	1
Number of fuel assemblies in radial array	748	193–241	380	493	364 driver, 233 blanket

[†]More detailed data and references are contained in App. IV.

[‡]The HTGR fuel geometry is different from that of the other reactors, leading to some slightly awkward classifications.

[§]All of the cores approximate right circular cylinders. Fuel assemblies are loaded and/or stacked lengthwise parallel to the axis of the cylinder.

Light-Water Reactors

The fuel assemblies for the two types of light-water reactor [LWR] are very similar. Slightly enriched uranium dioxide is fabricated into the form of cylindrical fuel *pellets*. The pellets are then loaded into long zirconium-alloy *cladding* tubes to produce *fuel pins* or *fuel rods*. A rectangular array of the pins forms the final *fuel assembly*, or *fuel bundle*.

The fuel assembly for the boiling-water reactor [BWR] is shown in Fig. 1-7. The individual fuel pins consist of the clad tube, the fuel pellet stack or "active" fuel region, a retention spring, and welded end caps.

Upper and lower tie plates plus interim spacers secure the fuel pins into a square array with eight pins on a side. The fuel channel encloses the fuel pin array, so that coolant entering at the bottom of the assembly will remain within this boundary as it flows upward between the fuel pins and removes the fission energy.

The fuel assembly for the pressurized-water reactor [PWR] is shown in Fig. 1-8. The fuel pins are quite similar to those for the BWR.

The PWR pin array, typically 16×16 or 17×17, is larger than that considered previously. The array is not enclosed by a fuel channel since the behavior of the non-boiling coolant is much more predictable than that of the BWR.

Pressurized-Heavy-Water Reactor

The fuel assembly shown in Fig. 1-9 is that of the pressurized-heavy-water reactor [PHWR] known as *CANDU-PHW* [Canada Deuterium Uranium-Pressurized Heavy Water]. PHWR and CANDU are often used interchangeably.

The CANDU fuel bundles are designed for insertion into pressure tubes through

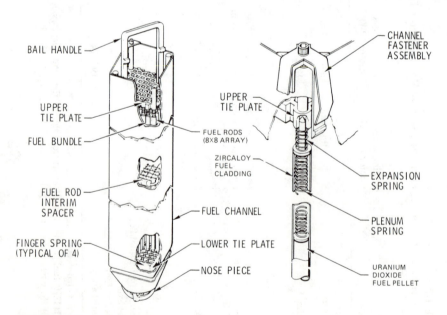

FIGURE 1-7

Fuel assembly for a typical boiling-water reactor. (Adapted courtesy of General Electric Company.)

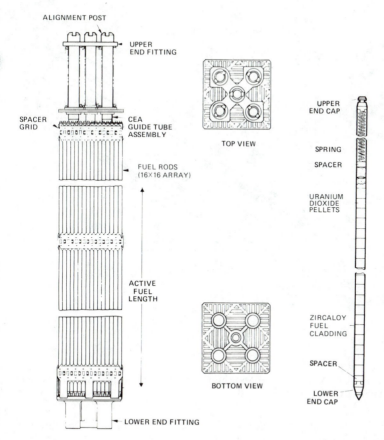

FIGURE 1-8

Fuel assembly for a typical pressurized-water reactor. (Adapted courtesy of Combustion Engineering, Inc.)

which the primary coolant flows. These tubes penetrate a large vessel, which contains the separate heavy-water moderator. (The specific design is treated in some detail in Chap. 13.)

The fuel pins consist of natural uranium in UO_2 pellets clad in zirconium alloy. Since these short pins do not have to be free-standing (as is the case for the LWR's), the clad is quite thin. The interelement spacers serve to separate the pins from each other, while the bearing pads separate the bundle from the pressure tube.

High-Temperature Gas-Cooled Reactor

The high-temperature gas-cooled reactor [HTGR] developed in the United States is a recent representative of a wide range of gas-cooled, graphite-moderated reactors around the world. One other state-of-the-art system, the pebble-bed reactor, is considered in Chap. 13.

The HTGR has a unique fuel design, as shown in Fig. 1-10. The basic units are tiny particles of uranium carbide or of thorium carbide surrounded by various coat-

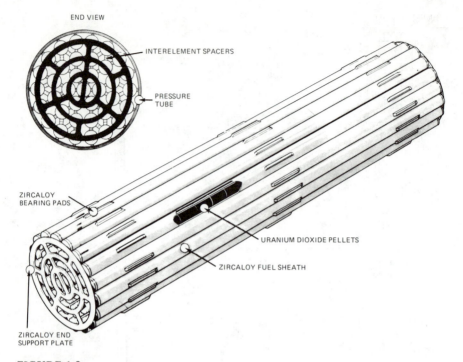

FIGURE 1-9
Fuel assembly for a typical CANDU pressurized-heavy-water reactor. (Adapted courtesy of Atomic Energy of Canada Limited.)

ings. Separate fissile and fertile particles, or *microspheres*, are used to facilitate the reprocessing separations described in Chap. 10.

As originally conceived, the fissile microsphere contains a small (approximately 0.2 mm) core of highly enriched ^{235}U in the form of uranium carbide. Since it employs three different types of coatings designed for fission produce retention, it carries the name *TRISO*.

The fertile microsphere has a core of thorium carbide. It is called the *BISO* because it is only doubly coated. Other types of microspheres may also be used, especially when fissile ^{233}U, produced from fertile ^{232}Th, becomes available.

The microspheres may be mixed to provide any desired effective fissile "enrichment." A composition of 5 wt % ^{235}U and 95 wt % ^{232}Th would be typical for a new HTGR. A graphite resin binder is used to form the microsphere mixture into small (roughly finger-sized) fuel rods.

The fuel rods are loaded into holes in a hexagonal graphite block to form the fuel assembly shown in Fig. 1-10. The block also contains holes for helium coolant flow. It should be noted that in the HTGR design the fuel and moderator are contained in an integral unit.

Liquid-Metal Fast-Breeder Reactor

A typical fuel assembly for a liquid-metal fast-breeder reactor [LMFBR] is shown in Fig. 1-11. Although the fuel pins have the same basic features as those for the water reactors, they have a much smaller diameter, are more closely spaced, and contain two different fuel pellet types.

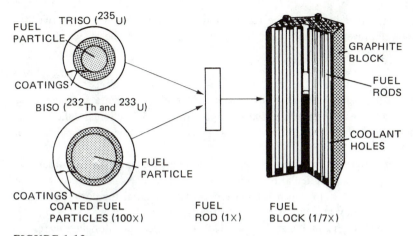

FIGURE 1-10
Fuel assembly for a typical high-temperature gas-cooled reactor. (Adapted courtesy of General Atomic Company.)

The primary fissile composition for LMFBR fuel is plutonium. The fertile composition is ^{238}U in the form of depleted uranium, the byproduct of the enrichment process. Some pellets contain a PuO_2–UO_2 combination called *mixed oxide*. Typical compositions have 10–20 wt % Pu with depleted uranium accounting for the remainder. Other pellets contain only depleted uranium.

Basic *driver* fuel pins contain a central stack of mixed-oxide pellets with stacks of depleted-uranium *blanket* pellets on both ends. Blanket pins consist entirely of depleted-uranium pellets. In both cases, the blanket material is employed to enhance breeding by absorbing neutrons that would otherwise escape from the system.

The LMFBR driver assembly shown in Fig. 1-11 has a fuel channel that encloses the hexagonal fuel-pin array. Liquid sodium enters at the bottom of the as-

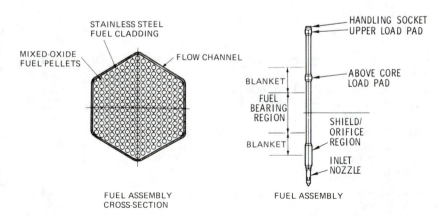

FIGURE 1-11
Fuel assembly for the Fast Flux Test Facility characteristic for a liquid-metal fast-breeder reactor. (Courtesy of U.S. Department of Energy.)

sembly, is distributed in the orifice region, and then flows through the active fuel region. The wall prevents the mixing of flow from adjacent assemblies.

Reactor Cores

Fuel assemblies loaded into the reactor vessel form the *core* wherein the fission process produces heat energy. Core configurations for each reactor are summarized in Table 1-2. In all cases the fuel is loaded to approximate a right circular cylinder with the fuel assemblies placed parallel to the axis. Thus, coolant flow is always *axial* (i.e., parallel to the axis).

The axis is vertical for four of the reactors and horizontal for the other. For each system a plane perpendicular to the axis defines a *radial* cross section of the core. (Examination of the fuel-management patterns in Chap. 7 and the core drawings in Chaps. 12–15 may aid visualization of the situation.)

In the LWR's, the fuel bundles are held in a vertical position in the reactor vessel. A typical BWR has 748 assemblies, while the PWR has between 193 and 241 larger assemblies.

Current CANDU designs call for 12 bundles end-to-end in each of the horizontal pressure tubes. An array of 380 such tubes makes up the reactor core.

The HTGR is the only reference system in which the fuel assemblies are stacked vertically. A typical core has 8 fuel blocks along the axis and 493 radially.

The LMFBR core contains two basic fuel assembly types. Assemblies of driver fuel pins constitute most of the central region of the core. The depleted uranium at their top and bottom forms an *axial blanket*. The assemblies made from blanket fuel pins are loaded around the outside of the central region to form the *radial blanket*. Through the combination of the axial blankets with the radial blanket, the central mixed-oxide core is essentially surrounded with depleted uranium for ^{239}Pu breeding.

One current LMFBR design calls for 364 driver assemblies and 233 external blanket assemblies. Although most of the blanket assemblies are placed around the core periphery, some are interspersed among the driver fuel in the central core region.

EXERCISES

Questions

1-1. Identify, explain briefly, and sketch the relationships among each step of the uranium fuel cycle as it currently exists in the United States. Include features that may not show explicitly on the sketch.

1-2. Expand on question 1-1 to include the conceptual closed fuel cycle (add to the sketch with a different color).

1-3. Identify the reactor system(s) characterized by fuel cycles with:
   a. slightly enriched uranium
   b. 93 wt % ^{235}U
   c. required plutonium recycle
   d. no enrichment facilities
   e. thorium use

1-4. Identify each of the five major reactor designs by full name and acronym. Use these as column headings for a table with row headings of:
   a. moderator

b. coolant
c. fuel assembly
  - fissile composition
  - fuel-pellet geometry
  - cladding
  - assembly geometry

Fill in the table. Explain the multiple barrier containment for each reactor.

Numerical Problems[†]

1-5. Using data from App. IV, make full-scale radial cross-section drawings of the outside dimension and shape of the five basic fuel assemblies. Center them on the same standard $[8\frac{1}{2} \times 11'']$ sheet of paper (except the largest, for which one edge of the assembly may be centered on the right-hand edge of the paper and as much drawn as will fit; also indicate where the fuel-assembly center would be located). In a convenient corner of each assembly, also draw a circle with the outside diameter of the corresponding fuel pin.

1-6. Calculate the masses of ^{235}U and ^{238}U per metric ton (t)[‡] of uranium ore, assuming ^{235}U is 0.25 wt % of the total uranium content.

1-7. Coal has a heat content of 19–28 GJ/t of mined material. Uranium as employed in an LWR has a heat content of 460 GJ/kg of natural uranium metal.
  a. Calculate the heat content of a ton of typical uranium ore (see the previous problem) and its ratio to that of the extreme coal values.
  b. Considering that electrical conversion efficiencies [MW(e)/MW(th)] are about 32 percent and 38 percent for an LWR and a coal plant, respectively, calculate electric energy ratios as in (a).

1-8. The ^{235}U and ^{238}U masses in natural uranium are split between enriched and depleted streams as a result of an enrichment process. If the input masses and output enrichments are specified, mass conservation determines the maximum (i.e., zero-loss) quantity of each isotope in the output streams. Considering a 1 kg input of natural uranium, a 3 wt % ^{235}U enriched stream, and a 0.3 wt % ^{235}U depleted stream:
  a. Calculate the masses of ^{235}U, ^{238}U, and total U in each of the two output streams.
  b. Calculate the fraction of the initial ^{235}U that ends up in each output stream.
  c. Repeat part (b) for ^{238}U and total U.

SELECTED BIBLIOGRAPHY[§]

General Nuclear Engineering Texts
      Connolly, 1978
      Foster & Wright, 1977
      Glasstone & Sesonske, 1967
      Lamarsh, 1975
      Murray, 1975

General References
      Babcock & Wilcox, 1980
      Etherington, 1958
      Lapedes, 1976

[†]Units and conversion factors are contained in App. II.
[‡]Metric ton is also sometimes abbreviated as te.
[§]Full citations are contained in the General Bibliography at the back of the book.

General Fuel Cycle Information
   APS, 1978
   Benedict, 1981
   Benedict & Pigford, 1957
   Elliot & Weaver, 1972
   WASH-1250, 1973

General Reactor Design Information
   CONF-740501, 1974
   ERDA-1535, 1975
   Hogerton, 1970
   Lish, 1972
   Nero, 1979
   Thompson & Beckerley, 1964

Reference Reactor Designs
   BWR
      GE/BWR-6, 1978
      GESSAR
   PWR
      B-SAR-241
      Babcock & Wilcox, 1975
      Babcock & Wilcox, 1978
      CESSAR
      Combustion Engineering, 1978
      RESSAR
      Westinghouse, 1975
      Westinghouse, 1979a
   PHWR
      Atomic Energy of Canada Limited, 1976
      Haywood, 1976
   HTGR
      GASSAR
      Nucl. Eng. & Des., 1974
   LMFBR
      Nucl. Eng. Int., May 1977
      Vendryes, 1977

Current Information Sources (General)
   EPRI Journal—published monthly by the Electric Power Research Institute; feature articles on the general status of alternative energy sources and on specific research projects; brief abstracts from recently-issued EPRI reports (see "Reports" section below)

   Nuclear Engineering International [Nucl. Eng. Int.]—published monthly in the United Kingdom; excellent coverage of worldwide developments in the nuclear fuel cycle and reactor technology; some issues devoted to a single reactor (often including a colored wall chart of the system), a reactor concept, a national program, or a fuel cycle step (specific topics of interest are included in the Bibliography); annual July supplement—"Power Reactors"—summarizes design parameters and operating history for the world's past, present, and planned systems.

   Nuclear Industry [Nucl. Ind.]—published monthly by the Atomic Industrial Forum [AIF], a U.S. organization representing electric utilities and other organizations involved in commercial nuclear power; summary coverage of current issues and status, especially concerned with federal regulatory policy and practice.

Nuclear News [Nucl. News] —published monthly by the American Nuclear Society [ANS], the major professional organization for nuclear engineers; summary coverage of current issues and status, plus feature articles on topics of general interest; updated "World List of Nuclear Power Reactors" appears in each February and August issue.

Nuclear Safety [Nucl. Safety] —published bimonthly by the U.S. Nuclear Regulatory Commission; feature articles on various aspects of the nuclear fuel cycle and reactor safety including material safeguards; summary of current status of regulatory actions.

Physics Today [Phys. Today] —published monthly by the American Institute of Physics [AIP], the major professional organization for physicists; summary coverage of current issues, plus occasional feature articles on nuclear energy and other energy sources.

Power—published monthly as "the magazine of power generation and plant energy systems"; frequent overview articles on various aspects of nuclear and other power sources written for a general engineering audience.

Power Engineering [Power Eng.] —published monthly as "the engineering magazine of power generation"; summary coverage of current issues in "Nuclear Power Engineering" section plus frequent overview articles on nuclear and other power sources written for a general engineering audience (often includes good color illustrations and pictures).

Science—published weekly by the American Association for the Advancement of Science [AAAS] for the general scientific and engineering community; coverage of current energy policies and issues including occasional feature articles on specific energy sources with some emphasis on cost–risk–benefit implications.

Scientific American [Sci. Am.] —published monthly; excellent feature articles on specific aspects of nuclear energy and other energy sources (usually several per year) written for a general technical audience.

Technology Review [Tech. Rev.] —published monthly by the Massachusetts Institute of Technology; good review articles on nuclear and other energy technologies with some emphasis on sociopolitical interactions.

The Bulletin of the Atomic Scientists [Bull. At. Sci.] —published monthly (except July and August) by the Educational Foundation for Nuclear Science as "a magazine of science and public affairs"; although each cover holds "the Bulletin clock, symbol of the threat of nuclear doomsday hovering over mankind" (in reference mainly to nuclear *weapons*), it has been a good forum for opinion on commercial nuclear power by proponents and some responsible critics; the overall thrust tends to be political rather than technical.

Current Information Sources (Professional Journals)

Nuclear Engineering and Design [Nucl. Eng. & Des.] —published monthly by the North Holland Publishing Company; reports on a range of engineering research related to nuclear energy, emphasizing structural and thermodynamic analysis and design; occasional issues are devoted to a specific topic area (e.g., HTGR systems in January 1974 and LWR fuel rod modeling in February 1980).

Nuclear Science and Engineering [Nucl. Sci. Eng.] and *Nuclear Technology* [Nucl. Tech.] — published monthly by the American Nuclear Society; the principal technical journals of the nuclear engineering profession for theoretical and applications-oriented topics, respectively.

Transactions of the American Nuclear Society [Trans. Am. Nucl. Soc.] —paper summaries from ANS meetings; up-to-date coverage of current research activities.

Current Information Sources (Reports)

Electric Power Research Institute [EPRI] –organization funded by the U.S. electric utility industry to conduct research in subject areas of practical interest; reports cover a wide range of topics related to nuclear power and other energy sources and are usually well written for a general engineering audience. (Many EPRI reports are referenced by author in the selected bibliographies.)

United States Government–reports of research supported by government funds from departments, agencies, and national laboratories; all such reports (plus some of foreign origin) are available from the National Technical Information Service [NTIS]. (Many of these reports are referenced by author or report number in the selected bibliographies.)

II

BASIC THEORY

Goals

1. To introduce the basic theoretical concepts of nuclear physics, radiation protection, reactor physics, reactor kinetics, fuel depletion, and energy removal
2. To develop fundamental calculational skills that can aid in understanding nuclear energy problems and solutions
3. To identify the bases for some of the "uniquely nuclear" features of the operations and systems described in the remaining parts of the book

Chapters in Part II

Nuclear Physics
Nuclear Radiation Environment
Reactor Physics
Reactor Kinetics and Control
Fuel Depletion and Related Effects
Reactor Energy Removal

2

NUCLEAR PHYSICS

Objectives

After studying this chapter, the reader should be able to:

1. Identify the constituent parts and describe the basic structure of an atom.
2. Define the following terms:
   a. isotope
   b. half-life
   c. mean free path
3. Explain the fundamental differences among fissile, fissionable, and fertile nuclides; identify members of each classification.
4. Identify the radiations associated with fission.
5. Sketch the relationship among the following reactor cross sections:
   a. total interaction
   b. scattering
   c. absorption
   d. fission
   e. capture
6. Differentiate between microscopic and macroscopic cross sections.
7. Write equations and perform simple calculations for radioactive decay and nuclear reactions.

Prerequisite Concepts

Energy Forms
Basic Chemistry
Basic SI Units   Appendix II

The characteristics of the radioactive decay and nuclear reaction processes have been the driving forces behind nearly every unique design feature of nuclear energy systems. Thus, familiarity with some of the fundamental principles of nuclear physics is essential to understanding nuclear technology.

Radioactive decay and nuclear reactions are unique in that they

1. provide clear evidence that mass and energy can be interconverted
2. involve a variety of particles and radiations, which often have discrete, or quantized, energies
3. require descriptive formulations based on the laws of probability

Such characteristics have prompted the development of many new experimental and analytical methods.

THE NUCLEUS

The atom is the basic unit of matter. As first modeled by Niels Bohr in 1913, it consists of a heavy central *nucleus* surrounded by orbital *electrons*. The nucleus, in turn, consists of two types of particles, namely *protons* and *neutrons*. Table 2-1 compares the atom and its constituents in terms of electric charge and mass.[†]

The proton and electron are of exactly opposite charge. A complete atom has the same number of protons and electrons, each given by the *atomic number Z*. The electrostatic [Coulomb] attractive forces between the oppositely charged particles is the basis for the electron orbits (in a manner similar to the way solar system orbits result from gravitational forces).

Although the atom is electrically neutral, the number and resulting configuration of the orbital electrons uniquely determines the chemical properties of the atom, and hence its identity as an *element*. It may be recalled that the atomic number is the basis for alignment in the period table of elements.

[†]Unless otherwise noted, data here and in the remainder of the book are based on the General Electric Chart of the Nuclides (GE/Chart, 1977). The definition for the "amu" is provided at the end of this section.

TABLE 2-1
Characteristics of Atomic and Nuclear Constituents

Constituent	Charge (e)[†]	Mass (amu)[‡]	Radius (m)
Electron	-1	5.5×10^{-4}	
Proton	$+1$	1.00727	
Neutron	0	1.008665	
Nucleus	$+Z$[§]	$\sim A$[¶]	$\sim 10^{-16}$
Atom	0	$\sim A$	$\sim 10^{-11}$

[†]$e = 1.6 \times 10^{-19}$ C.
[‡]amu $= 1.6605 \times 10^{-27}$ kg.
[§]Z = atomic number.
[¶]A = atomic mass number.

Structure

Very strong, short-range forces override the Coulomb repulsive forces to bind the positively charged protons (along with the uncharged neutrons) into the compact nucleus. The protons and neutrons in a nucleus are collectively referred to as *nucleons*. Since each nucleon has roughly the same mass, the nucleus itself has a mass that is nearly proportional to the *atomic mass number A*, defined as the total number of nucleons. The electrons are very light compared to the particles in the nucleus, so the mass of the atom is nearly that of the nucleus.

The characteristic dimensions of the nucleus and atom are listed in Table 2-1. The latter value is based on the effective radius of the outer electron orbits in arrays of atoms or molecular combinations.

A useful shorthand notation for nuclear species or *nuclides* is $^A_Z X$, where X is the chemical symbol, Z is the atomic number, and A is the atomic mass number. Since the subscript Z is actually redundant once the chemical element has been identified, its use is somewhat discretionary. (An alternative formulation, $_Z X^A$, is also found, although current practice favors retention of the upper right-hand location for charge-state information.)

Different nuclides of a single chemical element are called *isotopes*. For example, uranium isotopes $^{233}_{92} U$, $^{235}_{92} U$, and $^{238}_{92} U$ were mentioned in the last chapter. Each has 92 protons and electrons with 141, 143, and 146 neutrons, respectively. Another important isotope group is the hydrogen family—$^1_1 H$, $^2_1 H$, and $^3_1 H$. The latter two are often given separate names and symbols—deuterium [$^2_1 D$] and tritium [$^3_1 T$], respectively.

Binding Energy

One of the most startling observations of nuclear physics is that the mass of an atom is less than the sum of the masses of the individual constituents. When all parts are assembled, the product atom has "missing" mass, or a *mass defect* Δ, given by

$$\Delta = [Z(m_p + m_e) + (A - Z)m_n] - M_{atom} \qquad (2\text{-}1)$$

where the masses m_p, m_e, and m_n of the proton, electron, and neutron, respectively, are multiplied by the number present in the atom of mass M_{atom}.

It is now known that the defect mass is converted into energy at the time the nucleus is formed.[†] The conversion is described by the expression

$$E = mc^2 \qquad (2\text{-}2)$$

for energy E, mass m, and proportionality constant c^2, where c is the speed of light in a vacuum. This simple-appearing equation (one of the world's most famous!) was first derived by Albert Einstein.

The energy associated with the mass defect is called the *binding energy*. It is

---

[†]Mass changes also occur with chemical binding (e.g., electrons with a nucleus to form an atom and atoms with each other to form molecules), but they are so small as to defy measurement.

said to put the atom into a "negative energy state" since positive energy from an external source would have to be supplied to disassemble the constituents. (This is comparable to the earth–moon system, which could be separated only through an addition of outside energy.) The binding energy [BE] for a given nucleus may be expressed as

$$BE = [M_{atom} - Z(m_p + m_e) - (A - Z)m_n]c^2 = -\Delta c^2 \tag{2-3}$$

As the number of particles in a nucleus increases, the BE also increases. The rate of increase, however, is not uniform. In Fig. 2-1, the BE in MeV* per nucleon is plotted as a function of atomic mass number. The nuclides in the center of the range are more tightly bound on the average than those at either very low or very high masses.

The existence of the fission process is one ramification of the behavior shown in Fig. 2-1. Compared to nuclei of half its mass, the ^{235}U nucleus is bound relatively lightly. Energy must be released to split the loosely bound ^{235}U into two tightly bound fragments. A reasonably good estimate for the energy released in fission can be obtained by using data from the curve in Fig. 2-1.

Energy production in our sun and other stars is based on the fusion process, which combines two very light nuclei into a single heavier nucleus. As shown by Fig.

*The MeV is a convenient unit for energy, which is defined shortly.

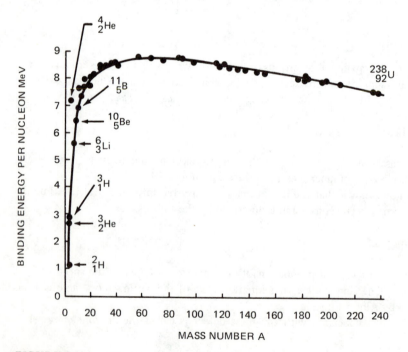

FIGURE 2-1

Binding energy per nucleon as a function of mass number.

2-1, two deuterium nuclei [${}^2_1$H or ${}^2_1$D], for example, could release a substantial amount of energy if combined to form the much more tightly bound helium [${}^4_2$He]. A number of fusion reactions are considered as potential terrestrial nuclear energy sources in Chap. 21.

Mass and Energy Scales

The masses and energies associated with nuclear particles and their interactions are extremely small compared to the conventional macroscopic scales. Thus, special units are found to be very useful.

The *atomic mass unit* [amu] is defined as 1/12 of the mass of the carbon-12 [${}^{12}_6$C] atom.[†] The masses in Table 2-1 are based on this scale.

When an electron moves through an electrical potential difference of 1 volt [V], it acquires a kinetic energy of 1 *electron volt* [eV]. This unit (equal to 1.602 $\times$ 10^{-19} J, as noted in App. II), along with its multiples keV and MeV, is very convenient for nuclear systems.[‡]

Mass and kinetic energy, as noted previously, may be considered equivalent through the expression $E = mc^2$. Thus, for example, it is not uncommon to express mass differences in MeV or binding energies in amu based on the conversion 1 amu = 931.46 MeV. (Other useful factors are contained in App. II.)

RADIOACTIVE DECAY

The interactions among the particles in a nucleus are extremely complex. Some combinations of proton and neutron numbers result in very tightly bound nuclei, while others yield more loosely bound nuclei (or do not form them at all).

Whenever a nucleus can attain a more stable (i.e., more tightly bound) configuration by emitting radiation, a spontaneous disintegration process known as *radioactive decay* may occur. (In practice this "radiation" may be actual electromagnetic radiation or may be a particle.) Examples of such processes are delayed briefly to allow for an examination of important basic principles in the following paragraphs.

Conservation Principles

Detailed studies of radioactive decay and nuclear reaction processes have led to the formulation of very useful *conservation principles*. For example, electric charge is conserved in a decay; the *total* amount is the same before and after the reaction even though it may be distributed differently among entirely different nuclides and/or particles. The four principles of most interest here are conservation of

1. charge
2. mass number or number of nucleons

[†]Two earlier mass scales defined the amu as 1/16 the mass of elemental oxygen and of oxygen-16, respectively. Since each of these differs from the current standard, caution is advised when using data from multiple sources [e.g., Kaplan (1962) is based on the ^{16}O scale, while the Chart of the Nuclides (GE/Chart, 1977) employs the ^{12}C scale].

[‡]eV is usually pronounced as "ee-vee," keV as "kay-ee-vee," etc.

3. total energy
4. linear and angular momentum

Conservation of electric charge implies that charges are neither created nor destroyed. Single positive and negative charges may, of course, neutralize each other. Conversely, it is also possible for a neutral particle to produce one charge of each sign.

Conservation of mass number does not allow a net change in the number of nucleons. However, the conversion of a proton to a neutron is allowed. Electrons follow a separate particle conservation law (which is beyond the scope of this discussion). By convention, a mass number of zero is assigned to electrons.

The total of the kinetic energy and the energy equivalent of the mass in a system must be conserved in all decays and reactions. This principle determines which outcomes are and are not possible.

Conservation of linear momentum is responsible for the distribution of the available kinetic energy among product nuclei, particles, and/or radiations. Angular momentum considerations for particles that make up the nucleus play a major role in determining the likelihood of occurrence of the outcomes that are energetically possible. (This latter consideration is substantially beyond the scope of this book.)

Natural Radioactivity

A wide range of radioactive nuclides, or *radionuclides*, exist in nature (and did so before the advent of the nuclear age). Artificial radionuclides produced by nuclear reactions are considered separately.

The naturally occurring radioactive decay processes may produce any of three radiations. Dating back to the time of their discovery and identification, the arbitrary names *alpha, beta,* and *gamma* are still employed.

Alpha Radiation

Alpha radiation is a helium nucleus, which may be represented as either ^4_2He or $^4_2\alpha$. An important alpha-decay process with $^{235}_{92}\text{U}$ is written in the form

$$^{235}_{92}\text{U} \rightarrow ^{231}_{90}\text{Th} + ^4_2\text{He}$$

where $^{231}_{90}\text{Th}$ [thorium-231] is the decay or *daughter* product of the $^{235}_{92}\text{U}$ *parent* nucleus. It may be noted that the reaction equation demonstrates conservation of mass number A and charge (or equivalently, atomic number Z) on both sides of the equation. Thus, once any two of the three constituents are known, the third may be determined readily.

Most alpha-emitting species have been observed to generate several discrete kinetic energies. Thus, to conserve total energy, the product nuclei must have correspondingly different masses. The discrete or *quantum* differences in energy are related to a complex *energy level* structure within the nucleus. (Explanation of such phenomena is delegated to the field of *quantum mechanics*, which is outside the scope of this book.)

Beta Radiation

Beta radiation is an electron of nuclear, rather than orbital, origin. Since, as noted earlier, the electron has a negative charge equal in magnitude to that of the proton and has a mass number of zero, it is represented as $_{-1}^{0}e$ or $_{-1}^{0}\beta$. A beta-decay reaction, which is important to production of plutonium in a breeder reactor, is

$$_{93}^{239}\text{Np} \rightarrow {}_{94}^{239}\text{Pu} + {}_{-1}^{0}\beta + {}_{0}^{0}\nu*$$

where $_{93}^{239}\text{Np}$ [neptunium-239] is the parent of the $_{94}^{239}\text{Pu}$, and $_{0}^{0}\nu*$ is an uncharged, massless "particle" called an *antineutrino*.[†] As required by conservation principles, the algebraic sums of both charge and mass number on each side of the equation are equal.

The nuclear basis for beta decay is

$$_{0}^{1}\text{n} \rightarrow {}_{1}^{1}\text{p} + {}_{-1}^{0}e + {}_{0}^{0}\nu*$$

where the uncharged neutron emits an electron and an antineutrino while leaving a proton (and a net additional positive charge) in the nucleus. The slight mass difference between the neutron and proton may be noted from Table 2-1 to be sufficient to allow for electron emission as well as a small amount of kinetic energy.

The nature of the antineutrino defies the human senses. Since it has neither charge nor mass, it does not interact significantly with other materials and is not readily detected. However, it does carry a portion of the kinetic energy that would otherwise belong to the beta particle. In any given decay, the antineutrino may take anywhere from 0 to 100 percent of the energy, with an average of about two-thirds for many cases.

Analogously to alpha decay, the beta process in a given radionuclide may produce several discrete transition energies based on the energy levels in the nucleus. Here, however, a range of beta energies from the transition energy down to zero are actually observed, due to the sharing with antineutrinos.

Gamma Radiation

Gamma radiation is a high-energy electromagnetic radiation that originates in the nucleus. It is emitted in the form of *photons*, discrete bundles of energy that have both wave and particle properties.

Gamma radiation is emitted by *excited* or *metastable* nuclei, i.e., those with a slight mass excess (which may, for example, result from a previous alpha or beta transition of less than maximum energy). In one gamma-decay reaction

$$^{60m}_{27}\text{Co} \rightarrow {}_{27}^{60}\text{Co} + {}_{0}^{0}\gamma$$

an excited nucleus is transformed into one that is more "stable" (although, in this case, still radioactive). Such processes increase the binding energy but do not affect either the charge or the mass number of the nucleus.

The gamma-ray energies represent transitions between the discrete energy levels in a nucleus. As a practical matter, nuclides can often be readily identified or differentiated from each other on the basis of their distinctive gamma energies.

[†]There is also a *neutrino* $_{0}^{0}\nu$, which is associated with *positron* and *electron capture* decay processes. Since neutrinos are not significant in the context of commercial nuclear energy, they are not considered here. The interested reader may consult nearly any textbook on nuclear physics.

Decay Probability

The precise time at which any single nucleus will decay cannot be determined. However, the average behavior of a very large sample can be predicted accurately by using statistical methods.

An average time dependence for a given nuclide is quantified in terms of a *decay constant* λ—the probability per unit time that a decay will occur. The *activity* of a sample is the average number of disintegrations per unit time. For a large sample, the activity is the product of the decay constant and the number of atoms present, or

$$\text{Activity} = \lambda n(t) \tag{2-4}$$

where $n(t)$ is the concentration, which changes as a function of the time t. Since λ is a constant, the activity and the concentration are always proportional and may be used interchangeably to describe any given radionuclide population.

It has been typical to quote activities in units of the *Curie* [Ci], defined as 3.7×10^{10} disintegrations per second, which is roughly the decay rate of 1 g of radium. The currently favored SI unit is the *Becquerel*, which is 1 disintegration per second. The rate at which a given radionuclide sample decays is, of course equal to the rate of decrease of its concentration, or

$$\text{Activity} = \text{rate of decrease}$$

Mathematically, this is equivalent to

$$\lambda n(t) = - \frac{dn(t)}{dt}$$

By rearranging terms,

$$\lambda = - \frac{dn(t)/n(t)}{dt} \tag{2-5a}$$

or $$\frac{dn(t)}{n(t)} = - \lambda \, dt \tag{2-5b}$$

where Eq. 2-5a shows that the decay constant λ is the fractional change in nuclide concentration per unit time.

The solution to Eq. 2-5b is

$$n(t) = n(0)e^{-\lambda t} \tag{2-6}$$

where $n(0)$ is the radionuclide concentration at time $t = 0$. As a consequence of the exponential decay, two useful times can be identified:

1. The *mean lifetime* τ—the average [statistical mean] time a nucleus exists before undergoing radioactive decay. Since it may be shown that

$$\tau = \frac{1}{\lambda}$$

this lifetime is also the amount of time required for the sample to decrease by a factor of e (see Eq. 2-6).

2. The *half-life* $T_{1/2}$ —the average amount of time required for sample size or activity to decrease to one-half of its initial amount.

The half-life, mean lifetime, and decay constant are found to be related by

$$T_{1/2} = \ln 2\, \tau = \frac{\ln 2}{\lambda}$$

or equivalently

$$T_{1/2} \approx 0.693\tau = \frac{0.693}{\lambda}$$

The basic features of decay of a radionuclide sample are shown by the graph in Fig. 2-2. Assuming an initial concentration $n(0)$, the population may be noted to decrease by one-half of this value in a time of one half-life. Additional decreases occur such that *whenever* one half-life elapses, the concentration drops to one-half of *whatever* its value was at the beginning of that time interval.

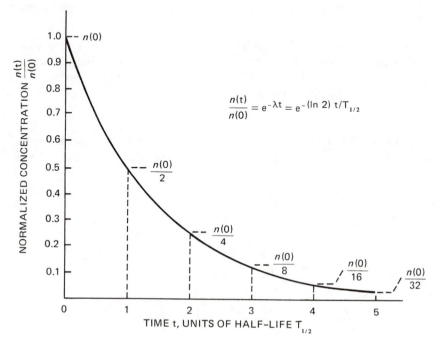

FIGURE 2-2
Radioactive decay as a function of time in units of half-life.

An example of an important application of radioactive decay is in the management of radioactive wastes (a subject considered in detail in Chap. 10). The fission products ^{85}Kr [krypton-85] and ^{87}Kr, which have half-lives of roughly 11 years and 76 min, respectively, are generally both present in LWR cooling water. In 10 half-lives, each would be reduced in population and activity to about 0.1 percent (actually 1/1024). Thus, ^{87}Kr would essentially disappear of its own accord in a little over one-half day. The ^{85}Kr, on the other hand, would be of concern for on the order of a hundred years. (The latter, for example, was a problem following the accident at the Three Mile Island reactor, discussed in Chap. 18.)

NUCLEAR REACTIONS

A majority of all known radionuclides are produced when nuclear particles interact with nuclei. The *"man-made"*[†] or *artificial nuclides* of interest in nuclear reactors span nearly all elements.

A simple reaction is depicted in Fig. 2-3. It may be represented in equation form as

$$X + x \rightarrow (C)^* \rightarrow Y + y$$

for target nucleus X, projectile particle x, compound nucleus $(C)^*$, product nucleus Y, and product particle y. Common shorthand notations for the reaction are

$$X(x, y)Y \quad \text{or simply} \quad X(x, y)$$

where conservation principles allow the latter simplification. The implied designations are actually quite arbitrary since, for example, the target and projectile may both be moving (and, occasionally, are even the same nuclear species) and the products may consist of several nuclei and/or particles.

Compound Nucleus

The compound nucleus temporarily contains all of the charge and mass involved in the reaction. However, it is so unstable in an energy sense that it only exists for on

†This term is still commonly used (with apologies to women).

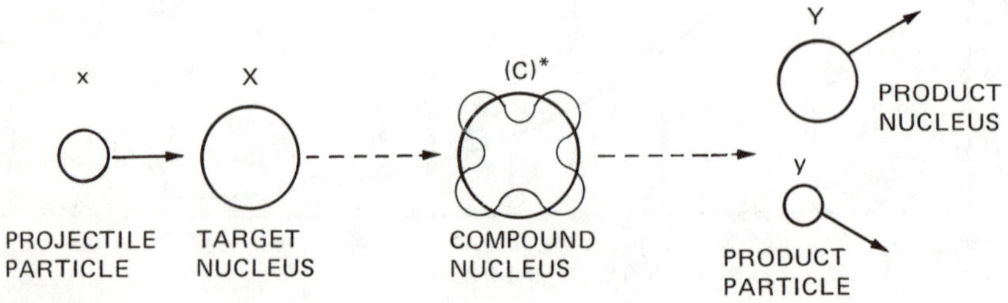

FIGURE 2-3
Generic nuclear reaction.

the order of 10^{-14} s (a time so short as to be insignificant, and undetectable, on a scale of human awareness). Because of its instability, a compound nucleus should never be considered equivalent to a nuclide that may have the same number of protons and neutrons.

Nuclear reactions are subject to the same conservation principles that govern radioactive decay. Based on conservation of charge and mass number alone, a very wide range of reactions can be postulated. Total energy considerations determine which reactions are feasible. Then, angular momentum (and other) characteristics fix the relative likelihood of occurrence of each possible reaction. Equations for a number of important reactions are considered at the end of this section. Reaction probabilities are the subject of the last section of this chapter.

Conservation of total energy implies a balance, including both kinetic energy and mass. The simple reaction in Fig. 2-3 must obey the balance equation

$$(KE)_X + M_X c^2 + (KE)_x + M_x c^2 = (KE)_Y + M_Y c^2 + (KE)_y + M_y c^2 \qquad (2\text{-}7)$$

where $(KE)_i$ and $M_i c^2$, respectively, are the kinetic and mass-equivalent energies of the ith participant in the reaction. Rearranging the terms of Eq. 2-7 shows that

$$[(E_Y + E_y) - (E_X + E_x)] = [(M_X + M_x) - (M_Y + M_y)] c^2 \qquad (2\text{-}8)$$

where the left-hand bracket is the *Q-value* for the reaction.

When $Q > 0$, the kinetic energy of the products is greater than that of the initial reactants. This would imply that mass has been converted to energy (a fact that may be verified by examining the right-hand side of Eq. 2-8). Such a reaction is said to be *exothermal* or *exoergic* because it produces more energy than that required to initiate it.

For cases with $Q < 0$, the reaction reduces the kinetic energy of the system and is said to be *endothermal* or *endoergic*. These reactions have a minimum *threshold energy*, which must be added to the system to make it feasible (i.e., to allow the mass increase required by Eq. 2-8).

A different type of energy threshold exists in reactions between charged particles of like sign (e.g., an alpha particle and a nucleus) due to the repulsive Coulomb forces. In this case, however, the reaction may still be exoergic as long as $Q > 0$ or, equivalently, as long as the product kinetic energy exceeds the energy required to override the electrostatic forces. The fusion reactions considered in Chap. 21 are an important example of threshold reactions of this latter type.

Reaction Types

A very wide range of nuclear reactions have been observed experimentally. Of these, the reactions of most interest to the study of nuclear reactors are the ones that involve neutrons.

When a neutron strikes $^{235}_{92}U$, for example, a compound nucleus $(^{236}_{92}U)^*$ is formed, as sketched in Fig. 2-4. The compound nucleus then divides in one of several possible ways. These reactions and several others are discussed in the remainder of this section. (It should be noted that charge Z and mass number A are conserved in each case.)

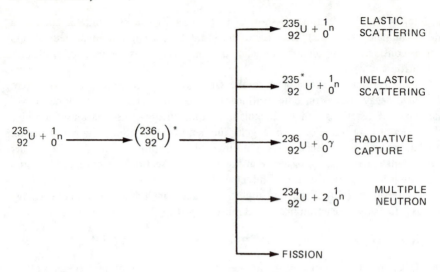

FIGURE 2-4
Possible outcomes from neutron irradiation of $^{235}_{92}$U.

Scattering

A *scattering* event is said to have occurred when the compound nucleus emits a single neutron. Despite the fact that the initial and final neutrons do not need to be (and likely are not) the same, the net effect of the reaction is as if the projectile neutron had merely "bounced off," or scattered from, the nucleus.

The scattering is *elastic* when the kinetic energy of the system is unchanged by the reaction. Although the equation for elastic scattering, ^{235}U(n, n), or

$$^{235}_{92}U + ^{1}_{0}n \rightarrow (^{236}_{92}U)^* \rightarrow ^{235}_{92}U + ^{1}_{0}n$$

looks particularly uninteresting, the fact that the neutron generally changes both its kinetic energy and its direction is significant. The former change, in particular, may substantially alter the probability for further reactions.

If the kinetic energy of the system decreases, the scattering is *inelastic*. The equation $^{235}_{92}U(n, n')$, or

$$^{235}_{92}U + ^{1}_{0}n \rightarrow (^{236}_{92}U)^* \rightarrow ^{235*}_{92}U + ^{1}_{0}n$$

represents the fact that kinetic energy is not conserved and that, thus, the product nucleus is left in an excited state. This extra loss of kinetic energy makes inelastic scattering significant for certain applications to slowing down neutrons [moderation]. The excited $^{235*}_{92}U$ nucleus quickly decays to its more stable form by emitting gamma radiation.

Radiative Capture

The reaction $^{235}_{92}U(n, \gamma)$ or

$$^{235}_{92}U + ^{1}_{0}n \rightarrow (^{236}_{92}U)^* \rightarrow ^{236}_{92}U + ^{0}_{0}\gamma$$

is known as *radiative capture* or simply n, γ. The *capture gamma* in this case has an energy of about 6 MeV (corresponding roughly to the binding energy per nucleon on Fig. 2-1 for this extra neutron).

The same reaction occuring in materials other than the fuel is often called *activation.* When sodium coolant in an LMFBR gains a neutron in the reaction

$$^{23}_{11}\text{Na} + ^{1}_{0}\text{n} \rightarrow (^{24}_{11}\text{Na})^* \rightarrow ^{24}_{11}\text{Na} + ^{0}_{0}\gamma$$

the $^{24}_{11}\text{Na}$ is radioactive (i.e., the sodium has been activated). The desire to eliminate the possibility of radioactive sodium contacting water is responsible for the introduction into the LMFBR steam cycle (Fig. 1-6) of an intermediate loop of "clean" sodium.

On the more positive side, the entire field of *activation analysis* is based on producing artificial radionuclides by neutron bombardment of an unknown sample. Then, by using gamma-ray information to determine the identity and quantity of each species, the composition of the initial material can be determined, often to a high degree of accuracy.

Multiple Neutron

The compound nucleus may also deexcite by emitting more than one neutron. The reaction $^{235}_{92}\text{U(n, 2n)}$ or

$$^{235}_{92}\text{U} + ^{1}_{0}\text{n} \rightarrow (^{236}_{92}\text{U})^* \rightarrow ^{234}_{92}\text{U} + 2^{1}_{0}\text{n}$$

evolves a pair of neutrons. Reactions that produce three or more neutrons are also possible.

The multiple neutron reactions are generally endoergic. For example, above a neutron threshold energy of about 6 MeV, the reaction

$$^{233}_{92}\text{U} + ^{1}_{0}\text{n} \rightarrow (^{234}_{92}\text{U})^* \rightarrow ^{232}_{92}\text{U} + 2^{1}_{0}\text{n}$$

is possible. The relatively short half-life of ^{232}U can have a significant impact on operations in ^{233}U–thorium fuel cycles (described in Chap. 6).

Fission

The typical fission reaction $^{235}_{92}\text{U(n, f)}$ or

$$^{235}_{92}\text{U} + ^{1}_{0}\text{n} \rightarrow (^{236}_{92}\text{U})^* \rightarrow F_1 + F_2 + ^{0}_{0}\gamma\text{'s} + ^{1}_{0}\text{n's}$$

yields two fission-fragment nuclei plus several gammas and neutrons. As considered in the next section, fission produces many different fragment pairs.

Charged Particles

Although not common for ^{235}U, there are many neutron-initiated reactions that produce light charged particles. An important example is the $^{10}_{5}\text{B(n, }\alpha)$ reaction

$$^{10}_{5}\text{B} + ^{1}_{0}\text{n} \rightarrow (^{11}_{5}\text{B})^* \rightarrow ^{7}_{3}\text{Li} + ^{4}_{2}\alpha$$

where boron-10 is converted to lithium-7 plus an alpha particle. On the basis of this reaction, boron is often used as a "poison" for removing neutrons when it is desired to shut down the fission chain reaction.

Other neutron-induced reactions yield protons or deuterons. Multiple charged-particle emissions, possibly accompanied by neutron(s) and/or gamma(s), have also been observed.

Neutron Production

Two other types of reactions are of interest because their product is a neutron. The reaction $^9_4\text{Be}(\alpha, n)$ or

$$^9_4\text{Be} + {}^4_2\alpha \rightarrow ({}^{13}_{6}\text{C})^* \rightarrow {}^{12}_{6}\text{C} + {}^1_0\text{n}$$

can occur when an alpha emitter is intimately mixed with beryllium. Plutonium-beryllium [Pu–Be] and radium-beryllium [Ra–Be] sources both employ this reaction to produce neutrons. Either may be used to initiate a fission chain reaction for startup of a nuclear reactor.

High-energy gamma rays may interact with certain nuclei to product *photoneutrons*. One such reaction with deuterium, $^2_1\text{D}(\gamma, n)$ or

$$^2_1\text{D} + {}^0_0\gamma \rightarrow ({}^2_1\text{D})^* \rightarrow {}^1_1\text{H} + {}^1_0\text{n}$$

occurs in any system employing heavy water. A similar reaction occurs with ^9_4Be in research reactors that have beryllium components. Both reactions have energy thresholds based on overcoming the "binding energy of the last neutron" (i.e., the binding energy difference between the isotopes with $A - Z$ neutrons and $A - Z - 1$ neutrons, respectively).

NUCLEAR FISSION

The fission reaction has been noted to be the basis for current commercial application of nuclear energy. Several of the important features of the process are discussed in this section.

Nature of Fission

According to a very simple qualitative model, a nucleus may be considered as a liquid drop that reacts to the forces upon and within it. The nucleus, then, assumes a spherical shape when the forces are in equilibrium. When energy is added, the nucleus is caused to oscillate from its initially spherical shape. If the shape becomes sufficiently elongated, it may neck down in the middle and then split into two or more fragments. A large amount of energy is released in the form of radiations and fragment kinetic energy (e.g., see Fig. 1-1).

Almost any nucleus can be fissioned if a sufficient amount of excitation energy is available. In the elements with $Z < 90$, however, the requirements tend to be prohibitively large. Fission is most readily achieved in the heavy nuclei where the threshold energies are 4–6 MeV or lower for a number of important nuclides. Certain of the heavy nuclides exhibit the property of *spontaneous fission* wherein an external energy addition is not required. In californium-252 [$^{252}_{98}\text{Cf}$], for example, this process occurs as a form of radioactive decay with a half-life of about 2.6 years.

Charged particles, gamma rays, and neutrons are all capable of inducing fission. The first two are of essentially no significance in the present context. As indicated in the previous chapter, neutron-induced fission chain reactions are the basis for commerical nuclear power.

Neutron-Induced Fission

When a neutron enters a nucleus, its mere presence is equivalent to an addition of energy because of the binding energy considerations discussed earlier. The binding energy change may or may not be sufficient by itself to cause fission.

A *fissile* nuclide is one for which fission is possible with neutrons of *any* energy. Especially significant is the ability of these nuclides to be fissioned by thermal neutrons, which bring essentially no kinetic energy to the reaction. The important fissile nuclides are the uranium isotopes $^{235}_{92}$U and $^{233}_{92}$U and the plutonium isotopes $^{239}_{94}$Pu and $^{241}_{94}$Pu. It has been noted that ^{235}U is the only naturally occurring member of the group.

A nuclide is *fissionable* if it can be fissioned by neutrons. All fissile nuclides, of course, must fall in this category. However, nuclides that can be fissioned only by high-energy, "above-threshold" neutrons are also included. This latter category includes $^{232}_{90}$Th, $^{238}_{92}$U, and $^{240}_{94}$Pu, all of which require neutron energies in excess of 1 MeV.

Since several of the fissile nuclides do not exist in nature, they can only be produced by nuclear reactions. The target nuclei for such reactions are said to be *fertile*. Figure 2-5 traces the mechanisms by which the three major fertile nuclides, $^{232}_{90}$Th,

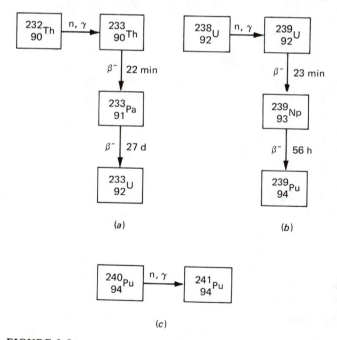

(a)

(b)

(c)

FIGURE 2-5
Chains for conversion of fertile nuclides to fissile nuclides: (*a*) $^{232}_{90}$Th to $^{233}_{92}$U; (*b*) $^{238}_{92}$U to $^{239}_{94}$Pu; (*c*) $^{240}_{94}$Pu to $^{241}_{94}$Pu.

$^{238}_{92}$U, and $^{240}_{94}$Pu, produce $^{233}_{92}$U, $^{239}_{94}$Pu, and $^{241}_{94}$Pu, respectively. The first two are each based on radiative capture followed by two successive beta decays. The last process is much simpler in being complete following the capture reaction.

It may be noted that the fertile nuclides are also fissionable. Thus, while neutrons below the threshold energy cannot cause fission, they do produce more fissile material. However, from the standpoint of neutron economy, it must be recognized that the latter process requires two neutrons for each fission.

As introduced in the previous chapter, a reactor will be classified as a converter or a breeder based on its production of new fissile from fertile nuclides. These concepts are considered in more detail in Chap. 6.

When a nucleus fissions, the major products are fission fragments, gamma rays, and neutrons. The fission fragments then undergo radioactive decay to yield substantial numbers of beta particles and gamma rays, plus a small quantity of neutrons.

Fission Fragments

A fissioning nucleus usually splits into two fragments. Because of some complex effects related to nuclear stability, the split does not usually produce equal masses. The asymmetric distribution, or characteristic "double-humped" distribution, for thermal-neutron fission of ^{235}U is shown in Fig. 2-6. It may be noted that equal-mass fragments ($A \approx 117$) are produced by only about 0.01 percent of the fissions. Fragment nuclides in the ranges of roughly 90–100 and 135–145 occur in as many as 7 percent of the fissions.

In about 1 case out of 4000, 3 fragments are produced by a *ternary fission.* One

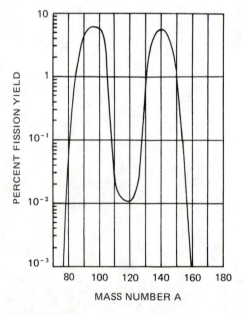

FIGURE 2-6

Fission yield as a function of mass number for thermal-neutron fission of ^{235}U. (Adapted from *Nuclear Chemical Engineering* by M. Benedict and T. H. Pigford. © 1957 by McGraw-Hill Book Company, Inc. Used by permission of McGraw-Hill Book Company.)

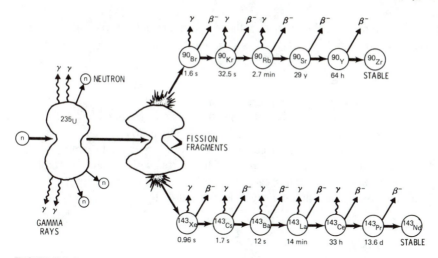

FIGURE 2-7
Two representative fission-product decay chains (from *different* fissions). (Adapted courtesy of U.S. Department of Energy.)

fragment is usually tritium $[^3_1 T]$, the beta-active isotope of hydrogen. As discussed in Chap. 10, the presence of tritium can be a significant problem in waste management.

The fission fragments tend to be neutron-rich with respect to stable nuclides of the same mass number. The related energy imbalance is generally rectified by successive beta emissions, each of which converts a neutron to a proton. Two beta-decay "chains" (from *different* fissions) are shown in Fig. 2-7. The gamma rays are emitted whenever a beta decay leaves the nucleus in an excited state. The antineutrinos that accompany the beta decays are not shown because they have no direct effect on nuclear energy systems.

The chain in Fig. 2-7, which contains strontium-90 $[^{90}_{38} Sr]$, is especially trouble-some in waste management because the relatively long 29-year half-life is coupled to the high fission yield shown in Fig. 2-6. Similar considerations apply to cesium-137 $[^{137}_{55} Cs]$, which has a 30-year half-life.

Another problem associated with the fission fragments is the presence within the decay chains of nuclides, which capture neutrons that would otherwise be avail-able to sustain the chain reaction or to convert fertile material. Two especially im-portant neutron "poisons" are xenon-135 $[^{135}_{54} Xe]$ and samarium-149 $[^{149}_{62} Sm]$. Each poses a slightly different problem for reactor operation (as discussed in Chap. 6).

Neutron Production
Thermal-neutron fission of $^{235} U$ produces (an average of) 2.5 neutrons per reaction. The majority of these are *prompt neutrons* emitted at the time of fission. A small fraction are *delayed neutrons*, which typically appear from seconds to minutes later.

The number of neutrons from fission depends on both the identity of the fissionable nuclide and the energy of the incident neutron. The parameter ν (referred to simply as "nu") is the average number of neutrons emitted per fission.

The energy distribution, or spectrum, for neutrons emitted by fission $\chi(E)$ is relatively independent of the energy of the neutron causing the fission. For many purposes, the expression

$$\chi(E) = 0.453e^{-1.036E} \sinh \sqrt{2.29E} \tag{2-9}$$

provides an adequate approximation to the neutron spectrum for ^{235}U shown by Fig. 2-8. The most likely neutron energy of about 0.7 MeV occurs where $\chi(E)$ is a maximum. The average neutron energy may be calculated to be nearly 2.0 MeV.

The neutron spectrum $\chi(E)$ is actually defined as a probability density, or the probability per unit energy that a neutron will be emitted within increment dE about energy E. Thus, neutron fractions can only be obtained by integration over finite energy intervals. As a probability density, it is required that

$$\int_0^\infty \chi(E)\, dE = 1$$

(where for actual data the infinite upper limit would be replaced by the maximum observed energy). The average or mean energy $\langle E \rangle$ of the distribution is then

$$\langle E \rangle = \int_0^\infty E\, \chi(E)\, dE$$

Almost all fission neutrons have energies between 0.1 MeV and 10 MeV. Thus, reactor concepts that are based on using the very-low-energy thermal neutrons ($E <$ 1 eV) must moderate or slow down the fission neutrons by a factor of approximately one million. (Chapter 4 considers this situation in more detail.)

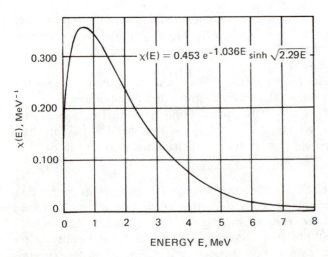

FIGURE 2-8
Fission-neutron energy spectrum $\chi(E)$ for thermal fission of ^{235}U approximated by an empirical expression.

TABLE 2-2
Representative Distribution of Fission Energy

Energy source	Fission energy (MeV)	Heat produced MeV	Heat produced % of total
Fission fragments	168	168	84
Neutrons	5	5	2.5
Prompt gamma rays	7	7	3.5
Delayed radiations			
Beta particles[†]	20	8	4
Gamma rays	7	7	3.5
Radiative capture gammas[‡]	—	5	2.5
Total	207	200	100

[†]Includes energy carried by beta particles and antineutrinos; the latter do not produce heat in reactor systems.

[‡]Nonfission capture reactions contribute heat energy in all real systems; design-specific considerations may change this number by about a factor of two in either direction.

Delayed neutrons are emitted by certain fission fragments immediately following the first beta decay. They enter the system on time scales characterized by the half-lives of the decay processes. Although delayed neutrons constitute a small fraction (typically less than 1 percent) of the neutron production attributed to the fission process, they play a dominant role in reactor control (as described in Chap. 5).

Energy Production

A typical fission produces nearly 200 MeV of energy. By comparison, 2-3 eV of energy is released for each carbon atom that is burned with oxygen. The per-reaction difference by a factor of 70-100 million is a major advantage for nuclear fission.

A representative distribution of energy among the various products of the fission process is provided in Table 2-2. Essentially, all of the kinetic energy from the fragments, neutrons, and gammas is deposited within the reactor vessel in the form of heat. Of the energy associated with fragment beta decays, however, only the fraction carried by the beta particles produces heat; the remainder leaves the system with the antineutrinos.

Radiative capture reactions occur in all reactor systems. Thus, while they are not fissions, they do absorb fission neutrons and add gamma energy. Their actual contribution depends on the composition of the reactor in terms of fissile, fertile, and other species.

The energy from the prompt radiations is deposited at the time of fission. That from the delayed radiations appears in a time-dependent manner characterized by the half-lives of the various fission-product species. In both cases, the energy appears as heat.

The rate of energy generation—i.e., the power—from fission product decay can be described by the equation

$$\frac{P(t)}{P_0} = 5.8 \times 10^{-3} \left[t^{-0.2} - (t + t_0)^{-0.2} \right] \tag{2-10}$$

for decay power $P(t)$ at t days after reactor shutdown and where t_0 is the number of days that the reactor has operated at a steady thermal power level of P_0. The heat generation rate falls off only as the one-fifth power of time, resulting in a sizeable heat load at relatively long times after shutdown. This decay-heat source has important impacts on spent fuel handling, reprocessing, waste management, and reactor safety (all of which are considered in later chapters).

Assuming that 200 MeV is available each time a fission occurs, the ultimate heat production may be quantified as

$$3.1 \times 10^{10} \text{ fission/s} \approx 1 \text{ W}$$

A gram of any fissile nuclide contains about 2.5×10^{21} nuclei. Thus, the energy production per unit mass may be expressed roughly as

$$\text{Fission energy production} \approx 1 \text{MW} \cdot \text{d/g}$$

In real systems where radiative capture reactions convert some nuclei before they can fission, somewhat larger amounts of fissile material are expended for a given energy production.

REACTION RATES

The rate at which fissions occur is directly proportional to the power output of a nuclear reactor. However, the rates for all reactions that produce or remove neutrons determine the overall efficiency with which fissile and fertile materials are employed. The ability to calculate such reaction rates is extremely important to the design and operation of nuclear systems.

When a single compound nucleus has more than one mode for deexcitation, it is not possible (with current knowledge) to predict with absolute certainty which reaction will occur. (This characteristic is analogous to the uncertainty in time of emission for radioactive decay.) The relative probability for each outcome, however, can be determined.

Because of the extreme complexity of the interactions that occur within nuclei, theoretical considerations alone are seldom adequate for predicting nuclear reaction rates. Instead, experimental measurements provide the bulk of available information.

Reaction rates are generally quantified in terms of two parameters—a *macroscopic cross section* related to the characteristics of the material in bulk and a *flux* characterizing the neutron population. The form is based, at least in part, on the historical development of the fields of nuclear physics and reactor theory.

Nuclear Cross Sections

The concept of a nuclear *cross section* σ was first introduced with the idea that the effective size of a nucleus should be proportional to the probability that an incident particle would react with it. The following steps are representative of the conceptual development:

1. Each spherical nucleus is pictured as representing a cross-sectional, or target, area to a parallel beam of neutrons traveling through the reference medium.
2. A "small" cylinder of area dA and thickness dx is constructed of a single-nuclide material with an *atom density* of n per cm^3 (as shown in Fig. 2-9).
3. A "point" neutron traveling perpendicular to the face of the cylinder sees each nucleus in the sample (i.e., the sample is "thin" enough that the nuclei do not "shadow" each other).
4. A neutron entering the disk at a random location on surface dA has a probability for a "hit," which is equal to the total area presented by the targets divided by the area of the disk.
5. If each nucleus is assumed to have a cross-sectional area σ, the total area of the targets is the number of nuclei in the disk (i.e., atom density times volume, $n\,dV = n\,dA\,dx$) multiplied by the area per nucleus.

$$\text{Total target area} = \sigma\,n\,dA\,dx$$

6. Thus, the probability that a neutron will interact in traveling a distance dx through the material is

$$\text{Interaction probability} = \frac{\text{total target area}}{\text{disk surface area}} = \frac{\sigma\,n\,dA\,dx}{dA} \tag{2-11}$$

$$\text{Interaction probability} = n\sigma\,dx$$

According to the original definition in Eq. 2-11, the interaction probability is the product of the atom density n, the cross-sectional area σ of the nucleus, and the distance of travel dx (the latter only for "short" distances where nuclei do not shadow each other). Measurements of interaction probabilities in materials of known density and thickness were employed to determine values for σ for various nuclei.

Later development showed, however, that there are many practical cases where the apparent area changes dramatically with neutron energy. These seemingly unphysical phenomena were eventually traced to the complex interactions of the nuclear forces and particles. Nuclei can have an especially great affinity for those neutrons that are capable of exciting their discrete energy levels.

This situation suggested that the connotation of cross-sectional area be dropped. Thus, σ is merely a *cross section* defined such that its product with the atom density n and distance dx is equal to the interaction probability. (It should be noted that although Eq. 2-11 is still valid, the emphasis has been shifted from defining the proba-

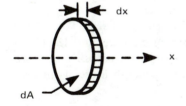

FIGURE 2-9
Basic geometry for developing the concept of a nuclear cross section: thickness $= dx$ (cm); area $= dA$ (cm^2); atom density $= n$ (at/cm^3).

bility to defining the cross section.) In the present context, Eq. 2-11 may be rearranged to

$$\sigma = \frac{\text{interaction probability}}{n \, dx} \tag{2-12}$$

which defines the cross section σ as the interaction probability per unit atom density per unit distance of neutron travel.

For reasons that become more apparent later, σ is more formally called the *microscopic cross section* (where the added word signifies application to describing the behavior on a nucleus-by-nucleus, or actually *sub*microscopic basis). Every nuclide can be assigned a cross section for each possible type of reaction and each incident neutron energy. The form $\sigma_r^j(E)$ may be employed to emphasize dependence on the nuclide j, reaction type r, and neutron energy E.

Many typical cross sections have been found to have values on the order of 10^{-24} cm^2. The tongue-in-cheek comment that this area is "as big as a barn door" led to defining the unit

$$1 \text{ barn } [b] = 10^{-24} \text{ cm}^2$$

A long-standing compilation of cross-section data is contained in the various editions and parts of BNL-325 (Hughes/BNL-325, 1955), collectively referred to as the "barn book" (having a sketch of a barn on the covers). BNL-325 and the more recent computer-based Evaluated Nuclear Data File [ENDF] (Honeck/ENDF, 1966) contain substantial cross-section data classified by nuclide, interaction type, and energy.

Interaction Types

One possible classification system for neutron cross sections is shown in Fig. 2-10. The *total cross section* σ_t represents the probability that *any* reaction will occur for the given nuclide and neutron energy. It consists of scattering (σ_s) and absorption (σ_a) components. The scattering, in turn, is split between the elastic (σ_e) and inelastic (σ_i) processes described earlier.

The *absorption cross section* σ_a includes contributions from all reactions except scattering. Each absorption, then, produces one or more new nuclei. Fission is treated as an absorption event based largely on useage in the various calculational methods. The neutrons produced by fission are treated separately from the initial absorption. (Chapter 4 discusses the role of the cross sections in reactor calculations.)

The nonfission absorption processes are referred to as *capture* [nonfission-capture] events. Radiative capture [(n, γ)], charged-particle [(n, p), (n, d), (n, α), (n, 2α), ...], multiple-neutron [(n, 2n), (n, 3n), ...], and charged-particle/neutron [(n, pn), (n, dn), ...] processes are the important constituents.

The detailed interaction spectrum—i.e., the far-right portion of Fig. 2-10—is generally used only to construct a "condensed" *library* of cross sections. The library is then employed to generate the total, scattering, absorption, and fission cross sections used in most reactor calculations. (The basic models and their use of cross sections are described in Chap. 4.)

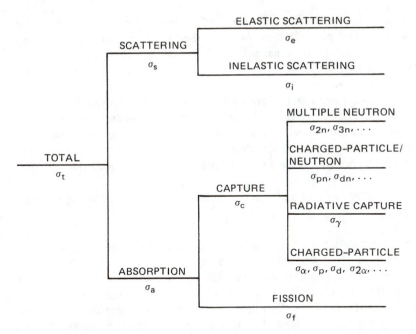

FIGURE 2-10
The hierarchy of microscopic cross sections used to calculate neutron interaction rates.

Fissile nuclei case to exist whenever they experience an absorption reaction. In fission the nucleus is split, while in capture some new species is produced. The *capture-to-fission ratio* α is defined as

$$\alpha = \frac{\sigma_c}{\sigma_f} \tag{2-13}$$

It has been found to provide a convenient measure of the probabilities of the two processes with respect to each other. By itself, a small value of α would be favored since it implies more fission (and, thus, more energy production) per unit mass of fissionable material.

The prospects for converting and breeding in fertile material depend on the number of neutrons produced per fission. However, a more important measure is the number of neutrons produced per nucleus destroyed η (referred to simply as *eta*),[†] defined by

$$\eta = \nu \, \frac{\sigma_f}{\sigma_a}$$

where ν is the average number of neutrons per fission and the ratio of cross sections is the fraction of absorptions that produce fission. Since one of the fission neutrons

[†] η for a *mixture* is defined in Chap. 4 for the "four-factor formula."

is always required to sustain the chain reaction, the remainder may be available for fertile conversion. The possibility for *breeding* (i.e., producing an amount of new fissile material that is greater than that expended) exists whenever η *exceeds two.*

Both the capture-to-fission ratio and eta vary with nuclide and neutron energy since they are based on cross sections with similar dependencies. The forms $\alpha^j(E)$ and $\eta^j(E)$ for nuclide j and energy E are appropriate for more detailed representation.

Cross sections, capture-to-fission ratios, and neutron production factors for the important fissile and fertile nuclides are compared in Table 2-3. All parameters have been evaluated at an energy of 0.025 eV (a standard reference that corresponds to the most probable neutron speed of a population in thermal equilibrium at room temperature).

Some observations based on the data in Table 2-3 (which have later significance) are:

- ^{233}U has the smallest fission cross section and the second lowest ν yet has the largest η and, thus, the best prospect for breeding.
- Although the fissile plutonium isotopes each produce about three neutrons per fission, their large capture-to-fission ratios result in fairly low values of η.
- The fertile nuclides ^{232}Th and ^{238}U have absorption cross sections that are on the order of only 1 percent of those of their respective conversion products ^{233}U and ^{239}Pu.
- Fertile ^{240}Pu has a large capture cross section for production of fissile ^{241}Pu.

It must be emphasized that the comments above apply strictly to 0.025 eV neutrons and may at most extend to thermal neutrons. In fact, as shown later in Chap. 6, several completely opposite conclusions must be drawn when fission-neutron energies are considered.

The fissionable nuclides may be identified by a shorthand notation (which actu-

TABLE 2-3

Thermal-Neutron Cross Sections and Parameters for Important Fissile and Fertile Nuclides

| Nuclide | Cross section (b)[†] | | | α | ν[‡] | η |
	σ_a	σ_c	σ_f			
^{232}Th	7.4	7.4	–	–	–	–
^{233}U	575	46	529	0.087	2.50	2.30
^{235}U	682	98	584	0.168	2.44	2.09
^{238}U	2.7	2.7	–	–	–	–
^{239}Pu	1012	270	742	0.364	2.90	2.13
^{240}Pu	290	290	0.05	–	–	–
^{241}Pu	1370	360	1010	0.356	3.00	2.21

[†]Cross sections from Chart of the Nuclides (GE/Chart, 1977) for 0.025 eV "thermal" neutrons; σ_c has been assumed to be equal to the radiative-capture cross section σ_γ.

[‡]ν values at 0.025 eV from Hughes/BNL-325 (1958).

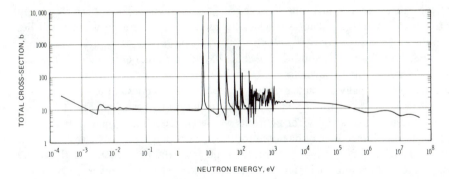

FIGURE 2-11
Total microscopic cross section for ^{238}U as a function of incident neutron energy. (Data from Hughes/BNL-325, 1955.)

ally dates back to the Manhattan Project and the need for code names for the weapons-useable materials). For example,

$$^{235}_{92}U \rightarrow 25$$

based on the last digits of the atomic number and the atomic mass number, respectively. The plutonium isotopes $^{239}_{94}$Pu, $^{240}_{94}$Pu, and $^{241}_{94}$Pu, which become 49, 40, and 41, respectively, may be noted to have identifiers that are not consecutive. Overall, the format can be quite useful, especially as a compact superscript for cross sections or other parameters (e.g., as nuclide identification j in σ_f^j or η^j).

Energy Dependence

The change in cross sections with incident neutron energy can be fairly simple or extremely complex, depending on the interaction and the nuclide considered. Only some of the more important general features are considered here.

The total cross section for ^{238}U plotted as a function of energy is shown in Fig. 2-11. Significant features are the "linear" region at low energies (on the doubly logarithmic scale) and the region of high, narrow peaks at intermediate energies.

The sloped, linear portion of the cross section curve in Fig. 2-11 is characteristic of a process referred to as *1/v-absorption* (or "one-over-v" absorption) for neutron speed v. Among neutrons passing close to a given nucleus, the slower ones spend more time near the nucleus and experience the nuclear forces for a greater period of time. The absorption probability, then, tends to vary inversely as the neutron speed and, thus, give rise to the $1/v$-behavior. When an absorption cross section σ_0 is known for energy E_0, values for other energies in the $1/v$-region may be readily calculated from

$$\sigma_a(E) = \sigma_0 \frac{v_0}{v}$$

or $\sigma_a(E) = \sigma_0 \dfrac{\sqrt{E_0}}{\sqrt{E}}$ (2-14) since the kinetic energy in terms of speed is just $E = \frac{1}{2}mv^2$. The "thermal" energy $E = 0.025$ eV (or $v = 2200$ m/s) is generally used as the reference whenever it lies within the range.

Almost all fissionable and other nuclides have regions that are roughly $1/v$, many to higher energies than shown for ^{238}U. In the important neutron poison ^{10}B, this regular behavior actually continues rather precisely up to energies in the keV range.

The very high, narrow *resonance peaks* in the center region of Fig. 2-11 are a result of the nucleus's affinity for neutrons whose energies closely match its discrete, quantum energy levels. It may be noted for the lowest-energy resonance that the total cross section changes by a factor of nearly 1000 from the peak energy to those just slightly lower. All fissionable materials exhibit similar resonance behavior.

Scattering, the other contributor to the total cross section, tends to take two forms. In *potential scattering*, the cross section is essentially constant (i.e., independent of energy) at a value somewhat near the effective cross-sectional area of the nucleus. *Resonance scattering*, like its absorption counterpart, is based on the energy-level structure of the nucleus.

The distinction between fissile and nonfissile isotopes is readily observed from the plot of the fission cross sections for ^{235}U and ^{238}U in Fig. 2-12. As neutron energy increases, the ^{235}U curve is characterized by large (near-$1/v$) values, resonance behavior, and slightly irregular low values. The nonfissile ^{238}U shows a fission threshold near 1 MeV followed by a maximum cross section of one barn. Fission in ^{235}U clearly dominates that for ^{238}U at all neutron energies of interest in reactor systems.

Figure 2-12 also shows that thermal-neutron fission of ^{235}U is more probable than fast-neutron ($E \gtrsim 0.1$ MeV) fission by over three orders of magnitude. This is one of the important factors that favor thermal reactors over fast reactors.

Interaction Rates

The cross section quantifies the relative probability that a nucleus will experience a neutron reaction. The overall rate of reaction in a system, however, also depends on the characteristics of the material and of the neutron population. The following simplified derivation identifies the important features of interaction-rate calculations.

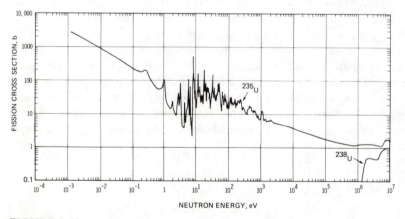

FIGURE 2-12

Microscopic fission cross section for fissile ^{235}U and fissionable ^{238}U. (Data from Hughes/BNL-325, 1955.)

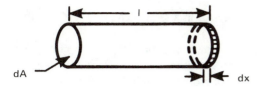

FIGURE 2-13
Basic geometry for developing the concept of nuclear reaction rate in terms of macroscopic cross section and neutron flux: $l = v\, dt$; $V = l\, dA = v\, dt\, dA$.

Considering first a parallel beam of monoenergetic neutrons with a speed v and a density of N per unit volume, it is necessary to determine the rate at which they pass through a sample like that of Fig. 2-9. If, as before, the disk has an area dA and a thickness dx, then:

1. When the beam is perpendicular to dA, only neutrons directly in front of and within a certain distance l can reach the surface in a given time dt.
2. The distance l traveled by the neutrons is equal to their speed v multiplied by the time dt, or $l = v\, dt$.
3. As shown by Fig. 2-13, only the neutrons within the cylinder of length l and area dA can enter the sample in time dt (all others will either not reach the target or will pass outside of its boundaries).
4. Since the number of neutrons in the cylinder is the density N times the volume dV [$= l\, dA$, or $v\, dt\, dA$ from (2)],

$$\frac{\text{Number of neutrons passing}}{dA \text{ per unit time}} = \frac{N\, dV}{dt} = \frac{Nv\, dt\, dA}{dt} = Nv\, dA \qquad (2\text{-}15)$$

The reaction rate must be equal to the product of the rate of entry of neutrons and the probability that a neutron will interact with a nucleus. Thus, combining the results of Eqs. 2-15 and 2-11,

Reaction rate $= (Nv\, dA)(n\sigma\, dx)$

for neutron density N, neutron speed v, nuclide density n, cross section σ, and disk area and thickness dA and dx, respectively. Rearranging terms and noting that $dA\, dx$ is just the disk volume dV,

Reaction rate $= (n\sigma)(Nv)\, dV \qquad (2\text{-}16)$

This result separates the effects of nuclide characteristics, neutron population, and sample volume.

The *macroscopic cross section* Σ is defined as

$\Sigma = n\sigma \qquad (2\text{-}17)$

Since n is the number of atoms per unit volume and σ is the area per atom, the macroscopic cross section is "per unit distance" or generally cm^{-1}. It is the probability per unit distance of travel that a neutron will interact in a sample characterized by atom density n and microscopic cross section σ.

It is, perhaps, unfortunate that both σ and Σ bear the title cross section since they have different units. The microscopic cross section is an "effective area" used to characterize a single nucleus. The macroscopic cross section is the probability that a neutron will interact in traveling a unit distance through a (macroscopic) sample of material. Although renaming Σ could eliminate some confusion, established practice dictates against doing so (at least in this book). Use of the common nicknames "micro" and "macro" without attaching the words "cross section" may provide a partial solution to the problem.

The *neutron flux* Φ is defined by

$$\Phi = Nv \tag{2-18}$$

for neutron density N and speed v. Since the density N is in terms of neutrons per unit volume, and the speed v is distance per unit time, neutron flux Φ is neutrons per unit area per unit time, or generally neutrons/cm$^2\cdot$s. Although the earlier derivation was based on a monodirectional, monoenergetic beam, the definition itself is completely general.

The reaction rate in Eq. 2-16 may be rewritten as

$$\text{Reaction rate} = \Sigma\Phi\,dV \tag{2-19}$$

for macroscopic cross section Σ and neutron flux Φ as defined by Eqs. 2-17 and 2-18, respectively, and for sample volume dV. An alternative form is

$$\text{Reaction rate per unit volume} = \Sigma\Phi \tag{2-20}$$

where now the volume dV is unspecified.

The macroscopic cross section Σ is constructed from microscopic cross sections, and thus varies with nuclide, interaction type, and neutron energy. In the form $n^j\sigma_r^j(E) = \Sigma_r^j(E)$, it represents the probability per unit distance of travel that a neutron of energy E will interact by mechanism r with nuclide j. Since these interaction probabilities for a given reaction and given nucleus are independent of other reactions and the other nuclei present, they may be summed to describe any desired combination. The total macroscopic cross section for any given mixture is

$$\Sigma_t^{\text{mix}} = \sum_{\text{all } j}\sum_{\text{all } r}\Sigma_r^j = \sum_{\text{all } j}\sum_{\text{all } r}n^j\sigma_r^j \tag{2-21}$$

where the summations are over all nuclides j in the mixture and all reactions r. In summing the reactions, care is advised in assuring that each is independent (e.g., Σ_a may not be added to Σ_f or Σ_c since $\Sigma_a = \Sigma_f + \Sigma_c$, as shown by Fig. 2-10).

Based on the inherent limitations of the earlier development of the concept of a microscopic cross section, it must be recognized that Eq. 2-11 and others based on it are valid only for "small" samples where the nuclides do not shadow or obscure each other from the neutron beam. "Thick" samples have the effect of removing neutrons and, therefore, changing the neutron flux seen by the more internal nuclei. If a beam of neutrons passes into a material sample, it is said to experience *attenuation* or a decrease of intensity.

If a beam of parallel monoenergetic neutrons crosses the surface of a sample, one neutron will be removed each time a reaction occurs. Thus, the rate of decrease of neutron flux $\Phi(x)$ with distance x will be equal to the reaction rate,

$$-\frac{d\Phi(x)}{dx} = \Sigma_t \Phi(x)$$

where use of the macroscopic total cross section implies that all reactions remove neutrons from the beam. (Note that even scattering does this by changing the energy and direction of the neutron.) Rearrangement of terms yields

$$\Sigma_t = -\frac{d\Phi(x)/\Phi(x)}{dx} \tag{2-22a}$$

or

$$\frac{d\Phi(x)}{\Phi(x)} = -\Sigma_t \Phi x \tag{2-22b}$$

where Eq. 2-22a shows that Σ_t is the fractional loss of flux (or neutrons, since $\Phi = Nv$) per unit distance of travel. Viewed another way, it restates the definition of Σ_t as the total interaction probability per unit distance of travel.

The solution to Eq. 2-22b is

$$\Phi(x) = \Phi(0)e^{-\Sigma_t x} \tag{2-23}$$

for entering flux $\Phi(0)$ at position $x = 0$. This exponential decay may be noted to parallel the decay of radionuclides considered in a previous section. Figure 2-14 depicts the effects of attenuation.

The neutron *mean free path* λ, or the average distance of travel before an interaction, may be shown to be

$$\lambda = \frac{1}{\Sigma}$$

For any given nuclide j and reaction r (or combination thereof), the mean free path $\lambda_r^j = 1/\Sigma_r^j$ is the average distance of neutron travel between reactions of the type considered.

Because of the exponential nature of the attenuation described by Eq. 2-21, the neutron flux never becomes identically zero, not even at arbitrarily large distances. There is, then, always some probability for very deep penetration.

Nuclear Data

Four of the more important sources[†] of nuclear data are:

1. Chart of the Nuclides [GE/Chart, 1977]
2. Table of Isotopes [Lederer & Shirley, 1978]

[†]All are subject to updating (frequently, it is hoped!).

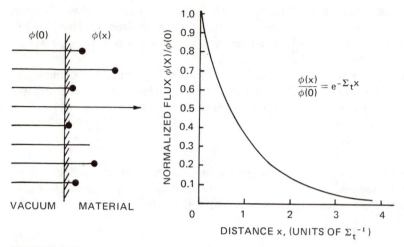

$\phi(0)$ $\phi(x)$

VACUUM   MATERIAL

$$\frac{\phi(x)}{\phi(0)} = e^{-\Sigma_t x}$$

NORMALIZED FLUX $\phi(x)/\phi(0)$

DISTANCE x, (UNITS OF Σ_t^{-1})

FIGURE 2-14

Exponential attenuation of a neutron beam in a material with total macroscopic cross section Σ_t.

3. Neutron Cross Sections [Hughes/BNL-325, 1955]
4. Evaluated Nuclear Data File (ENDF) [Honeck/ENDF, 1966]

The first of these is a *must* for virtually anyone doing work related to nuclear energy. The others, which are of somewhat more specialized interest, are described first.

The Table of Isotopes is essentially a compilation of experimental data on half-lives, radiations, and other parameters for almost all radionuclides. It contains detailed energy-level diagrams and identifies the energy and emission probability for each radiation. The data itself plus the extensive referencing make the Table of Isotopes virtually indispensable to anyone doing activation-analysis or other radionuclide measurements.

As mentioned previously, BNL-325 and ENDF contain cross-section data. Since BNL-325, or the "barn book," presents data in tabular and graphical form, it is of most use when dealing with a limited range of nuclides, reactions, and/or energies.

The computer-based ENDF is the workhorse for nearly all large-scale calculations. The ENDF/B set contains complete, evaluated sets of nuclear data for about 80 nuclides and for all significant reactions over the full energy range of interest. Revised versions are developed, tested, and issued periodically.

Chart of the Nuclides

The Chart of the Nuclides is very useful for performing preliminary or scoping calculations related to reactors. A portion containing important fissile and fertile nuclides is sketched in Fig. 2-15. Because of the vast amount of information, the figure shows the basic structure plus detail for only the ^{235}U, ^{238}U, and ^{239}Pu nuclides. The basic grid has the elements in horizontal rows with vertical columns representing neutron numbers. The block at the left end of each row gives the chemical symbol and, as appropriate, the thermal [0.025 eV] absorption and fission cross sections for the *naturally occurring* isotopic composition.

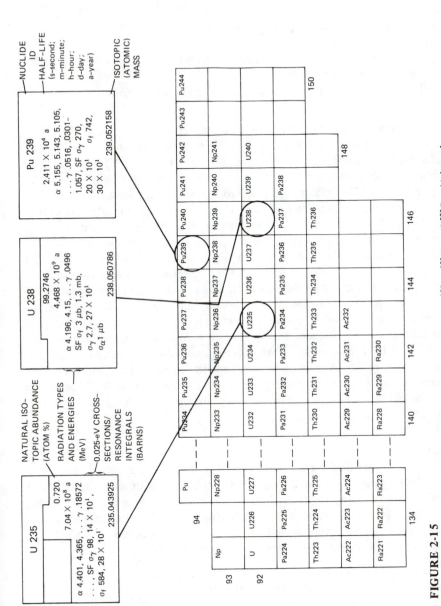

FIGURE 2-15

Structure and data presentation in the Chart of the Nuclides for ^{235}U, ^{238}U, and ^{239}Pu. (Adapted from the Chart of the Nuclides, courtesy of Knolls Atomic Power Laboratory, Schenectady, New York. Operated by the General Electric Company for the United States Department of Energy Naval Reactors Branch.

Basic data for individual nuclides includes (as appropriate):

- nuclide ID
- natural isotopic composition in atom percent [at %] [†]
- half-life
- type of radiation(s) emitted and energies in decreasing order ("..." implies additional lower energy values not included)
- thermal neutron cross sections and resonance integrals, both in barns
- isotopic (atomic) mass

Since the isotopic compositions are given in atom percent [at %], while fuel cycle mass flows are usually in weight percent [wt %], there is occasional confusion (e.g., the enrichment of natural uranium is 0.72 at % or 0.711 wt % ^{235}U). The resonance integral mentioned above is intended as a measure of an "effective" cross section that fission neutrons would see in slowing down through the resonance-energy region.

The masses used in the Chart of the Nuclides are for the entire atom rather than for the nucleus. Thus, they include the mass of the electrons less the contribution of electron binding energy. The electron binding energies are almost always too small to be significant in comparison with nuclear binding energies. Further, since it is nuclear mass differences (as opposed to the masses themselves) which are of most interest, the use of atomic masses is fully consistent. (It may be noted that charge conservation guarantees that electrons are handled appropriately. In beta decay, for example, the new electron can enter an orbit to balance the extra proton in the nucleus.)

The Chart of the Nuclides also contains a wealth of useful information not shown in Fig. 2-15. One interesting feature is the use of color codes for both half-life and cross-section ranges. A recent addition has been fission-product production data for typical power reactors. Fortunately, the introduction and guide to use of the Chart of the Nuclides are well written and easily understood. Currently, it is available in both wall-chart and booklet form.

Summary

Most of the important nuclides and many of the concepts of this chapter are directly or indirectly summarized in Fig. 2-16. Isotopes are in rows with arrows representing radiative capture or (n, γ) reactions. Beta decays are shown by downward-pointing arrows. Thermal-neutron fission is represented by diagonal dashed arrows. Decay half-lives and thermal cross sections are from the Chart of the Nuclides.

As a review, the reader should identify on Fig. 2-16:

- the four major fissile nuclides
- three chains for converting fertile nuclides to fissile nuclides
- an important (n, 2n) reaction
- nonfission capture events in fissile nuclides

Overall, the information on this figure should be of substantial value in a number of the later chapters.

[†]Atom percent is also sometimes abbreviated as a/o.

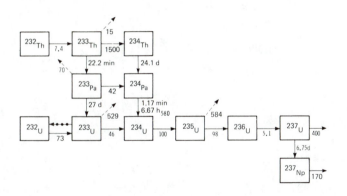

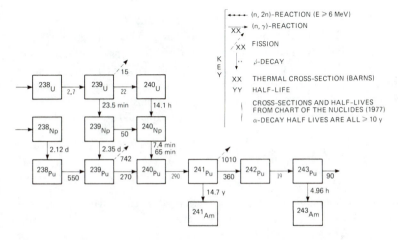

FIGURE 2-16
Neutron irradiation chains for heavy elements of interest for nuclear reactors.

Sample Calculations

The data in Figs. 2-15 and 2-16 plus the equations developed in this chapter can be used to determine a number of important reactor and/or fuel cycle characteristics. The following are representative examples:

1. The reaction and decay equations and total mass change for conversion of fertile $^{238}_{92}U$ to fissile $^{239}_{94}Pu$:

$$\left.\begin{array}{l} ^{238}_{92}U + ^{1}_{0}n \rightarrow \left(^{239}_{92}U\right)^{*} \rightarrow ^{239}_{92}U + ^{0}_{0}\gamma \\[2mm] ^{239}_{92}U \rightarrow ^{239}_{93}Np + ^{0}_{-1}\beta \\[2mm] ^{239}_{93}Np \rightarrow ^{239}_{94}Pu + ^{0}_{-1}\beta \end{array}\right\} \quad \text{(Fig. 2-15)}$$

Noting that charge and mass number are conserved,

$^{238}_{92}$U mass   238.050786 amu   (Fig. 2-15)

$^{1}_{0}$n mass   +   1.008665 amu   (Table 2-1)
                                    ─────────────
                                    239.059451 amu

$^{239}_{94}$Pu mass   − 239.052158 amu   (Fig. 2-15)
                                    ─────────────
Total change $\boxed{0.007293 \text{ amu}}$

In terms of kinetic energy,

$$0.007293 \text{ amu} \times 931.46 \text{ MeV/amu} = \boxed{6.79 \text{ MeV}}$$

2. The nuclide density n^{28} of ^{238}U. The definition of SI units (see App. II) provides that one mole [mol] of any element contains the same number of atoms as 0.012 kg [12 g] of ^{12}C. In the same manner that a mole of ^{12}C has a mass equivalent to its mass number ($A = 12.000$, as used to define the atomic mass unit [amu]) in grams, a mole of each other element consists of *its own* A-value in grams. A mole of substance, in turn, contains Avogadro's number A_0 of atoms ($\sim 6.02 \times 10^{23}$ at/mol as noted in App. II). Thus, nuclide density n^j may be expressed as

$$n^j = \frac{A_0 \text{ (at/mol)}}{A \text{ (g/mol)}} \times \rho \text{ (g/cm}^3) = \frac{A_0}{A} \rho \text{ (at/cm}^3)$$

for density ρ. Uranium-238 with nominal density $\rho^{28} \approx 19.1$ g/cm^3 would then have

$$n^{28} = \frac{6.02 \times 10^{23} \text{ at/mol}}{238 \text{ g/mol}} \times 19.1 \text{ g/cm}^3 = \boxed{4.83 \times 10^{22} \text{ at/cm}^3}$$

(Note that use of the integer value 238 compared to the actual atomic mass of 238.05 (Fig. 2-15) introduces a negligible error.) [Nuclide density calculations for mixtures are somewhat more complex since compositions are typically expressed as percentages or fractions by molar composition (or, equivalently, atom number), weight, or volume. Accordingly, the corresponding A- and ρ-values must be weighted averages of those for the constituent species.]

3. The activity of 1 g of ^{238}U:

$$\text{Activity} = \lambda^{28} n^{28}(t) \quad \text{(Eq. 2-4)}$$

$$T^{28}_{1/2} = 4.47 \times 10^9 \text{ years} \quad \text{(Fig. 2-15)}$$

$$\lambda^{28} = \frac{\ln 2}{T_{1/2}} = \frac{0.693}{4.47 \times 10^9 \text{ years}} = 1.55 \times 10^{-10} \text{ years}^{-1}$$

$$n^{28} = \frac{A_0}{A_{28}} = \frac{6.02 \times 10^{23} \text{ at/mol}}{238 \text{ g/mol}} = 2.53 \times 10^{21} \text{ at/g}$$

$$\text{Activity} = 1.55 \times 10^{-10} \text{ years}^{-1} \times 2.53 \times 10^{21}$$

$$\text{Activity} = 3.92 \times 10^{11} \text{ years}^{-1} \times \frac{1 \text{ year}}{3.15 \times 10^7 \text{ s}} \times \frac{1 \text{ Bq}}{1 \text{ s}^{-1}} = \boxed{1.24 \times 10^4 \text{ Bq}}$$

$$\text{Activity} = 1.24 \times 10^4 \text{ s}^{-1} \times \frac{1 \text{ Ci}}{3.7 \times 10^{10} \text{ s}^{-1}} = 3.37 \times 10^{-7} \text{ Ci} = \boxed{0.337 \text{ } \mu\text{Ci}}$$

4. Time required for ^{238}U to decay by 1 percent:

$$n(t) = n(0)e^{-\lambda^{28} t} \quad \text{(Eq. 2-6)}$$

$$n(t) = 0.99n(0)$$

$$\lambda^{28} = 1.55 \times 10^{-10} \text{ year}^{-1}$$

$$\frac{n(t)}{n(0)} = 0.99 = e^{-(1.55 \times 10^{-10} \text{ year}^{-1})t}$$

$$\ln (0.99) = -(1.55 \times 10^{-10} \text{ year}^{-1})t$$

$$t = 6.48 \times 10^7 \text{ year} = \boxed{64,800,000 \text{ years}}$$

5. Average neutron density corresponding to a typical LWR thermal flux $\Phi = 5 \times 10^{13}$ cm^{-2} s^{-1}, assuming an effective speed $v = 2200$ m/s:

$$\Phi = Nv \quad \text{(Eq. 2-18)}$$

$$N = \frac{\Phi}{v} = \frac{5 \times 10^{13} \text{ cm}^{-2} \text{ s}^{-1}}{2.2 \times 10^5 \text{ cm s}^{-1}} = \boxed{2.27 \times 10^8 \text{ cm}^{-3}}$$

6. Power produced by 1 g of ^{235}U fission in the LWR thermal flux in (5):

$$\text{Fission rate} = \Sigma_f^{25} \Phi \, dV = n^{25} \sigma_f^{25} \Phi \, dV \quad \text{(Eq. 2-19)}$$

$$n^{25} \, dV = n' = \text{number of atoms in the 1-g sample}$$

$$n' = \frac{6.02 \times 10^{23} \text{ at}}{235 \text{ g}} = 2.56 \times 10^{21} \text{ at/g}$$

$$\sigma_f^{25} = 584 \text{ b} \times \frac{10^{-24} \text{ cm}^2}{1 \text{ b}} = 584 \times 10^{-24} \text{ cm}^2$$

$$\text{Fission rate} = 7.48 \times 10^{13} \text{ fissions/s}$$

$$\text{Power} = 7.48 \times 10^{13} \text{ fissions/s} \times \frac{1 \text{ W}}{3.1 \times 10^{10} \text{ fissions/s}} = 2.41 \times 10^3 \text{ W}$$

$$\text{Power} = \boxed{2.41 \text{ kW}} \text{ per 1 g of } ^{235}\text{U}$$

7. For equal nuclide densities of ^{235}U and ^{239}Pu in a given reactor, find (a) fraction of fissions for each and (b) absorption mean free path for each nuclide and for the mixture. Assume each has an atom density of 10^{21} at/cm^3.

a. Fission fraction

$$F^j = \frac{\Sigma_f^j \Phi}{\Sigma_f^{mix} \Phi} = \frac{n^j \sigma_f^j}{n^{25} \sigma_f^{25} + n^{49} \sigma_f^{49}} \xrightarrow{n^{25} = n^{49}} \frac{\sigma_f^j}{\sigma_f^{25} + \sigma_f^{49}}$$

$$F^{25} = \frac{584 \text{ b}}{584 \text{ b} + 742 \text{ b}} = \boxed{0.44} \quad \text{in } ^{235}\text{U}$$

$$F^{49} = \frac{742 \text{ b}}{584 \text{ b} + 742 \text{ b}} = \boxed{0.56} \quad \text{in } ^{239}\text{Pu}$$

b. Mean free paths

$$\Sigma_a^{25} = n^{25}(\sigma_\gamma^{25} + \sigma_f^{25}) \quad n^{25} = 10^{21} \text{ at/cm}^3 \times \frac{10^{-24} \text{ cm}^2}{1 \text{ b}} = 10^{-3} \text{ at/b·cm}^\dagger$$

$$\Sigma_a^{25} = 10^{-3} \text{ at/b·cm } (98 \text{ b} + 584 \text{ b}) = 0.682 \text{ cm}^{-1}$$

$$\Sigma_a^{49} = 10^{-3} \text{ at/b·cm } (270 \text{ b} + 742 \text{ b}) = 1.012 \text{ cm}^{-1}$$

$$\Sigma_a^{mix} = \Sigma_a^{25} + \Sigma_a^{49} = 1.694 \text{ cm}^{-1}$$

$$\lambda_a^{25} = (\Sigma_a^{25})^{-1} = \boxed{1.47 \text{ cm}} \quad \text{for } ^{235}\text{U}$$

$$\lambda_a^{49} = (\Sigma_a^{49})^{-1} = \boxed{0.988 \text{ cm}} \quad \text{for } ^{239}\text{Pu}$$

$$\lambda_a^{mix} = (\Sigma_a^{mix})^{-1} = \boxed{0.590 \text{ cm}} \quad \text{for both}$$

EXERCISES

Questions

2-1. Identify the constituent parts and describe the basic structure of an atom.
2-2. Define the following terms: isotope, half-life, and mean free path.
2-3. Explain the fundamental differences among fissile, fissionable, and fertile nuclides; identify members of each classification.
2-4. Identify the radiations associated with fission.
2-5. Sketch the relationship among the following reactor cross sections:
    a. total interaction
    b. scattering
    c. absorption
    d. fission
    e. capture
2-6. Differentiate between microscopic and macroscopic cross sections.

Numerical Problems

2-7. Consider a thermal-neutron fission reaction in ^{235}U that produces two neutrons.
    a. Write balanced reaction equations for the two fission events corresponding to the fragments in Fig. 2-7.

---

†The unit at/b·cm is convenient and often used for atom densities employed in construction of macroscopic cross sections.

    b. Using the binding-energy-per-nucleon curve, estimate the total binding energy for ^{235} U and each of the fragments considered in (a).

    c. Estimate the energy released by each of the two fissions and compare the results to the accepted average value.

2-8. The best candidate for controlled nuclear fusion is the reaction between deuterium and tritium. The reaction is also used to produce high energy neutrons.

    a. Write the reaction equation for this D–T reaction.

    b. Using nuclear mass values in Table 2-4, calculate the energy release for the reaction.

    c. Calculate the reaction rate required to produce a power of 1 W, based on the result in (b).

    d. Compare the D–T energy release to that for fission on per-reaction and per-reactant-mass bases.

2-9. Gamma rays interacting with $^{9}_{5}$ Be or $^{2}_{1}$ H produce "photoneutrons." Write the reaction equation for each and calculate the threshold gamma energy. (Use mass data from Table 2-4.)

2-10. The nuclide $^{218}_{84}$ Po emits either an alpha particle or a beta particle with a half-life of 3.05 min.

    a. Write equations for each reaction.

    b. Calculate the decay constant λ and mean lifetime τ.

    c. Determine the number of atoms in a sample that has an activity of 100 μCi.

    d. Calculate the activity after 1, 2, and 2.5 half-lives.

2-11. Natural boron has a density of 0.128×10^{24} at/cm^3 and cross section $\sigma_a = 760$ b and $\sigma_s = 4$ b at an energy of $E = 0.025$ eV.

    a. Calculate the macroscopic cross sections at 0.025 eV for absorption, scattering, and total interaction.

    b. What fractional attenuation will a 0.025-eV neutron *beam* experience when traveling through 1 mm of the boron? 1 cm?

    c. Assuming the absorption cross section is "one-over-v" in energy, calculate the macroscopic cross sections for boron for neutrons of 0.0025-eV and 100-eV energies.

    d. What thickness of boron is required to absorb 50 percent of a 100-eV neutron beam?

2-12. Estimate the fraction of thermal neutron absorptions in natural uranium that causes fission. Estimate the fraction of fast-neutron fissions in natural uranium that occurs in ^{238} U. (Use data from tables and figures in this chapter.)

TABLE 2-4
Nuclear Mass Values for Selected Nuclides

Nuclide	Nuclear mass (amu)
$^{1}_{0}$n	1.008665
$^{1}_{1}$H	1.007825
$^{2}_{1}$H	2.014102
$^{3}_{1}$H	3.016047
$^{3}_{2}$He	3.016029
$^{4}_{2}$He	4.002603
$^{8}_{5}$Be	8.005305
$^{9}_{5}$Be	9.012182

SELECTED BIBLIOGRAPHY[†]

Nuclear Physics
    Burcham, 1963
    Evans, 1955
    Hyde, 1964
    Kaplan, 1963
    Kramer, in press.

Nuclear Data Sources
    GE/Chart, 1977
    Honeck/ENDF, 1966
    Hughes/BNL-325, 1955
    Lederer & Shirley, 1978

Other Sources with Appropriate Sections or Chapters
    Benedict, 1981
    Benedict & Pigford, 1957
    Cohen, 1974
    Connolly, 1978
    Duderstadt & Hamilton, 1976
    Etherington, 1958
    Foster & Wright, 1977
    Glasstone & Sesonske, 1967
    Henry, 1975
    Lamarsh, 1966, 1975
    Murray, 1975
    Rydin, 1977
    WASH-1250, 1973

[†]Full citations are contained in the General Bibliography at the back of the book.

3

NUCLEAR RADIATION ENVIRONMENT

Objectives

After studying this chapter, the reader should be able to:

1. Identify the basic mechanisms for radiation interaction with matter.
2. Calculate radiation dose from a mixed source.
3. Describe the basic principles for setting maximum permissible concentrations (MPC) of radionuclides and explain the ALARA concept.
4. Identify the basic features of radiation damage in biological tissue and reactor materials.
5. Describe the basic features of a composite reactor shield.
6. Calculate attenuation for gamma-ray and neutron beams.

Prerequisite Concepts

Nuclear Radiations	Chapter 2
Radioactive Decay	Chapter 2
Nuclear Reactions, Cross Sections, Flux	Chapter 2

Nuclear fission results in the production of many types of radiation by the direct and indirect processes considered in the previous chapter. The major potential hazard of commercial nuclear power is associated with the ability of the radiations to damage biological and material systems.

Public perception of radiation hazards is colored in part by the three following considerations.

1. Atomic and nuclear science is relatively new, dating back only as far as 1895 when W. K. Roentgen discovered the x-ray. The basis for nuclear energy applications was the discovery of fission announced in 1939.[†]
2. Radiation is not directly detectable by any of the human senses except at levels well above lethality.
3. The detonation of two nuclear weapons in Japan in 1945 provided the world with a dramatic and terrifying introduction to nuclear energy and radiation effects.

These have led to the concept of radiation as a new, invisible, silent, and deadly hazard.

In reality, radiation effects are not conceptually different from those known to occur from physical, chemical, and/or biological agents. Radiation effects have been very thoroughly studied and are better understood than the effects of many common environmental "insults" like the emissions from a coal-burning power plant.

The radiation environment associated with the fission process results in several unique problems in nuclear reactor design. Based on origin and impact, the following classifications are useful:

1. fission fragments, prompt neutrons, and gamma radiation emitted at the time of fission
2. activation-gamma radiation emitted as a result of (n, γ) reactions
3. delayed neutrons emitted by fission fragments and spontaneous-fission neutrons from transmutation products
4. delayed alpha, beta, and gamma radiations with half-lives from fractions of a second to millions of years emitted by fission fragments, activation products, and transmutation products

The first category consists of radiations that are emitted only while the fission chain reaction continues. As shown in Table 2-2, these radiations account for 90 percent of the energy associated with fission. Their major impact is the requirement for energy removal and radiation shielding at all times when a reactor is producing power.

Activation gammas are present only when there is a neutron source. Since most neutrons appear with fission, the impact of these gammas is comparable to that of the prompt fission gammas (except as noted below).

The delayed and spontaneous-fission neutrons are essentially insignificant in number and total energy compared to other neutron sources. However, they are the only neutrons that exist after shutdown of the chain reaction. They are also the only *delayed* radiations capable of causing nuclear reactions and thereby producing secondary radiations and radioactive species in a reactor.

Although the major significance of the small fraction of delayed fission neutrons is in reactor control (as considered in Chap. 5), they can also cause fission and activation outside of the reactor core in certain systems. The molten-salt-breeder reactor (MSBR) described in Chap. 14, for example, has a fluid core that carries the delayed

---

[†]Although the history of the "atomic and nuclear age" is fascinating, little of it is traced herein. Kaplan (1963) integrates history and theory well for the technically minded. Hiebert and Hiebert (1970) profile the leading personalities in interesting pamphlets directed to the general public.

neutrons through the entire primary coolant loop. Thus, extra shielding is required for these neutrons and the radiations produced by the reactions they induce. The direct effects of delayed neutrons disappear within a few minutes after reactor shutdown because of the short half-life of the phenomena. On the other hand, components activated by the neutrons may provide a radiation source of relatively long duration.

Certain of the new nuclides produced by transmutation undergo spontaneous fission and consequently emit neutrons. The isotope ^{240}Pu, for example, is a long-lived spontaneous-fission neutron source in all irradiated fuel that contains ^{238}U. Substantial concentrations of ^{240}Pu or other such nuclides lead to requirements for neutron shielding after reactor shutdown.

The final category of fission-related radiations includes the betas and gammas from fission products and the alphas, betas, and gammas from transmutation products. The fission-product radiations are responsible for $7\frac{1}{2}$ percent of fission energy (as per Table 2-2). According to Eq. 2-10, the power falls off only as roughly the one-fifth power of time following shutdown, e.g., to roughly 1 percent of operating power after one day and to 0.1 percent after two months. The long-lived activity leads to a requirement for virtually constant shielding, heat removal, and/or remote handling for reactor, spent-fuel storage, reprocessing, waste management, and related transportation operations.

The transmutation [transuranic-element] products as a group generate less power soon after shutdown, but have longer half-lives than their fission-product counterparts. For a typical LWR, their post-shutdown power is only 3–5 percent as great for on the order of a year. At very long times, however, the heat and radioactivity of the transuranics becomes the dominant problem of waste management (as described in Chaps. 6 and 10).

INTERACTION MECHANISMS

An alternative classification scheme for the radiations is based on the basic mechanisms by which they interact with various atoms and nuclei. The charged particles, electromagnetic radiation, and neutrons each behave in fundamentally different ways.

Charged Particles

Alpha particles [$^4_2$He^{2+}], beta particles [$^0_{-1}$e], and fission fragments each have one or more unpaired charges. The charges experience electrostatic [Coulomb] forces when they come close to the electrons of the atoms that compose the medium of interest. As a result of the forces, the charged particles lose energy with each interaction. This energy, in turn, ultimately appears in the system as heat.

The Coulomb forces are proportional to the product of the charges and inversely proportional to the square of the distance between them, or

$$F = k\,\frac{qq'}{r^2} \tag{3-1}$$

for force F, charges q and q', distance r, and proportionality constant k. According to the equation, the force decreases fairly rapidly with distance, but becomes negligible only at very large distances. This implies that at any given instant of time a charged particle experiences forces from a large number of electrons. The resulting energy losses are found to be rather well defined for each charged particle and each material medium.

The net macroscopic effect of charged-particle interactions is often characterized in terms of *range* and *linear energy transfer* [LET]. As the name implies, the range is the average distance traveled by a charged particle before it is completely stopped. The LET is the energy deposition per unit distance of travel.

Mathematically,

$$\text{LET} = \frac{dE}{dx}$$

for particle energy E and distance x and where the LET itself is generally a function of E. Since the range R is the total distance of travel for initial particle energy E_0 to be reduced to zero,

$$R = \int_0^{E_0} \frac{dE}{\text{LET}} = \int_0^{E_0} \frac{dE}{dE/dx}$$

Both the range and the LET of a specific radiation contribute to the effect they have on a material. The range determines the distance of penetration. The LET determines the distribution of energy deposited along the path.

Fission fragments generally have masses between 80 and 150 amu (Fig. 2-6) and charges of about $+20e$ (Evans, 1955) at the time of fission. The combination of large mass and high charge result in a range of only a few centimeters in air and a fraction of a millimeter in solid material. Thus, fission fragments generally stop very near their point of origin and deposit all of their energy within this short distance of travel. As a consequence, they have a high LET.

Alpha particles have a mass of 4 amu and are doubly charged. At typical energies, they have ranges only 3–6 times greater than fission fragments and LETs about an order of magnitude lower. As a point of reference, this sheet of paper is thick enough to stop any of the alphas or fission fragments produced by a nuclear reactor.

The combination of low mass and single charge gives electrons relative ranges about 100 times greater than those for alpha particles and LETs correspondingly reduced. Because of their low mass, the paths traced out by electrons deviate greatly from the roughly straight paths of the heavy charged particles. Both the total path length and net straight-line distance of travel (i.e., the range) vary substantially for individual electrons but have a well-defined "spread" of values.

Electromagnetic Radiation

Photons of electromagnetic radiation interact directly with electrons and more rarely with nuclei. Three important mechanisms shown by Fig. 3-1 are:

1. *The photoelectric effect*—photon energy is converted completely to kinetic energy of an orbital electron.

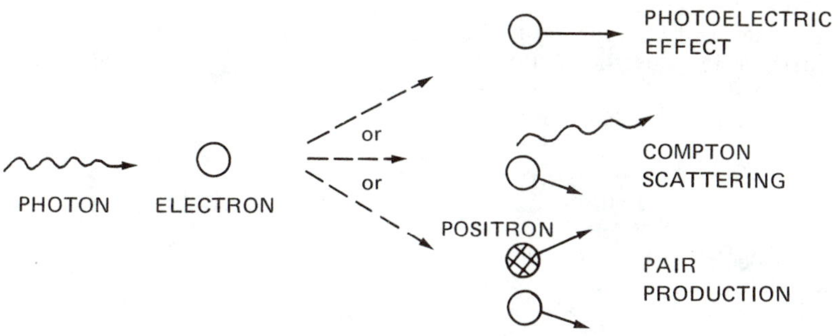

FIGURE 3-1
Interaction of electromagnetic radiation with an electron by photoelectric, Compton scattering, and pair-production mechanisms.

2. *Compton scattering*—photon transfers a portion of its energy to an electron and leaves the reaction at a correspondingly lower energy.
3. *Pair production*—photon energy is converted to mass and kinetic energy of an electron–positron pair.

Only the photoelectric effect results in the complete loss of an x- or gamma-ray photon. The Compton process reduces the energy and changes the direction of the incident photon. The energy imparted to the electron is dissipated to heat as for any charged particle.

The pair production interaction can occur only for a photon whose energy exceeds the mass of the two particles, i.e., twice the electron mass of 5.5×10^{-4} amu $= 0.511$ MeV. When the positron ultimately stops and contacts another electron, the combined mass is converted into two 0.511-MeV photons called *annihilation gammas*. Thus, the net effect of pair production is the conversion of one high-energy photon into two of 0.511 MeV.

Very short-range forces govern the electromagnetic mechanisms. A photon must essentially "hit" an electron for an interaction to occur. Thus, a parallel drawn to the neutron interactions described in Chap. 2 shows the same type of statistical behavior. Since for such a system an individual photon may travel essentially any distance, the concept of photon range may be defined only in terms of the average or mean of a large sample. This, of course, is in contrast to the well-defined range associated with charged-particle interactions.

A very rough comparison of relative range and LET for the three naturally occurring radiations is shown in Table 3-1. Since the important mechanisms each depend more on the density of electrons than on the specific atom composition, the relationships are relatively material-independent. These "rule of thumb" values demonstrate that gamma radiation is about 100 times as penetrating as beta particles. The betas, in turn, are more penetrating than alpha particles by about the same factor. As would be expected, the LET values are inversely related to the ranges.

Neutrons

A wide range of neutron interaction mechanisms were identified in Chap. 2. Of these, the absorption and scattering reactions are of most interest in the context of radiation effects.

TABLE 3-1
Comparison of Range and LET for Naturally
Occurring Radiations in a Specified Material

Radiation	Relative range	Relative linear energy transfer [LET]
Alpha	1	10,000
Beta	100	100
Gamma	10,000	1

Most absorption reactions result in the loss of a neutron coupled with production of a charged particle or a gamma ray. When the product nucleus is radioactive, additional radiation is emitted at some later time.

Scattering reactions result in the transfer of energy from a neutron to a nucleus. The latter then interacts in the system as a charged particle.

As was established previously, neutron interactions may be characterized by a mean free path λ as an average range. Generalized comparisons of neutron behavior with that of other radiations, however, are not readily made because of the extreme sensitivity of λ values to neutron energy and material composition.

RADIATION EFFECTS

Since nuclear radiations are energetic, they all have some potential for producing changes in biological tissue and other materials. Overall effects are determined by the type, energy, and intensity of the radiation and the detailed composition of the material medium.

Absorbed Dose

Historically, radiation *exposure* for x- and gamma-radiations was measured in units of the *roentgen* [R], where

1 roentgen = amount of radiation required to produce 1 esu of charge from
either part of an ion pair in 1 cm³ of air at standard temperature
and pressure

One roentgen is also equivalent to depositing about 88 ergs of energy in 1 g of air (88 ergs will move the point of a sharpened pencil about $1\frac{1}{2}$ mm across a piece of paper). Straightforward measurements may be made to determine radiation exposure in roentgens. However, the unit was not found to be very useful for comparing the effects of the various radiations on materials (i.e., effects on air do not necessarily correlate well to effects on other substances).

The concept of *radiation absorbed dose* [rad] † is based on energy deposition per unit mass of material, defined as

---

†Although the rad and the rem are not SI-derived units, their counterparts—1 gray [Gy] = 1 J/kg = 100 rads and 1 sievert [Sv] = 1 J/kg × QF = 100 rem—have not yet been generally accepted.

1 rad = 100 erg/g

This unit is applicable to any radiation and to any material. However, it is applied mostly to biological systems.

It has been found that the effect of radiation on a system depends on both the absorbed dose and the LET of the radiation. For biological systems, it is convenient to define a *relative biological effectiveness* [RBE] for the various types of radiation as

$$RBE = \frac{\text{dose of 250-keV x-rays producing given effect}}{\text{dose of reference radiation for same effect}}$$

The RBE depends on the effect studied, the dose, the dose rate, the physiological condition of the subject, and other factors.

The upper limit of RBE's for a specific type of radiation is called the *quality factor* [QF]. Currently recommended factors are provided in Table 3-2. It may be noted that fission fragments are included in the third entry, along with alpha particles.

The unit *rem*[†] (originally "roentgen equivalent man," but now simply rem) is defined by

$$\text{Dose (rem)} = QF \times \text{dose (rad)}$$

so that the *potential* effects (since QF is an *upper limit*) of all types of radiation can be considered from a common reference. Doses in rem are additive, independent of the specific radiation types involved.

A dose accumulated over a very short period of time is said to be *acute*, while one accumulated over an extended period is said to be *chronic*. For a given total dose, an acute dose has been considered more harmful because natural repair mechanisms can operate during the acquisition of a chronic dose. The *dose rate* is the absorbed dose per unit time. Conversely, the dose is the product of the dose rate and the time over which it is delivered.

[†]See footnote on page 70.

TABLE 3-2
Quality Factors Recommended by the International Commission on Radiological Protection[†]

Radiation	Quality factor
X-rays, gamma rays, and electrons	1
Neutrons, protons, and singly charged particles of rest mass greater than one atomic mass unit of unknown energy	10
Alpha particles and multiply charged particles (and particles of unknown charge) of unknown energy	20
Thermal neutrons	2.3

[†]Reprinted with permission from ICRP/26 "Recommendations," copyright © 1977, Pergamon Press, Ltd.

Radiation Damage

Radiation causes damage to various materials through three main mechanisms:

1. displacement of electrons and atoms
2. large energy release in small volumes
3. production of impurities

All types of radiation cause displacements and energy deposition. Only neutrons produce impurity nuclei. Such nuclei result from fission and activation reactions.

Heavy charged particles and neutrons can transfer large amounts of energy to cause displacement of "knock-on" atoms. These atoms, in turn, can cause ionization and produce a cascade of secondary knock-on atoms. In high-symmetry crystalline lattices, displacement atoms may leave lattice vacancies and lodge in interstitial locations or cause interchange of dissimilar atoms in the lattice structure.

Fission fragments are highly energetic (84 percent of the fission energy, as per Table 2-2), highly charged ions which cause considerable ionization, displacement of atoms, and heat deposition over their very short ranges. They also become impurities with respect to the lattice and may contribute further damage by emission of delayed beta and gamma radiations.

Like fission fragments, alpha particles cause ionization, displacement, and heat deposition over a very short range. Since alpha particles are helium nuclei, buildup of this inert gas may also cause pressurization problems, e.g., from (n, α)-reactions in ^{10}B control-rod material.

Beta radiation causes ionization and some displacement of atoms. The relatively short range leads to localized heat deposition. Gamma radiation also causes ionization but only rare displacements (the latter via nuclear Compton interactions). Gamma heating occurs over fairly substantial distances. The relatively lower damage potential of the betas and gammas is reflected in their quality factor of unity (see Table 3-2).

Fast neutrons generally cause multiple displacements through scattering interactions. Because of their great range, they present a biological hazard and cause most of the radiation damage experienced by ex-core reactor components.

Thermal neutrons cause radiation damage indirectly through absorption reactions. These reactions may lead to fission, charged-particle emission, or gamma emission. They may also produce lattice impurities.

Radiation tends to be increasingly damaging in the following order of molecular formation:

1. metallic bond
2. ionic bond
3. covalent bond
4. Van der Waals bond

largely due to the ability of ionization to disrupt the bonds. Since biological tissue is characterized by substantial covalent bonding, it is generally more susceptible to radiation damage than the metallic-bonded structural components. Other radiation-damage considerations include the observations that:

- Damage effects are generally less at elevated temperatures where enhanced diffusion may provide repair mechanisms.

- Low melting points enhance "annealing" (i.e., migration of dislocated atoms) and reduce radiation damage effects.
- Dose and dose rate are both important in determining overall damage levels.

Biological Effects

Biological cells are subject to radiation damage via direct and indirect mechanisms. Direct effects are thought to cause about 20 percent of the damage, while indirect effects account for the remainder.

Tissue is affected directly when radiation interacts with cell nuclei to break important molecular chains, e.g., the DNA required for cell reproduction. This type of damage is generally not repairable.

Indirect damage mechanisms break less critical molecules, like H_2O, into reactive parts, which in turn undergo detrimental chemical reactions with DNA, protein, or other important molecules. Since diffusion processes control such damage, natural defense mechanisms of the body have some opportunity to act to reduce the effects. Figure 3-2 shows three ways that radiation can break the covalent H_2O bonds. Recombination of components can then result in production of new species, of which H_2O_2 (peroxide) and $HO_2^{\cdot}$ are potentially most damaging.

Cell damage tends to be greatest in cells which are multiplying most rapidly. This establishes the following hierarchy of highest to lowest susceptibility:

1. lymph
2. blood
3. bone
4. nerve
5. brain
6. muscle

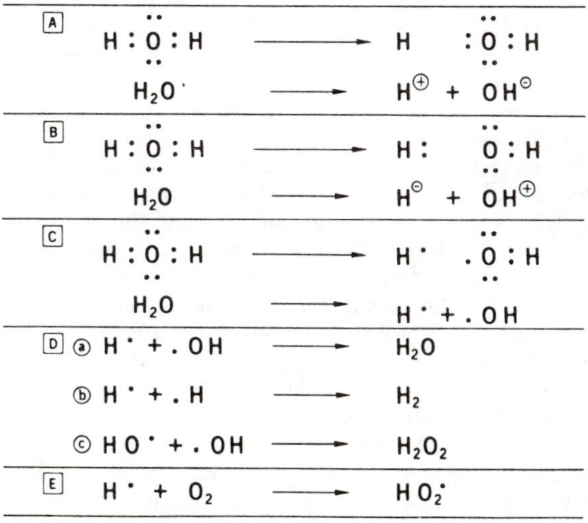

FIGURE 3-2
Primary (A, B, C) and secondary (D, E) products of radiation interaction with water molecules (courtesy of U.S. Department of Energy).

Human Response

Very large acute doses of radiation have readily identifiable effects on the human body. Although individual differences occur, there are general trends which are summarized in Table 3-3. It may be noted that even doses well above lethality cannot be detected by the human senses until the medical effects appear.

Lethal dose estimates are often expressed in the form LD 50/30, meaning "*Lethal Dose for 50 percent* of the population within *30 days.*" A whole body dose of about 450 rem is considered to be LD 50/30 for a general human population.

Delayed effects of large acute doses and of comparable chronic doses include:

• leukemia and cancers
• cataracts
• genetic effects
• blood disorders
• lifespan shortening

Although considerable controversy exists over the magnitude of delayed effects, especially for low-dose-rate chronic exposure, radiation effects are known substantially better than comparable "environmental insults" of other origins. The "Biological Effects of Ionizing Radiation" [BEIR (1980)] report, prepared by a committee of the National Academy of Sciences and National Research Council, is a definitive reference on radiation effects.

Reactor Materials

Essentially all radiation damage to materials in nuclear reactors results from neutron interactions. Fragments, fast neutrons, gammas, and delayed radiations from fission are major contributors. Radiations from activation- and transmutation-product nuclides

TABLE 3-3
Probable Effects of Acute Whole-Body Radiation Doses[†]

Acute dose (rem)	Probable clinical effect
0–75	No effects apparent. Chromosome abberations and temporary depression in white blood cell levels found in some individuals.
75–200	Vomiting in 5 to 50 percent of exposed individuals within a few hours, with fatigue and loss of appetite. Moderate blood changes. Recovery within few weeks for most symptoms.
200–600	For doses of 300 rem or more, all exposed individuals will exhibit vomiting within 2 h or less. Severe blood changes, with hemorrhage and increased susceptibility to infection, particularly at higher doses. Loss of hair after 2 weeks for doses over 300 rem. Recovery within 1 month to a year for most individuals exposed at lower end of range; only 20 percent survive at upper end of range.
600–1000	Vomiting within 1 h, severe blood changes, hemorrhage, infection, and loss of hair. From 80–100 percent of exposed individuals will succumb within 2 months; those who survive will be convalescent over a long period.

[†]From WASH-1250 (1973).

also cause damage. Thus, damage magnitude in a given system may be viewed as being related to the neutron flux history. The product of the neutron flux and the time over which it occurs is called *fluence* and serves as a convenient replacement for absorbed dose in measuring radiation damage in reactor materials. Since flux has units of neutrons per unit area per unit time [n/cm^2·s], fluence is expressed in terms of neutrons per unit area [n/cm^2]. Neutron *irradiation* is said to occur when a material is subjected to a neutron flux or, equivalently, when it accumulates neutron fluence.

Nonfuel compositions are damaged primarily by fast neutrons. For this purpose it is common to define a *fast fluence* as that based only on neutrons whose energies exceed a threshold value (e.g., neutrons above 1 MeV that cause damage to steel structural components).

In mathematical terms, fluence is defined as

$$\text{Fluence} = \int_0^t \Phi(t)\,dt \tag{3-2}$$

for time-dependent flux $\Phi(t)$. The flux-time integral reduces to

$$\int_0^t \Phi(t)\,dt = \Phi_0 t = Nvt \tag{3-3}$$

when a constant flux Φ_0 is present since from Eq. 2-16 the flux is equal to the product of the neutron density N and speed v. Based on Eqs. 3-2 and 3-3, the terms *fluence, flux-time,* and *nvt* are often used interchangeably. Typical fast fluences for steel components, or "epi-1-MeV" values, are determined from

$$\text{Fluence} \ (\geqslant 1 \ \text{MeV}) = \int_{1\,\text{MeV}}^{E_{\max}} \int_0^t \Phi(E, t)\,dt\,dE \tag{3-4}$$

for maximum neutron energy $E_{\max}$ and energy- and time-dependent flux $\Phi(E, t)$.

Essentially all materials are subject to radiation damage. Selection of reactor materials is influenced heavily by their stability in the anticipated neutron environment. For example, ceramic reactor fuels like UO_2 are generally favored over metallic forms (as considered in some detail in Chap. 9).

The water moderator and coolant employed in LWRs is subject to dissociation in a neutron environment. The hydrogen and oxygen produced in the process tend to enhance corrosion of cladding and other structures.

The graphite blocks used as moderators in the HTGR design tends to swell (i.e., increase in volume and decrease in density) with increasing fluence. Thermal resistance and stored or internal energy are also amplified by neutron irradiation.

Metal structural materials including the cladding, support fixtures, and pressure vessel are subject to a variety of neutron-induced changes. Important examples are:

- hardening and embrittlement due to disruption of initially symmetric lattice patterns
- swelling, or decreased density, caused by displacement of atoms from normal lattice sites and/or the presence of impurity atoms
- transformation of metallurgical phase (perhaps including a change of the overall lattice pattern)
- decreased corrosion resistance due to transmutation of alloying constituents
- changes in mechanical properties

In water-cooled systems, reduced corrosion resistance in structures combines with the increased corrosiveness of dissociated water to enhance the overall rate at which these chemical reactions occur.

The effects of irradiation on several mechanical properties of one particular type of steel are shown in Fig. 3-3. Although description of the general mechanisms involved is beyond the scope of this book, the significance of the shapes of the two yield curves may be readily inferred. Yield strength is related to the amount of force required for permanent sample deformation, ultimate strength to that required for sample fracture. Although both increase with fluence, they converge, providing successively smaller ranges over which deformation without fracture is possible.

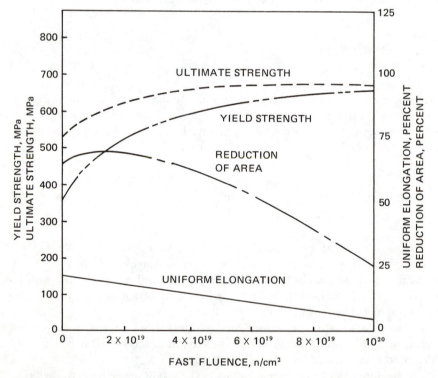

FIGURE 3-3
Dependence of mechanical properties of A212B carbon–silicon steel on fast-neutron fluence. (Adapted from *Engineering Materials Science*, by C. W. Richards. Copyright © 1961 by Wadsworth, Inc. Reprinted by permission of the publisher, Brooks/Cole Publishing Company, Monterey, California. Data from Wilson and Berggren, Am. Soc. Test. Mats. *Proceedings* 55, 702, 1955.)

Changes in structural properties often determine the maximum useful lifetimes of various structural components. Steel pressure vessels in LWRs, for example, have estimated lifetimes of 40-50 years in typical fast-neutron environments.

Electronic components are another important class of materials subject to radiation damage. Since semiconductor devices like transistors and integrated circuits depend on very closely controlled lattice compositions and impurity levels, they are especially susceptible to neutron damage via displacements and activation. Thus, control and safety circuitry must either be radiation-resistant or well protected from radiation exposure.

DOSE ESTIMATES

It is often possible to make rough estimates of absorbed radiation doses from knowledge of radiation type and energy plus material composition. For charged particles, energy deposition occurs over a short range. Electromagnetic radiations and neutrons, by contrast, are generally characterized by widely distributed energy deposition.

Radiation dose from a given source may be reduced by limiting exposure time or increasing distance. The use of shielding material is often the most effective means for restricting personnel doses.

Alpha and Beta Radiation

None of the charged particles produced in a reactor core have ranges great enough to be of concern outside of the reactor vessel. However, radionuclides that emit alpha and/or beta particles are produced by fission and other neutron reactions. When such species leak or are otherwise removed from the core, they may come in contact with biological tissue. The resulting absorbed dose depends on particle energy, nuclide activity or decay rate, length of time, and tissue density.

If radionuclides enter the body, essentially all of the charged-particle energy is deposited in organ tissue. External radioactivity affects only the skin. Alpha-particle emitters must be in direct contact to have any impact. Beta-particles, on the other hand, may penetrate some amount of clothing, with a resulting decrease in energy. In either case, only the particles which actually strike the skin contribute to absorbed dose (e.g., since the particles have random directions, half of those emitted from a flat surface travel outward from the skin and are of no further concern).

If an internal source deposits all particle energy in an organ of mass m, the time-dependent dose rate $R_p(t)$ is approximately

$$R_p(t) = \frac{Q(t)\bar{E}}{m} \tag{3-5}$$

for activity $Q(t)$ and average energy $\bar{E}$ per particle. The energy $\bar{E}$ is the full transition energy for alpha decay, but is only about one-third of the transition energy for beta decay (because of the sharing with antineutrinos noted in Chap. 2). For external doses, the rates may be modified by factors which account for surface and/or clothing effects as appropriate.

Typical applications of Eq. 3-5 are based on the replacement

$$Q(t) = Q_0 e^{-\lambda t}$$

from Eq. 2-6 for initial activity Q_0 and decay constant λ. Conversion of $R_p(t)$ to units of rad/s and integration over the desired time interval provides the total absorbed dose. If the dose equivalent in rem is desired, the result is multiplied by a quality factor of 1 or 20 for beta or alpha particles, respectively.

Gamma Radiation

Gamma radiation is subject to electromagnetic interactions with atomic electrons and nuclei. Since these occur on a "one-shot," statistical basis, there is some probability for very great distances of travel, even through dense materials. Primary gamma radiation associated with fission or the secondary radiation produced by its interactions may escape from a reactor core and the surrounding structures.

The probability per unit distance of travel that a gamma ray photon will interact with nuclei of a given element is the *linear attenuation coefficient* μ. It is fully analogous to the neutron macroscopic cross section Σ, which is also an interaction probability per unit path length. Thus, Eq. 2-21 for narrow-beam neutron attenuation may be modified directly to

$$\Phi(x) = \Phi(0)e^{-\mu x} \tag{3-6}$$

for gamma flux Φ, linear attenuation coefficient μ, and distance of travel x. As was the case for neutrons, the form of the equation dictates that the mean free path of a gamma ray is $\lambda = 1/\mu$ and that there is a finite (though perhaps extremely small) probability of a photon penetrating to any arbitrarily large distance.

The coefficient μ is dependent on the gamma-ray energy and on the density and elemental composition of the material. Since the interactions occur predominantly with orbital electrons, the isotopic make-up of a sample has no significant effect (e.g., all uranium enrichments "look the same" to gamma rays).

The density dependence of μ may be removed by defining a mass attenuation coefficient μ/ρ for elemental density ρ. Equation 3-6 is then modified to

$$\Phi(x) = \Phi(0)e^{-(\mu/\rho)\rho x} \tag{3-7}$$

where the product ρx, the areal density (with typical units g/cm^2), replaces distance x in the formulation. The mass attenuation coefficient as a function of energy for lead is shown in Fig. 3-4. Contributions from the photoelectric, Compton-scattering and pair-production mechanisms are identified on the figure.

Although the total attenuation coefficient is appropriate for describing narrow-beam attenuation, it does not allow direct calculation of energy deposition and absorbed dose. Compton-scattering and pair-production interactions result in partial energy conversion plus emission of secondary radiation. Thus, a total absorption coefficient is defined to include all photoelectric interactions but only the appropriate contributions of the other two mechanisms. Figure 3-4 has divided the Compton interaction into absorption and scattering components with only the former summed into the curve for energy absorption. As a practical matter, the *total* pair-production coefficient is added into the total absorption coefficient since even the potential loss of one or more of the .511-MeV annihilation photons has a minimal effect when viewed in the context of the other processes that occur simultaneously (Evans, 1955).

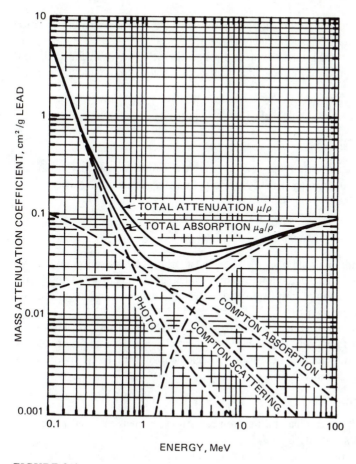

FIGURE 3-4

Mass attenuation coefficients for total interaction and absorption of electromagnetic radiation in lead, including contributions from photoelectric, Compton scattering and absorption, and pair-production effects. (Adapted from *The Atomic Nucleus*, by R. D. Evans, © 1955 by McGraw-Hill, Inc. Used by permission of McGraw-Hill Book Company.)

Gamma radiation is generally so penetrating that both internal and external sources can contribute to the net flux and ultimately to the absorbed dose. The dose rate R_γ is approximately

$$R_\gamma(t) = \Phi_\gamma(t)E_\gamma \frac{\mu_a}{\rho} \tag{3-8}$$

for time-dependent gamma flux Φ_γ, gamma energy E_γ, and mass absorption coefficient μ_a/ρ for tissue density ρ. The total absorbed dose in rad is obtained by converting units as necessary and integrating over all energies and over the desired time interval.

Neutron Radiation

Fast-neutron radiation is highly penetrating in biological tissue and is, thus, potentially hazardous as an external source. As a practical matter, internal neutron sources are rare and are associated with much more highly damaging charged-particle emission (e.g., ^{240}Pu neutrons are from spontaneous fission).

Fast-neutron scattering generates heat through kinetic energy transfer to charged nuclei. The fast-neutron dose rate R_{fn} in a single-constituent medium may be approximated by

$$R_{fn}(t) = \frac{\Phi_n(t)E_n\Sigma_s f}{\rho} \tag{3-9}$$

for time-dependent neutron flux Φ_n, neutron energy E_n, macroscopic scattering cross section Σ_s, density ρ, and average fractional energy transfer per collision f where

$$f = \frac{2A}{(A + 1)^2}$$

for a nuclide of atomic mass number A. When considering actual tissue, contributions in the form of Eq. 3-9 must be computed for all nuclides. Total absorbed dose in rad then depends on conversion of units, summation over all nuclides, and integration over fast-neutron energies and over time.

Absorbed dose from thermal neutron radiation is an indirect result of absorption reactions. Activation radiations and induced radioactivity are the important sources. Dose calculations require detailed descriptions of the neutron flux, tissue composition, and interaction and decay mechanisms.

Sample Calculations

Consider dose rates from gamma and neutron radiations. Each is assumed to have an energy of 1 MeV and a flux of 10^8 cm$^{-2}\cdot$s^{-1}.

For 1-MeV gamma rays in water (or, roughly, hydrogenous material like most human tissue), $\rho \approx 1$ g/cm^3 and $\mu_a \approx 0.03$ cm^{-1}. The dose rate R_γ (Eq. 2-8) is thus

$$R_\gamma = \frac{\Phi_\gamma E_\gamma \mu_a}{\rho} = \frac{10^8 \text{ } \gamma/\text{cm}^2 \cdot \text{s} \times 1 \text{ MeV} \times 0.03 \text{ cm}^{-1}}{1 \text{ g/cm}^3} \times \frac{1.60 \times 10^{-13} \text{ J}}{1 \text{ MeV}}$$

$$\times \frac{1 \text{ erg}}{1 \times 10^{-7} \text{ J}} \times \frac{1 \text{ rad}}{100 \text{ erg/g}} = \boxed{0.048 \text{ rad}}$$

where the energy conversion factors are obtained from App. II. Since QF = 1 for gamma radiation,

$$R_\gamma = 0.48 \text{ rad} \times 1 = \boxed{0.048 \text{ rem}}$$

Fast neutron dose rates are obtained from Eq. 3-9:

$$R_{fn} = \frac{\Phi_n E_n \Sigma_s f}{\rho}$$

Since neutrons in water react primarily with the hydrogen atoms,

$$f = \frac{2A}{(A+1)^2} = \frac{2(1)}{(1+1)^2} = 0.5$$

Assuming $\Sigma_s \approx 0.1 \text{ cm}^{-1}$,

$$R_{\text{fn}} = \frac{10^8 \text{ n/cm}^2 \cdot \text{s} \times 1 \text{ MeV} \times 0.1 \text{ cm}^{-1} \times 0.5}{1 \text{ g/cm}^3} \times \frac{1.60 \times 10^{-13} \text{ J}}{1 \text{ MeV}}$$

$$\times \frac{1 \text{ erg}}{1 \times 10^{-7} \text{ J}} \times \frac{1 \text{ rad}}{100 \text{ erg/g}} = \boxed{0.08 \text{ rad}}$$

or since QF = 10 for fast neutrons,

$$R_{\text{fn}} = 0.08 \times 10 = \boxed{0.80 \text{ rem}}$$

If both fluxes existed simultaneously,

$$R = R_\gamma + R_{\text{fn}} = 0.048 \text{ rem} + 0.80 \text{ rem} = \boxed{0.848 \text{ rem}}$$

(The attenuation of either the gamma rays or the neutrons can be calculated in the manner shown for the latter at the end of Chap. 2.)

Dose Reduction

Radiation levels from operating reactors and from irradiated fuel can be extremely large. Three basic principles for reducing personnel dose from such radiation sources are to:

1. restrict the *time* of proximity
2. increase the *distance* from the source
3. use *shielding* material to attenuate the radiation

Time and Distance

Time restriction is generally valid only in situations where distance and shielding cannot be used and where short-term exposure will result in performing the necessary task. In high radiation environments, workers may accumulate maximum weekly or quarterly doses very quickly (i.e., they are "burned out"), and then be excluded from further exposure for a specified period of time.

The decrease of dose rate with distance is most readily observed by considering a point radiation source. Since the photons or particles from the source S_0 "spread out" to progressively larger $4\pi r^2$ spherical-areas, the flux becomes

$$\Phi(r) = \frac{S_0}{4\pi r^2}$$

a familiar "inverse square" or "one-over-r-squared" attenuation. The use of this "geometrical" attenuation is of value mainly where radiation levels are reasonably low and where substantial amounts of unused space are available.

Shielding Principles

Since time restrictions have limited application and the cost of facility floor-space dictates against distance, shielding plays the dominant role in dose reduction for nuclear facilities. Relationships among the important radiations associated with the fission process are depicted in Fig. 3-5. With one exception, all constituents in the figure have been described in this or the previous chapter. *Bremsstrahlung*, an electromagnetic radiation produced by deceleration of electrons, makes a negligible contribution to radiation dose from reactor systems.

Shielding for charged particles is readily accomplished. Even longer range electrons are stopped by a few millimeters of metal, e.g., the walls of a typical liquid-waste handling tank.

Gamma and neutron radiations create very complex shielding problems because of the potential long ranges for both primary and secondary radiations. The principles embodied in Eqs. 2-23 and 3-6 are always valid, but must be applied separately for each energy. Because of the highly energy-dependent nature of cross sections—e.g., as for ^{238}U in Fig. 2-11—neutron calculations are very complex. Even the more regular variation of attenuation coefficients—e.g., in Fig. 3-4—make gamma calculations difficult.

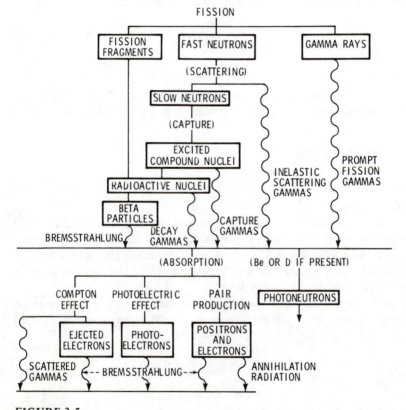

FIGURE 3-5

Radiation produced as a result of fission. (Adapted from *Nuclear Reactor Engineering* by Samuel Glasstone and Alexander Sesonske, © 1967 by Litton Educational Publishing, Inc. Reprinted by permission of Van Nostrand Reinhold Company.)

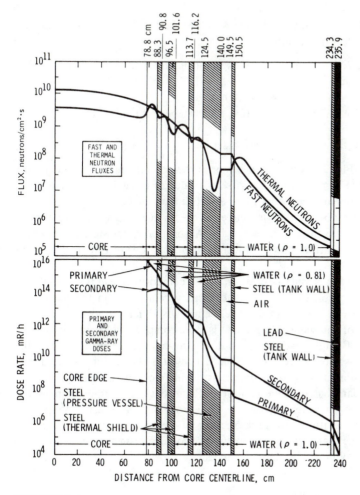

FIGURE 3-6

Neutron-flux and gamma-radiation dose profiles in a composite shield for a 70-MW reactor designed for use aboard the S. S. Savannah. (Adapted from *Nuclear Reactor Engineering* by Samuel Glasstone and Alexander Sesonske, © 1967 by Litton Educational Publishing, Inc. Reprinted by permission of Van Nostrand Reinhold Company.)

Scattering reactions for both radiations produce changes in energy and direction which in turn determine new reaction probabilities and escape path lengths. The secondary radiations shown on Fig. 3-5 complicate the picture further.

Calculational procedures have been developed to describe the transport of neutrons, of gamma rays, and of the two together. The basic principles of neutron calculations are described briefly in Chap. 4. Applications to photon and coupled neutron–photon transport employ conceptually similar methods.

Reactor Shields

Effective reactor shields must attenuate both gammas and neutrons, including the secondary radiations they produce. Figure 3-6 shows an example of a composite shield designed to minimize the total weight of a 70-MW reactor system aboard the S.S. Savannah. Important features include:

1. steel *thermal shields* to remove heat energy from the neutrons and gammas to adjacent cooling water
2. steel and lead for gamma and neutron attenuation
3. water to thermalize fast neutrons so that they can be absorbed by the water, steel, or lead
4. a slab arrangement to handle secondary radiations

The curves on Fig. 3-6 show the behavior of the neutron and gamma radiations. In the top figure, the fast-neutron flux is seen to decrease regularly, with inelastic scattering in the steel being somewhat more effective than the elastic scattering in the water at causing attenuation. Moderation increases the thermal flux when it first enters a water layer, but capture then reduces it again. Absorption in the steel decreases the thermal flux, but fast-neutron slowing down compensates toward the outside of the slab.

The behavior of the gamma population in the composite shield is shown by the curves in the lower portion of Fig. 3-6. The primary gammas from the reactor are attenuated somewhat regularly, with the steel being substantially more effective than water. The behavior of the secondary gammas is more complicated because it includes not only effects of Compton scattering, but is also coupled to neutron radiative-capture and inelastic-scattering reactions. The net result is that the dose rate due to secondary neutrons dominates that of the primaries for most of the shield thickness.

Conventional land-based reactors have little incentive for minimizing shield weight. Thus, massive amounts of concrete can provide a more easily fabricated, lower cost shield. The relatively low mass of concrete is good for neutron moderation, while at the same time many of its constituents are good neutron absorbers. Required gamma attenuation is obtained by sheer thickness.

RADIATION STANDARDS

Ideal standards, or limits, for radiation exposure are those that assure the health of operations personnel and the public, while allowing for reasonable efficiency in operations with radioactive materials. To determine appropriate standards, two types of exposure must be considered:

1. exposure to external radiation
2. exposure from internal radiation

The practical basis for setting standards considers natural background radiation levels and experimental data on dose–effect relationships.

Natural background radiation varies throughout the world. For the United States, typical contributions are:

1. cosmic or extraterrestrial radiation, 40–130 mrem per year
2. terrestrial radiation (mainly U, Th, and ^{40}K in the soil), 30–115 mrem per year
3. internal radiation (mainly U, Th, ^{40}K, ^{14}C, and ^{3}T in the human body), 25 mrem per year

The general range is 100–250 mrem per year with an average of about 130 mrem per year. Medical and other radiation sources contribute an extra 40 mrem per year to bring the "best estimate" average to 170 mrem per year.

Experimental dose–effect data on human subjects has been limited to post-exposure studies of groups of individuals who have received very large acute or chronic radiation exposures. Nuclear-weapon detonations, accidents, other inadvertent exposures (e.g., the radium-dial painters mentioned later in this chapter), and medical radiation treatments provide the bulk of the information base. Detailed, controlled experiments have been performed only with animal populations.

Available human data and that inferred from animal experiments is generally combined to prepare estimates of low-level radiation consequences as a function of population dose. Such doses are quantified in terms of *person-rem* (formerly "man-rem")—the product of the number of individuals and the dose received summed over the population of interest. It is estimated, for example (BEIR, 1972), that a general population dose of 10^6 person-rems would cause 90–460 excess cancers (e.g., over and above normal incidence).

External Radiation

Careful analyses of available data have helped to identify external radiation doses which may be expected to produce no statistically significant medical effects. Organizations like the International Commission on Radiological Protection [ICRP] and the National Council on Radiation Protection and Measurements [NCRP] have pioneered such efforts. Current recommended dose limits[†] from the latter are tabulated in Table 3-4. The recommendations are divided roughly into occupational, general public, and special-case categories.

Radiation workers in the nuclear industry, hospitals, and elsewhere are required to wear *dosimeters*, or devices that monitor radiation exposure. Composite devices may be employed to make separate measurements of charged-particle, gamma, and/or neutron doses as appropriate to a given facility. Conversion of measured parameters to doses in rems allows calculation of the total dose equivalent.

The most general occupational dose limit is the 5-rem annual whole-body value. It is intended to handle nonspecific situations in a manner conservative enough to cover very sensitive tissues like the gonads, lens of the eye, and red bone marrow. Under more controlled conditions, where only particular tissues are exposed, the higher limits noted in Table 3-4 may be applied.

Whole body exposure of 5 rems is a prospective or target limit. Since dosimeters typically require processing time, however, it is not uncommon for some excess to be accumulated. A retrospective annual limit of 10–15 rems is suggested on this basis. However, there is also a total long-term accumulation limit of $(N - 18) \times 5$ rems to an age of N years (with an inherent assumption that no occupational dose should be received before age 18). If either the retrospective or long-term limit is exceeded in normal operations, the worker should be prohibited from further exposure for some period of time.

Exposure limits for individual nonemployees and student trainees are set at

[†]These are reasonably consistent with the federal regulations in 10CFR20 (as described in Chap. 19).

TABLE 3-4
Dose-Limiting Recommendations from the National Council on Radiation Protection
and Measurements[†]

Maximum permissible dose equivalent for occupational exposure	
Combined whole-body occupational exposure	
Prospective annual limit	5 rems in any 1 year
Retrospective annual limit	10-15 rems in any 1 year
Long-term accumulation to age N years	$(N - 18) \times 5$ rems
Skin	15 rems in any 1 year
Hands	75 rems in any 1 year (25/qtr)
Forearms	30 rems in any 1 year (10/qtr)
Other organs, tissues, and organ systems	15 rems in any 1 year (5/qtr)
Fertile women (with respect to fetus)	0.5 rem in gestation period

Dose limits for the public, or occasionally exposed individuals	
Individual or occasional	0.5 rem in any 1 year
Students	0.1 rem in any 1 year

Population dose limits	
Genetic	0.17 rem average per year
Somatic	0.17 rem average per year

Emergency dose limits—life-saving	
Individual (older than 45 years if possible)	100 rems
Hands and forearms	200 rems, additional (300 rems, total)

Emergency dose limits—less urgent	
Individual	25 rems
Hands and forearms	100 rems, total

[†]From NCRP/39 (1971).

lower levels of 0.5 and 0.1 rem per year, respectively. Maximum limits for a general population exposure are 0.17 rem per year based on both genetic and somatic considerations.

Special circumstances are recognized by provisions for preplanned emergency radiation doses. The recommended whole-body limits are 100 rem for saving a life and 25 rem for less urgent actions such as limiting radiation release or fighting fires. In both cases, the NCRP recommends use of volunteers as well as certain other actions and restrictions.

All of the recommended dose limits, and especially the general population value of 170 mrem per year, should be recognized as maxima rather than targets. The ICRP, NCRP, and current federal regulations all call for exposures to be *"as low as reasonably achievable"* [ALARA]. This suggests the desirability of reduced radiation doses (individual as well as person-rem) tempered by cost-versus-benefit considerations.

Historically, reactor emissions have been on the order of a few mrem per year, and only a small fraction of radiation workers have received as much as 5 rem in a given year.

Internal Radiation

Once radionuclides are deposited in organ tissues, little can be done to reduce the resulting radiation dose. The primary method for controlling internal exposure is to limit potential uptake by specifying *maximum permissible concentrations* [MPC] for radionuclides in air and water discharges from nuclear facilities.

In practice, MPCs are established according to the following concepts that:

1. A 0.1 μCi *body burden* of ^{226}Ra has no discernable health effects (as discerned from studying "radium dial painters" who had ingested substantial quantities of radium), so an equivalent dose from one or more other radionuclides should also be considered "safe."
2. Evaluation of the effect of a particular radionuclide must consider
   - radiation type
   - radiation energy
   - radioactive half-life
   - biological clearance rate
   - critical organ† (i.e., that organ damage to which is most detrimental to the entire organism)
   - fraction of ingested nuclide deposited in organ
3. An *effective half-life* $T_{1/2}^{\text{eff}}$ for a radionuclide in the human body can be calculated from the radioactive half-life $T_{1/2}$ and the biological clearance rate $T_{1/2}^{\text{bio}}$ according to the expression

$$\frac{1}{T_{1/2}^{\text{eff}}} = \frac{1}{T_{1/2}} + \frac{1}{T_{1/2}^{\text{bio}}} \tag{3-10}$$

4. The following assumptions apear to be conservative
   - dose-versus-effect data can be extrapolated linearly to very low doses (i.e., do not assume that a threshold exists below which no damage will occur).
   - there is no dose-rate effect (i.e., neglect the fact that repair mechanisms may reduce the effect of a dose received at low dose rates over a long period of time).
5. Nuclide ingestion rates can be determined from intake patterns for air and water.
6. The data can be used to determine the maximum concentration of a nuclide that will lead to a dose equivalent no greater than that of the reference 0.1 μCi source of ^{226}Ra.
7. MPC's can be established by considering the above result, correcting for the effects of other nuclides, and including a "safety margin."

Table 3-5 contains MPC and other data for five nuclides of interest in the nuclear fuel cycle.

---

†As noted earlier, the "critical organs" for external radiation are the gonads, lens of the eye, and the red bone marrow.

TABLE 3-5

Data Related to Maximum Permissible Concentrations [MPC] for Several Radionuclides of Interest in the Nuclear Fuel Cycle

	$^{3}_{1}$H	$^{51}_{24}$Cr	$^{90}_{38}$Sr	$^{226}_{88}$Ra	$^{239}_{94}$Pu
Form of nuclide	H_2O	Soluble	Soluble	Soluble	Soluble/insoluble
Physical half-life	12.3 y	27 d	28 y	1622 y	24,360 y
Biological half-life	19 d	110 d	11 y	44 y	120 y/360 d
Effective half-life	19 d	22 d	9 y	44 y	120 y/360 d
Radiation (energy)	β(18 keV)	γ(.33 MeV)	β(.5/2.3 MeV)	α(14.5 MeV)	α(5.2 MeV)
Critical organ	Whole-body	Kidney	Bone	Bone	Bone/lungs
MPC[†] (μCi/ml) water	3×10^{-3}	2×10^{-3}	3×10^{-7}	3×10^{-8}	$5 \times 10^{-6}/3 \times 10^{-5}$
air	2×10^{-7}	4×10^{-7}	3×10^{-11}	2×10^{-12}	$6 \times 10^{-14}/1 \times 10^{-12}$

[†]MPC values prescribed by 10CFR (1979) for general effluent release.

When several radionuclides are under consideration, each concentration must be reduced to a fraction of its MPC value. The limiting expression that must be satisfied is

$$\sum_{\text{all } j} \frac{C_j}{(\text{MPC})_j} \leqslant 1.0$$

for concentration Cj and (MPC)j of the nuclide identified by subscript j.

EXERCISES

Questions

3-1. Explain why charged particles can be assigned definite ranges while neutrons and gamma rays cannot.

3-2. State the three mechanisms by which electromagnetic radiations interact with electrons and identify the secondary radiation(s) produced by each.

3-3. Describe the basic principles for calculating maximum permissible concentrations [MPC]. Explain the ALARA concept.

3-4. Discuss the basic differences between gamma rays and neutrons in terms of the damage they could produce in a steel pressure vessel.

Numerical Problems

3-5. Calculate the individual and total radiation doses in rem for each radiation and in total for a 2-h exposure to:
a. 20-mrad/h gamma
b. 15-mrad/h alpha
c. 5-mrad/h fast neutron
d. 25-mrad/h thermal neutron
Which radiation is potentially most harmful according to the calculations?

3-6. Repeat the previous problem using the SI units of Gy and Sv.

3-7. Consider a 15,000-Ci ^{60}Co source (typical of several university campuses) which gives off gamma rays of 1.17 MeV and 1.33 MeV from each decay. Assume that the dose rate at a distance R cm from a source of strength C Curies emitting gamma energy E MeV for each disintegration is given by

$$D \text{ (mrad/h)} = 4.6 \times 10^6 \frac{CE}{R^2}$$

a. Calculate the average hourly dose associated with a 5-rem yearly limit (assume a 50-week year of 40-h weeks).

b. Calculate the dose rate at 1 m from the source and the time to receive the "emergency, life-saving dose."

c. Neglecting the effect of air and other materials, calculate the distance from the ^{60}Co source required to achieve the dose rate in (a).

d. Based on the mass attenuation coefficient curve in Fig. 3-3, estimate the thickness of lead ($\rho = 11.35$ g/cm^3) necessary to attenuate a beam with the dose rate in (b) to the acceptable level in (a). Neglect secondary radiations and geometry effects.

3-8. Since the Compton-interaction probabilities depend only on the density of electrons, the mass attenuation coefficient is material-independent for this mechanism. The total coefficient, then, is roughly material-independent for the energy range where the Compton effect is dominant (e.g., μ/ρ has the same energy dependence from 0.3–3 MeV for water and elements up to $Z \approx 26$ (iron); all elements up to lead have roughly the same μ/ρ over the 1–2 MeV range).

  a. Estimate the fractional absorption of ^{60}Co gammas passing through your chest, assuming that the density of the human body is that of water.

  b. Calculate the thickness of water and of concrete ($\rho \approx 4$ g/cm^3), which provide the same attenuation as the lead in Prob. 3-7(d). Assuming that each shield is a sphere surrounding a (point) source, calculate the required masses of lead, concrete, and water, respectively.

3-9. Boron-10 is often used as a thermal-neutron shield material because of its high absorption cross section (3837 b at 0.025 eV). Noting that full-density boron has an atom density of 0.128×10^{24} per cm^3, calculate for a beam of 0.025-eV neutrons:

  a. the absorption mean free path for ^{10}B

  b. the fraction transmitted through a 1-mm-slab of ^{10}B

  c. the relative (fractional) density of ^{10}B required for the 1-mm slab to attenuate the beam to 1 percent of its initial strength

3-10. The mean free path of a neutron beam from the D–T fusion reaction is about 135 m. What is the total macroscopic cross section of air?

3-11. Compute the effective half-life of ^{90}Sr in the human body. Also determine its maximum allowable discharge concentration in water if it must be mixed with 1×10^{-3} μCi/ml tritium.

3-12. The expression for the effective half-life in a biological system has the same form as that for parallel resistors in an electric circuit. Explain why.

3-13. Assuming that human tissue has the same heat capacity as water and that a 1°C temperature change is just noticeable, calculate the absorbed gamma dose in rad necessary to reach this "threshold of feeling." Compare the result to LD 50/30.

3-14. A pressure vessel is fabricated from a material whose properties become unacceptable after receiving a fast fluence of 10^{21}/cm^2. Calculate the expected lifetime in years for such a vessel subject to a 5×10^{11}/cm$^2 \cdot$s^{-1} fast neutron flux.

SELECTED BIBLIOGRAPHY[†]

Radiation Effects and Protection
  Azimov & Dobzhansky, 1966
  BEIR, 1972, 1980
  Billington & Crawford, 1961
  Cember, 1969
  Eisenbud, 1973
  EPA, 1974
  Frigerio, 1967
  Glasstone & Jordan, 1980 (ch. 5–7)
  Goldstein, 1959
  Henry, 1969
  Hiebert & Hiebert, 1970, 1973, 1974
  ICRP/26, 1977
  Kelly, 1966

[†]Full citations are contained in the General Bibliography at the back of the book.

NCRP/39, 1971
Profio, 1979
Robertson, 1969

Controversy on Low-Level Radiation
Archer, 1980
Bull. At. Sci. (current)
Lapp, 1979
Morgan, 1978
Rotblat, 1978
Science (current)

Other Sources with Appropriate Sections or Chapters
Burcham, 1963
Cohen, 1974
Connolly, 1978
Etherington, 1958
Evans, 1955
Foster & Wright, 1977
Glasstone & Sesonske, 1967
HEW, 1970
Holden, 1958
Kaplan, 1963
Lamarsh, 1975
Murray, 1975
Sagan, 1974
Schaeffer, 1973
WASH-1250, 1973
Wilkinson & Murphey, 1958

4

REACTOR PHYSICS

Objectives

After studying this chapter, the reader should be able to:

1. Differentiate between the infinite multiplication factor k_∞ and the effective multiplication factor k_{eff}.
2. Explain why neutrons experience greater average energy losses from collisions with hydrogen than from collisions with any other nuclides.
3. Identify each term in the four-factor formula and explain its significance.
4. Explain several basic neutron-balance controls.
5. Describe the effect of uniform dilution with water on the minimum critical mass for an unreflected, homogeneous plutonium system.
6. Differentiate among the diffusion-theory, transport-theory, and Monte Carlo calculational procedures.
7. Perform basic calculations of average neutron cross sections, multiplication factors, and average collision energy losses.

Prerequisite Concepts

Cross Sections	Chapter 2
ν, η, and α	Chapter 2
Flux	Chapter 2
Reaction Rates	Chapter 2

The theory of neutron chain reacting systems is called *reactor physics* or *reactor theory*. Time-independent and time-dependent phenomena are considered under the classifications of *reactor statics* and *reactor kinetics*, respectively. The former is the subject of this chapter, the latter of the next chapter.

In a *reactor* (implying, for now, *any* collection of materials in which fission can occur), the neutron population may be described by the equation

Rate of increase   rate of   rate of   rate of
in the number   = production − absorption − leakage
of neutrons   of neutrons   of neutrons   of neutrons

Accumulation = production − absorption − leakage   (4-1)

Accumulation = production − losses

This *neutron balance* equation represents the fact that neutrons must be conserved, i.e., neither created nor destroyed.

When the neutron population is steady at a nonzero level, the fission chain reaction is exactly self-sustaining and the system is said to be *critical*. Criticality may occur at *any* fission rate (or, equivalently, at any power level) as long as neutron losses are exactly balanced by neutron production.

Systems in which neutron production exceeds the losses are *supercritical* and characterized by increasing power levels. *Subcritical* systems have neutron losses in excess of production and, therefore, decrease in power level until a shutdown condition occurs.

Power reactors require provisions for adjusting the neutron balance. The reactors must be critical to produce steady-state power, supercritical to increase power, and subcritical to reduce power and/or shutdown. By contrast, nuclear materials in fuel-cycle facilities must be kept subcritical at all times.

INFINITE SYSTEMS

The idealized concept of an infinite system with homogeneous material properties is a useful starting point for developing reactor theory. Such systems can exhibit no spatial variations of the neutron population because the material composition is everywhere the same and neutrons would not leak from an infinite system.

Nonleakage is the functional basis for defining an infinite system. The behavior of neutrons in the central region of a very large system may often be approximated by that of an infinite system.

Neutron Multiplication

A critical system experiences no change in neutron level. Thus, the balance in Eq. 4-1 reduces to

Rate of production   rate of absorption
of neutrons   = of neutrons

Production = absorption   (4-2)

This result indicates that where leakage is not possible, all neutrons are eventually absorbed.

Recalling the definitions introduced in Chap. 2, the production rate (per unit volume) may be represented as

Production rate = number of neutrons per fission $\times$ fission rate

or   Production rate = $\nu \Sigma_f \Phi$ $\hfill$ (4-3)

for average number of neutrons per fission ν, macroscopic fission cross section Σ_f, and neutron flux Φ. The cross section and flux as employed here are independent of position and represent the *average* behavior of the entire (*energy-dependent*) neutron population. In a similar manner, the absorption rate (per unit volume) is

Absorption rate = $\Sigma_a \Phi$ $\hfill$ (4-4)

for macroscopic absorption cross section Σ_a.

The criticality condition for an infinite system is obtained by combining Eqs. 4-3 and 4-4 with the result that

$$\nu \Sigma_f \Phi = \Sigma_a \Phi \hfill (4\text{-}5)$$

Since the flux is the same on both sides of the equation,

$$\nu \Sigma_f = \Sigma_a \hfill (4\text{-}6)$$

or the product of the average number of neutrons per fission ν and the macroscopic fission cross section Σ_f is equal to the macroscopic absorption cross section Σ_a for an infinite critical system.

Since the product $\nu \Sigma_f$ is used routinely, it is conveniently considered as a unit. It may be viewed as the macroscopic cross section for fission-neutron production rather than as the product of the two terms. Therefore, $\nu \Sigma_f$ and $\nu \sigma_f$, respectively, are called the macroscopic and microscopic *neutron production cross sections*[†] in the remainder of this book.

The tendency of the neutron population in an infinite system to change is often quantitifed in terms of the *infinite multiplication factor k_∞* [*k-infinity*], defined as

$$k_\infty = \frac{\nu \Sigma_f}{\Sigma_a} \hfill (4\text{-}7)$$

the ratio of the macroscopic cross sections for neutron production and absorption.

[†]This nomenclature and a few other examples may be somewhat unique to the author. However, since the use of the mathematical symbols is completely unaffected, the reader should experience no difficulty in applying concepts among other reference documents.

This ratio is the same as the ratio of the average number of neutrons produced to the average number of neutrons absorbed. According to the time sequence, the neutrons in one "generation" are absorbed, cause fissions, and produce the neutrons of the next generation. Thus, k_∞ is a measure of the multiplication between neutron generations.

A critical infinite system has a precise balance between neutron production and absorption and, thus, has a multiplication of unity. Equivalently, it has a k_∞-value of unity as demonstrated by comparison of Eqs. 4-6 and 4-7.

Energy Dependence

The definition of the infinite multiplication factor k_∞ in Eq. 4-7 is deceptively simple, since the average macroscopic cross sections are not easily determined. The extreme energy dependence of fissionable-nuclide cross sections (e.g., as shown by Figs. 2-11 and 2-12) must be considered. Then, since reaction rates depend on both the cross sections and the neutron fluxes, the interaction between the latter two must be determined. The net result is that average cross sections for a given system are highly dependent on the detailed material composition and its arrangement.

Typical neutron spectra for an LWR and an LMFBR are shown in Fig. 4-1. The reactor designs, as described in Chap. 1, have slightly enriched uranium fuel with water moderator/coolant and mixed-oxide fuel with sodium coolant, respectively. The energy distribution of neutrons from fission is essentially the same for both systems. Thus, the shapes of the curves are dependent on the material

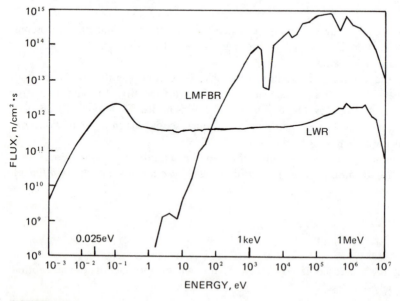

FIGURE 4-1
Typical neutron-flux spectra for an LWR and an LMFBR. (LWR data from a calculation for typical LWR fuel; LMFBR data from J. A. Rawlins, "Calculation of Passive Sensor Perturbations in FFTF," HEDL–TME 77–59, November 1977.)

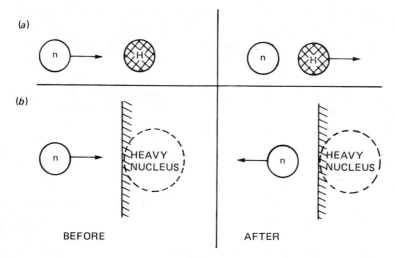

FIGURE 4-2
Effects of head-on "billiard-ball" collisions between (*a*) a neutron and a hydrogen nucleus (two balls) and (*b*) a neutron and a heavy nucleus (a ball and the table "bumper").

compositions and their geometric arrangements. The differences can be attributed to the neutron moderation or slowing-down effects which are discussed next. (A description of the methods for calculating detailed neutron fluxes, average cross sections, and reaction rates is deferred to the end of the chapter.)

Moderation

The relative probability of fission, or equivalently, the fission cross section, is smaller for the high-energy neutrons produced by fission than it is for very low energy neutrons, e.g., as shown for ^{235}U in Fig. 2-12. Thus, the ability to slow down neutrons may be employed to good advantage.

A neutron involved in a scattering reaction with a stationary nucleus loses some of its initial energy. When the scattering is elastic (i.e., when kinetic energy is conserved as described in Chap. 2), energy changes are governed by the same laws of physics used to describe macroscopic "billiard-ball" collisions.

At this point, it may be instructive to consider an actual set-up for billiards. The balls can represent a neutron and a proton [a hydrogen nucleus], both of which have roughly one unit of mass. The table can represent a very massive nucleus—e.g., of a uranium or plutonium isotope.

When a moving ball strikes a stationary ball in a direct, head-on collision, the former comes to a complete stop while the latter moves off in the same direction and with the same speed that initially characterized the other ball. This is equivalent to a neutron transferring all of its energy to a stationary hydrogen nucleus as shown in Fig. 4-2.

On the other hand, when a moving ball strikes the table "bumper," it is reflected back in the opposite direction but still has essentially the same speed. This is equivalent to a neutron striking a very heavy nucleus and losing almost none of its energy.

These examples illustrate the fact that a neutron can give up all of its energy in a collision with hydrogen and very little in a collision with a heavy nucleus. For nuclei of intermediate masses, the maximum energy transfer is somewhere between the two extremes. Collisions that are "glancing," i.e., not head-on, result in neutron energy losses of less than the maximum value.

Quantitatively, elastic collisions conserve both kinetic energy and momentum. This results in a maximum energy transfer ΔE_{max} from a moving neutron to a stationary nucleus of

$$\Delta E_{max} = E\left[1 - \left(\frac{A-1}{A+1}\right)^2\right] = E(1-\alpha) \tag{4-8}$$

where E is the initial energy of the neutron, A is the atomic mass number of the target nucleus (i.e., the ratio of its mass to that of the neutron), and α is defined by the equation.

Since the energy change in a collision with a given nucleus can vary from zero for a "near miss" to ΔE_{max}, the final neutron energy varies from its initial value E down to a minimum value E_{min} calculated from

$$E_{min} = E - \Delta E_{max} = \alpha E \tag{4-9}$$

Rearranging terms to

$$\frac{E_{min}}{E} = \alpha \tag{4-10}$$

shows that α is the smallest fractional energy a neutron can have after a single collision with a specified nucleus. Table 4-1 contains data of interest to neutron moderation, including maximum fractional energy change $\Delta E_{max}/E$ and α as defined above.

The fractional nature of the energy change in elastic scattering suggests the

TABLE 4-1
Parameters Related to Neutron Moderation by Nuclides of Interest in
Power Reactors

Nuclide	$\dfrac{\Delta E_{min}}{E}$	$\alpha = \dfrac{E_{min}}{E}$	$(\Delta u)_{max}$[†]	ξ	Number of collisions[‡]
$^1_1 H$	1.000	0.000	–	1.000	18
$^2_1 D$	0.889	0.111	2.198	0.725	24
$^{12}_6 C$	0.284	0.716	0.334	0.158	111
$^{23}_{11} Na$	0.160	0.840	0.174	0.085	206
$^{238}_{92} U$	0.017	0.983	0.017	0.0084	2084

[†] Lethargy change associated with maximum energy change.
[‡] Average requirement for a 1-MeV fission neutron to be reduced in energy to 0.025 eV.

use of a logarithmic scale for many neutron slowing-down calculations. A new variable, *lethargy, u*, is defined as

$$u(E) = \ln \frac{E_0}{E} = \ln E_0 - \ln E \qquad (4\text{-}11)$$

for (corresponding) energy E and reference energy E_0 selected to be the maximum under consideration. Lethargy has a zero value at the maximum energy and then increases with decreasing energy, i.e., as the neutrons get more "lethargic" or lazy. Thus, lethargy and energy scales run in opposite directions. The lethargy scale also has the effect of expanding the lower energies (equivalent, of course, to plotting energy on a logarithmic scale rather than a linear one). From Eq. 4-11, it may be noted that the lethargy change Δu associated with an energy change between any two arbitrary energies $E_1 > E_2$ is just $\ln (E_2/E_1)$. Table 4-1 shows lethargy changes corresponding to the maximum fractional energy changes. For all cases, $(\Delta u)_{max} = \ln (1/\alpha)$; this results in an infinite value for hydrogen (since its final energy can be zero and $u(0)$ is undefined) and finite values for all other nuclides.

The maximum lethargy change is a much less useful concept than the *average* lethargy change, or the *average logarithmic energy decrement* ξ. The latter can be employed in a number of calculations aimed at determining typical neutron-scattering behavior. For example, an estimate of the number of collisions needed to slow a fission neutron to thermal energies may be made by dividing the total required lethargy change Δu by the average per-collision ξ. A typical case starts with a fission neutron at about 1 MeV and ends with it at about 0.025 eV, for which Δu is 17.5. Table 4-1 includes values of ξ and estimated collisions for each nuclide; e.g., for hydrogen, $\xi = 1$ with roughly 18 collisions required for thermalization.

The number of collisions required to thermalize a fission neutron is only one factor employed in selecting a moderator. Scattering and absorption cross sections are also important. The basic goal is to slow the neutrons to thermal energies while losing as few as possible to absorption or leakage along the way.

Hydrogen in the form of ordinary water is an effective moderator in the LWR design. Because few collisions are required for thermalization, the neutrons do not travel as far as they might otherwise and allow for a relatively compact system. The absorption cross section, however, is large enough that some neutrons are lost. Deuterium and carbon, on the other hand, have very low absorption, but because of the larger number of collisions require larger cores. Design features of the CANDU and HTGR systems reflect this latter concern.

Liquid sodium was selected as a coolant for the fast-neutron-based LMFBR in part because it is *not* an effective moderating material. However, even the small energy changes with each collision can have an important impact on system safety as described in the next chapter.

Elastic scattering from heavy elements is responsible for very little slowing down of neutrons. Inelastic scattering, however, does remove some neutron energy (and reemits it in the form of gamma radiation). Mathematical modeling of the latter process is substantially more complicated than that for its elastic counterpart.

Four-Factor Formula

Thermal reactors by definition rely primarily on thermal neutrons to cause fissions. At these low energies, the fission cross section for fissile nuclides is quite large (e.g., see Fig. 2-10 for ^{235}U). Thermal-neutron fission produces fast neutrons, making slowing-down a crucial part of the overall process. If a system is relatively "large," it can be assumed to approximate an infinite one, i.e., to be leakage-free.

One of the earliest attempts to describe and calculate the behavior of thermal reactor systems is the *four-factor formula*. As implied by the name, the formula breaks neutron behavior into four parts. The *fast fission factor* ϵ is the ratio of total fissions to those caused by thermal neutrons only. The *resonance escape probability* p is the ratio of thermal-neutron absorption to that for neutrons of all energies. The *thermal utilization factor* f is the fraction of the total thermal-neutron absorption that occurs in *fissionable* nuclides. The *thermal "eta" factor* η is the average number of fission neutrons produced per thermal neutron absorbed in *fissionable material*. (It may be recalled that this latter concept was introduced in Chap. 2 for individual nuclides.)

The factors as defined seem somewhat disjointed at first. However, the following scenario puts them in context. Considering a large population of fast neutrons from fission, it is found that:

1. A fraction p reach thermal energies after *escaping* capture while slowing down through the *resonance* energies (i.e., those corresponding to the highly peaked resonance cross sections shown, for example, in Figs. 2-11 and 2-12).
2. A fraction f of the thermal neutrons are absorbed in fissionable material (while the remainder are absorbed elsewhere).
3. For each thermal neutron absorbed in fissionable material, an average of η fast neutrons are produced by fission.
4. The fast neutrons are augmented by a factor ϵ from those resulting from the fissions caused directly by fast neutrons (actually by some of the $(1 - p)$-fraction that did *not* reach thermal energies).

The relationship among the four factors is shown in a slightly different manner by Fig. 4-3. The number of initial fast neutrons is decreased by both fast and resonance absorption, although the effects are lumped into the resonance escape probability p. Part of the absorption results in the fissions that are characterized by the fast fission factor ϵ. Since by definition $\eta = \nu(\sigma_f/\sigma_a)$, the term in parentheses—the fraction of absorptions that lead to fission—is just η/ν (accounting for the temporary presence of the average number of neutrons per fission ν as a "fifth" factor). The thermal utilization f is treated in a straightforward manner. As must be true, the system is critical only if the product of the four factors is unity (i.e., it exactly replicates the initial neutron).

The ratio of the number of "second-generation" fission neutrons to the initial number for these examples is the product of the four terms. Since this ratio matches the definition of the infinite multiplication k_∞,

$$k_\infty = pf\eta\epsilon \tag{4-12}$$

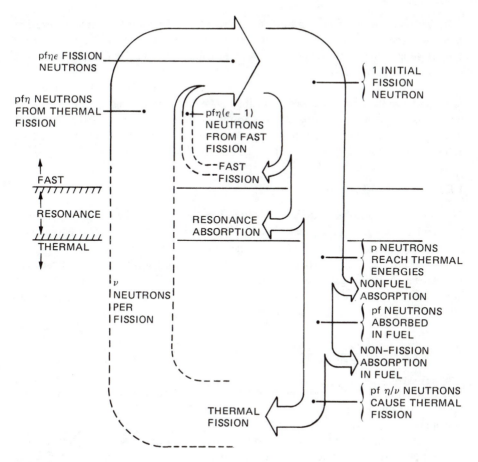

FIGURE 4-3
Flow diagram for neutrons in a thermal reactor from the standpoint of the four-factor formula.

The formulation was useful primarily because the thermal terms η and f could be calculated (e.g., based on a $1/v$-absorption cross section and a thermal flux with a regular behavior like that in Fig. 4-1). The parameters ϵ and p could be determined experimentally. Further modifications, of course, were required to treat finite systems (as considered briefly later in this chapter).

The four-factor formula also proves to be helpful for examining the difference between homogeneous and heterogeneous arrangements of fissionable materials. A uniform solution or mixture of natural uranium and moderator, for instance, cannot be made critical because neutrons are absorbed in the ^{238}U resonances (i.e., the resonance escape probability p is too small). However, a distribution of "lumped" natural uranium fuel distributed in a heavy water or graphite moderator can sustain a neutron chain reaction. This latter case increases p by allowing fission neutrons to leave the fuel lumps, to thermalize in moderator where no ^{238}U is present for absorptions, and to reenter the fuel at energies where ^{235}U fission is highly probable.

The separation of fuel and moderator tends to decrease the thermal utilization f because the likelihood of absorption in nonfuel constituents is enhanced. With careful design, the larger value of p can overshadow the smaller f to produce a net increase in the multiplication of the system.

The CANDU reactor system relies on separation of natural-uranium fuel and heavy-water moderator to attain criticality. The LWR designs use fuel-pin arrangements, as introduced in Chap. 1, to form a heterogeneous pattern that enhances neutron multiplication. Uranium enrichment, pin size, and pin spacing can all be adjusted to achieve the desired goal (e.g., as described in Chap. 11).

FINITE SYSTEMS

The abstraction of an infinite system is useful for eliminating the spatial dependence in preliminary or scoping calculations. However, all real systems have finite dimensions and experience the effects of neutron leakage.

The infinite multiplication factor k_∞ is really only significant as a measure of the general multiplying properties of a specific material or group of materials. The multiplication of a real system as a whole, of course, must account for leakage effects. The *effective multiplication factor* k_{eff} [*k-effective*] is defined as

$$k_{\text{eff}} = k = \frac{\text{neutron production}}{\text{neutron losses}}$$

where the losses result from both absorption and leakage. Since all systems are finite, it is common practice to drop the subscript "eff" such that k is the effective multiplication factor.

Unfortunately, the leakage term cannot be expressed as readily as the production and absorption terms. Approximate quantitative expressions for k are considered later in this chapter.

In a finite critical system, k must be unity. This may be expressed as

$$k = 1 = \frac{\text{production}}{\text{absorption} + \text{leakage}} \tag{4-13}$$

or
$$k = 1 = \frac{k_\infty^c}{1 + \text{leakage/absorption}}$$

where the latter is based on division by the absorption term and substitution of $k_\infty^c = $ production/absorption—the infinite multiplication factor of this *finite, critical* system. Since

$$k_\infty^c = 1 + \frac{\text{leakage}}{\text{absorption}}$$

for criticality, the ratio of leakage to absorption is essentially the excess multiplication required to compensate for the finite extent of the system.

Typical reactor systems contain regions of varying composition. Each individual region has the same value of k as the system as a whole. However, each region will have its own k_∞ characteristic of the material composition. In a critical reactor, for example, the physical leakage of neutrons that reduces the effective multiplication of a k_∞-greater-than-one region is a neutron source ("negative leakage") to a region with k_∞-less-than-one.

Even a region with $k_\infty = 0$ has an effective multiplication of unity in a critical system. Such a situation is common in *reflector* regions that contain no fissile material but serve the useful purpose of returning some of the neutrons that would otherwise leak completely away from the reactor core. The difference between the in-leakage from the core *to* the reflector and the out-leakage *from* the reflector is equal to the absorption rate. Thus, since the net leakage *from* the region is negative, its sum with the absorption term is zero (i.e., *net* losses *from* the region are nil). The zero production and zero net loss then balance to result in $k = 1$ when the reflector is an integral part of an otherwise critical system.

Criticality Control

It is instructive to examine some of the general features of the neutron balance before considering mathematical solutions. Methods of criticality control and the concept of critical mass are the topics here.

Criticality for a finite system may be determined by comparing its k-effective value to:

$k = 1$ critical

$k > 1$ supercritical

$k < 1$ subcritical

According to Eq. 4-13, the multiplication may be adjusted by changing production, absorption, and/or leakage. Typical procedures may concentrate on one aspect but may end up changing all three to some extent.

Even though steady power production may be desirable, no reactor is built to be exactly and constantly critical. Provisions must be made to accommodate power-level changes and to compensate for fuel depletion and related effects.

Active neutron-balance controls in reactors operate on one or more of the three terms in Eq. 4-13. Examples include:

1. production—adjust the amount of fissile material in the active core region
2. absorption—
   - use solid, moveable absorbers [*control rods*]
   - dissolve absorbing material in the coolant [*soluble poisons*]
   - employ solid, fixed absorbers [*burnable poisons*] which "burn out" gradually with neutron reactions over the lifetime of the reactor core
3. leakage—change system dimensions and density and/or modify the effectiveness of neutron reflection

All of these methods have been employed in research, testing, and training reactors.

Most power-reactor designs rely on one or more of the neutron-absorption procedures. However, neutron-production control is exercised by on-line refueling in the CANDU and in two advanced-reactor concepts. The light-water-breeder reactor [LWBR] system relies on in-core movement of driver fuel with respect to the surrounding blanket region to adjust the fission rate. Further discussions of basic control principles are contained in the next three chapters, while reactor-system design details are discussed in Chaps. 12-15.

Neutron-balance control is also important in fuel cycle operations outside of nuclear reactors. Here, it is desirable to maintain all fissionable materials in the subcritical state. Basic nuclear criticality safety depends on geometric and/or administrative controls applied to the neutron balance in one or more of the following ways:

1. production—limit fissile mass and/or concentration
2. absorption—use fixed or soluble neutron poisons
3. leakage—
   - employ high-leakage, "favorable-geometry" equipment and containers
   - limit neutron reflection
   - separate equipment and/or storage containers to limit neutronic interaction

Although it does not show explicitly in simple neutron-balance equations, moderation can have a major impact on each of the three other factors. Dilution of fissile material with water moderator, for example, enhances fission when neutron energies are reduced to the thermal range where the fission cross section is so large (e.g., as shown by Fig. 2-12 for ^{235}U). Similarly, the energy of neutrons determines the effectiveness of absorbers as, for example, with the large $1/v$-absorption cross section of ^{10}B. Moderation also affects leakage because the resulting dilution makes fissile nuclei "harder for a neutron to find." Reflectors constructed of moderating materials reduce leakage by returning neutrons and may also enhance multiplication by returning the fast neutrons at thermal or otherwise-reduced energies.

The interplay among production, absorption, leakage, and moderation is often extremely complex. The example that follows is intended to provide some perspective on the subject.

Illustrative Example

In the course of nuclear criticality safety studies, Clayton (1979) has identified a number of "anomalies of nuclear criticality" where systems of fissile materials behave in ways that are completely foreign to nonnuclear activities. The "anomaly" considered here is the criticality of plutonium as it is diluted uniformly with water.

Pure plutonium-239 metal in the metallurgical form known as "α-phase" has a density of 19.6 kg/liter. If it were formed into a sphere of 5-cm radius and left completely unreflected [*bare*], the resulting 9.8 kg of plutonium would be just critical or, equivalently, would be a *critical mass*. Dilution of the plutonium with water changes its density, critical mass, and critical dimensions as shown by the uppermost curve on Fig. 4-4. The figure (including a second curve for reflected critical mass) actually represents the results of calculations for idealized ^{239}Pu—water mixtures maintained in spherical geometries with and without water reflection.

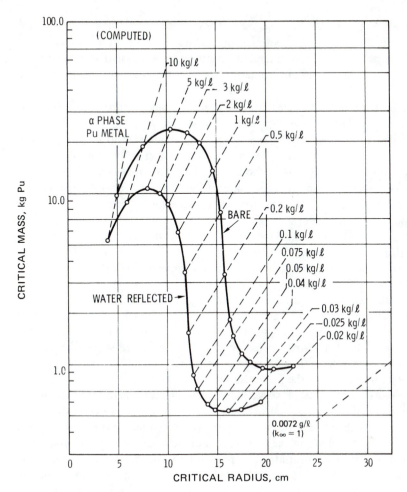

FIGURE 4-4
Computed mass and radius of critical [239]Pu–water spheres. (Adapted courtesy of E. D. Clayton, Battelle Pacific Northwest Laboratories.)

(Reality dictates multiisotope plutonium, dissolution of plutonium in nitric acid rather than water, and some kind of outer container(s); however, such differences do not affect the general trends of and conclusions drawn from the data on the figure.)

Criticality in the pure-metal sphere depends entirely on fast neutrons since moderation by this heavy metal is negligible. The dimensions of the system are comparable to the fission mean free path for the fast neutrons.

Initial dilution of the plutonium results in an increase in critical mass and critical size as shown in Fig. 4-4. The slight amount of moderation reduces some fission-neutron energies to the extent that resonance absorption is increasingly favored over fission. At the same time, the decreased fissile density enhances leakage.

Increasingly effective moderation allows more and more neutrons to escape capture at resonance energies. The effect, as shown in Fig. 4-4, is that the upper curve first levels off at a maximum of about 22 kg and 5 kg/liter. The critical mass then decreases, but the size increases as neutrons less easily "find the dispersed plutonium nuclei." The trends continue to a density of 32 g/l where *optimum moderation* results in an overall minimum critical mass of about 0.9 kg. This value is more than an order of magnitude less than that for the pure metal, but the associated volume is 75 times greater.

Although hydrogen in water always absorbs some neutrons, its effect is overshadowed by other mechanisms until dilution is very large, i.e., until the macroscopic fission cross section Σ_f is no longer large compared to the macroscopic absorption cross section Σ_a^H for hydrogen. The increase in critical mass at lower-than-optimum concentrations shown in Fig. 4-4 is caused by hydrogen absorption. This increase continues and approaches ever more closely the dashed line defining the 7.2-g/liter-concentration of an infinite critical system (i.e., the concentration for which $k_\infty = 1$).

A critical plutonium-metal sphere with water reflection has a critical mass of a little more than 5 kg, as shown by the lower curve on Fig. 4-4. This value of just over half of the bare critical mass results primarily from the ability of the reflector to return the fast neutrons at substantially lower energies. The moderating effect of the reflector reduces the critical mass and volume at all concentrations. However, at greater dilutions the reflector has successively less impact as its moderating properties more closely approach those of the core and as large core size results in a smaller fraction of total neutrons becoming available for reflection. On these bases, the curve for water reflection approaches increasingly more closely to its bare counterpart and ultimately to the k_∞-equals-one line.

Conventional safety wisdom suggests that dilution will generally reduce overall hazard. This is certainly true for most toxic chemicals. However, Fig. 4-4 indicates that dilution may actually encourage criticality within certain concentration ranges.

According to the curves in Fig 4-4, plutonium has the same critical mass at as many as three different concentrations—representing high, intermediate, and dilute (approaching $k_\infty = 1$) values respectively. Dilution of a fixed mass to a concentration between the high and intermediate points would leave the system subcritical. However, further dilution would lead to supercriticality.

Consider the highly hypothetical situation of a solid metal sphere that dissolves in a very large tank of acid such that the resulting solution is always homogeneous and expands spherically. This situation is applied to a plutonium sphere of about 5.1 kg; a horizontal line drawn at this mass on Fig. 4-4 can be used to analyze the situation. First, the sphere placed in the center of the acid is reflected enough to be just critical (assuming the acid reflects like water). However, as soon as any dilution occurs, the mass is insufficient to sustain the chain reaction, so the system becomes subcritical. Somewhere slightly below 1 kg/l, the solution will become just critical for a second time. Continued dissolution will produce supercriticality since the critical mass will be exceeded. Thus, until the very low concentration and large volume approaching the k_∞-equals-one condition is reached, the system would produce fissions, heat, and radiation at an ever-escalating rate. (That such chain reactions are generally self-limiting or self-terminating is addressed in the next chapter).

The discussions based on Fig. 4-4 point out the complexities that can be associated with processing operations that involve fissile solids and solutions. Although safety could be assured by limiting plutonium mass to 0.5 kg or less (i.e., to the overall minimum critical mass), realistically large fuel cycle requirements dictate against this. Instead, criticality safety must be implemented through equipment design and administrative procedures.

COMPUTATIONAL METHODS

The neutron population of any chain reacting system is difficult to model because it is characterized by a wide range of energies and directions. There are a variety of reactions, some with complex cross sections and secondary neutron emissions.

The first calculational techniques were based on highly simplified models. Then as digital computer technology has developed, successively more sophisticated methods have been developed and employed. State-of-the-art procedures and computers are now capable of performing highly accurate reactor physics calculations.

Diffusion Theory

In the simplest representation of a finite system, all neutrons are treated as if they have the same energy and have a net flow through the material. These assumptions liken the situation to that incurred in chemical diffusion, so the method is called *diffusion theory*. Although originally developed as "one-speed diffusion theory" in reference to the assumed monoenergetic neutrons, the basic method can be modified to be substantially more general (as considered later in this section).

The one-speed diffusion theory method describes neutron leakage through Fick's law of diffusion [*Fick's law*],

$$J(r) = -D \nabla\Phi(r) \tag{4-14}$$

for neutron current J, diffusion coefficient D, and neutron flux Φ. The current J represents the net *vector* flow of neutrons through a surface in the material of interest. Equation 4-14 shows that J is proportional to the gradient of the neutron flux but oppositely directed. Stated another way, the flow of neutrons increases with decreasing neutron flux.[†]

The actual volumetric leakage in terms of the current J is just

$$\text{Leakage} = \nabla \cdot J(r)$$

The substitution of Eq. 4-14 yields

$$\text{Leakage} = \nabla \cdot (-D \nabla\Phi(r)) = -D \nabla^2\Phi(r) \tag{4-15}$$

[†]The definitions of current and flux as applied to neutrons are somewhat unique. In most other engineering applications, "flux" is a vector quantity that plays much the same role as J in Eq. 4-14. For example, in the heat transfer equation, $q(r) = k \nabla T(r)$, the heat *flux* q and the temperature T assume roles comparable to the neutron *current* J and the neutron *flux* Φ, respectively.

(This result may be noted to have essentially the same form as the leakage terms describing a wide range of phenomena considered in physics, chemistry, and engineering.) A simple one-dimensional, slab geometry in coordinate x, for example, is characterized by

$$J(x) = -D \frac{d\Phi}{dx}$$

and   Leakage $= -D \dfrac{d^2 \Phi}{dx^2}$

Combining Eq. 4-15 with Eqs. 4-3 and 4-4 for a critical homogeneous medium with position-dependent neutron flux $\Phi(\mathbf{r})$, the neutron balance is

Production = absorption + leakage

$$\nu\Sigma_f \Phi(\mathbf{r}) = \Sigma_a \Phi(\mathbf{r}) - D \, \nabla^2 \Phi(\mathbf{r})$$

(4-16)

(It may be noted that the term $-D\nabla^2\Phi$ is defined in such a manner that it does represent a *positive* leakage of neutrons.) Rearranging terms in Eq. 4-16 yields

$$\nabla^2 \Phi(\mathbf{r}) + \frac{\nu\Sigma_f - \Sigma_a}{D} \, \Phi(\mathbf{r}) = 0$$

or $\nabla^2 \Phi(\mathbf{r}) + B_m^2 \Phi(\mathbf{r})$

(4-17)

where the *material buckling* B_m^2 is defined from the equation as

$$B_m^2 = \frac{\nu\Sigma_f - \Sigma_a}{D}$$

Constructing general solutions to Eq. 4-17 is a nontrivial problem (to which a sizeable portion of any introductory reactor theory course is likely to be devoted). For the purpose of this chapter, it is sufficient to recognize that a critical system should have a neutron flux that is stable, everywhere positive, and zero at the external boundaries of the material system. These restrictions result in the flux solutions shown in Table 4-2 for five geometries of interest.

Criticality for a finite homogeneous system requires that $\nu\Sigma_f > \Sigma_a$ to allow for some leakage. This in turn assures that $B_m^2 > 0$ and that Eq. 4-17 has oscillatory solutions, e.g., $\sin B_m x$ and $\cos B_m x$ for a "thin" slab of infinite area. The restrictions noted above, however, require that only the positive half of one cycle represent the flux. Thus, the buckling B_m^2 in Eq. 4-17 is restricted to the one specific value B_g^2 that provides a match to the given geometry. Values for this *geometric buckling* B_g^2 are provided in Table 4-2. The flux shapes normalized to unit magnitude and a unit boundary distance are shown in Fig. 4-5 for the three geometries that are characterized by a single finite dimension. All may be noted to have "cosine-like" shapes (i.e., the Bessel function J_0 and the $(1/r) \sin$ function are peaked in the center and fall smoothly to zero at the boundaries).

TABLE 4-2
Diffusion Theory Fluxes and Bucklings for Bare Critical Systems of Uniform Composition

Geometry	Dimensions	Normalized flux $\dfrac{\Phi(r)}{\Phi(0)}$	Geometric buckling B_g^2
Sphere	Radius R	$\dfrac{1}{r}\sin\left(\dfrac{\pi r}{R}\right)$	$\left(\dfrac{\pi}{R}\right)^2$
Finite cylinder	Radius R, height H (centered about $z = 0$ and extending to $z = \pm H/2$)	$J_0\left(\dfrac{2.405r}{R}\right)\cos\left(\dfrac{\pi z}{H}\right)$	$\left(\dfrac{2.405}{R}\right)^2 + \left(\dfrac{\pi}{H}\right)^2$
Infinite cylinder	Radius R	$J_0\left(\dfrac{2.405r}{R}\right)$	$\left(\dfrac{2.405}{R}\right)^2$
Rectangular parallelepiped [cuboid][†]	$A \times B \times C$ (centered about $x = y = z = 0$ and extending to $x = \pm A/2$, etc.)	$\cos\left(\dfrac{\pi x}{A}\right)\cos\left(\dfrac{\pi y}{B}\right)\cos\left(\dfrac{\pi z}{C}\right)$	$\left(\dfrac{\pi}{A}\right)^2 + \left(\dfrac{\pi}{B}\right)^2 + \left(\dfrac{\pi}{C}\right)^2$
Infinite slab	Thickness A (centered about $x = 0$ and extending to $x = \pm A/2$)	$\cos\left(\dfrac{\pi x}{A}\right)$	$\left(\dfrac{\pi}{A}\right)^2$

[†]The term *cuboid*—a synonym for rectangular parallelepiped used in the KENO Monte Carlo code—is commonly employed in the field of nuclear criticality safety (and may catch on elsewhere).

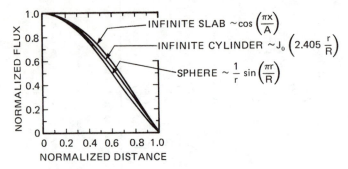

FIGURE 4-5
Normalized flux shapes for critical spheres of radius R, infinitely-long cylinders of radius R, and infinite-area slabs of thickness A predicted by diffusion theory.

As could be inferred from the discussions surrounding Fig. 4-4, material composition and geometry cannot be specified separately if a system is to be critical. This is why the material buckling B_m^2 must exactly match the geometric buckling B_g^2 for criticality. If $B_m^2 > B_g^2$, the multiplying properties of the material overpower the geometric leakage to make the system supercritical; by contrast, $B_g^2 > B_m^2$ results in a subcritical system. The *buckling* B^2 may generally be assumed to be the critical value where the material and geometric components balance.

Mathematically, the ratio of the second derivative of a function to the function itself is a measure of curvature. The buckling serves this role for the neutron flux, as shown by rearrangement of Eq. 4-17 to

$$B^2 = -\frac{\nabla^2 \Phi}{\Phi}$$

An infinite system has $B^2 = 0$ and a correspondingly uncurved or flat flux. Positive values of B^2 associated with finite dimensions may be viewed as requiring the flat flux to be bent or "buckled" to match the zero-flux boundary conditions. The larger the magnitude of B^2, the greater is the curvature required for criticality (i.e., the smaller is the geometry it must occupy).

Elimination of the $\nabla^2 \Phi$ terms in Eqs. 4-16 and 4-17 results in the neutron balance

$$\nu \Sigma_f \Phi(\mathbf{r}) = \Sigma_a \Phi(\mathbf{r}) + DB^2 \Phi(\mathbf{r}) \tag{4-18}$$

for critical buckling B^2. The form of the equation suggests that DB^2—the product of the diffusion coefficient D and the buckling B^2—assumes the role of a "macroscopic leakage cross section" since its product with the flux Φ is the volumetric leakage rate.

The definition of k for a critical system plus Eq. 4-18 show that

$$k = 1 = \frac{\nu \Sigma_f}{DB^2 + \Sigma_a}$$

or since $k_\infty = \nu\Sigma_f/\Sigma_a$,

$$k = 1 = \frac{k_\infty}{DB^2/\Sigma_a + 1}$$

and $\quad k_\infty = 1 + \frac{DB^2}{\Sigma_a}$ \hfill (4-19)

This last result shows that DB^2/Σ_a is the excess multiplication required to compensate for leakage in a finite system.

One empirical formulation for neutron leakage treats fast and thermal neutrons separately. The effective multiplication factor is considered as the product

$$k = k_\infty P_{fnl} P_{tnl} \hfill (4-20)$$

for *fast nonleakage probability* P_{fnl} and *thermal nonleakage probability* P_{tnl} where each is the probability that neutrons will *not* leak from the system in the respective energy ranges. The fast-neutron component may be treated according to the *age* [*Fermi age*] approximations where neutron slowing-down is considered as continuous rather than the discrete-step process which actually occurs. The formulation results in the expression

$$P_{fnl} = e^{-B^2\tau} \hfill (4-21)$$

for age τ. Enrico Fermi selected the latter name because τ derives from its differential equation in the same manner that time-dependence is derived from a heat transfer equation of the same form. Actually, τ is proportional to the mean square distance $\langle r^2 \rangle$ that a neutron travels in slowing from fission energies to thermal energies. Equation 4-21 indicates that neutron nonleakage is reduced (i.e., leakage is enhanced) by large τ and B^2. Since these imply a combination of a long path length and a small system, leakage should indeed be large.

The behavior of thermal neutrons may be modeled approximately using diffusion theory. Thus, the thermal nonleakage probability becomes

$$P_{tnl} = \frac{1}{1 + L^2 B^2} \hfill (4-22)$$

for *thermal diffusion length L*, defined by

$$L = \sqrt{\frac{D_t}{\Sigma_{at}}}$$

with thermal diffusion coefficient D_t and macroscopic absorption cross section Σ_{at}. The square of the diffusion length is proportional to the mean square distance traveled by a thermal neutron before it is absorbed. Thus, τ and L^2 have comparable significance for fast and thermal neutrons, respectively.

The combination of Eqs. 4-20, 4-21, and 4-22 results in the *age-diffusion approximation*

$$k = \frac{k_\infty \, e^{-B^2 \tau}}{1 + L^2 B^2} \tag{4-23}$$

for effective and infinite multiplication factors k and k_∞, respectively, age τ, and thermal diffusion length L. For very large systems, Eq. 4-23 may be approximated by

$$k \approx \frac{k_\infty}{1 + (L^2 + \tau)B^2} = \frac{k_\infty}{1 + M^2 B^2} \tag{4-24}$$

for neutron *migration area* M^2, defined by the equation. The migration area concept simply combines fast and thermal leakage effects into a single term. Comparison of Eqs. 4-24 and 4-19 shows that the migration area M^2 assumes the same role as the ratio D/Σ_a in determining the incremental multiplication that is necessary for a critical system.

The four-factor formula (Eq. 4-12) and Eqs. 4-23 and/or 4-24 are not very powerful calculational tools compared to others available. However, they can be particularly useful in providing insight into physical phenomena. Computer calculations often include "effective" values of some or all of the parameters in their outputs for exactly this reason.

In another application, the migration area is applied to evaluating the multiplication of fissionable materials in the fuel cycle. Extensive correlations of experimental data allow calculation and tabulation of k_∞ and "effective" M^2 values. Thus, the appropriate k_∞ and M^2 can be "looked up" for a given material and the geometric factors handled completely by the B^2 calculation with Eq. 4-24 yielding a surprisingly accurate value of k.

Sample Calculations

The four-factor formula may be used to estimate the multiplication factor for a "large" ($P_{nl} \sim 1.0$) system. The fast fission factor ϵ generally takes on values in the 1 to 1.05 range. It and the resonance escape probability p (which varies between zero and unity depending on the composition and heterogeneity of the system) must generally be found from some reference. The other two factors—η and the thermal utilization f—can be estimated from *thermal* cross sections:

$$f = \frac{\Sigma_a^{\text{fuel}}}{\Sigma_a^{\text{fuel}} + \Sigma_a^{\text{nonfuel}}}$$

$$\eta = \frac{\nu \Sigma_f}{\Sigma_a^{\text{fuel}}}$$

Then knowing the nuclide densities, or their ratios, η and f can be calculated.

For the case where the fuel and the nonfuel nuclide densities are equal and

$\sigma_f^f = 100$ b, $\sigma_c^f = 10$ b, $\nu = 2.5$, and $\sigma_a^n = 20$ b (where the superscripts f and n indicate fuel and nonfuel compositions, respectively),

$$f = \frac{n(100 \text{ b} + 10 \text{ b})}{n(100 \text{ b} + 10 \text{ b}) + n(20 \text{ b})} = \frac{110 \text{ b}}{130 \text{ b}} = \boxed{0.846}$$

$$\eta = \frac{2.5n(100 \text{ b})}{n(100 \text{ b} + 10 \text{ b})} = \frac{250 \text{ b}}{110 \text{ b}} = \boxed{2.27}$$

Values of $\epsilon = 1.03$ and $p = 0.60$ would then result in

$$k \approx k_\infty = \epsilon p \eta f = (1.03)(0.60)(2.27)(0.846) = \boxed{1.19}$$

If the material were placed in an unreflected spherical system 4 m in diameter and with an effective migration area of 60 cm^2, the nonleakage probability (Eq. 4-24) would be

$$P_{nl} = \frac{1}{1 + M^2 B^2}$$

The buckling B^2 for a sphere (Table 4-2) is

$$B^2 = \left(\frac{\pi}{R}\right)^2 = \left(\frac{\pi}{2.0 \text{ m}}\right)^2 = 2.47 \text{ m}^{-2} \times \left(\frac{1 \text{ m}}{100 \text{ cm}}\right)^2 = 2.47 \times 10^{-4} \text{ cm}^2$$

so that

$$P_{nl} = \frac{1}{1 + (60 \text{ cm}^2)(2.47 \times 10^{-4} \text{ cm}^2)} = 0.985$$

or leakage would be about $1\frac{1}{2}$ percent. For the material considered above, the multiplication factor would be

$$k = k_\infty P_{nl} = (1.19)(0.985) = \boxed{1.17}$$

and the system would still be supercritical.

Neutron Transport[†]

The behavior of individual neutrons and nuclei cannot be predicted. However, the average behavior of a statistically large population of neutrons can be described quite accurately by extending the concepts of neutron fluxes, nuclear cross sections, and reaction rates.

[†]The remainder of this chapter is provided as an overview of the mathematically sophisticated computations used by the nuclear industry. Since it is *not* necessary to any later chapters, omission is recommended for those who are not already familiar with the field of reactor physics.

A complete description of each "representative" neutron requires knowledge of seven variables:

1. position in space **r** (three coordinates, e.g., x, y, z or r, θ, ϕ for rectangular and spherical systems, respectively)
2. velocity **v** (three coordinates), usually broken into energy E (since $E \approx v^2$) and direction Ω (with the latter consisting of components θ and ϕ)
3. time t, for which the coordinates **r**, E, and Ω are appropriate

The simple position-dependent neutron flux $\Phi(\mathbf{r})$ considered to this point must now be replaced by its multivariable counterpart $\Phi(\mathbf{r}, E, \Omega, t)$. The neutron balance known as the *Boltzman neutron transport equation* (or some subset of these terms, e.g., "transport equation") may be written as:

$$\frac{1}{v(E)} \frac{d\Phi\,(\mathbf{r}, E, \Omega, t)}{dt}^{①} = -\Omega\cdot\nabla\Phi(\mathbf{r}, E, \Omega, t)^{②} - \Sigma_t(\mathbf{r}, E, \Omega)\Phi(\mathbf{r}, E, \Omega, t)^{③}$$

$$+ \chi(E) \int_{E'}\int_{\Omega'} v\Sigma_f(\mathbf{r}, E', \Omega')\Phi(\mathbf{r}, E', \Omega', t)\, d\Omega'\, dE'^{④}$$

$$+ \int_{E'}\int_{\Omega'} \Sigma_s(\mathbf{r}; \Omega' \to \Omega; E' \to E)\Phi(\mathbf{r}, E', \Omega', t)\, d\Omega'\, dE'^{⑤}$$

$$(4\text{-}25)$$

where each term represents a rate (per unit volume, per unit energy, and per unit direction to be precise) involving neutrons with the specified coordinates. Term 1 is simply the rate of accumulation of such neutrons. The second is the leakage term. Term 3 is the total interaction rate or the rate of removal of neutrons due to absorption and scattering interactions (since these latter "out-scatters" change neutron energy and direction).

The last two terms in Eq. 4-25 represent the production phenomena where neutrons at arbitrary energy E' and direction Ω' react with nuclei to generate those with the reference parameters E and Ω, respectively. The integrals sum over all initial energies and directions. Specifically, the double integral in term 4 is just the total fission rate; then given that a fission has occurred, the neutron spectrum $\chi(E)$ represents the fission-neutron distribution. (It may be recalled, for instance, that in a thermal reactor the slow neutrons *cause* most of the fissions, but that fast neutrons are *produced*).

Term 5 in Eq. 4-25 is based on differential scattering of neutrons from initial energy E' to final energy E and from initial direction Ω' to final direction Ω. The cross section $\Sigma_s(\mathbf{r}; E' \to E; \Omega' \to \Omega)$ accounts for the relative probabilities of all possible combinations (recalling, for example, that fast neutrons can only *lose* energy in collisions with stationary nuclei). This "in-scatter" term may be noted to be the *only* source of neutrons at energies below those of the fission neutrons. Thermal neutrons, for example, *all* result from scattering reactions.

By simply remembering the complex energy dependence of $\Sigma_t(E)$ for ^{238}U (Fig. 2-11), it is apparent that Eq. 4-25 cannot be solved in closed form (or anything close to it!). The methods considered in the remainder of this chapter, however, provide generally acceptable, approximate solutions.

Averaged Cross Sections

The first step in attempting to solve Eq. 4-25 must be to obtain appropriate reaction cross sections. In many cases, averaging the cross sections over one or more parameters results in valuable simplification.

The continuous energy dependence of the neutron flux, $\Phi(E)$, for example, may be divided into intervals or "groups" according to

$$\Phi_{\Delta E} = \int_{\Delta E} \Phi(E)\,dE \tag{4-26}$$

where $\Phi_{\Delta E}$ is the total flux within energy range ΔE. The corresponding average cross section $\Sigma_{r\,\Delta E}$ is defined as

$$\Sigma_{r\,\Delta E} = \frac{\int_{\Delta E} \Sigma_r(E)\Phi(E)\,dE}{\int_{\Delta E} \Phi(E)\,dE} \tag{4-27}$$

for reaction r and cross section $\Sigma_r(E)$. This latter definition is based on preserving the reaction rate so that

$$\Phi_{\Delta E}\Sigma_{r\,\Delta E} = \int_{\Delta E} \Sigma_r(E)\Phi(E)\,dE$$

i.e., assuring that the product of the incremental flux and average cross section faithfully reproduce the actual reaction rate. When the procedure in Eq. 4-27 is applied, *flux-averaged* or *flux-weighted cross sections* result.

If the averaging procedure is carried out over the entire energy spectrum, the resulting parameters

$$\Phi = \int_0^{\infty} \Phi(E)\,dE \tag{4-28}$$

and $\quad \Sigma_r = \dfrac{\displaystyle\int_0^{\infty} \Sigma_r(E)\Phi(E)\,dE}{\displaystyle\int_0^{\infty} \Phi(E)\,dE} \tag{4-29}$

are referred to as the *one-energy-group* flux and cross section, respectively. Cross sections computed in this manner may be substituted for the relatively simple-minded "*one-speed*" values introduced earlier so that Eq. 4-16 has exactly the same form, namely

$$\nu\Sigma_f\Phi(\mathbf{r}) = \Sigma_a\Phi(\mathbf{r}) - D\,\nabla^2\Phi(\mathbf{r}) \tag{4-16}$$

Under these conditions, however, this *one-energy-group* expression can be quite accurate. It is also rather difficult to obtain, since the energy-dependent flux $\Phi(E)$ must be known (or, in practice, adequately approximated) before the average cross sections $\nu\Sigma_f$, Σ_a, and D can be computed from Eq. 4-27. (Certain other difficulties with this diffusion-theory model are identified shortly.)

The process of generating cross sections for a calculational model with a limited number of energy groups usually begins by considering an infinite homogeneous system of a representative composition. Such a system has no leakage and cannot support any position- or direction-dependent flux variation. By selecting a critical $[k_\infty = 1]$ composition, there will also be no time dependence so that Eq. 4-25 reduces to

$$\Sigma_t(E)\Phi(E) = \chi(E)\int_0^\infty \nu\Sigma_f(E')\Phi(E')\,dE'$$

$$+ \int_0^\infty \Sigma_s(E' \to E)\Phi(E')\,dE' \tag{4-30}$$

where the terms have been arranged to equate losses to production. The total-interaction rate consists of absorption and out-scatter reactions, each of which removes neutrons from reference energy E. The two source terms rely on fission and scattering reactions, respectively, in which neutrons at any initial energy E' produce neutrons at the reference energy E.

The energy dependence of the cross sections prevents closed-form solution of Eq. 4-30. However, breaking the energy range into groups can provide a useful simplification. This *multigroup method* is based on an energy structure like that shown in Fig. 4-6. Because of the characteristics of the neutron slowing-down

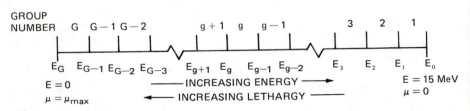

FIGURE 4-6
Partitioning of the energy axis for multigroup calculations.

process considered previously, the energy groups are numbered according to increasing *lethargy*, i.e., from highest to lowest energy. Equation 4-30 may be intergrated over each energy group $\Delta E_g = E_{g-1} - E_g$ and thereby converted to a set of equations of the form

$$\Sigma_{tg}\phi_g = \chi_g \sum_{g'=1}^{G} \nu\Sigma_{fg'}\phi_{g'} + \sum_{g'=1}^{G} \Sigma_{g'\to g}\phi_{g'} \tag{4-31}$$

for each g from 1 to G where

$$\phi_g = \int_{\Delta E_g} \Phi(E)\, dE$$

$$\Sigma_{tg} = \frac{1}{\phi_g} \int_{\Delta E_g} \Sigma_t(E)\Phi(E)\, dE$$

$$\nu\Sigma_{fg'} = \frac{1}{\phi_{g'}} \int_{\Delta E_{g'}} \nu\Sigma_f(E')\Phi(E')\, dE'$$

$$\chi_g = \int_{\Delta E_g} \chi(E)\, dE$$

$$\Sigma_{g'\to g} = \frac{1}{\phi_{g'}} \int_{\Delta E_g} \int_{\Delta E_{g'}} \Sigma_s(E' \to E)\Phi(E')\, dE'\, dE$$

It must be emphasized that Eq. 4-31 is actually a set with one for each of the G energy groups in Fig. 4-6. The energy groups g' correspond to the energies E' in Eq. 4-30 and their summations replace the integrals for constructing the fission- and scattering-source terms. The cross section $\Sigma_{g'\to g}$ represents the probability that neutrons in energy group g' will scatter to an energy within group g.

Equations 4-31 are a set of G algebraic equations, from which the unknown fluxes may be determined in a relatively straightforward manner. Again, however, the cross sections do depend on the neutron flux, so nothing is gained unless they can be calculated accurately. The basic procedure is to have enough energy groups that the flux will not change greatly over the energy range of interest. In the limiting case where ΔE_g is "extremely small" (i.e., essentially the same width as the data point that represents the cross section), the flux is constant at ϕ_g,

$$\int_{\Delta E_g} \Sigma_r(E)\Phi(E)\, dE \approx \Sigma_r(E_g)\phi_g$$

and $\dfrac{1}{\phi_g} \displaystyle\int_{\Delta E_g} \Sigma_r(E)\Phi(E)\,dE \approx \Sigma_r(E_g)$

i.e., the average cross section is flux-independent. Alternatively, accurate knowledge of the flux shape in a given energy region (e.g., from theoretical considerations) may allow use of a wider energy interval.

Computer programs that employ both methods are used to calculate very detailed energy-dependent fluxes. These fluxes, in turn, may be used to calculate new energy-group cross sections for a coarser structure.

With the infinite-homogeneous-medium cross sections in hand, sophistication can be added through modifications that account for the leakage effects and heterogeneities characteristic of all real systems. Directional dependencies can also be included by decomposing direction Ω into discrete components.

Multigroup Diffusion Theory

The diffusion theory approximation that was developed earlier, on the basis of some simple assumptions, may also be derived more rigorously from the Boltzmann transport equation (Eq. 4-25). In this latter case, an assumption of a well-behaved flux allows the leakage to be approximated by Fick's law according to Eq. 4-14. The current $\mathbf{J}$ is represented by the same general expression

$$\mathbf{J}(\mathbf{r}) = -D(\mathbf{r})\,\nabla\Phi(\mathbf{r}) \tag{4-15}$$

but with the diffusion coefficient being position-dependent and defined in terms of parameters developed from the transport equation. The one-energy-group diffusion equation then becomes

$$\nu\Sigma_f(\mathbf{r})\Phi(\mathbf{r}) = \Sigma_a(\mathbf{r})\Phi(\mathbf{r}) - \nabla\cdot D(\mathbf{r})\,\nabla\Phi(\mathbf{r}) \tag{4-32}$$

for appropriately flux-averaged cross sections, which may also vary with position.

Since criticality requires a very precise match between material properties and geometry, the balance implied by Eq. 4-32 is relatively difficult to obtain (and certainly not likely to be "guessed" on the first try). Current practice treats this problem by noting that

$$k = \frac{\nu\Sigma_f\Phi}{\Sigma_a\Phi - \nabla\cdot D\,\nabla\Phi}$$

for all situations and that modifying Eq. 4-32 to

$$\frac{1}{k}\,\nu\Sigma_f(\mathbf{r})\Phi(\mathbf{r}) = \Sigma_a(\mathbf{r})\Phi(\mathbf{r}) - \nabla\cdot D(\mathbf{r})\,\nabla\Phi(\mathbf{r}) \tag{4-33}$$

will guarantee a (mathematical) balance. If after $\Phi(\mathbf{r})$ and k are computed the latter is not unity (or a close approximation thereto), the composition and/or size of the system may be varied and the calculations repeated until a balance is obtained.

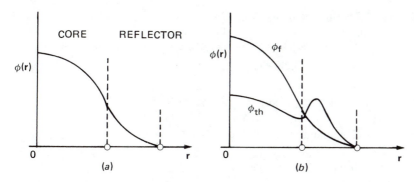

FIGURE 4-7

Comparison of (*a*) one- and (*b*) two-energy-group (fast and thermal) fluxes in core and reflector regions of a spherical reactor. (Adapted from *Reactor Analysis*, by R. V. Megrehblian and D. K. Holmes, © 1960 by McGraw-Hill Book Company, Inc. Used by permission of McGraw-Hill Book Company.)

The one-group diffusion formulation suffers severely in accuracy by treating all neutrons as having the same average behavior. The multiple-energy-group approach can be extended directly to rectify this situation. For example, Fig. 4-7*a* shows the average one-group flux for adjacent core and reflector regions. Figure 4-7*b* shows the fluxes of fast and thermal neutrons, respectively, for the same system. The reflector peak of the thermal flux is especially significant since it results in a net leakage *into* the core of these neutrons (at the same time both the total flux and fast flux exhibit net out-leakage). The potential accuracy of calculations is generally enhanced by employing additional energy groups.

With a leakage term of the form $-\nabla \cdot D_g \nabla \phi_g$, Eq. 4-31 becomes

$$-\nabla \cdot D_g \nabla \phi_g + \Sigma_{tg} \phi_g = \frac{1}{k} \chi_g \sum_{g'=1}^{G} \nu \Sigma_{fg'} \phi_{g'} + \sum_{g'=1}^{G} \Sigma_{g' \to g} \phi_{g'} \tag{4-34}$$

for each of the G groups included in the model. The $1/k$ term has been applied to this balance as it was to Eq. 4-33. These represent one formulation of the *multigroup diffusion equations*.

A two-group model based on Eq. 4-34 is written as

$$-\nabla \cdot D_1 \nabla \phi_1 + \Sigma_{t1} \phi_1 = \frac{1}{k} \chi_1 (\nu \Sigma_{f1} \phi_1 + \nu \Sigma_{f2} \phi_2) + \Sigma_{1 \to 1} \phi_1 + \Sigma_{2 \to 1} \phi_2 \tag{4-35a}$$

and $$-\nabla \cdot D_2 \nabla \phi_2 + \Sigma_{t2} \phi_2 = \frac{1}{k} \chi_2 (\nu \Sigma_{f1} \phi_1 + \nu \Sigma_{f2} \phi_2) + \Sigma_{1 \to 2} \phi_1 + \Sigma_{2 \to 2} \phi_2 \tag{4-35b}$$

where the equations balance the populations of higher-energy ($g = 1$) and lower-energy ($g = 2$) neutrons respectively. One of the first features noted is that the formulation includes self-scatter terms in the form $\Sigma_{g \to g}$. The latter represent the scattering events that do not change the neutron energy enough to have it leave the group. However, for this example, $\Sigma_{tg} = \Sigma_{ag} + \Sigma_{g\to 1} + \Sigma_{g\to 2}$, so the self-scatter terms can be eliminated from both sides of the equation.

Equations 4-35a and 4-35b can be modified to represent fast and thermal neutrons in a reactor system by noting that:

- $\chi_1 = 1$ and $\chi_2 = 0$ since all fission neutrons are fast [regardless of whether they result from fast fission ($\nu\Sigma_{f1}\phi_1$) or thermal fission ($\nu\Sigma_{f2}\phi_2$)]
- $\Sigma_{2\to 1} = 0$ since "up-scattering" from thermal to fast energies is precluded (by conservation of energy and momentum considerations)

Employing information from this and the previous paragraph, Eq. 4-35 can be rewritten as

$$-\nabla \cdot D_1 \ \nabla\phi_1 + \Sigma_{a1}\phi_1 + \Sigma_{1\to 2}\phi_1 = \nu\Sigma_{f1}\phi_1 + \nu\Sigma_{f2}\phi_2 \qquad (4\text{-}36a)$$

$$-\nabla \cdot D_1 \ \nabla\phi_1 + \Sigma_{a2}\phi_2 = \Sigma_{1\to 2}\phi_1 \qquad (4\text{-}36b)$$

where each equation balances the neutron population for its own energy group. The only source of fast neutrons is fission, while the only source of thermal neutrons is fast-neutron scattering.

Solution of Eq. 4-36 or any other set based on Eq. 4-34 is most readily accomplished by using iterative computer methods that handle both the energy and position dependencies. The latter generally employ finite-difference methods based on spatial grids (similar in concept to the energy-group structure shown by Fig. 4-6) of regular geometry.

The validity of the multigroup diffusion equations is limited by the inherent assumptions contained in Fick's law. The equations are generally not valid:

1. at exterior (vacuum boundaries)
2. near or within strong absorbers
3. at interfaces between dissimilar materials

Although these restrictions would seemingly leave them unavailable for most real applications, the diffusion equations can be modified. The discrepancy at external boundaries, for example, has been found to be predictable enough that an *extrapolation distance* δ can be subtracted from the predicted dimensions (e.g. radius r_{DT}) to yield a good estimate of actual critical values, i.e.,

$$r = r_{\text{DT}} - \delta \qquad (4\text{-}37)$$

The problems with strong absorbers and interfaces are most readily overcome by using more accurate calculational methods to determine "effective" cross sections (i.e., those that result in correct reaction rates). Reactor design often relies very heavily on such procedures for analysis of their highly heterogeneous fuel-pin lattices.

Discrete Ordinates Method

The limitations on diffusion theory may be removed only by adding explicit consideration of the directional properties of the neutron population. One method is to approximate Ω in Eq. 4-25 by a limited set of directions known as *discrete ordinates*. The procedure is also known as S_n, since it evolves from an nth order approximation of the directional scattering *source*. In fact, the general term *transport theory* as applied to neutrons has come to be synonymous with the discrete ordinates or S_n method.

The discrete ordinates method has the complication of direction "groups" in addition to the energy groups and spatial meshes of diffusion theory. Although this new feature increases computational time, the general principles of iterative solution are relatively unaffected. The overall complexity of the method precludes further discussion in this book.

Monte Carlo Method

There is an approach to neutron transport that appears to be entirely different from the methods considered previously. It is based on conceptual tracking of individual neutrons through a material medium. Random numbers are selected and correlated to various possible reaction events in accordance with the underlying physical principles. The method is called *Monte Carlo*, in reference to the European city by the same name known for its "games of chance."

The Monte Carlo procedure seems far removed from the iterative solutions to the integrodifferential form of the Boltzmann equation in Eq. 4-25. However, the two methods are completely consistent. The Monte Carlo method is actually equivalent to evaluating the source integrals by a "random-sampling" process.

The most fundamental requirement for developing a Monte Carlo procedure is to relate a random number to a physical event in an unambiguous manner that is firmly tied to the governing laws of probability. This is generally accomplished by computing a *probability density function* $p(x)$ for each possible interaction parameter x defined by

$$p(x) \, dx = \begin{array}{l} \text{probability that the outcome of an interaction will} \\ \text{result in the parameter being between } x \text{ and } x + dx \end{array} \tag{4-37}$$

and normalized such that

$$\int_{-\infty}^{\infty} p(x) \, dx = 1$$

In practice, the function may be continuous as defined above, or it may be discrete when only a finite number of outcomes are possible. The fission-neutron energy spectrum $\chi(E)$, for example, is a continuous probability density function. An example of a discrete function may be constructed by noting that Σ_a/Σ_t and Σ_s/Σ_t are the probabilities for neutron absorption and scattering, respectively, in a material with total macroscopic cross section Σ_t.

The *cumulative probability distribution function P(x)* is defined by

$$P(x) = \int_{-\infty}^{x} p(x') \, dx' \tag{4-38}$$

It is single-valued, ranges from zero to unity, and preserves probability. Thus, a number selected randomly on the 0-1 interval can be assigned unambiguously to event *x*. Most importantly, the probability of occurrence of any given event is identical to the probability that the corresponding random number will be selected.

The basic components of a Monte Carlo code are related to random numbers, geometry, tracking, and scoring. Random numbers are selected on the 0-1 interval. In practice, a mathematical algorithm which generates "pseudorandom" numbers is often preferred for simplicity and the ability to repeat calculations in an exact manner.

The geometry identifies the location of each material type in the system. Macroscopic cross sections describe the nuclear characteristics of each region. The boundaries which separate regions often serve as convenient neutron "scoring" locations. For example, neutrons leaking from one material to another or from the system as a whole may be counted, or "scored," as they cross boundaries.

Neutrons are tracked through the system on the basis of random selection of reaction parameters. Procedures for tracking and scoring are generally based on the nature of the desired result. Shielding calculations, for example, are concerned with the neutrons that leave external material boundaries. On the other hand, multiplication and radiation dose calculations depend on counting reaction rates.

The features of a multiplication calculation are shown schematically in Fig. 4-8. The procedure, a form of *analog Monte Carlo*, follows "generations" of neutrons and compares the number started to the number produced to compute k_{eff}. In an arbitrary generation, the locations for starting individual neutron histories are selected from those of the previous generation (the first generation starts neutrons from some arbitrary distribution). The energy and direction are selected randomly from appropriate cumulative distribution functions. Neutron path lengths between collisions depend, of course, on the total macroscopic cross section $\Sigma_t(E)$. The geometry determines whether a neutron leaks or experiences a collision at the end of its path length. Collision types are selected randomly in accordance with the appropriate reaction cross sections. Scattering events change the energy and direction of the neutron before it continues through the system. Leakage, capture, and fission terminate the history and signal the start of the next fission neutron. For fission reactions, the number of neutrons is randomly selected with the resulting number and the location of the event stored for use in starting neutrons of the next generation. Since it is typical to start a fixed number of neutrons in each generation *n*, the number of fission points in generation $n-1$ may be adjusted upward or downward by random duplication or elimination, respectively.

By considering "neutron weight" rather than whole particles, the *biased Monte Carlo* methods enhance computational efficiency. The situation sketched by Fig. 4-8, for example, can be modified by considering each collision to be a partial absorption and a partial scatter in proportion to the respective cross sections. Likewise, fission can produce a neutron weight equivalent to the average number of neutrons per fission rather than an integral number of neutrons. Calculations other than those for k_{eff} can be biased by other procedures to enhance efficiency.

Applications

Neutron transport methods are generally applied to reactor and fuel-cycle design problems by using large computer codes developed by government laboratories

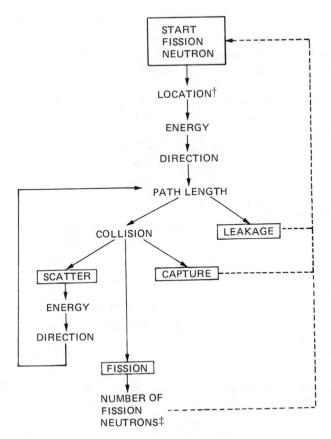

FIGURE 4-8
Flow diagram for an analog Monte Carlo method used to calculate the effective multiplication factor k. †Locations for generation N based on fission points from generation $N - 1$; ‡record neutron number and fission location for generation $N + 1$ starting locations.

and/or individual companies. The selection and use of such codes is usually problem-specific.

Although a few packaged cross-section sets are available, most high-level design work ties back to the Evaluated Nuclear Data File [ENDF] described in Chap. 2. The detailed cross sections are used to generate reduced *libraries* for representative compositions. Then, the group structure is condensed further to produce composition-dependent microscopic cross sections that can be used directly in the neutron-transport models.

Multigroup diffusion theory codes are used for the bulk of the calculations related to reactor fuel management, especially in light-water reactors. Transport theory, Monte Carlo, and/or other procedures must be employed to "homogenize" the physical heterogeneities and generate "effective" cross sections consistent with the limitations of the diffusion-theory approximation. Typical LWR applications, for example, use a cross-section library with 50–60 fast groups and about 30 thermal groups to calculate the four-group parameters employed in the spatial calculations.

Diffusion-theory codes perform calculations in a relatively inexpensive manner, but are limited to large, regular (usually rectangular-mesh) geometries.

Transport-theory codes are employed to perform many calculations that are not possible with diffusion theory. They are especially valuable for systems that can be modeled in one or two dimensions, since three-dimensional calculations are prohibitively expensive. They employ multigroup cross section sets, which, of course, include the provisions necessary for the discrete-ordinates formulation. Transport codes are also limited to regular-mesh geometries.

Computer time requirements for both diffusion-theory and transport codes are roughly proportional to the product of their numbers of spatial mesh points, energy groups, and ordinates. Thus, there is always strong incentive to identify arrangements that optimize "accuracy-per-unit-computer-time." Since diffusion theory is less expensive than transport theory in rough proportion to the number of ordinates used by the latter, it is preferred for all calculations within its range of applicability.

The Monte Carlo method has a major advantage in being able to represent almost any geometric configuration. Thus, it is especially useful for irregular geometries like those found in fuel-cycle processing and storage equipment. The Monte Carlo method may also use cross sections in any number of energy or angular groups (up to and including the essentially continuous representation of the ENDF set) without major impact on computational time. The latter occurs because the cross sections are used in ratios of reaction probabilities and *not*, for example, in the integrodifferential formulation of Eq. 4-25.

The running time for Monte Carlo codes is not nearly as strong a function of either geometry or cross-section representation as it is for the other two methods. As a rule of thumb, two-dimensional Monte Carlo and transport calculations of the same system use nearly equivalent amounts of computer time. Transport has a clear advantage when a one-dimensional representation is adequate. Monte Carlo is the only real option for three-dimensional problems and those with particularly complex geometries.

After some straightforward modifications, the transport and Monte Carlo methods are equally applicable to the transport of electromagnetic photons in reactor shields and materials. The methods can also be applied to electrons and ions, e.g., in the controlled fusion systems described in Chap. 21. Even coupled neutron–photon or electron–photon transport can be handled by some of the more sophisticated models.

EXERCISES

Questions

4-1. Differentiate between the infinite multiplication factor k_∞ and the effective multiplication factor k_{eff}.

4-2. Explain why neutrons experience greater average energy losses from elastic collisions with hydrogen than from elastic collisions with any other nuclides.

4-3. Describe the effect of uniform dilution with water on the minimum critical mass for the unreflected, homogeneous plutonium-water system in Fig 4-4.

4-4. List three different methods used to control absorption in the neutron balance for a reactor. List two methods for leakage control in fuel-processing facilities.

4-5. Differentiate between one-speed and one-group-diffusion theory calculational models.[†]

4-6. List the major random decisions necessary for a Monte Carlo calculation of k_{eff}.[†]

Numerical Problems

4-7. For a PWR with a roughly cylindrical shape (3.8-m diameter $\times$ 4.1-m height):
   a. Calculate the geometrical buckling.
   b. Calculate k_∞ for a critical system where the migration area, $M^2 = D/\Sigma_a$, is 60 cm^2.

4-8. A very large homogeneous reactor is to be built using a solution of ^{235}U-salt and water. Based on the following thermal parameters:

$$\sigma_f^{25} = 584 \text{ barns} \qquad n^{25} = 10^{21} \text{ at/cm}^3$$

$$\sigma_c^{25} = 98 \text{ barns} \qquad n^{salt} = 4 \times 10^{21} \text{ at/cm}^3$$

$$\sigma_c^{salt} = 25 \text{ barns} \qquad \nu = 2.5$$

$$\sigma_c^{H_2O} \approx 0 \text{ barns}$$

and using a four-factor-formula model with $\epsilon = 1.03$:
   a. Calculate η and f.
   b. Estimate the value of p required for criticality.

4-9. Verify that for a slab geometry, $\Phi(x) = A \cos (\pi x/A)$ is a solution to Eq. 4-17 which matches the boundary condition $\Phi(\pm A/2) = 0$.

4-10. Using the average logarithmic energy decrement, estimate the average fractional energy loss per collision for H, D, Na, and U. Calculate the ratio between the average and maximum energy loss per collision for each.

4-11. Verify that Eq. 4-34 for $G = 1$ is equivalent to Eq. 4-32.[†]

4-12. The distribution of neutron energies in thermal equilibrium with reactor material may be approximated by the *Maxwellian distribution*

$$N(E) \approx \sqrt{E}\, e^{-E/kT}$$

for absolute temperature T and Boltzmann constant k ($= 8.62 \times 10^{-5}$ eV/K).
   a. Determine the most probable neutron energy in terms of kT.
   b. For $T = 25°$C, calculate the most probable energy and the energies where $N(E)$ is 1 percent of its maximum value.

4-13. An extreme example of the dependence of average absorption cross sections on the neutron flux spectrum may be considered on the basis of the following assumptions:

 • for a thermal reactor

[†]Questions and problems followed by a dagger refer to the optional section *Neutron Transport*.

$$\Phi(E) = \begin{cases} \phi_{th} & 0 \leqslant E \leqslant 1 \text{ eV} \\ 0 & \text{other energies} \end{cases}$$

- for a fast reactor

$$\Phi(E) = \begin{cases} \phi_f & 0.1 \text{ MeV} \leqslant E \leqslant 10 \text{ MeV} \\ 0 & \text{other energies} \end{cases}$$

- the absorption cross section σ_a^X for material X has a value of 100 barns at 0.025 eV and a one-over-v dependence at all energies.

For this example:
a. Calculate the average absorption cross section in the thermal reactor.
b. Calculate the average cross section in the fast reactor.
c. Identify the system in which material X is the better neutron "poison" and determine the factor by which the absorptions differ.[†]

4-14. In the diffusion-theory model, leakage from a *critical* system of any regular geometry is characterized by the *same* critical buckling B^2. A method called *buckling conversion* employs this characteristic to compare a known critical geometry to other shapes by:

- converting the actual critical dimensions to diffusion-theory-predicted values by adding an extrapolation distance δ to radii and *each* flat surface for cylindrical and cuboidal geometries
- calculating B^2 from the adjusted dimensions
- equating the result to the B^2-expression for the desired alternate shape and then calculating dimensions for the latter
- adjusting the new dimensions by reversing the procedure of the first step

Considering a bare, critical sphere of α-phase plutonium with the radius noted on Fig. 4-4:
a. Calculate its buckling assuming an extrapolation distance of 2.0 cm.
b. Calculate the dimensions for critical cylinders with height-to-diameter ratios of 1.0 and 3.0, respectively.
c. Determine the critical volume for the sphere and the two cylinders.

SELECTED BIBLIOGRAPHY[†]

Basic Reactor Theory
    Duderstadt & Hamilton, 1976
    Henry, 1975
    Lamarsh, 1966
    Onega, 1975
    Rydin, 1977
    Zweifel, 1973

Transport Theory
    Bell & Glasstone, 1970
    Duderstadt & Martin, 1979

[†]Full citations are contained in the General Bibliography at the back of the book.

Megrehblian & Holmes, 1960
Weinberg & Wigner, 1958

Monte Carlo
Carter & Cashwell, 1975
Schaeffer, 1973 (Ch. 5)

Numerical Methods for Reactor Calculations
Clark & Hansen, 1964
Greenspan, 1968

Other Sources with Appropriate Sections or Chapters
Clayton, 1979
Connolly, 1978
Etherington, 1958
Foster & Wright, 1977
Glasstone & Sesonske, 1967
Graves, 1979
Knief, 1981
Lamarsh, 1975
Murray, 1975
O'Dell, 1974
Sesonske, 1973
Thompson & Beckerley, 1964
WASH-1250, 1973

5

REACTOR KINETICS AND CONTROL

Objectives

After studying this chapter, the reader should be able to:

1. Define the concepts of reactivity, prompt neutron lifetime, delayed neutron precursor, and delayed and prompt criticality.
2. Describe the behavior of a reactor following a small step insertion of reactivity with and without negative temperature feedbacks.
3. Calculate reactivity defects for given reactivity coefficients and temperature changes.
4. Sketch the general behavior of a pulse reactor following a large (prompt supercritical) reactivity insertion.

Prerequisite Concepts

Neutron Balance, k_{eff}	Chapter 4
One-Group Diffusion Theory	Chapter 4
Four-Factor Formula	Chapter 4

A reactor core could be loaded with just enough fuel to make it critical. However, it would be capable of producing power only until there were enough fissions (i.e., destruction of fissile atoms and production of fission product poisons) to reduce the effective multiplication below unity. For this reason, all power reactors have a substantial "excess" multiplication to allow for multiyear operation.

Energy production in a nuclear power reactor is based on four time-dependent features that have no direct counterparts in more conventional systems:

1. Excess multiplication must be "held down," or compensated, by various control systems so that the system remains well below "prompt critical" at all times.
2. Control response times are based on the characteristics of the delayed-neutron population (born 0.2 s–1 min after fission and constituting <1 percent of all neutrons).
3. Several temperature-related feedback mechanisms are present which can produce very small to very large power changes on extremely short time scales.
4. Long-term changes occur in the fuel as nuclear reactions cause destruction of some nuclides and production of others.

All are important considerations to the control of reactor power. The first three features have short-term impacts that are considered in this chapter. The effects of the fourth are described in the next chapter.

Time-dependent behavior of neutron chain-reacting systems is the purview of *reactor kinetics* or *reactor dynamics*. Important features include characteristic system response times, automatic feedback mechanisms, and reactor control strategies. Response times are found to depend on the amount of excess multiplication and on the effective lifetimes of both prompt and delayed neutrons. Temperature-based feedbacks cause changes in cross sections and/or geometries and thereby modify system multiplication. External procedures and devices must function in concert with the inherent mechanisms to provide effective control of reactor systems.

NEUTRON MULTIPLICATION

A critical system exists when an average of one neutron from each fission goes on to cause another fission. Under such conditions, the effective multiplication factor k is unity, the fission rate is constant, and the power level is steady. Any adjustment in multiplication causes the power level to change in a time-dependent manner.

The state of criticality of a system depends on all of the neutrons that result from the fission process. Prompt neutrons appear at the time of fission. Delayed neutrons from fission-product decay enter the system at substantially later times. In most instances, neutrons of both origins affect overall behavior of the general neutron population and the reactor power level.

Prompt Neutrons

Prompt neutrons may be expected to undergo a few scattering interactions before they are absorbed or leak. The average amount of time that these neutrons exist in the system determines the minimum time between one generation of neutrons and the next.

The neutron balance for a general chain reacting system may be written as

Rate of change = rate of production − rate of absorption − rate of leakage

If as a starting point it is assumed that one-group diffusion theory is valid, the balance may be rewritten as

$$\frac{dN}{dt} = \nu\Sigma_f\Phi - \Sigma_a\Phi - DB^2\Phi$$

where dN/dt is the time rate of change of the neutron density N and $\nu\Sigma_f\Phi$, $\Sigma_a\Phi$, and $DB^2\Phi$ are the fission-production, absorption, and leakage rates (per unit volume), respectively. The equation may be modified to

$$\frac{1}{\nu}\frac{d\Phi}{dt} = \nu\Sigma_f\Phi - \Sigma_a\Phi - DB^2\Phi \tag{5-1}$$

since the neutron flux Φ is the product of the neutron density N and speed ν. Equation 5-1 contains the assumption that *all* fission neutrons are *prompt* since delayed neutrons are not represented.

If Eq. 5-1 is divided by $\nu\Sigma_f$, the result is

$$\frac{1}{\nu\nu\Sigma_f}\frac{d\Phi}{dt} = \left(1 - \frac{\Sigma_a + DB^2}{\nu\Sigma_f}\right)\Phi$$

or, by rearranging terms and noting that $k_{\text{eff}} = k = \nu\Sigma_f/(\Sigma_a + DB^2)$,

$$\frac{1}{\nu\nu\Sigma_f}\frac{d\Phi}{\Phi} = \frac{k-1}{k}\,dt \tag{5-2}$$

for a system with only prompt neutrons.

Since k is the multiplication of a particular system and $k = 1$ always implies criticality, the difference $k-1$ is the "excess multiplication" with respect to a critical condition. The term $(k-1)/k$ in Eq. 5-2 then represents the "fractional excess multiplication." More commonly this is the *reactivity* ρ, defined as

$$\rho = \frac{k-1}{k} = \frac{\Delta k}{k} \tag{5-3}$$

Table 5-1 compares the multiplication and reactivity scales as they describe the state of criticality of a system. The term $\Delta k/k$ is also used for reactivity.

Since $\nu\Sigma_f$ may be considered as a "macroscopic neutron production cross section," it is also the probability per unit distance of neutron travel that a *single fission neutron* will be produced (by contrast, Σ_f is the probability per unit distance

TABLE 5-1
Relationship between the Effective Multiplication Factor k, the Reactivity ρ, and the State of Criticality of a System

Multiplication	Reactivity	State of criticality
$k > 1$	$\rho > 0$	Supercritical
$k = 1$	$\rho = 0$	Critical
$k < 1$	$\rho < 0$	Subcritical

that a *fission* will occur and thereby produce v neutrons). The reciprocal of $v\Sigma_f$ is the mean free path for fission-neutron production λ_{vf}. The term $1/vv\Sigma_f$ in Eq. 5-2 may be rewritten as

$$\frac{1}{vv\Sigma_f} = \frac{\lambda_{vf}}{v} = l^* \tag{5-4}$$

where l^* (pronounced "ell-star") has units of time. As the ratio of a mean distance of travel and the average speed of the neutron, l^* is an average time that the neutron exists. It is often called the *mean neutron generation time*[†] since it is the time required for neutrons from one generation to cause the fissions that produce the next generation of neutrons. It is usual to break this characteristic time into two parts—the times spent as fast and thermal neutrons, respectively. Fast neutrons slow to thermal energies or leak out in $\approx 10^{-7}$ s. Thermal neutrons tend to exist for $\approx 10^{-4}$ s before they are captured. Thus, fast reactors have $l^* \approx 10^{-7}$ s, while thermal reactors have $l^* \approx 10^{-7}$ s $+ 10^{-4}$ s $\approx 10^{-4}$ s.

Equations 5-3 and 5-4 may be substituted in Eq. 5-2 to yield

$$l^* \frac{d\Phi}{\Phi} = \rho \, dt$$

$$\frac{d\Phi}{\Phi} \, dt = \frac{\rho}{l^*} \tag{5-5a}$$

or $\quad \dfrac{d\Phi}{\Phi} = \dfrac{\rho}{l^*} \, dt \tag{5-5b}$

For constant ρ and l^*, Eq. 5-5a indicates that the fractional change in the neutron flux per unit time is constant and equal to the ratio of the two. Equation 5-5b has the same, now-familiar form as Eq. 2-5b and 2-22b for radioactive decay and beam attenuation, respectively. The solution to Eq. 5-5b is

$$\frac{\Phi(t)}{\Phi(0)} = e^{\rho t/l^*} = e^{t/T} \tag{5-6}$$

for time-dependent flux $\Phi(t)$ with a value $\Phi(0)$ at time $t = 0$ in a system containing only *prompt* neutrons. The *period* T, defined as l^*/ρ, is the amount of time required for the flux to change by a factor of e.

If only prompt neutrons were present, the small excess multiplication associated with $k = 1.001$ in a thermal reactor with $l^* = 10^{-4}$ s would result in

$$\rho = \frac{1.001 - 1}{1.001} \approx 10^{-3}$$

[†]Some references call this *same* term the "prompt neutron lifetime" and/or represent it by the symbol Λ. Since they also define other lifetimes, care is advised when using multiple references.

$$T = \frac{10^{-4} \text{ s}}{10^{-3}} = 0.1 \text{ s}$$

and $\frac{\Phi(t)}{\Phi(0)} = e^{t/0.1 \text{ s}}$

An elapsed time of only 1 s would cause the flux to increase by a factor of $e^{10} \approx 20{,}000$. Fast systems where $l^* \approx 10^{-7}$ s would, of course, exhibit even more rapid power increases. Since 1 s or less would hardly allow for action of either an operator or an automatic mechanical system, positive control under the circumstances in the example would be nearly impossible. This is especially true since statistical variations of the neutron population routinely produce small variations in k.

Delayed Neutrons

Somewhat more than 99 percent of all neutrons from fission appear at the time of fission. The remainder—the delayed neutrons—are a small but very significant contributor to the time-dependent behavior of the population.

Each delayed neutron is emitted immediately following the first beta decay of a fission fragment known as a *delayed-neutron precursor*. About 20 different fragment nuclides have been identified as precursors. Since the neutrons appear on a time scale characteristic of precursor decay rather than at the time of fission, they have *effective* generation times that are very long compared to l^*.

For most applications it is convenient to combine the known precursors into groups with appropriately averaged properties. Table 5-2 contains six-precursor-group data for thermal fission of ^{235}U. The half-lives, ranging from about one-quarter of a second to nearly a minute, are much larger than $l^* \approx 10^{-4}$ s. The *delayed fractions* β_i are the ratios of the number of delayed neutrons from precursor group i to the total number of fission neutrons ν. The *total delayed fraction* β is the sum of the six β_i. For other fissionable nuclides, the half-life structures are similar, but the delayed fractions may be quite different.

The presence of delayed neutrons requires that the fission-neutron source term in Eq. 5-1 be modified. Since β is the fraction of delayed neutrons the prompt source must be

$$\text{Prompt source} = (1 - \beta)\nu\Sigma_f$$

The delayed source, predicated on the decay of each of the six precursor groups, is

$$\text{Delayed source} = \sum_{i=1}^{6} \lambda_i C_i(t)$$

for decay constant $\lambda_i [= \ln 2/(T_{1/2})_i]$ and time-dependent precursor concentration $C_i(t)$. Each $C_i(t)$, in turn, depends on the balance

$$\frac{dC_i(t)}{dt} = \beta_i \nu \Sigma_f \Phi(t) - \lambda_i C_i(t)$$

TABLE 5-2
Six-Precursor-Group Half-Lives and Delayed Neutron
Fractions for Thermal Fission of ^{235}U†

Group	Half-life $T_{1/2}$ (s)	Delayed fraction β_i
1	55.0	0.00021
2	23.0	0.00142
3	6.2	0.00127
4	2.3	0.00257
5	0.61	0.00075
6	0.23	0.00027
Total	–	0.0065

†Data from G. R. Keepin, T. F. Wimett, and R. K. Zeigler, "Delayed Neutrons from Fissionable Isotopes of Uranium, Plutonium, and Thorium," *Phys. Rev.*, vol. 107, 1957, pp. 1044–1049.

where the term $\beta_i \nu \Sigma_f \Phi$ may be noted to represent both the (eventual) production rate of group i neutrons and the rate of formation of precursor i. (Since not all decays of a given precursor nuclide actually lead to neutron emission, $C_i(t)$ is really an *effective* concentration weighted to reflect the actual production of delayed neutrons.)

The time-dependent neutron balance for six delayed groups consists of the following set of seven coupled differential equations

$$\frac{1}{v}\frac{d\Phi(t)}{dt} = (1 - \beta)\nu\Sigma_f\Phi(t) + \sum_{i=1}^{6} \lambda_i C_i + \Sigma_a\Phi(t) + DB^2\Phi(t)$$

$$\frac{dC_i(t)}{dt} = \beta_i\nu\Sigma_f\Phi(t) - \lambda_i C_i(t) \qquad i = 1, 2, \ldots, 6$$

(5-7)

where all terms have been defined previously. Although not stated explicitly, these equations treat the neutron flux as if the time dependence at each spatial location were the same. This is equivalent to averaging all of the spatial dependence so that the reactor system may be considered position-independent or as being represented by a "point model."

Point Kinetics Equations

The form of Eqs. 5-7 may be modified by substituting the definitions of neutron flux Φ in Eq. 2-18, reactivity ρ in Eq. 5-3, and neutron generation time l^* in Eq. 5-4, with the results

$$\frac{dN(t)}{dt} = \frac{\rho - \beta}{l^*}N(t) + \sum_{i=1}^{6} \lambda_i C_i(t)$$

$$\frac{dC_i(t)}{dt} = \frac{\beta_i}{l^*}N(t) - \lambda_i C_i(t) \qquad i = 1, 2, \ldots, 6$$

(5-8)

where $N(t)$ = neutron density

$C_i(t)$ = effective precursor concentration for group i

ρ = reactivity

l^* = mean neutron generation time

β = total effective delayed neutron fraction ($\beta = \Sigma_{i=1}^6 \beta_i$)

β_i = effective delayed neutron fraction of group i

λ_i = effective decay constant of group i

These equations are generally known as the *point kinetics equations*. Since neutron density is proportional to neutron flux, fission rate, and power, the ratios of any of these parameters may be substituted for $N(t)/N(0)$ in Eqs. 5-8.

Although they were developed herein on the basis of one-group diffusion theory, Eqs. 5-8 do not contain any explicit remnants of their origin. The kinetics parameters ρ, l^*, and β_i, in fact, may be redefined to account for energy, position, and composition effects so that the same equations can be applied with substantial generality. Reactivity ρ is still defined by Eq. 5-3 with k replaced by a more accurate value than that from one-group diffusion theory. Although the generation time l^* must be determined by an entirely different procedure, it is still found to be approximately equal to the neutron lifetime when the system is near critical.

The delayed neutron fractions β for the three fissile nuclides of most interest are shown in Table 5-3. Delayed neutrons are emitted with an average energy only about one-half that of prompt neutrons. For this reason, they are somewhat more effective in producing fissions and their presence results in *effective delayed fractions* β_{eff}, which are larger than the physical values. Typical β_{eff} values for LWR systems are included in Table 5-3. When more than one fissile nuclide is present, the effective delayed fraction is a combination of separate values. LWR systems, for example, experience a decrease in β_{eff} as the core burns out ^{235}U and produces ^{239}Pu.

The point kinetics formulation in Eqs. 5-8 is a very powerful tool for describing the time-dependent behavior of neutron chain reacting systems. However, the presence of the seven coupled differential equations results in rather complicated solutions. Only some important general features of the solutions are considered here.

Reactivity Insertions

A critical system operates at a steady power level based on equilibrium of *all* neutrons, i.e., of both the prompt and delayed populations. When reactivity is

TABLE 5-3

Delayed Neutron Fractions and Effective Delayed Neutron Fractions for ^{233}U, ^{235}U, ^{239}Pu

Nuclide	Delayed fraction β	Effective delayed fraction B_{eff}[†]
^{233}U	0.0026	0.003
^{235}U	0.0065	0.0070
^{239}Pu	0.0021	0.0023

[†]Typical for LWR systems.

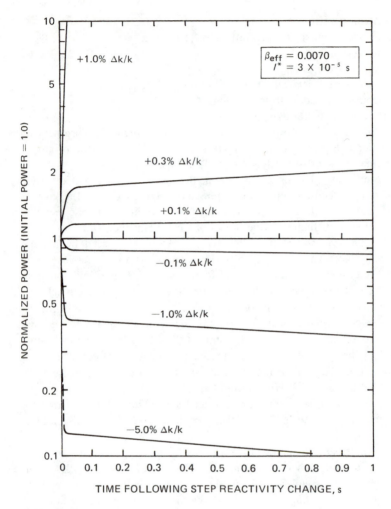

FIGURE 5-1
Time-dependent power behavior following various reactivity insertions in a typical reactor using slightly enriched uranium fuel.

added, the power level changes in a time-dependent manner in accordance with the point kinetics equations. Figure 5-1 shows the effect of various reactivity insertions. In all cases, it is assumed that the insertion is instantaneous (a "step" insertion) and sustained. (The latter neglects the presence of feedback mechanisms that tend to cause the reactivity to change with time.)

A negative insertion of reactivity into an initially critical system causes subcriticality and a decreasing power level. As shown on Fig. 5-1, the power first experiences a rapid drop and then continues downward on a more gentle linear slope. The initial *prompt drop* occurs because the multiplication has been decreased and there are now fewer neutrons in each succeeding generation. However, the existing delayed neutron precursors, which are unaffected by the reactivity change, continue to add neutrons and prevent the power from falling too rapidly. The *asymptotic period* (which is a straight line on the logarithmic scale of Fig. 5-1) results from the entry of the delayed neutrons coupled with the secondary fissions, prompt neutrons, and delayed precursors they produce. In the limit of very large negative reactivity insertions where no secondary fissions occur, the power falls off on a period of ≈80 s (corresponding to the precursor group with the 55 s half-life). All other cases result in longer decay periods.

Positive insertions of reactivity result in a supercritical system and an increasing power level. Figure 5-1 shows the effect of such insertions in two different behavior regimes. When $\rho < \beta$, the initial *prompt jump* is caused by the increased multiplication of prompt neutrons. Less than one *prompt* neutron results from each fission (since the delayed fraction β exceeds the excess multiplication $\rho \approx k - 1$). Thus, it is necessary that some delayed neutrons enter the system for criticality (and supercriticality) to be sustained. The positive asymptotic period results from the presence of the long-lived delayed neutrons. As was also true for the previous example, the delayed neutrons decrease the rate of change of the power level over that which would occur if only prompt neutrons were produced by fission.

At successively larger positive reactivity insertions, the prompt jump becomes larger and the asymptotic period shorter as fewer and fewer delayed neutrons are required for a self-sustaining chain reaction. *No* delayed neutrons are required at the *prompt critical* condition of $\rho = \beta$. For $\rho > \beta$, the power rises on a time-scale characteristic of the prompt neutron generation time, as shown in Fig. 5-1 (1% $\Delta k / k$).

The prompt critical condition does not signal a dramatic change in neutron behavior, i.e., the characteristic period changes in a regular manner between reactivities below and above this reference. Prompt critical is, however, as convenient as any other condition for marking the transition from delayed-neutron to prompt-neutron time scales. On this basis, the unit of the *dollar* [$] is defined as

$$\text{Reactivity in dollars} = \frac{\rho}{\beta}$$

The relationship to the U.S. currency is also carried over to the decimal fractions, with each 0.01 being equivalent to 1¢ of reactivity. It should be noted that the scale is nuclide dependent because the reference β differs among the fissile nuclides as shown by Table 5-3.

Inhour Equation

The point kinetics formulation in Eqs. 5-8 has solutions of the form

$$N(t) \approx C_i(t) \approx e^{\omega t}$$

where each inverse period ω is related to the kinetics parameters by the characteristic equation

$$\rho = \omega l^* + \sum_{i=1}^{6} \frac{\omega \beta_i}{\omega + \lambda_i} \tag{5-9}$$

which is called the *inhour equation*. (The name of the latter derives from early application to systems which had long periods that were measured "*in* (units of) *hours*.")

The inhour formulation in Eq. 5-9 provides a relationship between the reactivity ρ and the inverse periods ω, since the parameters l^*, β_i, and λ_i have fixed positive values for most applications. It is found that for each ρ there are seven real values of ω such that when

$\rho > 0$, one is positive and the others are negative
$\rho = 0$, one is zero and the others are negative
$\rho < 0$, all are negative

Each value of ω contributes to the behavior of the system according to an expression of the form

$$e^{\omega t} = e^{t/T}$$

for period T defined as $T = 1/\omega$. The latter suggests that the terms related to the largest negative values of ω will die out quickly, leaving the least negative (or most positive) root to dominate. The transient behavior of the six negative roots gives rise to the prompt drop and prompt jump discussed previously and shown by Fig. 5-1.

The most positive root of the inhour equation may be labeled ω_0 and its corresponding period T_0. The asymptotic period, which follows a step reactivity insertion, is this period T_0.

The inhour equation reduces to a relatively simple form in two limiting cases. For small reactivity insertions ($\rho \leqslant 0.1\beta$),

$$T_0 \approx \frac{\langle l \rangle}{\rho} \tag{5-10}$$

for an average neutron lifetime $\langle l \rangle$ represented by

$$\langle l \rangle = (1 - \beta)l^* + \sum_{i=1}^{6} \frac{\beta_i}{\lambda_i} \approx \sum_{i=1}^{6} \frac{\beta_i}{\lambda_i}$$

The average lifetime computed in this manner weights the neutron generation time [prompt lifetime] l^* and the mean lifetime τ_i for each delayed group (recall from Chap. 2 that $\tau_i = 1/\lambda_i$) by the fractions $1 - \beta$ and β_i, respectively. The generation time is usually small enough to be neglected in comparison to the summation (for a typical LWR, $\langle l \rangle \approx 0.08$ s, while $l^* \approx 10^{-4}$ s).

Large reactivity insertions ($\rho \gtrsim 1.5\beta$) result in periods that can be approximated from Eq. 5-9 as

$$T_0 = \frac{l^*}{\rho - \beta} \tag{5-11}$$

Thus, the period for very large insertions depends on the reactivity *above prompt critical* and the *prompt* neutron generation time. By contrast, the period for a very small insertion is characterized by the *total* reactivity and the average lifetime for *delayed* neutrons.

FEEDBACKS

The preceding analyses of the point kinetics equations considered system behavior after a stable reactivity level was established. In real systems, however, various feedback mechanisms are present which cause reactivity to change as the neutron level changes. These mechanisms operate on short time scales and are very important to reactor operation and safety.

For purposes of further discussion, feedbacks may be classified as positive or negative. A positive feedback tends to enhance the condition that produced it, while a negative feedback tends to diminish the condition. For example, a positive reactivity–power feedback proceeds as follows:

1. Increased reactivity leads to a greater rate of power increase.
2. Increased power raises core temperatures.
3. Higher temperatures increase reactivity.

The cycle could continue (in principle) until the temperatures were sufficiently high to destroy the system. On the other hand, with negative feedback the same reactivity–power–temperature cycle would reduce the reactivity and tend to stabilize the system.

Feedback Mechanisms

The feedback mechanisms of most interest in nuclear reactor systems are based on the changes in nuclear and/or physical characteristics that accompany variations in power level. Fuel temperature, coolant/moderator conditions, and fuel motion may each be responsible for a portion of the overall effect of a power level change.

Fuel Temperature Feedback
An increase in fuel temperature generally affects the neutron balance by decreasing fuel density and by changing the characteristics of the absorption of resonance-energy neutrons. A fuel density decrease has a negative reactivity effect in virtually all systems since leakage increases. This feedback is a relatively minor contribution in power reactors where the ceramic fuel pellets expand little. On the other hand, in certain test reactors made of metal (e.g., Sandia Laboratories' SPR-III and Los Alamos' Godiva), the fuel density feedback is strong enough to bring the system subcritical following a large reactivity insertion.

A second fuel-temperature feedback depends on increased neutron absorption in the high, narrow resonance peaks which are characteristic of fissionable nuclides (e.g., as shown in Fig. 2-11 for ^{238}U). The phenomenon known as the *Doppler effect* is caused by an apparent broadening of the resonances due to thermal motion of nuclei, as indicated on Fig. 5-2.

Each resonance corresponds to a quantum energy level in a nucleus. The level may be assumed to absorb neutrons of one specific energy E_0. Since the *relative*

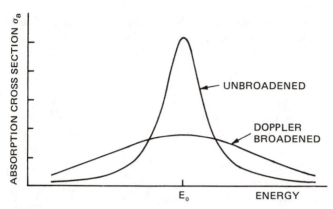

FIGURE 5-2
Effect of temperature on the effective shape of a resonance absorption cross section.

energy between the neutron and the nucleus is the appropriate reference, stationary nuclei absorb only neutrons of energy E_0, while those in motion can absorb at slightly different energies. Thermal motion of nuclei characterized by a wide distribution of energies can lead to absorption within a range about E_0 (as in Fig. 5.2). A neutron slowing down in a thermal reactor, for example, loses large amounts of energy in each collision. Although its final energy is not likely to be exactly E_0, the probability of absorption increases as the effective range of the resonance about E_0 is expanded. This "effective broadening" of the energies absorbed in resonances is the principle behind the Doppler feedback mechanism.

The Doppler effect can produce either a positive or negative reactivity feedback. In fuel that is primarily fissile, the increased absorption is likely to be in fission resonances (e.g., as shown in Fig. 2-12) and likely to enhance multiplication for a positive feedback effect. The opposite is true for the parasitic absorption in fertile material.

The Doppler feedback is strongly negative in the thermal reactor systems because their effective fissile content is low. LMFBR designs limit effective enrichment to 15–20 wt% to assure that this feedback effect is negative.

Moderator/Coolant Feedback

All power reactor systems have a moderator and/or a coolant which slows down neutrons via scattering interactions. An increase in temperature generally decreases both the density and the effectiveness for slowing down. This effect may produce a positive or negative feedback, depending on the design of the system.

Decreased neutron thermalization in LWRs reduces reactivity, since the higher energy neutrons are more subject to resonance absorption in the ^{238}U (i.e., the resonance escape probability p is decreased). The BWR design depends heavily on changes in water density (or equivalently, steam-bubble volume or void content) for power level control.

A PWR uses dissolved boric acid for reactivity control. Since boron density and thus absorption decreases with the water density, there is a positive feedback that may offset the effect of reduced thermalization.

The PHWR employs separate heavy-water supplies for coolant and moderator, respectively. A reduction in coolant density has little effect on the neutron thermalization or resonance escape (i.e., the factor p) associated predominately with the moderator. However, the reduced moderation in the fuel bundle itself can lead to slightly more fast fission (i.e., a higher value for ϵ). The net result is generally a positive coolant-temperature feedback. The magnitude of the feedback varies depending on complex fuel-depletion effects that occur over the lifetime of the core.

Graphite expansion with temperature in the HTGR reduces thermalization to provide a negative moderator feedback. Changes in the helium coolant density have a negligible effect on the system.

Although not intended as a moderator, the sodium coolant in the LMFBR reduces neutron energies, or "softens" the spectrum. The soft spectrum is more subject to resonance absorption, has a lower average η (as discussed in the next chapter) and, thus, has a lower reactivity. Since sodium voiding (density decrease) causes the spectrum to harden again, it constitutes a positive feedback. This void effect, in removing neutrons from the resonance energy range, may also serve to negate an otherwise-negative Doppler feedback effect.

Fuel Motion Feedback

When the neutron flux causes thermal gradients in fuel rods, differential expansion can produce bowing of the rods. This fuel motion, in turn, changes overall or local fuel densities. It may produce either positive or negative feedback in thermal reactors depending on the detailed design of the system. On the other hand, fuel bowing which increases overall or local densities tends to be a positive feedback mechanism in fast reactors like the LMFBR.

A limiting condition for the fuel motion feedback is an accident situation where there is substantial deformation and melting of fuel. As considered in some detail in Chaps. 16–17, the effect is a strong negative feedback in thermal reactors, but may provide a strong positive feedback in the LMFBR.

Reactivity Coefficients and Defects

The temperature-related feedbacks considered above are most readily quantified by a *reactivity coefficient* α—the rate of change of reactivity with respect to the feedback variable—and the *reactivity defect* $\Delta\rho$—the reactivity change associated with a macroscopic change of the feedback variable. Both parameters can be very useful for modeling the dynamic behavior of nuclear systems.

The *fuel-temperature coefficient* α_f is defined as

$$\alpha_f(T_f) = \frac{\partial \rho}{\partial T_f} \tag{5-12}$$

for "effective" (i.e., average) fuel temperature T_f. Since the coefficient is temperature-dependent, the reactivity defect associated with a particular fuel-temperature change must be obtained by integration. It is often useful to know the reactivity change associated with taking a reactor from hot, zero power [HZP] to hot, full power [HFP] as calculated with the *fuel-temperature defect*

$$\Delta\rho_f(\text{HZP} \to \text{HFP}) = \int_{T(\text{HZP})}^{T(\text{HFP})} \alpha_f(T_f)\, dT_f$$

for initial and final temperatures corresponding to the respective fuel conditions.

Coefficients for coolant and/or moderator feedbacks in specific systems take one or more of the following forms:

- *moderator temperature coefficient* [MTC]

$$\alpha_m(T_m) = \frac{\partial\rho}{\partial T_m} \qquad (5\text{-}13a)$$

for moderator temperature T_m
- *moderator density coefficient* [MDC]

$$\alpha_d(d_m) = \frac{\partial\rho}{\partial d_m} \qquad (5\text{-}13b)$$

for moderator density d_m
- *void coefficient*

$$\alpha_v(f_v) = \frac{\partial\rho}{\partial f_v} \qquad (5\text{-}13c)$$

for void fraction (or void percentage) f_v

The corresponding reactivity defects are calculated as for the fuel temperature case. The *power coefficient* α_P—defined as

$$\alpha_P = \frac{\partial\rho}{\partial P} \qquad (5\text{-}14)$$

for power P (typically expressed as a fraction of full power)—consists of multiple parts since fuel and moderator/coolant feedbacks accompany all power escalation. If the major contributors are fuel-temperature and moderator-temperature coefficients,

$$\alpha_P = \frac{\partial\rho}{\partial P} = \frac{\partial\rho}{\partial T_f}\frac{dT_f}{dP} + \frac{\partial\rho}{\partial T_m}\frac{dT_m}{dP} = \alpha_f\frac{dT_f}{dP} + \alpha_m\frac{dT_m}{dP} \qquad (5\text{-}15)$$

where the derivatives are the rates of change of fuel temperature and moderator temperature with system power, respectively.

For some purposes it is convenient to determine average coefficients. The average fuel-temperature coefficient $\langle\alpha_f\rangle$, for example, is just

$$\langle\alpha_f\rangle = \frac{\Delta\rho_f(\Delta T_f)}{\Delta T_f}$$

for defect $\Delta\rho_f$ and temperature change ΔT_f. The average power coefficient is based on the sum of the individual defects. If only the fuel and moderator coefficients are considered, $\langle\alpha_p\rangle$ from hot, zero power to hot, full power (i.e., 0 percent to 100 percent power) is

$$\langle\alpha_p\rangle = \frac{\Delta\rho_f(\text{HZP} \to \text{HFP}) + \Delta\rho_m(\text{HZP} \to \text{HFP})}{100\%}$$

Coefficients and defects may be represented in a variety of ways. Reactivity is expressed as either a fraction or a percentage and as "$\Delta\rho$" or the equivalent "$\Delta k/k$" (e.g., 0.01 $\Delta\rho$, 1% $\Delta\rho$, 0.01 $\Delta k/k$, and 1% $\Delta k/k$ are all comparable). Temperatures are expressed in degrees F or C, densities and voids in fractions or percentages, and power in percentage. Careful examination of the units can usually prevent confusion when parameters are extracted from different references.

Sample Calculations

If a reactor is found to experience a 0.1 percent $\Delta k/k$ reactivity reduction with a $100°\text{C}$ moderator temperature increase, its average MTC would be

$$\langle\alpha_m\rangle = \frac{\Delta\rho}{\Delta T_m} = \frac{-0.001\ \Delta\rho}{+100°\text{C}} = -10^{-5}\ \frac{\Delta\rho}{°\text{C}}$$

A decrease in temperature from $350°\text{C}$ to $325°\text{C}$, then, might be expected to produce a reactivity change of

$$\Delta\rho = \langle\alpha_m\rangle(T_{m2} - T_{m1}) = -10^{-5}\ \frac{\Delta\rho}{°\text{C}} \times (325°\text{C} - 350°\text{C})$$

$$= -10^{-5}\ \frac{\Delta\rho}{°\text{C}} \times -25°\text{C} = +2.5 \times 10^{-4}\ \Delta\rho = \boxed{0.00025\ \frac{\Delta k}{k}}$$

This "small" positive reactivity change would then make the system slightly supercritical. According to Eq. 5-10, if the average neutron lifetime were 0.08 s, the asymptotic period T_0 for the power would be approximately

$$T_0 \approx \frac{\langle l \rangle}{\rho} = \frac{0.08\ \text{s}}{0.00025} = \boxed{320\ \text{s}}$$

The power would increase by a factor of e in something over 5 min.

Noting the similarity in form between the neutron kinetic behavior in Eq. 5-6 and that of radioactive decay in Eq. 2-6, a *doubling time* can be defined in the same way as a decay half-life with the result

Doubling time = $(\ln 2)T_0$

The period calculated above would then correspond to

$$\text{Doubling time} = (\ln 2)(320 \text{ s}) = \boxed{222 \text{ s}}$$

(As mentioned in the next section, doubling times are often used in measurements of reactivity worth.)

CONTROL APPLICATIONS

Real reactor systems may be modeled by the point kinetics equations and other methods considered in this chapter. However, it is ultimately necessary to translate theoretical results into system design and practice. Joint goals are to maximize energy production while preventing overpower conditions that could ultimately lead to release of hazardous fission products. Certain aspects of meeting these goals are addressed in the remainder of this chapter, and others in Chap. 7.

Limitations and Correlations

It has been noted that positive reactivity insertions result in positive periods and power increase. Near or above the prompt critical condition, the periods can become so short as to eliminate the possibility for control by human operators or mechanical systems. Thus, it is appropriate to limit reactivity changes in power reactors to well below prompt critical.

Most reactors include integrated reactivity-control features of the following types:

1. control rods with limited drive speeds and limited individual reactivity worths
2. control system producing "scram" or "trip" (i.e , full control rod insertion and shutdown) for overpower or excess period
3. negative temperature feedbacks to mitigate the consequences of any unintentionally large reactivity insertions

When negative feedbacks cannot be readily obtained, the other mechanisms become even more important.

The CANDU system, for example, has a positive coolant-temperature feedback. It compensates for this, at least in part, by maintaining a relatively small excess multiplication. The system's unique ability to exchange fuel assemblies while operating at full-power conditions (as described in Chaps. 9 and 13) allows the multiplication to be kept low at all times in the lifetime of the core.

Reactivity is a rather abstract concept which quantifies the tendency of a neutron chain reacting system to undergo change. However, physical changes are ultimately responsible for causing reactivity to exist. Correlations between system conditions and reactivities must be developed for reactor operations to proceed in an orderly fashion.

The inhour formulation in Eq. 5-9 provides one useful method for determining the "reactivity worth" of nearly any of the neutron balance controls described in Chap. 4 (e.g., control rods and soluble poisons). The method is also applicable to measuring certain of the reactivity defects described in the previous section of this chapter.

The basic procedure for applying the inhour equation depends on knowledge of the parameters l^*, β_i, and λ_i so that tables of ρ vs. T_0 can be generated from Eq. 5-9. Next, a reactivity insertion is performed, e.g., by a specified control-rod movement, and T_0 is measured. (In practice, the latter may be performed by measuring doubling time—the time required for the power to increase by a factor of two—with a stopwatch and dividing by ln 2 to obtain the period T_0.) Comparison of the measured period to values in the $\rho - T_0$ table provides a value for the reactivity. This result is the reactivity *worth* of the initial change in system geometry or other physical condition. The inhour method is limited to small reactivities and/or low power levels where feedbacks do not cause the asymptotic period to change with time.

Integrated Response

More general situations consist of initial physical changes followed by feedback responses. Figure 5-3 is a reactivity feedback diagram that can be used to describe and/or model system behavior. For example, if an initially critical reactor experiences a positive external reactivity insertion ρ_{ext} (e.g., by control rod motion or soluble boron dilution), the "feedback loop" has the following features:

1. The rate of increase of power $P(t)$ is determined by the kinetics equations.
2. The output power changes system temperatures and densities resulting in a feedback reactivity ρ_F (the power defect for the given changes).
3. At the summation point Σ, the external and feedback reactivities combine to produce the net reactivity, which returns to the kinetics section.
4. If the feedback is negative (i.e., if ρ_{ext} and ρ_F have opposite sign), the cycle will continue until the system stabilizes at $\rho(t) = 0$ with power level and temperatures above the initial values.
5. If the feedback is positive, the power will tend to increase until ρ_{ext} is removed or the core disassembles (a strong negative feedback!).

Integrated dynamic responses like that above are especially important to analysis of potential reactor accidents in which external reactivity is introduced by direct or indirect methods. Poison-rod withdrawal or ejection and soluble poison dilution (the latter in PWR systems only) are the most direct means of reactivity insertion.

While negative temperature coefficients are always desirable from the standpoint of control stability, they result in a loss of reactivity as temperatures and the power level rise. This requires that the core have more excess reactivity than it would otherwise require (e.g., roughly 4 percent $\Delta k/k$ for a PWR, as considered more fully in Chap. 12). The other side of the picture is that decreasing temperatures restore the "lost" reactivity and produce a net positive insertion. Thus, the so-called *cold-water accidents* produce indirect reactivity insertions merely by lower core temperatures. Inadvertent startup of a "cold" loop of a multiloop coolant system is one example of such an accident. Another is a steam-line break where the rapid depressurization (of the BWR vessel or a steam generator in other designs) results in substantial cooling. As indirect reactivity mechanisms, they can be considered as part of ρ_{ext} in Fig. 5-3 or as a separate, parallel contribution acting directly on the feedback block.

The basic features of the response of an HTGR to a sudden, sustained decrease of 68°C in helium inlet temperature are shown by Fig. 5-4. The behavior of the power and temperature curves may be explained as follows:

1. The cooler helium at the inlet causes an initial reduction in the fuel, moderator, and helium-outlet temperatures on a time scale of tens of seconds.
2. Since both the fuel and moderator temperature feedbacks are negative, the lower temperatures lead to a positive reactivity insertion and a resulting power escalation.
3. The increased power then causes the fuel temperature to rise, followed by a similar change in the temperature of the graphite moderator.
4. This time the temperature effect is in the direction of a reactivity decrease, which causes the power level to stabilize.
5. In less than 3 min, the system approaches an equilibrium state at approximately 112 percent of the initial power level with about a 20°C reduction in helium-outlet temperature, a 2-3°C increase in fuel temperature, and a 5°C decrease in moderator temperature.

This example provides a good illustration of the type of feedback interactions that occur in all power reactors.

Reactivity Transients

A small positive reactivity insertion produces a prompt jump and then initial power increase on an asymptotic period, as shown by the appropriate curves on Fig. 5-1. However, when the power reaches high enough levels for significant temperature increases, the feedback mechanisms modify this behavior. The systems designed with

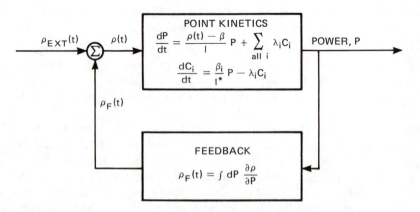

FIGURE 5-3
Reactivity feedback diagram (adapted from *The Technology of Nuclear Reactor Safety*, T. J. Thompson and J. G. Beckerley (eds.), Vol. 1, by permission of MIT Press, Cambridge, Massachusetts. Copyright © 1964 by the Massachusetts Institute of Technology).

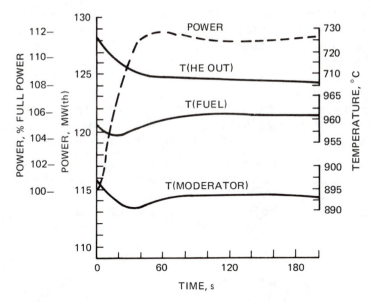

FIGURE 5-4
Response of the Peach Bottom HTGR to a 68°C decrease in helium inlet temperature (adapted from *The Technology of Nuclear Reactor Safety*, T. J. Thompson and J. G. Beckerley (eds.), Vol. 1, by permission of The MIT Press, Cambridge, Massachusetts. Copyright © 1964 by the Massachusetts Institute of Technology).

negative temperature responses experience a gradual lengthening of the asymptotic period followed ultimately by a leveling-off of the power at some constant value.

Larger reactivity insertions may produce similar effects or may result in a reactivity pulse. In the latter case, very rapid, supercritical power escalation produces a feedback response that actually drives the system subcritical and shuts it down neutronically.

Sandia Laboratories' Annular Core Research Reactor [ACRR] employs 35 wt% ^{235}U fuel mixed with BeO for routine pulsed operation. Based on the Doppler and density feedback mechanisms, the system has an overall fuel-temperature coefficient that is strongly negative. Figure 5-5 shows the time-dependent behavior of the power, fuel temperature, and net reactivity following the insertion of 3$ (or 2$ *above* prompt critical) of reactivity. Important features of the sequence are:

1. At time $= 0$, when the system is just critical at a very low power level, a signal from the reactor console begins the rapid hydraulic withdrawal of a poison "pulse rod" which has been calibrated to a worth of 3$.
2. By ≈ 55 ms the system is increasing power on a 0.87-ms period which, in the absence of feedback, would follow the dashed curve.
3. As the power increases to several percent of the ultimate maximum, the energy deposition increases the fuel temperature and, through feedback, decreases the reactivity.
4. The lower reactivity causes the period to lengthen, but since the power level is substantial, the temperature continues to rise. The fuel temperature ultimately

rises enough for the feedback to drive the system subcritical. (It is actually subcritical at $\rho = 1\$$, since the millisecond time scale is too short for many delayed neutrons to have been produced.)
5. With negative reactivity, the system is shut down except for the effect of the remaining delayed-neutron precursors (the effect of which is the power "tail" shown on the logarithmic plot).
6. The net result in this particular case is a pulse with a peak power slightly greater than 30,000 MW and with an FWHM (i.e., full-width-at-half-maximum or the time for which the power exceeds one-half of the peak value) of about 4 ms.

In typical operations, the pulse rod is dropped back into the core to assure final subcriticality. If it is not replaced, the system would become slightly supercritical as soon as the temperature dropped. The feedbacks would ultimately cause the power to become steady at a higher level (perhaps after one or more oscillations).

Power reactors would respond to a large reactivity insertion in the same

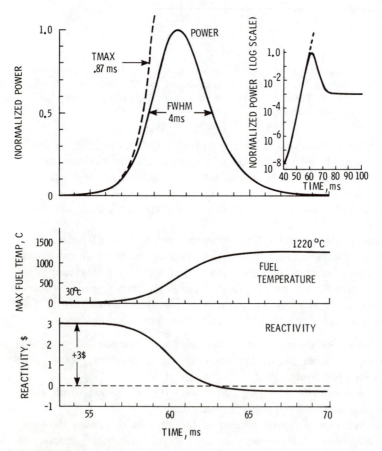

FIGURE 5-5
Typical response of Sandia Laboratories' ACRR pulse reactor to a 3\$ reactivity insertion. (Data courtesy of J. Philbin, Sandia National Laboratories.)

qualitative manner demonstrated by the example. The size of the insertion and the characteristics of the system determine the height and width of the pulse. If the resulting energy production (i.e., the integral of the power curve) is too great for the system to tolerate, core disassembly may become part of the negative feedback mechanism. Chapters 7 and 16 address these concerns further.

EXERCISES

Questions

5-1. Identify and define the parameters ρ, l^*, β, β_i, and λ_i.

5-2. Sketch and describe the basic features of the power, fuel temperature, and net reactivity histories for a large power pulse in a reactor like ACRR.

5-3. Explain the terms prompt jump, prompt drop, asymptotic period, and prompt critical.

Numerical Problems

5-4. An approximate solution to the point kinetics equations for a single group of delayed neutrons is

$$\frac{P(t)}{P(0)} = \frac{\beta}{\beta - \rho}\, e^{\lambda \rho t/(\beta - \rho)} - \frac{\rho}{\beta - \rho}\, e^{-(\beta - \rho)t/l^*}$$

for ρ at least 50¢ from prompt critical. Consider a typical LWR system with parameters $\beta_{\text{eff}} = 0.0070$, $l^* = 3 \times 10^{-5}$ s, and $\lambda = 0.084$ s^{-1}.

a. Plot $P(t)/P(0)$ and log $[P(t)/P(0)]$ for $\rho = 0.05\beta$ on a linear time scale up to $t = 10$ s.

b. Repeat (a) for a logarithmic time scale.

c. Using only the above results, estimate the fractional increase in power associated with the prompt jump and estimate the magnitude of the asymptotic period. (Hint: note the effect of each of the two terms in the equation.)

d. Repeat the plots and period calculation for $\rho = 1.5\beta$.

5-5. Estimate the reactivity insertions that will result in periods of 1 h and 1 day, respectively, for a system characterized by the parameters in Prob. 5-4.

5-6. Consider a PWR with the following average reactivity coefficients:

$$\langle \alpha_f \rangle = -1.0 \times 10^{-5}\, \frac{\Delta \rho}{°F}$$

$$\langle \alpha_m \rangle = -2.0 \times 10^{-4}\, \frac{\Delta \rho}{°F}$$

a. Calculate the reactivity defect in going from hot, zero power ($T_f = T_m = 530°F$) to hot, full power ($T_f = 1200°F$, $T_m = 572°F$) for each feedback mechanism.

b. Calculate the average power coefficient for 0 to 100 percent power.

5-7. The moderator temperature coefficient [MTC] in Prob. 5-6 is characteristic of the PWR at end of (core) life [EOL] when the soluble boron poison concentration is minimum. At beginning of life [BOL] when the concentration is high,

$$\langle \alpha_m \rangle = -0.2 \times 10^{-4} \frac{\Delta \rho}{{}^\circ F}$$

a. Identify the mechanism that is responsible for the difference.
b. Calculate the average power coefficient for BOL conditions.

SELECTED BIBLIOGRAPHY†

Reactor Kinetics
    Ash, 1979
    Hetrick, 1971
    Keepin, 1965
    Lewins, 1978
    Stacey, 1969

Other Sources with Appropriate Sections or Chapters
    Connolly, 1978
    Duderstadt & Hamilton, 1976
    Etherington, 1958
    Foster & Wright, 1977
    Glasstone & Sesonske, 1967
    Graves, 1979
    Henry, 1975
    Lamarsh, 1966
    Lamarsh, 1975
    Murray, 1975
    Onega, 1975
    Rydin, 1977
    Sesonske, 1973
    Thompson & Beckerley, 1964

†Full citations are contained in the General Bibliography at the back of the book.

6

FUEL DEPLETION AND RELATED EFFECTS

Objectives

After studying this chapter, the reader should be able to:

1. Describe the buildup of plutonium isotopes by neutron irradiation of ^{238}U.
2. Explain the relationship of eta $[\eta]$ to breeding and conversion.
3. Identify the mechanism by which the thorium fuel cycle experiences a radiation problem in recycle fabrication operations.
4. Explain the effects of ^{135}Xe and ^{149}Sm on thermal reactor design and control.
5. Define hazard index and describe the behavior of fission-product and transuranic wastes in terms of it.
6. Calculate fuel depletion and production effects for simple systems.

Prerequisite Concepts

Radioactive Decay	Chapter 2
Reaction Rates	Chapter 2
Eta $[\eta]$	Chapter 2
Transmutation (Fig. 2-16)	Chapter 2
Dose Limits	Chapter 3
Maximum Permissible Concentrations [MPC]	Chapter 3
Reactivity	Chapter 5
Sodium-void Feedback	Chapter 5
Doppler Feedback	Chapter 5

The time-dependent phenomena that occur in reactors over the long term are grouped in the category of *fuel depletion* because they are associated with the fission reactions directly or with reactions initiated by the fission neutrons. Burnup of fissile nuclides, fission-product formation, and transmutation are the major contributors. The reactivity effects of these slow-acting feedbacks must be compensated by appropriate neutron-balance control strategies.

FUEL BURNUP

Each nuclide in a reactor system obeys a simple balance equation of the form

$$
\begin{array}{l}
\text{Net rate of} \\
\text{production}
\end{array}
=
\begin{array}{l}
\text{rate of} \\
\text{creation}
\end{array}
-
\begin{array}{l}
\text{rate of} \\
\text{loss}
\end{array}
$$

Creation processes are based on nuclear reactions with and radioactive decay of other nuclides. When the nuclides are constrained from movement, the rate of loss is equivalent to the rate of neutron absorption, where the latter includes all nuclear processes which result in the loss of the initial nucleus (i.e., all reactions except elastic and inelastic scattering as shown by Fig. 2-10).

If there are no significant creation mechanisms for a particular nuclide, the absorption rate is the only contributor. The balance for this pure *depletion* or *burnup* case may be written as

$$
\frac{\partial n(\mathbf{r},\ t)}{\partial t} = -n(\mathbf{r},\ t)\sigma_a(E)\Phi(\mathbf{r},\ E,\ t) \tag{6-1}
$$

for concentration n, microscopic cross section σ_a, and flux Φ, which may depend on position $\mathbf{r}$, energy E, and/or time t as shown. Averaging the parameters over space and energy (as was done for the point kinetics model in Eqs. 5-8) results in a "point" depletion equation of the form

$$
\frac{dn(t)}{dt} = -n(t)\sigma_a\Phi(t)
$$

with solution

$$
n(t) = n(0)e^{-\sigma_a\int_0^t \Phi(t)\,dt} \tag{6-2}
$$

for initial concentration $n(0)$ at time $t = 0$ and where the integral term is neutron *fluence, flux-time,* or *nvt,* as defined by Eqs. 3-2 and 3-3. A constant (or time-averaged) flux Φ_0 provides the simplified expression

$$
n(t) = n(0)e^{-\sigma_a\Phi_0 t} \tag{6-3}
$$

Comparison with Eq. 2-6 shows that the nuclide concentration "decays" exponentially with a decay constant equivalent to the "microscopic absorption rate" ["absorption rate per nucleus"] $\sigma_a\Phi_0$. In current water-reactor applications, for example, the behavior of fissile ^{235}U and fertile ^{238}U are described well by Eq. 6-2 since there are no significant production mechanisms and both have extremely long decay half-lives.

Another measure of reactor core burnup does not depend on any knowledge of cross sections or fluxes. It merely considers *thermal* energy output per unit mass of fuel—usually MWD/T, *megawatt-days* per *ton* (metric ton) of fuel. According to convention, fuel is considered to be the heavy metal content (total Th, U, and Pu), exclusive of alloy or compound constituents.

The concepts of fluence and burnup in MWD/T are both fairly general measures of depletion effects. They were originally applied to natural-uranium systems where a particular value of either would imply specific depletion of ^{235}U and ^{238}U, production of ^{239}Pu, damage to internal structural components, etc. Since flux, power, and fission rate tend to be proportional, fluence and MWD/T can be employed as measures of energy deposition and radiation damage for any reactor types. The concepts still are most meaningful for comparisons among similar systems.

TRANSMUTATION

All of the neutron absorption reactions which do not result in fission do lead to the production of new nuclide species through *transmutation*. These can, in turn, be transmuted or may undergo radioactive decay to produce still more species. Because of the interrelationships among the nuclides created by absorptions and decays, all are commonly referred to as *transmutation products*. Figure 2-16 shows the interrelationships among many of the important nuclides in this category.

The production rate for any specific nuclide ^A_ZX is based on a balance equation of the form

Net rate of production $=$ rate of creation by (n, γ) reactions in $^{A-1}_Z$X $+$ rate of creation by other reactions r in nuclides j

$+$ rate of creation by decay of nuclides i $-$ rate of loss by absorption $-$ rate of loss by radioactive decay

$$\frac{dn(t)}{dt} = n^{A-1}\sigma_\gamma^{A-1}\Phi + \sum_{\text{all } j} n^j \sigma_r^j \Phi + \sum_{\text{all } i} n^i \lambda^i - n\sigma_a \Phi - n\lambda \qquad (6\text{-}4)$$

for nuclide concentrations n, decay constants λ, microscopic cross sections σ, and neutron flux Φ. The second and third creation terms include only those reactions and decays, respectively, which result in production of the reference nuclide.

The transmutation products which are fissile, fertile, or parasitic absorbers have impacts on criticality and are, thus, very important to reactor design. These and other nuclides are also of interest to fuel cycle applications.

Conversion and Breeding

Among the most important neutron interactions are those which convert the fertile nuclides ^{232}Th, ^{238}U, and ^{240}Pu to fissile nuclides ^{233}U, ^{239}Pu, and ^{241}Pu, respectively, as shown by Fig. 2-5 and on Fig. 2-16. The amounts of new fissile material that can be produced are based largely on the number of neutrons in the system which are not required to sustain the chain reaction.

The parameter η, the average number of neutrons produced per neutron absorbed in fuel, serves as a useful reference. Figure 6-1 shows η as a function of

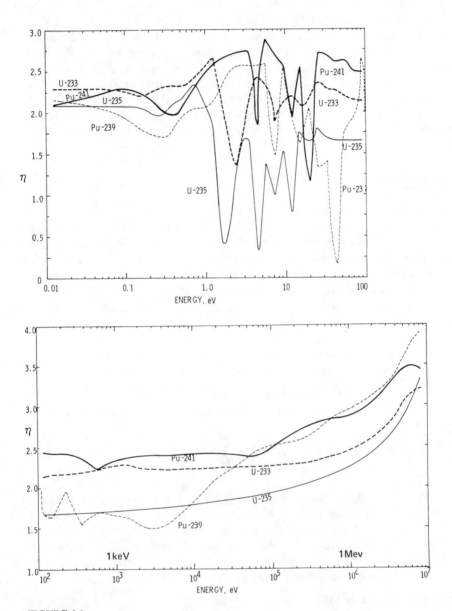

FIGURE 6-1
Values of eta [η] for fissile nuclides as a function of energy. [Courtesy of Electric Power Research Institute (Shapiro, 1977).]

TABLE 6-1
Average Conversion or Breeding Ratios for Reference Reactor Systems

Reactor	Initial fuel[†]	Conversion cycle[†]	Conversion ratio	Breeding ratio
BWR	2–4 wt% ^{235}U	^{238}U–Pu	0.6	–
PWR	2–4 wt% ^{235}U	^{238}U–Pu	0.6	–
CANDU	Natural U	^{238}U–Pu	0.8	–
HTGR	≈5 wt% ^{235}U	^{232}Th–^{233}U	0.8	–
LMFBR	10–20 wt% Pu	^{238}U–Pu	–	1.0–1.6

[†]All plutonium in power reactors is an isotopic mixture based on initial conversion of ^{238}U to ^{239}Pu and followed by transmutation to the "higher" isotopes.

energy for the important fissile nuclides. Since one neutron is required to sustain the chain reaction, $\eta - 1$ is an upper limit on the number of neutrons available for producing new fuel. For $\eta > 2$, the possibility exists for *breeding,* or producing more fissile nuclei than are used in the chain reaction. The same process is called *converting* when $\eta < 2$ (*or* when a potential breeding cycle produces less fuel than is used because of inherent neutron loss mechanisms). Since ^{235}U is the only fissile nuclide that exists in nature, the others *must* be generated by a transmutation process. The production of ^{233}U is based on neutron irradiation of ^{232}Th as shown by Fig. 2-5a. The fissile plutonium isotopes ^{239}Pu and ^{241}Pu are produced together in systems containing ^{238}U (as described in the next section).

According to Fig. 6-1, ^{233}U has the largest η for a thermal neutron spectrum with $E < 0.1$ eV. The ^{232}Th–^{233}U cycle, thus, offers the best possibility for breeding in a thermal reactor. A modified CANDU system, the light-water breeder reactor [LWBR], and the molten-salt breeder reactor [MSBR] (described in Chaps. 13 and 14) are all candidates for "thorium breeders."

A fast neutron spectrum with $E > 10^5$ eV favors breeding with the plutonium isotopes ^{239}Pu and ^{241}Pu. The LMFBR employs the ^{238}U–^{239}Pu cycle on this basis. Since the value of η increases with neutron energy, there is an incentive to have as little neutron moderation as possible, i.e., to maintain a "hard" neutron spectrum. (The slowing-down caused by sodium "softens" the spectrum slightly. Thus, sodium voiding and the resultant rehardening of the spectrum produces an increase in η and a positive feedback as mentioned in Chap. 5.)

The *instantaneous breeding ratio* and *conversion ratio* are both defined as the rate of creation of new fissile material divided by the rate of destruction of existing fissile. If the ratio exceeds unity, it is a breeding ratio; otherwise, it is a conversion ratio. Typical average values for the reference reactors are shown in Table 6-1. The values for the two LWR reactors are essentially the same because of design similarities. The HTGR ratio is somewhat higher because of the more favorable nature of the ^{232}Th–^{233}U cycle. The CANDU system, although dependent on ^{235}U and bred Pu like the LWRs, has a relatively high conversion ratio mainly because on-line refueling and the resultingly small requirements for neutron poisons to control excess reactivity allow the CANDU system to have a higher conversion ratio than the LWRs. In fact, conversion in both the LWR and HTGR designs suffers from the need for the poisons.

The LMFBR has the range of possible breeding ratios in Table 6-1 depending on the neutron energy spectrum. As noted from Fig. 6-1, a very hard spectrum results in a large value of η and a correspondingly large breeding ratio. A softer spectrum, however, is generally favored from a safety standpoint. The lower-energy neutrons, which are subject to resonance absorption, allow a negative Doppler feedback to enhance the stability of the system's response to reactivity or temperature transients, as described in Chap. 5. Final LMFBR designs are based on a trade-off between breeding ratio and favorable reactivity feedbacks.

The breeding ratio changes with time as the various effects of fuel depletion occur. A more useful concept for describing the process is *doubling time*—the time in years for the core to contain twice its initial fissile content. In principle, this is the amount of time required for enough fuel to be available to refuel the reactor itself and to provide an initial loading for another reactor of the same type and power level.

Other Effects

The concentrations of all fissionable nuclides and transmutation products change whenever neutron irradiation occurs. Such changes have many significant effects on design and operation of reactor and fuel systems. A few representative examples are considered in the following paragraphs.

Reactivity Penalties

The production of plutonium isotopes from irradiation of ^{238}U may be traced on Fig. 2-16. Following initial formation of ^{239}Pu, successive capture reactions can produce ^{240}Pu, ^{241}Pu, ^{242}Pu, and so forth. The concentration buildup of the four major plutonium isotopes with burnup is shown by Fig. 6-2 for a typical LWR fuel.

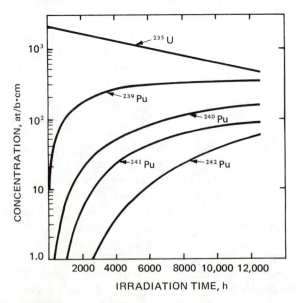

FIGURE 6-2
Buildup of plutonium isotopes with burnup for a typical LWR fuel composition.

TABLE 6-2
Reactivity Penalty from Selected Transmutation Products
for Recycle of BWR Fuel[†]

End of cycle number	Reactivity penalty at discharge, $\%\Delta k$			
	^{236}U[‡]	^{237}Np[§]	^{242}Pu	^{243}Am[§]
1	0.62	0.13	0.65	0.36
2	0.90	0.59	1.53	0.57
3	1.12	0.73	2.04	0.89

[†]From A. Sesonske, *Nuclear Power Plant Design Analysis*, TID-26241, 1973.
[‡]The ^{236}U concentration is assumed not to decrease in the diffusion plant.
[§]Neptunium and americium are removed by reprocessing on each recycle.

If plutonium is recycled with depleted uranium, ^{239}Pu production continues and the buildup of the higher isotopes continues toward an equilibrium level. Most of the fissile ^{239}Pu and ^{241}Pu fission, fertile ^{240}Pu produces ^{241}Pu, but ^{242}Pu acts only as a parasitic absorber. Since plutonium isotopes cannot be readily separated with current technology, the *reactivity penalty* associated with the ^{242}Pu content must be compensated by loading a larger total mass of plutonium in succeeding cycles. Table 6-2 shows the reactivity penalty associated with ^{242}Pu at the end of an initial core loading and each of the first two recycles for a typical BWR. Since the presence of ^{243}Am depends on radiative capture by ^{242}Pu and a beta decay (see Fig. 2-16), its reactivity penalty is directly tied to the presence of the plutonium isotope. It may be noted that even though ^{243}Am can be removed between each recycle, the increased presence of ^{242}Pu results in larger penalties with each successive cycle.

A similar problem exists when ^{236}U is built up from radiative capture in ^{235}U. Since these two uranium isotopes differ by only one mass unit, they are not as readily separated from each other as they are from the heavier ^{238}U (at least with the current technologies described in Chaps. 1 and 8). The reactivity penalties for ^{236}U and ^{237}Np in a BWR are also shown in Table 6-2. The situation for ^{237}Np may be noted to be quite analogous to that of ^{243}Am. The HTGR design which employs 93 wt% ^{235}U has even more severe reactivity penalties (the impacts of which are considered in Chap. 9 and illustrated by Table 9-4).

The reactivity penalties are especially bothersome when recycle fuel must be valued with respect to fresh fuel. For both the uranium and plutonium, isotopic compositions determine the amount of energy that can be extracted from a given mass of fuel.

Plutonium recycled in an LWR, or other thermal reactor, has an increasingly large reactivity penalty which may eventually dictate against its further use or, at least, may require that it be mixed with "fresher" plutonium. An alternative may be to employ LWR plutonium in an LMFBR. For this latter system, the presence of the higher isotopes can actually be an advantage because they will undergo fast-neutron fission. Both ^{240}Pu and ^{242}Pu have fission thresholds in the 1-2 MeV range (e.g., like the behavior of ^{238}U shown by Fig. 2-12). A symbiotic LWR–LMFBR cycle could lead to overall enhancement of energy production from recycled plutonium.

The current enrichment technologies employ continuous rather than batch processes (e.g., as described for gaseous diffusion in Chap. 8). Thus, a system normally handling only natural uranium input would experience some "contamination" from the ^{236}U content of each charge of recycle uranium. The enriched uranium output would need to be revalued (from the standpoint of ultimate energy production) to compensate for the ^{236}U reactivity penalty. With the advent of large-scale reprocessing and recycle, however, all recycle uranium might be confined to a dedicated enrichment facility to minimize such concerns.

Radioactivity

The radioactivity of the transmutation products can cause problems in the nuclear fuel cycle. General waste disposal is, of course, the primary concern. It is considered somewhat further later in this chapter, and in Chap. 10. One unique aspect of thorium fuel cycles bears some attention below.

The presence of ^{232}U in thorium-based fuel cycles poses a potentially serious radiation hazard. As shown by Fig. 6-3, ^{232}U may be produced by the ^{233}U (n, 2n) reaction with high-energy neutrons or by less direct routes from ^{230}Th and ^{232}Th, respectively. The major significance of the ^{232}U nuclide is due to its relatively short half-life of 72 years (compared to 10^5-10^8 years for the other important uranium isotopes) and the decay chain shown in Fig. 6-4, which includes high-energy gamma rays. Since all ^{233}U contains some amount of ^{232}U, gamma activity is always present. Freshly separated ^{233}U from reprocessing experiences a buildup of gamma dose as shown by Fig. 6-5 for one cycle and for equilibrium recycle. These plots are based on initial use of thorium, which is predominantly ^{232}Th. If there is a

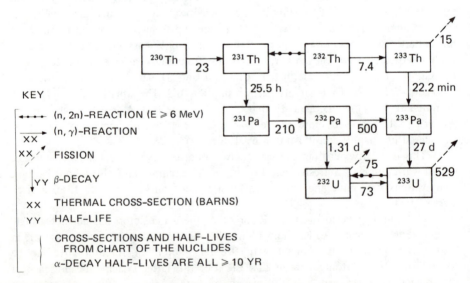

FIGURE 6-3
Transmutation chains for the production of ^{232}U.

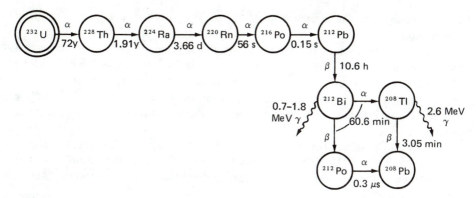

FIGURE 6-4
Radioactive decay chain for ^{232}U.

substantial amount of ^{230}Th (from decay of ^{238}U where thorium and uranium are found together), the gamma dose rates can be higher.

As a practical matter, fabrication of ^{233}U will require special procedures. Freshly separated product can be handled in the open for a relatively short time. After about 30 days, shielding and remote operations may be required for many fabrication processes.

Safeguards

The transmutation processes described above have some interesting impacts on nuclear material safeguards (considered further in Chap. 20). The isotopic content

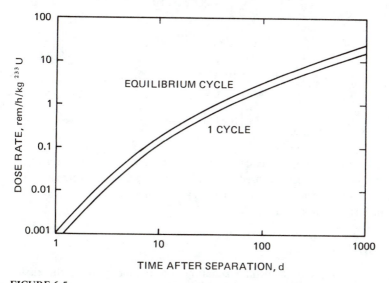

FIGURE 6-5
Dose rate at 1 m from a 1-kg sphere of ^{233}U as a function of time after separation. Curves correspond to material discharged after one cycle and after an equilibrium cycle, respectively. [Data courtesy of Electric Power Research Institute (Shapiro, 1977).]

of plutonium and the presence of ^{232}U in the thorium cycle are the major contributors.

The best plutonium for a nuclear weapon is generally considered to be that with at least 90–95 wt% ^{239}Pu. Buildup of ^{240}Pu and ^{242}Pu (Fig. 6-2) has the effect of diluting the fissile concentration and providing a stronger source of spontaneous-fission neutrons. Since each of these is detrimental to a weapon, fuels in the commercial nuclear fuel cycle become less desirable diversion targets with increasing burnup.

It is also possible to employ ^{233}U for nuclear weapons. However, the gamma radiation from the ^{232}U-decay chain makes it somewhat difficult to handle. Heating and potential radiation damage effects may also decrease the usefulness of ^{233}U for weapons applications.

Sample Calculation

A nuclide with a half-life of 100 days and an absorption cross section of 100 b could either decay or be transmuted in a reactor neutron beam. At any given time the fraction decaying f_d, for example, would be

$$f_d = \frac{\text{decay rate}}{\text{decay rate} + \text{absorption rate}} = \frac{n\lambda}{n\lambda + n\sigma\Phi} = \frac{\lambda}{\lambda + \sigma\Phi}$$

for decay and absorption rates inferred from Eqs. 2-4 and 6-1, respectively. The decay constant λ is

$$\lambda = \frac{\ln Z}{T_{1/2}} = \frac{0.693}{100 \text{ d}} \times \frac{1 \text{ d}}{24 \text{ h}} \times \frac{1 \text{ h}}{3600 \text{ s}} = 8.02 \times 10^{-8} \text{ s}^{-1}$$

Thus, if the neutron flux is 5×10^{13} n/cm$^2 \cdot$s,

$$\sigma\Phi = 100 \text{ b} \times 5 \times 10^{13} \frac{n}{\text{cm}^2 \cdot \text{s}} \times \frac{10^{-24} \text{ cm}^2}{1 \text{ b}} = 5 \times 10^{-9} \text{ s}^{-1}$$

and $\quad f_d = \dfrac{8.02 \times 10^{-8}}{8.02 \times 10^{-8} + 5 \times 10^{-9}} = \boxed{0.941}$

so that about 94 percent of the nuclides decay before they can capture a neutron.

This simple procedure can be used to estimate the relative importance of the loss mechanisms. If one mechanism is found to have a negligible effect, it may often be deleted from further consideration.

FISSION PRODUCTS

The other major products of neutron irradiation result directly from the fission process. *Fission fragments* generated at the time of fission decay to produce *fission products* (as shown, for example, by Fig. 2-7).

The buildup of fission products follows a balance equation similar to that of Eq. 6-4 for transmutation products except that there is an additional source term

Rate of creation due to fission $= \gamma \Sigma_f \Phi$ (6-5)

where $\Sigma_f \Phi$ is the fission rate and γ is the *fission yield*—the average number of the nucleus produced per fission. Since there are usually two fragments per fission, the sum of the yields is very close to two.

The balance for fission fragment nuclides is

$$\frac{dn(t)}{dt} = \gamma \Sigma_f \Phi + n^{A-1} \sigma_\gamma^{A-1} \Phi + \sum_{\text{all } j} n^j \sigma_r^j \Phi + \sum_{\text{all } i} n^i \lambda^i - n\sigma_a \Phi - n\lambda \quad (6\text{-}6)$$

where the parameters are as described previously for Eqs. 6-4 and 6-5. One such equation describes the behavior of each of the several hundred fission products which result from an initial fragment distribution like that in Fig. 2-6.

Fission products are of concern in reactors primarily as parasitic absorbers of neutrons and as long-term heat sources. Although several fission products have significant absorption cross sections, ^{149}Sm and ^{135}Xe have the most substantial impact on reactor design and operation. Fuel-cycle facilities are affected by the heat and radiation associated with the long beta-decay chains of some of the fission products.

Samarium-149

The ^{149}Sm nuclide is important in thermal reactors because it has a 4.1×10^3 b absorption cross section for 0.025-eV neutrons and a large resonance near thermal energies. It is produced from the ^{149}Nd [neodymium-149] fission fragment as shown by Fig. 6-6. The figure also has values of the fission yield γ for the three major fissile nuclides.

For the purpose of examining the behavior of ^{149}Sm, the 2-h half-life of ^{149}Nd is enough shorter than the 53-h value for ^{149}Pm [promethium-149] that the latter may be considered as if it were formed directly by fission. This assumption plus knowledge that the nuclides are produced and destroyed *only* by the mechanisms shown in Fig. 6-6 allows the situation to be described by the following pair of equations

$$\frac{dP}{dt} = \gamma^{Nd} \Sigma_f \Phi - \lambda^P P \tag{6-7a}$$

$$\frac{dS}{dt} = \lambda^P P - S\sigma_a^S \Phi \tag{6-7b}$$

where the time-dependent nuclide concentrations for promethium and samarium are represented by the letters P and S, respectively.

The buildup of the ^{149}Sm fission product poisons the system by absorbing neutrons that could otherwise be used to produce fissions. Since the product $S\sigma_a^S$ is proportional to the samarium absorption rate, $S\sigma_a^S / \Sigma_a$ is the fractional increase in the total absorption rate that must be compensated by an increased fission rate to maintain criticality (e.g., as may be inferred from Eq. 4-16). This ratio is a rough

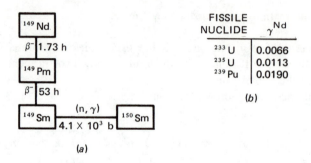

(a)

(b)

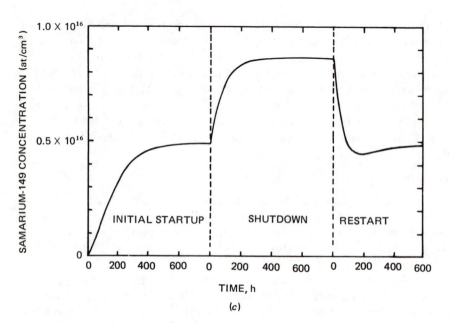

(c)

FIGURE 6-6
Behavior of ^{149}Sm in typical LWR fuel: (*a*) decay and reaction chain, (*b*) fission yields, (*c*) concentration vs. time.

measure of *poisoning* and the *negative* reactivity worth of the samarium compared to an initially critical system.[†] (A similar definition may be applied to other fission products, including the ^{135}Xe considered later.)

Solution of Eqs. 6-7 using parameters typical of an LWR reveals the following features shown by Fig. 6-6:

1. ^{149}Sm reaches an equilibrium concentration which is independent of the flux level. The time to reach equilibrium is flux dependent, being about 500 h ($\approx$10 half-lives) for many thermal reactors.

[†]See Glasstone & Sesonske (1967) for a more precise definition of poisoning and related concepts.

2. After shutdown, when the flux is no longer present to remove ^{149}Sm via absorption, the concentration builds to a higher level as the ^{149}Pm decays.
3. Since ^{149}Sm is stable, the concentration remains at this level until a neutron flux is present.
4. When the system is restarted, the original equilibrium level eventually reestablishes itself.

Although ^{149}Sm has a constant poisoning [negative reactivity] effect during long-term sustained operation, its behavior during initial startup and during post-shutdown and restart periods requires special consideration in designing control strategies.

Xenon-135

The ^{135}Xe nuclide has a 2.6×10^6 b absorption cross section. It is produced directly by some fissions but is more commonly a product of the ^{135}Te [tellurium-135] decay chain, as shown in Fig. 6-7.

For the purpose of examining the behavior of ^{135}Xe, the very short half-life of ^{135}Te allows the assumption that ^{135}I [iodine-135] is produced directly by fission. The additional knowledge that the nuclides are produced and destroyed only by the mechanisms shown in Fig. 6-7 allows the situation to be represented by the equations

$$\frac{dI}{dt} = \gamma^{Te}\Sigma_f\Phi - \lambda^I I \qquad (6\text{-}8a)$$

$$\frac{dX}{dt} = \gamma^X\Sigma_f\Phi + \lambda^I I - X\sigma_a^X\Phi - \lambda^X X \qquad (6\text{-}8b)$$

where the time-dependent nuclide concentrations for ^{135}I and ^{135}Xe are represented by the symbols I and X, respectively. Solution of these equations using typical LWR parameters reveals the following features shown by Fig 6-7:

1. The equilibrium ^{135}Xe level, determined by the absorption rate and both the Xe and I decay rates, is reached in about 50 h.
2. When the reactor is shut down, the absorption by ^{135}Xe ceases and leaves the two decay processes to compete. In thermal power reactors where the iodine concentration usually exceeds that for xenon, the shorter half-life of the iodine will cause an initial buildup of xenon.
3. As the ^{135}I atoms decay, the rate of decay decreases until it equals and then is less than that for ^{135}Xe. This produces a maximum which is followed by a drop-off of the ^{135}Xe level.
4. If the system is restarted when the ^{135}Xe concentration is above the equilibrium level, renewed absorption accelerates the losses until the equilibrium is re-established.

The operating neutron flux level determines the maximum post-shutdown poisoning. This behavior is shown by Fig. 6-8 for several fluxes in a typical LWR system. Since the amount of excess core reactivity available to "override" the

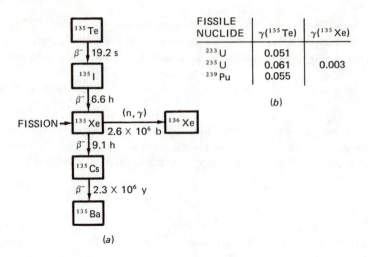

(a)

FISSILE NUCLIDE table:

FISSILE NUCLIDE	$\gamma(^{135}\text{Te})$	$\gamma(^{135}\text{Xe})$
^{233}U	0.051	
^{235}U	0.061	0.003
^{239}Pu	0.055	

(b)

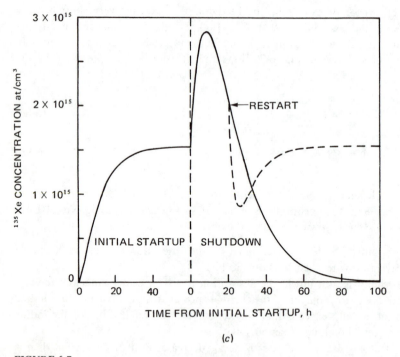

(c)

FIGURE 6-7
Behavior of ^{135}Xe in typical LWR fuel: (a) decay and reaction chain, (b) fission yields, (c) concentration vs. time.

negative reactivity of the xenon is usually well less than 10 percent $\Delta k/k$, thermal power reactors are limited to fluxes of about 5×10^{13} n/cm²·s so that timely restart can be assured.

Thermal reactors may also experience spatial power oscillations because of the presence of ^{135}Xe. The mechanism is based on the following scenario:

1. An initial asymmetry in the core power distribution (e.g., by control rod misalignment) causes an imbalance in fission rates and, thus, in the ^{135}I buildup and ^{135}Xe absorption.
2. In the high-flux region, ^{135}Xe burnout allows the flux to increase further while in the low-flux region the reverse occurs. The iodine concentration increases where the flux is high and decreases where the flux is low.
3. As soon as the ^{135}I levels build up sufficiently, decay to xenon reverses the initial situation.
4. Repetition of these patterns can lead to xenon oscillations with periods on the order of tens of hours.

With little overall reactivity change, the oscillations can change local power levels by a factor of three or more. In a system with strongly negative temperature feedbacks, ^{135}Xe oscillations are damped quite readily. This is one important reason for designing reactors to have negative moderator-temperature coefficients. If the feedback effects are relatively weak, it may also be necessary to implement a somewhat complicated control-rod insertion program to reduce the amplitude of the oscillations.

   The ^{135}Xe and ^{149}Sm mechanisms are dependent on the very large thermal cross sections and, thus, affect only thermal-reactor systems. In fast reactors neither these nor any other fission products have a major poisoning influence.

OPERATIONAL IMPACTS

The major impacts of fuel depletion and its related effects accrue to long-term reactivity control and radioactive-waste management. The reactivity control is implemented by a number of different mechanisms in the various reactor designs.

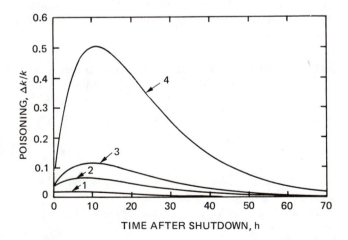

FIGURE 6-8
Poisoning of ^{135}Xe as a function of time after shutdown for a typical LWR fuel composition at various neutron flux levels. Curve 1: $\Phi = 1 \times 10^{13}$ n/cm$^2 \cdot$s; Curve 2: $\Phi = 5 \times 10^{13}$ n/cm$^2 \cdot$s; Curve 3: $\Phi = 1 \times 10^{14}$ n/cm$^2 \cdot$s; Curve 4: $\Phi = 5 \times 10^{14}$ n/cm$^2 \cdot$s.

Radioactive wastes include both the generally long-lived transmutation products and the shorter-lived fission products.

Calculations

All of the depletion-related phenomena occur on a time scale that is very long compared to that considered for reactor kinetics in the last chapter. All calculations can be performed by quasistatic procedures as opposed to the explicit kinetics formulations like Eqs. 5-8.

A typical sequence begins with calculations of core power and flux distributions with a static model, e.g., multigroup diffusion as per Chap. 4. The fluxes are then used in formulations like Eqs. 6-4 and 6-6 to determine nuclide concentration changes for a given time interval or "time step." The new concentrations can then be reintroduced for another power and flux calculation as the procedure continues in this stepwise fashion to the desired time endpoint.

This simple time-step depletion method is found to give reasonably accurate results for many applications. In thermal reactors, for example, good results can be obtained with a model based on initial time steps of 50 h and 500 h (corresponding to ^{135}Xe and ^{149}Sm near-equilibrium conditions, respectively) followed by routine steps of about 800 h, or approximately a month.

Reactor-design models consider only those nuclides which have an impact on reactivity. These include the fissile, fertile, and parasitic-absorber nuclides, e.g., including transmutation products shown on Fig. 2-16. Initial composition and neutron-flux-level information coupled with a comparison of related absorption and decay parameters often allows for a reduction in the number of nuclides that require explicit representation. Typical depletion models for thermal reactors use only 15–20 such compositions. Fission products are generally handled by representing ^{135}Xe and ^{149}Sm explicitly and "lumping" the effects of the remaining species into one or more "dummy" nuclides. In the more sophisticated computer codes, the nuclide concentrations can be depleted separately for each spatial mesh point (where 10–100,000 such points are not uncommon for final design calculations).

Since spatial calculations are not necessary for waste management concerns, essentially all transmutation and fission-product chains may be employed. Output of typical codes includes activity and heat generation for each nuclide and each chemical element at various times. For spent fuel applications, times are usually measured from reactor shutdown.

Long-Term Reactivity Control

Many of the reactivity-control methods described earlier are especially well-suited to the relatively long time scale that characterizes the depletion-related effects. In most systems, such control is accomplished by one or more of the following:

- programmed control-rod motion
- use of fixed, burnable poisons
- use of soluble poisons
- on-line refueling

Some control rods are used to change power level and to provide for shutdown when safety limits are exceeded. Others may be employed to compensate long-term reactivity changes. Following large insertion at the beginning of core lifetime, the poison rods are gradually withdrawn as fuel-burnup, transmutation, and fission-product effects reduce the capability of the core to maintain a neutron chain reaction. They are also used to compensate the reactivity effects of xenon and samarium on initial and other startups of the system.

Control rods may be employed to damp xenon oscillations. This generally requires pre-programmed action because of the long time delay between ^{135}I production and its decay to ^{135}Xe. In some reactors, control rods which contain neutron poison material only on a fraction of their total length—part-length rods—are used to damp oscillations at selected axial locations.

Burnable poisons are designed to have negative reactivity that decreases with irradiation. Conceptually, these poisons should decrease their negative reactivity at the same rate the core's excess positive reactivity is depleted. Burnable poisons are generally used in the form of separate lattice pins or of additives in fuel. Since they can usually be distributed more uniformly than control rods, these poisons are less disruptive to the core power distribution.

Soluble poisons, also called "chemical shim," produce a very spatially uniform neutron absorption when dissolved in pressurized-water or heavy-water coolant. The most common soluble poison is boric acid, which is often referred to as "soluble boron" or simply "solbor." The flexibility for concentration and dilution allows control rod use to be minimized. Xenon and samarium effects, for example, are handled by trading control rod insertion and soluble poison concentration in a manner that limits the use of the rods.

On-line refueling is, perhaps, the ultimate means of controlling the long-term reactivity effects. By loading and discharging fuel of appropriate composition, it is possible to reduce excess reactivity to a minimum and thereby limit the need for the other control methods.

Most reactors employ control rods plus one or more of the other methods. The specific choices are generally based on overall system design considerations. Power capability impacts are addressed in the next chapter. Features for specific reactor types are described in Chaps. 12–15.

Radioactive Wastes

Fission products and transmutation products are the major contributors to the radioactive wastes generated in the nuclear fuel cycle. Each category has a somewhat different effect on waste management.

The fission product wastes emit beta and gamma radiation, as is shown in the example in Fig. 2-7. Although they constitute the largest heat and radiation source at the time of reactor shutdown, these products decay to relatively low levels in a few hundred years. This latter effect is verified by Fig. 6-9, which shows a plot of *hazard index*—defined in terms of the quantity of water required to dilute the waste to the current maximum permissible concentration [MPC]—for the fission products versus their age. Comparison with the hazard index for the uranium ore from which these wastes were ultimately produced shows the products to be much less hazardous after several hundred years of decay.

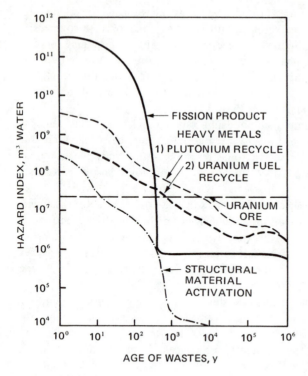

FIGURE 6-9

Hazard index as a function of time for spent LWR fuel constituents. (From *Oceanus*, vol. 20, no. 1, winter 1977, published by Woods Hole Oceanographic Institution.)

The transmutation products consist of those members of the periodic table known as the *actinide* elements. Equivalently, they are *transuranic* in the sense of being above uranium in atomic number. Thus, the terms actinide and transuranic are often used synonymously to describe the transmutation-product wastes. Figure 6-9 indicates that these products constitute a relatively low component of the initial waste hazard, but then become more dominant with time. After the fission-product hazard drops off, the transuranics become the major concern for an indefinite time, although their effect also drops below that of the initial uranium ore on the 1000–10,000 year time span. Figure 6-9 shows the effects of both uranium and plutonium recycle, each of which enhances the inventory of actinide nuclides with the latter doing more so (as might be inferred, for example, from the chains shown by Fig. 2-16).

The radionuclide content of spent-fuel or reprocessing wastes depends on the initial fuel composition and its burnup. Table 6-3 contains activities for selected nuclides that could be expected in typical LWR and LMFBR systems. The differences between the yields for ^{235}U and ^{239}Pu as shown by Fig. 6-10 account for the relatively increased presence of nuclides like ^{90}Sr and ^{134}Cs for the LWR and ^{95}Zr and ^{155}Eu for the LMFBR. Higher burnup plus having Pu as a starting material is responsible for the large actinide inventories for the LMFBR.

TABLE 6-3
Radionuclide Content of Typical LWR Spent Fuel at Discharge and 180 Days and of Typical LMFBR Fuel at Discharge and 30 Days[†]

Nuclide	Half-life $T_{1/2}$	Radia- tions[‡]	Activity, Ci/t heavy metal			
			LWR fuel		LMFBR fuel	
			Discharge	180 d	Discharge	30 d
^{3}H	12.3 y	β	5.744×10^2	5.587×10^2	1.648×10^3	1.640×10^3
^{85}Kr	10.73 y	β, γ	1.108×10^4	1.074×10^4	1.473×10^4	1.466×10^4
^{89}Sr	50.5 d	β, γ	1.058×10^6	9.603×10^4	1.333×10^6	8.939×10^5
^{90}Sr	29.0 y	β, γ	8.425×10^4	8.323×10^4	9.591×10^4	9.572×10^4
^{90}Y	64.0 h	β, γ	8.850×10^4	8.325×10^4	1.214×10^5	9.572×10^4
^{91}Y	59.0 d	β, γ	1.263×10^6	1.525×10^5	1.794×10^6	1.269×10^6
^{95}Zr	64.0 d	β, γ	1.637×10^6	2.437×10^5	3.215×10^6	2.340×10^6
^{95}Nb	3.50 d	β, γ	1.557×10^6	4.689×10^5	3.149×10^6	2.954×10^6
^{99}Mo	66.0 h	β, γ	1.875×10^6	3.780×10^{-14}	4.040×10^6	2.108×10^3
^{99m}Tc	6.0 h	γ	1.618×10^6	3.589×10^{-14}	3.487×10^6	2.002×10^3
^{99}Tc	2.1×10^5 y	β, γ	1.435×10^1	1.442×10^1	3.278×10^1	3.293×10^1
^{103}Ru	40.0 d	β, γ	1.560×10^6	6.680×10^4	4.617×10^6	2.730×10^6
^{106}Ru	369.0 d	β, γ	4.935×10^5	3.519×10^5	2.248×10^6	2.125×10^6
^{103m}Rh	56.0 min	γ	1.561×10^6	6.686×10^4	4.619×10^6	2.733×10^6
^{111}Ag	7.47 d	β, γ	5.375×10^4	3.005×10^{-3}	2.294×10^5	1.422×10^4
^{115m}Cd	44.6 d	β, γ	1.483×10^3	9.042×10^1	7.041×10^3	4.418×10^3
^{125}Sn	9.65 d	β, γ	1.081×10^4	2.624×10^{-2}	3.404×10^4	3.946×10^3
^{124}Sb	60.2 d	β, γ	4.147×10^2	5.219×10^1	2.329×10^3	1.649×10^3
^{125}Sb	2.73 y	β, γ	9.525×10^3	8.498×10^3	5.251×10^4	5.171×10^4
^{125m}Te	58.0 d	γ	1.976×10^3	2.031×10^3	1.121×10^4	1.144×10^4
^{127m}Te	109.0 d	β, γ	1.384×10^4	4.595×10^3	4.969×10^4	4.265×10^4
^{127}Te	9.4 h	β, γ	9.920×10^4	4.500×10^3	3.247×10^5	4.308×10^4
^{129m}Te	33.4 d	β, γ	8.508×10^4	2.041×10^3	2.316×10^5	1.249×10^5
^{129}Te	70.0 min	β, γ	3.211×10^5	1.296×10^3	8.454×10^5	7.932×10^4
^{132}Te	78.0 h	β, γ	1.486×10^6	3.159×10^{-11}	3.473×10^6	5.783×10^3
^{129}I	1.59×10^7 y	β, γ	3.219×10^{-2}	3.268×10^{-2}	1.033×10^{-1}	1.040×10^{-1}
^{131}I	8.04 d	β, γ	1.028×10^6	1.933×10^{-1}	2.602×10^6	2.020×10^5
^{132}I	2.285 h	β, γ	1.511×10^6	3.254×10^{-11}	3.546×10^6	5.956×10^3
^{133}Xe	5.29 d	β, γ	2.098×10^6	1.612×10^{-4}	4.414×10^6	1.076×10^5
^{134}Cs	2.06 y	β, γ	2.718×10^5	2.303×10^5	8.283×10^4	8.058×10^4
^{136}Cs	13.0 d	β, γ	6.962×10^4	4.719×10^0	2.577×10^5	5.204×10^4
^{137}Cs	30.1 y	β, γ	1.115×10^5	1.102×10^5	2.522×10^5	2.518×10^5
^{140}Ba	12.79 d	β, γ	1.953×10^6	1.133×10^2	3.636×10^6	7.153×10^5
^{140}La	40.23 h	β, γ	2.019×10^6	1.303×10^2	3.698×10^4	8.238×10^5
^{141}Ce	32.53 d	β, γ	1.784×10^6	3.876×10^4	3.730×10^6	1.979×10^6
^{144}Ce	284.0 d	β, γ	1.229×10^6	7.925×10^5	2.148×10^6	1.996×10^6
^{143}Pr	13.58 d	β	1.657×10^6	1.887×10^2	3.044×10^6	7.349×10^5
^{147}Nd	10.99 d	β, γ	7.902×10^5	9.278×10^0	1.513×10^6	2.283×10^5
^{147}Pm	2.62 y	β, γ	1.031×10^5	9.859×10^4	6.344×10^5	6.353×10^5
^{149}Pm	53.1 h	β, γ	3.919×10^5	1.326×10^{-19}	9.842×10^5	8.451×10^1
^{151}Sm	93.0 y	β^+, β^-, γ	8.658×10^2	8.696×10^2	9.693×10^3	9.703×10^3
^{152}Eu	13.4 y	β^+, β^-, γ	7.838×10^0	7.635×10^0	4.759×10^1	4.738×10^1
^{155}Eu	4.8 y	β, γ	2.540×10^3	2.365×10^3	4.305×10^4	4.255×10^4

(*See footnotes on p. 170.*)

TABLE 6-3
Radionuclide Content of Typical LWR Spent Fuel at Discharge and 180 Days and
of Typical LMFBR Fuel at Discharge and 30 Days[†] (*Continued*)

Nuclide	Half-life $t_{1/2}$	Radia-tions[‡]	Activity, Ci/t heavy metal			
			LWR fuel		LMFBR fuel	
			Discharge	180 d	Discharge	30 d
^{160}Tb	72.3 d	β, γ	1.418×10^3	2.525×10^2	4.880×10^3	3.661×10^3
^{239}Np	2.35 d	β, γ	2.435×10^7	2.050×10^1	5.990×10^7	8.727×10^3
^{238}Pu	87.8 y	α, γ	2.899×10^3	3.021×10^3	2.770×10^4	2.820×10^4
^{239}Pu	2.44×10^4 y	α, γ, SF	3.250×10^2	3.314×10^2	6.247×10^3	6.263×10^3
^{240}Pu	6.54×10^3 y	α, γ, SF	4.842×10^2	4.843×10^2	8.323×10^3	8.323×10^3
^{241}Pu	15.0 y	α, β, γ	1.098×10^5	1.072×10^5	7.280×10^5	7.252×10^5
^{241}Am	433.0 y	α, γ, SF	8.023×10^1	1.657×10^2	9.091×10^3	9.186×10^3
^{242}Cm	163.0 d	α, γ, SF	3.666×10^4	1.717×10^4	8.467×10^5	7.489×10^5
^{244}Cm	17.9 d	α, γ	2.772×10^3	2.720×10^3	8.032×10^3	8.007×10^3

[†]Calculated by SANDIA-ORIGEN code, courtesy D. E. Bennett, Sandia National Laboratories.
[‡]Radiations: α–alpha [$^4_2$He] ; β, β^- –beta [$^{\,0}_{-1}$e] ; β^+–positron [$^{\,0}_{+1}$e] ; γ–gamma [$^0_0\gamma$] ; SF–spontaneous fission.

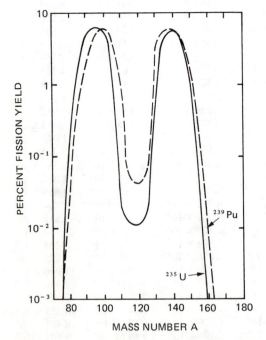

FIGURE 6-10
Fission yields for ^{235}U and ^{239}Pu. (Adapted from *Nuclear Chemical Engineering* by M. Benedict and T. H. Pigford. Copyright © 1957 by McGraw-Hill Book Company, Inc. Used by permission of McGraw-Hill Book Company.)

In the LMFBR fuel cycle, plutonium is likely to be reprocessed as quickly as possible because of its high economic value. LWR recycle, on the other hand, may proceed more slowly because it makes a fractionally lower contribution. It has been suggested on this basis that LMFBR and LWR fuels be allowed to "cool," i.e., decay, for 30 days and 150 days, respectively. Thus, the already "hot" LMFBR fuel would be processed at the very high radionuclide concentrations shown in Table 6-3. The table also includes the appropriate values for LWR reprocessing.

EXERCISES

Questions

6-1. Describe the buildup of plutonium isotopes by neutron irradiation of ^{238}U.

6-2. Define breeding ratio, conversion ratio, and doubling time. Discuss the relationship of the first two to η and their application to the reference reactor designs.

6-3. Explain how the presence of ^{135}Xe and ^{149}Sm affect reactor design and operation (a description of their behavior is *not* necessary).

6-4. Define hazard index and describe the behavior of fission-product and trans-uranic wastes in terms of it.

Numerical Problems

6-5. Based on the 0.025-eV cross sections and the decay constants in Fig. 2-16 and a thermal-neutron flux of $5 \times 10^{13}/\text{cm}^2 \cdot \text{s}$:
   a. Estimate the fractional depletion of ^{235}U during 1 year of operation.
   b. Estimate the fraction of ^{238}U that would be converted to ^{239}U in 1 year.
   c. Calculate the instantaneous decay and neutron absorption rates for ^{239}U nuclei and determine their fraction of the combined loss rate.
   d. Repeat (c) for ^{239}Np.
   e. Estimate from (b) and (c) the fraction of ^{239}U absorptions that result in the production of ^{239}Pu.

6-6. Calculate the burnup in MWD/T for a system operating for 1 year at a power level of 1000 MW(e) with a 32 percent thermal efficiency (i.e., MW(e)/MW(th)) and an initial loading of 100 t of heavy metal.

6-7. Define a "depletion half-life" which is analogous to the radioactive-decay half-life. Calculate this half-life for depletion of ^{233}U, ^{235}U, and ^{239}Pu in a 0.025-eV neutron flux of $5 \times 10^{13}/\text{cm}^2 \cdot \text{s}$.

6-8. Calculate the fraction of ^{239}Pu nuclei that would be converted to ^{240}Pu in a 0.025-eV neutron flux of $5 \times 10^{13}/\text{cm}^2 \cdot \text{s}$. Repeat the process for ^{241}Pu conversion to ^{242}Pu.

6-9. Identify the decays and half-lives by which ^{238}U converts to ^{230}Th. What effect does the presence of the latter have on a thorium fuel cycle?

6-10. Explain how the presence of ^{242}Pu causes a reactivity penalty. How is it responsible for the penalty associated with ^{243}Am?

6-11. Estimate the thickness of lead required to absorb 99 percent of the most energetic gamma rays from the ^{232}U decay chain.

6-12. Consider a 10-kg lot of equilibrium-recycled ^{233}U 30 days after separation. How long could one work at 1 m from the lot without exceeding the annual whole-body dose limit?

6-13. Using Fig. 6-2, estimate the maximum burnup in an LWR for which the plutonium would still be considered "weapons grade" (>90 wt% ^{239}Pu).

6-14. Consider the ^{238}Pu isotope.

    a. Calculate the rate of power generation in a 1-kg sample. (Use $T_{1/2} = 87.7$ years and $E_\alpha = 5.5$ MeV.)

    b. It has been suggested that such heating would make ^{238}Pu undesirable for a nuclear weapon. Explain how reactor use of recycle uranium could enhance ^{238}Pu production.

SELECTED BIBLIOGRAPHY†

Radioactive Wastes
    ORNL-4451, 1970
    Tonnessen & Cohen, 1977

Other Sources with Appropriate Sections or Chapters
    Benedict, 1981
    Benedict & Pigford, 1957
    Connolly, 1978
    Duderstadt & Hamilton, 1976
    Etherington, 1958
    Foster & Wright, 1977
    Glasstone & Jordan, 1980
    Glasstone & Sesonske, 1967
    Graves, 1979
    Henry, 1975
    Lamarsh, 1966
    Lamarsh, 1975
    Lapedes, 1976
    Murray, 1975
    Onega, 1975
    Rydin, 1977
    Sagan, 1974
    Sesonske, 1973
    Shapiro, 1977
    WASH-1250, 1973

†Full citations are contained in the General Bibliography at the back of the book.

7

REACTOR ENERGY REMOVAL

Objectives

After studying this chapter, the reader should be able to:

1. Identify the three reasons why heat removal in nuclear reactor systems may be more complex than in conventional systems.
2. Define hot-spot and hot-channel factors and relate them to core operating limits of linear heat rate and critical heat flux, respectively.
3. Explain how reflection and enrichment zoning can enhance reactor power capability.
4. Describe the effects of control rods, burnable poisons, and soluble poisons on reactor power peaking.
5. Perform fundamental calculations for conductive and convective heat transport and for peaking factors.

Prerequisite Concepts

Reference Reactor Designs	Chapter 1
Flux Shapes for Bare Reactors	Chapter 4
Feedback Mechanisms	Chapter 5
Reactor Control Methods	Chapters 4-6

Steady-state power generation in a reactor system depends directly on maintaining a critical neutron balance at the desired fission rate. The interaction between neutron multiplication and temperatures, however, make thermal-energy removal an integral component of any such balance.

In comparison to more conventional systems, energy removal in nuclear reactors is a more complex problem because:

- some designs incorporate very high power densities
- fuel is not "consumed" in the usual sense, but instead must maintain a fixed geometry for a number of years of operation
- the inherent nuclear-radiation fields limits selection of fuel and structural materials

There is always some incentive to optimize power output per unit mass of fissionable fuel and to minimize core size for reduction of containment and shielding volumes. Both goals often lead to the high power densities shown by the comparisons in Table 7-1. All reactors have fuel values higher than the average for a fossil plant and comparable to or greater than that for an aircraft turbine. The net effect of the reactors' high power densities is to produce large temperature gradients and resultingly large thermal stresses on core components.

Since typical fuel assemblies stay in the core for several years, they must maintain their geometrical shape through all expected routine operations and accident conditions. Temperatures must be limited to prevent any general geometry changes up to and including melting of fuel and clad. Extensive chemical reactions among the fuel, clad, moderator, and coolant are also to be avoided.

The effects of activation, parasitic capture, and radiation damage often dictate the use of different clad and structural materials than would result from a selection process based only on thermal, physical, and mechanical properties. This often leads to the use of expensive, unconventional materials which impose restrictive thermal limitations on core operations. Zirconium, for example, is favored as a clad material for water-reactor fuel because it leads to lower parasitic neutron absorption than would stainless steel. The zirconium, however, has the relative disadvantages of lower structural strength and a larger exothermal energy release with high-temperature oxidation (as considered further in Chaps. 9 and 16, respectively).

Safe reactor operation under these constraining conditions depends on

TABLE 7-1
Power Densities for the Reference Reactors and Other Systems

System	Power density (kW/liter)		
	Core average	Fuel average[†]	Fuel maximum[†]
Fossil-fuel plant	10	–	–
Aircraft turbine	45	–	–
Rocket	20,000	–	–
HTGR	8.4	44	125
CANDU	12	110	190
BWR	56	56	180
PWR	95–105	95–105	190–210
LMFBR	280	280	420

[†]Includes interspersed-coolant volume for systems with fuel-pin lattices; includes only fuel sticks for HTGR.

assurance that *local* conditions will not endanger the integrity of the core at *any* location. Thus, system limits must be determined by *maximum* temperatures and/or coolant flow conditions rather than by the average values that may be adequate for nonnuclear applications.

Reactor thermal-hydraulic analysis is based on very detailed modeling of core power distributions, including the effects of fuel-temperature, coolant/moderator, and other feedbacks. Correlation of local power densities to fuel-pin temperature distributions and coolant flow conditions provides the basis for establishing general operating limits for the system as a whole.

POWER DISTRIBUTIONS

Since most of the energy is deposited very near the site of each fission event, the power density in a reactor has essentially the same spatial distribution as the fission rate. The position-dependent power density $P(\mathbf{r})$ may be represented by

$$P(\mathbf{r}) = E_f R_f(\mathbf{r}) = E_f \Sigma_f(\mathbf{r}) \Phi(\mathbf{r}) \tag{7-1}$$

for energy per fission E_f, fission-rate density R_f, and one-energy-group macroscopic fission cross-section Σ_f and neutron flux Φ.

The total power output P_{tot} of a system of volume V is

$$P_{\text{tot}} = \int_V P(\mathbf{r}) \, dV = E_f \int_V \Sigma_f(\mathbf{r}) \Phi(\mathbf{r}) \, dV \tag{7-2}$$

and the average power density $\langle P(\mathbf{r}) \rangle$ is

$$\langle P(\mathbf{r}) \rangle = \frac{\displaystyle\int_V P(\mathbf{r}) \, dV}{\displaystyle\int_V dV} = \frac{P_{\text{tot}}}{V} \tag{7-3}$$

Since the ultimate power capability of a reactor system is limited by maximum local conditions, it is common practice to normalize the power-density distribution to the core-average value by using Eq. 7-3 to obtain

$$\frac{P(\mathbf{r})}{\langle P(\mathbf{r}) \rangle} = \frac{P(\mathbf{r})}{P_{\text{tot}}/V} \tag{7-4}$$

This is, of course, equivalent to setting the core-average power-density to unity and considering all other values as ratios thereto.

Homogeneous Systems

In homogeneous reactor systems, the material properties would be independent of position (i.e., $\Sigma_f(\mathbf{r}) = \Sigma_f$, a constant) such that the spatial power-density distribution from Eq. 7-1 would be proportional to the neutron flux according to

$$P(\mathbf{r}) = E_f \Sigma_f \Phi(\mathbf{r}) = A\Phi(\mathbf{r}) \tag{7-5}$$

for proportionality constant A. Normalizing to the core-average as in Eq. 7-4 and calling the resulting factor $F(\mathbf{r})$ yields

$$F(\mathbf{r}) = \frac{P(\mathbf{r})}{\langle P(\mathbf{r}) \rangle} = \frac{\Phi(\mathbf{r})}{\langle \Phi(\mathbf{r}) \rangle} \tag{7-6}$$

for spatially averaged flux $\langle \Phi(\mathbf{r}) \rangle$. The position-dependent factor $F(\mathbf{r})$ (which is redefined somewhat later in this chapter) represents both the power density and the neutron flux normalized to a core-average value of unity.

Considering a diffusion theory model of a bare reactor, the flux shapes for a sphere, infinite cylinder, and infinite slab, respectively, are shown in Fig. 4-5. Power capability for such systems would be limited by the peak value of the flux since it corresponds to the maximum power density and temperature. Table 7-2 shows maximum values of $F(\mathbf{r})$ for several geometric arrangements. These values are generally referred to as *power peaking factors*, although $F(\mathbf{r})$ itself is sometimes called by the same name. The peak-to-average power is quite large for the sphere and decreases for the infinite cylinder and slab, respectively. However, since these latter two are idealizations, their finite counterparts are the geometries of interest. The finite cylinder has a peaking factor comparable to the product of the factors for an infinite cylinder (the *radial* dimension) and a slab (the *axial* dimension). The

TABLE 7-2
Power Peaking Factors for Reactors of Various
Geometric Shapes

Geometry	Peaking factor	
	Total	Constituents
Sphere, bare	3.29	
Infinite slab, bare	1.57	
Cuboid,[†] bare	3.87	$x = 1.57$
		$y = 1.57$
		$z = 1.57$
Infinite cylinder, bare	2.32	
Cylinder, bare	3.64	$r = 2.32$
		$z = 1.57$
Cylinder, fully reflected	2.03	$r = 1.50$
		$z = 1.35$
Cylinder, fully reflected, enrichment-zoned radially	1.62	$r = 1.20$
		$Z = 1.35$

[†]A cuboid is a rectangular parallelepiped (see note on Table 4-2).

peaking factor for the cuboid is the product of one slab value for each of the three dimensions.

The significance of the peak-to-average power density may be established by rearranging Eq. 7-6 to

$$\langle P(\mathbf{r}) \rangle = \frac{P(\mathbf{r})}{F(\mathbf{r})}$$

or $\quad \langle P(\mathbf{r}) \rangle = \dfrac{P_{\max}}{F_{\max}}$

for power-peaking factor $F_{\max}$ corresponding to the maximum power density $P_{\max}$. $P_{\max}$ is generally fixed by the material composition of the reactor core, so that the average power density varies inversely with the peaking factor. Since the net energy output of a reactor core depends on the average power density (rather than on the maximum), a low peaking factor is highly favored. Power reactors are all roughly cylindrical (as described in Chap. 1) rather than cuboidal because of the enhanced power capability of the latter shape associated with the peaking factors in Table 7-2. (A spherical shape, which may be noted to have lower peaking also would be very difficult to design, build, and operate with solid fuel elements. The situation is somewhat different (for a number of reasons) in liquid-core systems like the molten-salt-breeder reactor [MSBR] described in Chap. 14.)

Reflectors and power shaping techniques can be used to reduce the power peaking factors. Figure 7-1 shows the general effect of reflection in a thermal reactor system (where the core power is proportional to the thermal flux, as shown, for example, by Fig. 4-7*b*). Without changing the peak power, a reflector can raise the power density at the core periphery and thus increase the core-average power level. Table 4-2 shows how axial and radial reflection can reduce the net peaking factor by 40-45 percent.

Power shaping may be accomplished most readily by varying the effective fuel enrichment across the core. The simple example in Fig. 7-2 shows the conceptual effect of employing a lower enrichment in the central region and a higher

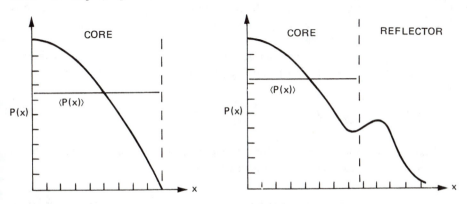

FIGURE 7-1
Flux shapes and average power densities for bare and reflected slab geometries.

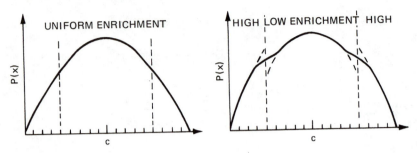

FIGURE 7-2
Power distributions for one- and two-batch fuel-management patterns in a bare-slab geometry.

enrichment in the outer regions of a core. This is the basis for the *multiple-batch fuel management* procedure which reduces power peaking and enhances core power capability (as described in more detail in Chap. 9). The final entry in Table 7-2 indicates that the use of two enrichment zones can reduce the radial peak in a reflected core by another 20 percent.

Peaking Factors

Reactor core design is often based on calculations of peaking factors which represent the maximum or limiting conditions for operation of the system. Two factors are in general use—one related to fuel-pin temperatures and the other related to coolant conditions.

In the regular-lattice configurations of the reference reactor designs, power density has several different but comparable formulations. First, the *volumetric heat rate* q''',[†] defined as the rate of heat production per unit volume, is precisely the same as the power density.

Under steady-state conditions, all energy produced in a fuel pellet must flow out through the cladding to the coolant. The (areal) *heat flux* q'' and the *linear heat flux* q' are defined as the rate of heat flow per unit surface area of the cladding and rate of heat flow per unit length of the fuel pin, respectively. Each is proportional to the power density for steady-state conditions in a regular-lattice fuel geometry. The *heat rate* q is the energy generation rate for a defined region and is just the power produced by that region.

The equivalence of power density and the various heat rates allows the *heat flux factor* F_Q to be defined as in Eq. 7-6 by any of the following:

$$F_Q(\mathbf{r}) = \frac{P(\mathbf{r})}{\langle P(\mathbf{r})\rangle} = \frac{q'''(\mathbf{r})}{\langle q'''(\mathbf{r})\rangle} = \frac{q''(\mathbf{r})}{\langle q''(\mathbf{r})\rangle} = \frac{q'(\mathbf{r})}{\langle q'(\mathbf{r})\rangle} \qquad (7\text{-}7)$$

In practice, the linear heat flux [or *linear heat rate*] formulation tends to be the most useful. Equation 7-7 may be rewritten as

[†]Each "prime" as used here and below is a common shorthand notation for one spatial dimension, e.g., q''' is heat rate per unit volume, q'' is heat rate per unit area, etc.

$$F_Q(\mathbf{r}, z) = \frac{q'(\mathbf{r}, z)}{\langle q'(\mathbf{r}, z)\rangle} \tag{7-8}$$

where now the radial position $\mathbf{r}$ is a two-component vector which typically identifies an entire fuel pin, and the axial position z is measured along the length of that fuel. The core-averaged linear heat rate $\langle q'\rangle$ may be determined by setting up an integral—e.g., like that in Eq. 7-3—or from the relatively simple relationships

$$\langle q'(\mathbf{r}, z)\rangle = \frac{\text{total core thermal power}}{\text{total length of fuel}}$$

or $\quad \langle q'(\mathbf{r}, z)\rangle = \dfrac{\text{total core thermal power}}{(\text{number of fuel pins})(\text{length per fuel pin})} \tag{7-9}$

Although the linear heat rate q' may be readily quantified in SI units like kW/m, current practice in this country still employs kW/ft based on measurement of electric (and the related thermal) power in watts and lengths in feet.

The second peaking factor is concerned with the enthalpy [or heat energy] content of liquid coolant as it flows through the channel formed by pins of a fuel assembly. The *enthalpy rise factor* $F_{\Delta H}$ is defined as

$$F_{\Delta H}(\mathbf{r}) = \frac{\text{enthalpy rise in the channel at } \mathbf{r}}{\text{enthalpy rise in the core-average channel}} \tag{7-10}$$

where the enthalpy increase depends on the average heat flux from the surrounding fuel pins and the identity, pressure, inlet temperature, heat capacity, and other parameters of the liquid coolant.

The heat flux factor F_Q is applicable to all of the reference reactors. The enthalpy rise factor $F_{\Delta H}$ is appropriate to all but the HTGR (since the latter has a nonliquid coolant). Both factors are determined (as appropriate) by the computer codes used for reactor physics calculations. Their correlation to limiting fuel temperatures and coolant conditions is required for the reactor design applications considered later in this chapter.

FUEL-PIN HEAT TRANSPORT

The temperatures within a fuel pin depend on the fission source distribution, the heat transport properties of the pin, and the ability of the coolant to act as a heat sink. Both conductive and convective heat transfer mechanisms operate to determine the temperature profile.

An idealized fuel pin with uniform axial composition and the cross section in Fig. 7-3 serves as the basis for discussion of heat transport properties in the following paragraphs. The fuel region is representative of cylindrical pellets. A gap between the pellet and the cladding generally exists because of necessary manufacturing tolerances. Initially, it may be filled with an inert gas, but later in core lifetime the gap contains gaseous fission products released from the fuel. The

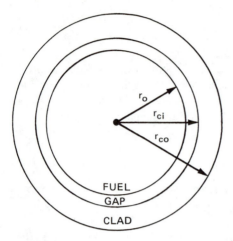

FIGURE 7-3
Cross section of a typical fuel pin (not drawn to scale).

cylindrical cladding tube should isolate the fuel and coolant chemically, but not thermally.

Conductive Heat Transfer

In the fuel and clad regions of a fuel pin, heat transport occurs by conduction. The fundamental balance equation for conduction is the Fourier equation,

Net rate of energy = rate of energy − rate of
accumulation   production   energy loss

or $\rho c \dfrac{\partial T}{\partial t} = S + \nabla \cdot k \, \nabla T$ (7-11)

for temperature T [$^\circ$C], heat source S [W/cm^3], and material density ρ [g/cm^3], heat capacity c [W·S/g·$^\circ$C], and thermal conductivity k [W/cm·$^\circ$C].

    To first approximation, fission may be considered to produce a spatially uniform volumetric heat rate q''' in a fuel pellet. A cylindrical fuel region with uniform properties will have a radial temperature profile $T(r)$ described by

$$T(r) = T(r_o) + \frac{q'''}{4k_f} \, (r_o^2 - r^2)   \qquad 0 < r < r_o$$ (7-12)

for thermal conductivity k_f and temperature $T(r_o)$ at the pellet outer radius r_o. $T(r_o)$ is determined by the interaction of the pellet with the remainder of the system.

    There are no heat sources or sinks in the cladding. Pure conduction occurs according to

$$T(r) = T(r_{co}) + \frac{q'}{2\pi k_{clad}} \, \ln\left(\frac{r_{co}}{r}\right)   \qquad r_{co} > r > r_{ci}$$ (7-13)

for linear heat rate q', thermal conductivity k_{clad}, temperature $T(r_{co})$ at the clad outer radius r_{co}, and clad inner radius r_{ci}. The linear heat rate q' is determined from the pellet heat source q''' as

$$q' = q''' A_f$$

for fuel-pellet cross-sectional area A_f. $T(r_{co})$ depends on the heat-transport properties of the coolant.

Convective Heat Transport

Heat transport from the clad to the coolant is by convection. The fundamental equation, sometimes called Newton's law of cooling, is

$$q = hA(T_S - T_B) \tag{7-14}$$

for heat rate q [W] through heat transfer area A [cm^2], surface temperature T_S [°C], bulk fluid temperature T_B [°C], and convective heat transfer coefficient h [W/cm$^2 \cdot$°C]. The coefficient h depends on geometry, surface conditions, and fluid properties (e.g., temperature, flow velocity, and pressure). Because of these complex dependencies, it is often more appropriate to rewrite the equation as

$$h = \frac{q}{A(T_S - T_B)}$$

to emphasize that h is a proportionality factor that is most often determined from experiments which measure q, A, and $T_S - T_B$.

For convective heat transfer from a cylindrical clad tube to a coolant, the temperature drop $(\Delta T)_{cool}$ is

$$(\Delta T)_{cool} = T(r_{co}) - T_B = \frac{q'}{2\pi r_{co} h_{cool}} \tag{7-15}$$

for clad surface temperature $T(r_{co})$, clad radius r_{co}, linear heat rate q', bulk coolant temperature T_B, and convective heat transfer coefficient h_{cool}. Given the linear heat rate of the pellet, the clad surface temperature can be calculated from the clad radius, the coolant bulk temperature, and the convective coefficient.

Heat transfer across a gap between the fuel pellet and the clad is also treated as a convective process. The temperature difference is

$$(\Delta T)_{gap} = T(r_o) - T(r_{ci}) = \frac{q'}{2\pi r_o h_{gap}} \tag{7-16}$$

for pellet outer radius r_o, clad inner radius r_{ci}, linear heat rate (pellet) q', and gap convective heat transfer coefficient h_{gap}. The coefficient, referred to as *h-gap*, is an empirical quantity inferred from various analyses and experiments. In practice it is also called the *gap conductivity* despite the fact that it it is a convective rather than a conductive heat transfer coefficient.

The maximum temperature difference between the coolant and the fuel centerline $(\Delta T)_{max}$ can be calculated directly by combining Eqs. 7-12, 7-13, 7-15, and 7-16. The resulting expression is

$$(\Delta T)_{max} = \frac{q'}{2\pi}\left[\frac{1}{r_{co}h_{cool}} + \frac{1}{k_{clad}}\ln\frac{r_{co}}{r_{ci}} + \frac{1}{r_{o}h_{gap}} + \frac{1}{2k_f}\right] \qquad (7\text{-}17)$$

where all parameters are as defined previously.

Temperature Profiles

The position-dependent temperature profile in the idealized fuel pellet can be constructed from Eqs. 7-12, 7-13, 7-15, and 7-16 according to the following steps:

1. Set the dimensions of the fuel pin, the heat transfer coefficients (h_{cool}, h_{gap}, k_f, k_{clad}), the volumetric heat source (q'''), and the bulk coolant temperature (T_B).
2. Calculate the linear heat rate q'.
3. Calculate the center-line temperature or temperature profiles, as appropriate.

The results of such calculations are temperature distributions which have the qualitative features shown in Fig. 7-4. The pellet center-line temperature $[T_c = T(0)]$ is especially important because it represents the maximum value and may ultimately limit core power capability.

All fuel pins have radii which are small compared to their lengths. Thus, axial conduction through the pellets is a relatively minor effect compared to radial heat conduction from the pellet to the coolant. A variation in fuel power density along the axial length of the pin results in a series of temperature distributions of the type shown in Fig. 7-4, each with different center-line values.

If the coolant were maintained at an everywhere-constant temperature, a given

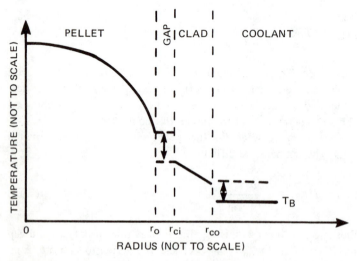

FIGURE 7-4

Basic features of the temperature profile across a clad fuel pin.

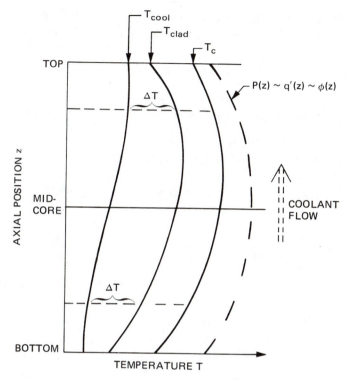

FIGURE 7-5
Axial temperature profiles for the fuel pellet center line, the clad, and the coolant in a reactor with a cosine flux distribution.

power density would result in specific fuel and clad temperatures. A cosine-shaped flux, for example, would result in similarly distributed fuel and clad temperatures.

In a more realistic situation, the coolant temperature increases as it removes heat from the fuel pins that form its flow channel. Equation 7-14 shows that for a given q' and a constant h, the clad and coolant temperatures would maintain a constant *difference* at each axial position. This also states that the clad temperature increases as the coolant temperature increases. The temperature profiles for the fuel, clad, and coolant in Fig. 7-5 are representative of a system which has a cosine-shaped flux. Thus, while the power density and the linear heat rate follow the flux shape, the temperature distributions are skewed by the changing capacity of the coolant to remove the heat energy. Axially symmetric locations, for example, have the same value of q' and the same ΔT between the coolant and the cladding. Since the coolant increases in temperature as it flows up the channel, the clad and, thus, the fuel temperature are higher in the upper axial region of the core.

Sample Calculations

Consider a 1-cm diameter fuel pin in direct contact with cooling water (i.e., no separate clad or gap). Assume an axial maximum 12 kW/m linear heat rate, thermal conductivity 0.05 W/cm·°C, convective heat transfer coefficient 2.0 W/cm²·°C, and bulk coolant temperature $T_B = 300°C$. In more convenient units, the heat rate is

$$q' = 12 \text{ kW/m} \times \frac{1 \text{ m}}{100 \text{ cm}} \times \frac{1000 \text{ W}}{1 \text{ kW}} = 120 \text{ W/cm}$$

The temperature difference between the coolant and the fuel surface (Eq. 7-15) is

$$\Delta T = \frac{q'}{2\pi r_{co} h_{cool}} = \frac{120 \text{ W/cm}}{(2\pi)(\frac{1}{2} \times 1 \text{ cm})(2.0 \text{ W/cm}^2 \cdot °C)} = 19.1°C$$

The maximum fuel-pin temperature (Eq. 7-12) occurs at the fuel center-line where $r = 0$, so

$$T_{max} = T_c = T(0) = T(r_o) + \frac{q'''}{4k_f}(r_o^2 - 0^2)$$

Since for this simple example,

$$T(r_o) = T_B + \Delta T = 300°C + 19.1°C = 319.1°C$$

and the pin cross-sectional area is just πr_o^2

$$q''' = \frac{q'}{A_f} = \frac{q'}{\pi r_o^2} = \frac{120 \text{ W/cm}}{\pi(\frac{1}{2} \times 1 \text{ cm})^2} = 153 \text{ W/cm}^3$$

$$T_{max} = 319.1°C + \frac{153 \text{ W/cm}^3}{(4)(0.05 \text{ W/cm} \cdot °C)}(\tfrac{1}{2} \times 1 \text{ cm})^2$$

$$T_{max} = 319.1°C + 191°C = 510°C$$

The process could have been shortened by using Eq. 7-17

$$(\Delta T)_{max} = \frac{q'}{2\pi}\left[\frac{1}{r_{co} h_{cool}} + \frac{1}{2k_f}\right] = \frac{120 \text{ W/cm}}{2\pi}\left[\frac{1}{(0.5 \text{ cm})(2 \text{ W/cm}^2 \cdot °C)}\right.$$
$$\left. + \frac{1}{(2)(0.05 \text{ W/cm} \cdot °C)}\right] = 210°C$$

The result, of course, is the same

$$T_{max} = T_B + \Delta T = 300°C + 210°C = \boxed{510°C}$$

If the axial distribution is assumed to be cosine in shape,

$$q'(z) = q'(0) \cos \frac{\pi z}{H}$$

for axial position z measured from the core center and for core height H. The temperature differences between the coolant and the fuel center-line (using the earlier result) would then be

$$\Delta T(0) = (\Delta T)_{max} = 210°C$$

$$\Delta T \left(\pm \frac{H}{2} \right) = 0$$

$$\Delta T \left(\pm \frac{H}{4} \right) = (\Delta T)_{max} \cos \frac{\pi z}{H} = (210°C) \cos \frac{\pi}{4} = 148°C$$

NUCLEAR LIMITS

The simplistic fuel-pin-temperature models considered previously provide some insight into the interactions among volumetric source rate, conductive and convective heat-transfer mechanisms, and temperature distributions. In actual design practice, more complicated modeling is required to account for spatial and temporal variations.

The effective thermal conductivity of fuel pellets is generally temperature- and position-dependent. It also tends to decrease with fuel burnup as fission-product buildup causes restructuring (described in Chap. 9). As shown by Eq. 7-12, smaller values of k tend to increase the center-line temperature for given heat rate and coolant conditions.

Burnup also causes fuel swelling, which decreases the thickness of the gap between the pellet and the cladding. This and the buildup of gaseous fission products (mainly Xe and Kr) in the gap tend to increase heat transfer and, thus, result in a larger value of h_{gap}. A decrease in the temperature drop across the gap (e.g., as shown by Eq. 7-16) tends to counteract the effect of reduced fuel thermal conductivity.

Calculations of axial temperature distributions are complicated by the temperature dependence of the convective heat transfer coefficients. Flux shape changes due to fuel and moderator temperature feedbacks, burnup, and control-rod motion also change the temperature profiles.

The results of analyses of radial and axial temperature distributions provide information which must be used to set limiting conditions for safe reactor operation. The heat flux factor F_Q and the enthalpy rise factor $F_{\Delta H}$ serve as convenient references for defining operating limits.

Hot-Spot Factor

The maximum value of F_Q identifies the maximum local power density or linear heat rate. It is referred to as the *hot-spot factor* because of its relationship to maximum temperatures in a fuel pin.

The hot-spot factor limits are set to prevent:

1. melting of fuel
2. cladding-coolant interactions following loss of cooling capability

Since the former is most likely to occur where the temperature is a maximum, it is commonly referred to as the *center-line melt* criterion in accordance with distributions like that shown by Fig. 7-4.

The *loss of cooling* condition may result from either an actual loss of coolant accident [LOCA] or from a loss of flow accident [LOFA]. In the latter, heat removal tends to decrease substantially as the coolant heats. Evaluation of LOCA scenarios is an important part of the design of the water reactors, while that of LOFA scenarios is important to all of the reference reactor designs. Mechanisms associated with such accidents are described in more detail in Chaps. 16-17.

A steady-state temperature profile like that in Fig. 7-4 is established in each fuel pin during normal operation. The pin contains a certain amount of *stored energy* based on its temperature distribution and its inherent heat capacity. If cooling capacity is suddenly lost (i.e., the pin becomes insulated), the stored energy in the pin will redistribute until the temperature profile is relatively flat. A final temperature level exceeding that required for an exothermal chemical reaction between the clad and the coolant may result in severe damage to the fuel.

The extent of fuel damage following loss of cooling will depend on the magnitude of the energy source and on the time history of cooling restoration. If the system remains critical or becomes supercritical following loss of coolant (as it may in LMFBR designs), fission provides a large heat source. Even in the LWR designs where the negative moderator feedback produces neutronic shutdown, the fission-product-decay heat source is sufficient to cause fuel melting if cooling is not restored.

In current reactor designs the loss of cooling criterion tends to result in the more restrictive limit—i.e., the temperatures needed for fuel melting require substantially higher power densities than can be allowed from the standpoint of LOCA or LOFA safety. A *peak clad temperature* is identified such that the extent of chemical reaction will be within an acceptable range. Then, a determination is made of the power density or linear heat rate that would be associated with this peak temperature (under certain *specified* accident conditions). The result provides a *design target hot-spot factor* $(F_Q)_{max}$ defined by

$$(F_Q)_{max} \geqslant \frac{[P(\mathbf{r})]_{max}}{\langle P(\mathbf{r}) \rangle} = \frac{[q'(\mathbf{r}, z)]_{max}}{\langle q'(\mathbf{r}, z) \rangle} \tag{7-18}$$

for maximum and core-averaged power densities P and linear heat rates q' as in Eq. 7-7. Whenever possible, the design is adjusted to limit $(q')_{max}$ such that the inequality is satisfied. The other option, reduction of $\langle q' \rangle$, is much less desirable, since it implies a lower core power level and reduced energy-generation capability.

Hot-Channel Factor

The *hot-channel factor* is the maximum enthalpy rise factor $(F_{\Delta H})_{max}$. It is associated with the possibility that the heat flux from the cladding may be sufficiently mismatched to the conditions of the coolant that energy transfer ceases to occur. Since it is possible for this to occur *without* a general loss of coolant and at a location other than that for the maximum heat flux, the phenomenon is quite different from the one considered previously.

The basic nature of the problem may be understood by considering a heated rod in a pool of liquid. If the bulk liquid is maintained at saturation conditions (i.e., at the temperature where vapor and liquid phases coexist at the given pressure), the heat flux and the relative surface temperature of the rod will be related in a manner shown by the curve in Fig. 7-6. The fundamental physical phenomena associated with the numbered regions are:

1. Heat is transferred by free convection of the liquid when the wall temperature T_{wall} is relatively close to the bulk temperature T_B of the saturated liquid.
2. Nucleate boiling results in a high rate of heat transfer as vapor bubbles are formed on the rod's surface, leave with convection and agitation, and are replaced by (cool) liquid.
3. Bubbles are formed at progressively greater rates until they coalesce to form a vapor column that prevents liquid from coming in contact with the surface of the rod. This point corresponds to the *dryout, burnout,* or *departure from nucleate boiling* [DNB] condition where the vapor blanket causes a severe reduction in heat transfer at the *critical heat flux* $(q'')_{crit}$.
4. Boiling is unstable with the curve being followed only if the heat flux is decreased according to a very precise program.
5. Film boiling occurs with radiation transport of energy across the vapor blanket to the bulk coolant.
6. Since the programmed temperature reduction in the unstable boiling region is very difficult to accomplish (and is impossible under accident conditions!), it is most likely that the temperature will jump from that of the DNB point to a value in the film boiling region.
7. Once the transition has been made to the film-boiling region, liquid can contact the rod surface only when the temperature is reduced at least as far as the *Leidenfrost temperature.*

The very large temperature change associated with the transition from the DNB point to the film-boiling region could lead to chemical reaction or melting of the rod.

The behavior of boiling liquid coolant in a fuel-assembly channel is conceptually similar to that described by the example. The critical heat flux must be avoided at all core locations to prevent cladding oxidation and/or melting. In pressurized systems, the formation of a layer of individual bubbles may lead to coalescence and the isolation of a section of the clad from the coolant, as shown by Fig. 7-7. The boiling coolant in a BWR often leads to a central vapor core with an annular region of liquid for cooling. When the annular region is thinned excessively by a high heat flux, dryout may occur, as shown in Fig. 7-7. (Gaseous coolants like the helium in the HTGR design, of course, are not subject to the burnout phenomenon.)

The hot-channel factor $(F_{\Delta H})_{max}$ identifies the coolant flow channel with the

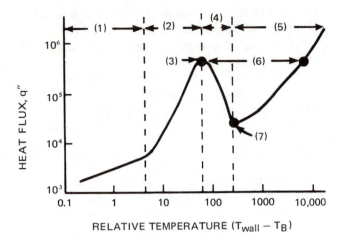

FIGURE 7-6
Heat flux versus surface temperature for a heated pin in a pool of water at saturation temperature.

maximum enthalpy rise. Since this channel should also have the highest temperature, the coolant tends to have the least heat-removal capability and the highest probability for departure from nucleate boiling.

Design Considerations

Power density calculations are quite complicated because reactors are heterogeneous in three dimensions, and their neutron population is time dependent through both kinetics and fuel-depletion mechanisms. Current design practice usually depends on three different types of calculational models:

1. one-dimensional axial with fuel- and moderator-temperature feedback
2. two-dimensional radial
3. point kinetics

The two spatial models each include stepwise depletion capability as described in the last chapter.

Since each of the calculational models relies on reducing the complexity of the actual system by from one to three dimensions, it is necessary that appropriate data and parameters be exchanged among them to assure consistency. The results must be combined or synthesized to provide estimates of the overall power density profile in space and time. Experience is required to develop synthesis methods which can be assured to be conservative (i.e., overpredict rather than underpredict the F_Q and $F_{\Delta H}$).

Limiting Factors

Based on the use of the several models, the hot-spot factor is usually expressed as:

$$\begin{matrix} \text{Calculated hot-} \\ \text{spot factor} \end{matrix} = \begin{bmatrix} \text{radial hot-} \\ \text{spot factor} \end{bmatrix} \times \begin{bmatrix} \text{axial hot-} \\ \text{spot factor} \end{bmatrix} \times \begin{bmatrix} \text{engineering} \\ \text{factor} \end{bmatrix}$$

$$(F_Q)^c_{\text{max}} = (F^r_Q)_{\text{max}} (F^z_Q)_{\text{max}} F^E_Q \qquad (7\text{-}19)$$

for factors calculated from the radial and axial models, respectively, and for the *heat flux engineering factor* F^E_Q, which is employed as an allowance for calculational uncertainties and manufacturing tolerances. Comparison of calculated values with the design-target factor in Eq. 7-18 determines the acceptability of specific design configurations.

The hot-channel factor may be expressed as

$$\begin{bmatrix} \text{Calculated hot-} \\ \text{channel factor} \end{bmatrix} = \begin{bmatrix} \text{radial hot-} \\ \text{channel factor} \end{bmatrix} \times \begin{bmatrix} \text{engineering} \\ \text{factor} \end{bmatrix}$$

$$(F_{\Delta H})^c_{\text{max}} = (F^r_{\Delta H})_{\text{max}} F^E_{\Delta H}$$

for a radial factor, which is the average of the heat fluxes from the fuel pins that form the coolant flow channel and for the *enthalpy rise engineering factor* $F^E_{\Delta H}$, which is employed as an allowance for calculational uncertainties and manufacturing tolerances. This hot-channel factor is generally used with the axial power density to construct the linear heat flux $q'(z)$ for the channel. The critical heat flux $q'_c(z)$, which depends on the detailed conditions in the coolant, is computed from empirical correlations. The ratio of the two terms is the *departure from nucleate boiling ratio* [DNBR] or the *minimum critical heat flux ratio* [MCHFR], defined as

$$\text{DNBR} = \text{MCHFR} = \frac{q'_c(z)}{q'(z)} \qquad (7\text{-}20)$$

The characteristic relationships among the core-average, average-channel, hot-

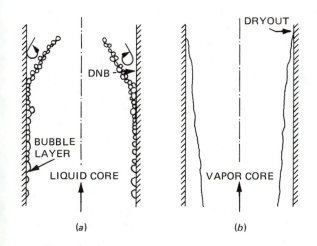

FIGURE 7-7
Critical heat flux effects for (a) pressurized and (b) boiling coolants. (Adapted from L. S. Tong, *Boiling Crises and Critical Heat Flux*, U.S.A.E.C., TID-25887, 1972.)

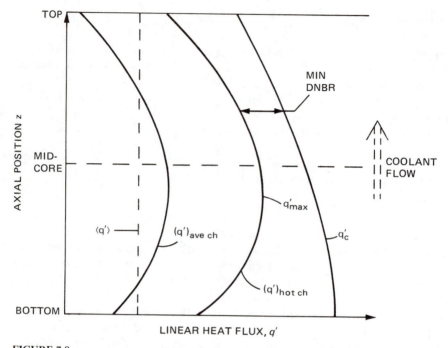

FIGURE 7-8

Characteristic relationship between the core average $\langle q' \rangle$, average channel $(q')_{\text{ave ch}}$, hot channel $(q')_{\text{hot ch}}$, and critical q'_c linear heat rates along the core axis of a PWR.

channel, and critical linear heat fluxes for a PWR are shown by Fig 7-8. The channel fluxes are peaked toward the bottom of the core, while the critical linear heat flux decreases toward the top of the core as the coolant becomes successively hotter. The result is a minimum value of DNBR at the position shown. The essential difference between hot-spot (i.e., q'_{max}) and DNB limits is emphasized by the distance between their respective locations on Fig. 7-8.

The linear heat fluxes in Fig. 7-8 are shown as being peaked toward the bottom of the core. Although a uniform material composition would tend to produce a cosine-shaped distribution, an axial temperature profile as in Fig. 7-5 would also result. The fuel [Doppler] and moderator feedbacks cause the local reactivity to decrease with increasing coolant temperature. Thus, the linear heat flux in fresh PWR fuel is peaked toward the region of lower coolant temperatures in the bottom of the core.

Reactor Control

Hot-spot and hot-channel limitations must generally be met for all normal operating conditions and all anticipated transients. Since from Eq. 7-18 the peaking factors and the core-average power density are both necessary considerations, the limits must accommodate reactivity control methods that affect either the shape or the level of the power distribution.

Control-rod insertion into a reactor core can be expected to increase local power peaking,[†] and may also increase the local heat flux until such time as the negative reactivity decreases the overall power level. Figure 7-9 shows the effect of control-rod insertion on a typical axial power distribution. A corresponding radial power profile is provided in Fig. 7-10. The overall power peaking effect is a combination of the radial and axial components.

Power-level increases due to reactivity insertions of any type (as described in Chap. 5) raise the heat fluxes throughout the core. Other effects like spatial xenon oscillations (as per Chap. 6) can increase peaking without a general change in power level. For both situations, it is important that automatic mechanisms cause termination before F_Q or $F_{\Delta H}$ limits are exceeded. Mechanical systems and inherent feedbacks can each be of help.

[†]Because the peaking-factor formulation holds the *effective* power density to unity (independent of the actual power *level*), it is sometimes helpful to visualize the system as a water-filled balloon. Every local perturbation (e.g., by "squeezing or poking" with a control rod) merely results in a shift of power to another region. The more severe the power depression, the more likely it is that there will be large peaks elsewhere.

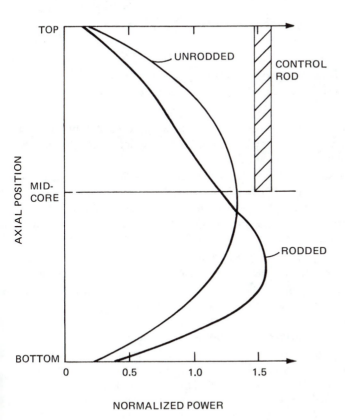

FIGURE 7-9
Effect of control-rod group insertion on PWR power shape axially for the core as a whole.

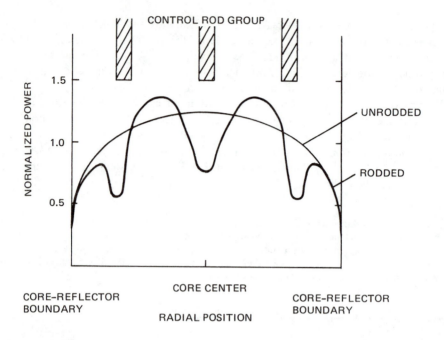

FIGURE 7-10
Effect of control-rod group insertion on PWR power shape radially in a plane through the control rods.

A scram or trip of the control rods will provide automatic insertion of negative reactivity. Since the rod motion may produce severe transient peaking effects, some care is required in system design.

Reactivity feedbacks may also serve to limit reactivity insertions and power peaking. A negative fuel-temperature [Doppler] feedback operates directly to reduce the linear heat rate where the fuel temperature is highest. Feedbacks have a somewhat more complex effect on the likelihood of DNB (since it relates to both fuel and coolant conditions).

Control-rod patterns are generally selected to provide operational reactivity control with reasonably low peaking factors. Burnable poisons within the fuel lattice that are used to hold down some of the excess core reactivity can also shape the power toward lower peaking.

Soluble poison is used in the PWR specifically to limit the need for control rod insertion under routine operating conditions. Thus, the power distribution in this system tends to be very uniform and the reactivity available for accidental insertion via rod withdrawal is small. Inadvertent boron dilution, however, could have a similar, if somewhat slower-acting, effect. The tendency of soluble poison to reduce the magnitude of the negative moderator temperature feedback (as per Chap. 5) is another drawback.

EXERCISES

Questions

7-1. Identify the three reasons why heat removal in nuclear reactor systems may be more complex than in conventional systems.

7-2. Define hot-spot and hot-channel factors and relate them to core operating limits of linear heat rate and critical heat flux, respectively.

7-3. Explain how reflection and enrichment zoning can enhance reactor power capability.

7-4. Describe the effects of control rods, burnable poisons, and soluble poisons on reactor power peaking.

Numerical Problems

7-5. Consider the Maine-Yankee Reactor, a Combustion Engineering PWR rated at 2440 MW(th). The fuel pins are characterized by:

UO_2 pellet diameter, .964 cm
Zr clad inside diameter, .986 cm
Zr clad outside diameter, 1.118 cm
Active fuel length, 3.66 m

The core contains 217 fuel assemblies each with a 14×14 lattice of fuel pins (less 20 per assembly for control element insertion, similar to Fig. 12-11). Assuming all fission energy is deposited uniformly in the fuel pellet, calculate the following core-averaged parameters:
a. heat rate per fuel pin q [kW]
b. linear heat rate q' [kW/m and kW/ft]
c. heat flux from fuel pellet q'' [kW/cm^2]
d. volumetric heat rate in fuel pellet q''' [W/cm^3]

7-6. Using the factors below, calculate the core-averaged fuel-pin temperature profile for the reactor in Prob. 7-5. Plot the results.

$$T_B = 298°C \qquad k_{UO_2} = 0.050 \frac{W}{cm \cdot °C}$$

$$h_{cool} = 1.8 \frac{W}{cm^2 \cdot °C} \qquad h_{gap} = 0.57 \frac{W}{cm^2 \cdot °C}$$

$$k_{clad} = 0.12 \frac{W}{cm \cdot °C}$$

7-7. For the temperature profile in Prob. 7-6, calculate the new center line temperature for:
a. a reduction of k_{UO_2} to one-half its initial value as a result of thermal and burnup effects in the pellet
b. an increase of h_{gap} by a factor of two due to fuel swelling and fission gas buildup
c. the combined net effect of the two mechanisms in (a) and (b)

7-8. Assuming a LOCA-limited design-target linear heat rate of 3.93 kW/m for the

Maine-Yankee core, compute the maximum radial peaking factor for an axial factor of 1.35 and an engineering factor of 1.05. Estimate the peak center-line temperature in the core using the analysis in Prob. 7-6.

7-9. Repeat the calculations in Prob. 7-6 and 7-7 for the following systems whose parameters are contained in App. IV:

   a. CE System 80 PWR
   b. BWR-6
   c. CANDU

SELECTED BIBLIOGRAPHY†

Reactor Heat Transport
   El-Wakil, 1979
   Ginoux, 1978
   Tong, 1972

Other Sources with Appropriate Sections or Chapters
   Babcock & Wilcox, 1980
   Connolly, 1978
   Duderstadt & Hamilton, 1976
   Etherington, 1958
   Foster & Wright, 1977
   Glasstone & Sesonske, 1967
   Graves, 1979
   Lamarsh, 1975
   Lewis, 1977
   Murray, 1975
   Rydin, 1977
   Sesonske, 1973
   Thompson & Beckerley, 1964

†Full citations are contained in the General Bibliography at the back of the book.

THE NUCLEAR FUEL CYCLE

Goals

1. To identify the most important features of utility and nuclear power economics.
2. To describe the important fuel cycle operations with uranium from exploration through enrichment.
3. To identify the basic principles of design, fabrication, use, storage, and transportation of reactor fuels.
4. To describe the fundamental aspects of spent-fuel reprocessing and fuel-cycle waste management.

Chapters in Part III

Economics and Uranium Processing
Reactor Fuels
Reprocessing and Waste Management

8

ECONOMICS
AND URANIUM PROCESSING

Objectives

After studying this chapter, the reader should be able to:

1. Explain each of the three components of electric energy costs and the importance of plant reliability.
2. Describe potentially harmful environmental aspects of uranium mining and milling.
3. Identify the basic requirements for uranium milling, processing, and enrichment in the nuclear fuel cycle.
4. Describe the general features of a solvent-extraction process.
5. Explain the unique role of UF_6 in the nuclear fuel cycle.
6. List the five basic methods for uranium enrichment and identify unique attributes of each.
7. Perform calculations of estimated fuel-cycle costs and enrichment requirements based on data contained in the chapter.

Prerequisite Concepts

Nuclear Fuel Cycle   Chapter 1

Nuclear energy is employed almost exclusively for generation of electrical power. Thus, it must compete economically with alternative energy sources and is subject to the basic constraints of the electric utility industry as a whole. The nuclear fuel cycle itself has a number of unique features that include long processing times and complex operations.

ECONOMICS OF NUCLEAR POWER

The United States has been the world leader in commercial development of nuclear energy. Until recently it produced as much electricity from this source as the rest of the world put together. Most of the world's current generating capacity is still based on reactors designed and fabricated in the United States. Canada, Japan, the Soviet Union, and the nations of Western Europe are becoming increasingly active in the international nuclear marketplace.

At the present time, nuclear energy accounts for about 11 percent of the electricity and 2 percent of the total energy used in the United States. Other nations with present and near-term commitments to nuclear power are identified in Table 8-1. A number may be noted to have substantially larger commitments than the United States. Long-term economic considerations (including security of energy supplies, as noted in App. III) are the underlying bases of the decision to depend on nuclear power.

Electric-Utility Economics

Electric utilities are charged with supplying enough electricity to match the demand at *all times* and at *all levels of demand*. Since this energy is a nonstorable, noninventory commodity, it must be generated as needed. These considerations suggest the need for a very high degree of reliability in generating systems.

Whether a utility is operated by a government or a private concern, it is a monopoly and is subject to some form of external regulation. A nuclear facility is generally subject to additional regulation (e.g., as described in Chap. 19). The combination tends to have an important impact on the conduct of utility activities, especially as related to uncertainty in future operational guidelines and economic constraints.

Cost Components

The cost of generating electricity may be divided into the following three categories:

1. *capital*
2. *fuel*
3. *operating and maintenance* [*O & M*]

In the case of a power plant, capital charges are related to facilities and equipment, while fuel and O & M costs are related to facility operation.

A fundamental economic principle is that capital [money] has a "time value." Simply stated, an investment or loan of current funds depends on the possibility that they will be enhanced in value at some future date. In general, the greater the risk of the venture, the higher must be the potential return. These principles are the bases for the *carrying charges* that are associated with any investment.

Capital for a new project is available only if the prospective economic incentives are large enough to encourage investment. Typical forms of investment include (but are not necessarily limited to) the following:

1. A company uses its *retained earnings* in expectation of earning a *profit.*
2. An investor buys *stock* in the company in expectation of earning *dividends* (or

TABLE 8-1
Nuclear Generating Capacity Installed in 1979 and Planned for 1985 by Various
Countries[†]

Country	Nuclear electric capacity in 1979		Nuclear electric capacity planned for 1985	
	Power, Mw(e)	Percentage of total capacity[‡]	Power, Mw(e)	Percentage of total capacity
Argentina	334	3 (2.67)	944	8.5
Belgium	1,667	16 (1.53)	5,427	38.0
Brazil	0	0	626	–
Bulgaria	880	–	1,760	–
Canada	5,514	8.0 (1.10)	9,700	10.0
China (Taiwan)	1,212	16.6 (0.81)	4,928	31.0
Cuba	0	0	880	–
Czechoslovakia	112	–	4,952	17.0
Finland	1,080	10.0	2,160	20.0
France	8,830	10.0 (1.25)	33,125	50.0
Germany (Dem. Rep.)	1,400	–	5,360	–
Germany (Fed. Rep.)	8,887	12.0 (1.26)	19,534	–
Hungary	0	0	1,760	–
India	596	2.5 (0.92)	1,676	4.0
Indonesia	0	0	1,600	–
Italy	1,412	3.2 (0.76)	2,434	4.5
Japan	14,952	12.2 (1.02)	30,000	16.7
Korea (Rep. S.)	587	8.5 (1.35)	3,815	20.0
Luxembourg	0	0	1,300	–
Mexico	0	0	1,308	5.0
Netherlands	505	3.3 (2.12)	505	2.7
Pakistan	125	–	725	–
Philippines	0	0	620	10.5
Poland	0	0	440	–
Romania	0	0	440	–
South Africa (Rep.)	0	0	1,844	7.5
Spain	1,082	3.9 (2.0)	–	–
Sweden	3,700	14.6 (1.7)	8,380	28.0
Switzerland	1,926	17.5 (1.69)	2,871	21.9
U.S.S.R.	9,905	4.3	34,135	10.0
United Kingdom	6,426	10.0 (1.3)	10,196	11.0
United States [§]	50,218	9.1 (1.2)	–	–
Yugoslavia	0	0	632	–

[†]Data from Atomic Industrial Forum [AIF] 2/6/80.
[‡]Since nuclear plants are usually base-loaded (for reasons described later in this chapter), the fraction of energy generated often exceeds this fraction of capacity. The number in parentheses is the ratio of generation to capacity percentages for 1978.
[§]AIF data 3/10/80.

of having the value of the stock increase due to profits and enhancement of retained earnings).
3. An investor buys *bonds* from the company in expectation of receiving fixed *interest* payments.

These profits, dividends and interest payments are the carrying charges associated with the investment.

Each of the three investment methods has drawbacks, especially at the present time. Retained earnings in most utilities can cover only 10–25 percent of the cost of the (large) plant which has the lowest incremental costs. Historically, half to two-thirds of the cost of a plant might have come from this source. Sale of stock can dilute the holdings of the original owners, especially when the regulatory process holds down profits to the point where the investment is attractive only at a very low per-share price. Sale of bonds places the company in debt. Further complications for this method can include governmental limits on utility borrowing and high interest rates engendered by competition for the very large sums of money associated with plant investments. Typical plant construction is financed by a combination of the three methods. Thus, capital costs include the investment in facilities and equipment plus *all* of the associated carrying charges.

Resource, processing, and transportation expenditures are all considered to be part of the fuel costs of an electric power plant. Waste disposal and even terminal decommissioning charges may also be included in this category. Expenses for on-site handling of fuel and wastes are considered to be part of operating and maintenance [O & M] costs. The latter also includes charges for administration, personnel, supplies, and the other day-to-day expenditures that are necessary to assure the proper functioning of the power plant.

The sum of the capital, fuel, and O & M costs determines the cost of the electricity generated at a particular facility. For proposed facilities, total projected costs are the bases for selection among alternative concepts. Consideration of economic impacts of contingencies like cost escalations and delays is an especially important component of the supporting analyses. Because the carrying charges continue to accrue until both they and the investment are paid off, prolonged delays caused by construction problems, regulation-related actions, or any other means tend to be very costly.

Demand Considerations

With an existing set of facilities, capital charges must be paid independent of the energy production. O & M costs are also relatively insensitive to plant output since administrative and staffing costs are a major component. As might be expected, the O & M costs are likely to be greatest when the plant is shut down for maintenance. Fuel costs, by contrast, are essentially proportional to the electrical energy production.

The nature of the cost contributions suggests that unused generating capacity can be very expensive. It is also apparent that *existing* facilities are best used in a manner that will minimize fuel costs.

One of the most expensive problems facing an electric utility is the unevenness of the demand for electrical energy. In the example shown by Fig. 8-1,

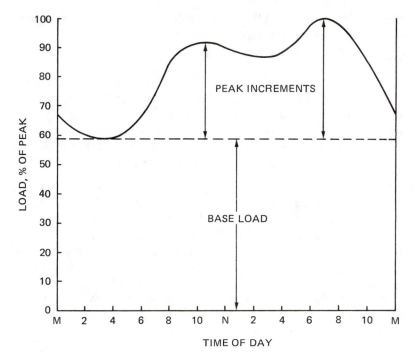

FIGURE 8-1
Typical load variation for an electric utility.

this demand changes by almost 50 percent between its minimum and peak values. Just as importantly, the demand is not stable for any significant period of time during a given day and changes from day to day. Generating capacity built to meet the overall maximum demand is thus underutilized most of the time.

The minimum demand shown on Fig. 8-1 may be met by *base-load* plants which produce power continuously. As the demand increases, other plants provide the incremental generation required to "fill in" the peak. Each operates initially in a *spinning reserve* mode, i.e., the turbine is turning, but not producing power. Then, as the demand increases, needed power is supplied, while another unit assumes a spinning-reserve role (and so on until the peak demand is met).

Energy-production costs are minimized by using the plant(s) with the lowest fuel cost to meet base-load requirements. Additional increments are added at successively greater fuel costs until the peak demand is satisfied. For this scenario, plant reliability is of extreme importance. This is especially true for the base-load plants, whose outage would require use of units with higher fuel costs or the still more expensive purchase of energy from another utility.

All costs for the generation of electrical energy must be covered by sales to customers in the utility's service area and to other electric utilities. The charge rate (typically in cents per kW·h) is a system-wide average that consists of contributions from each facility and from all three cost categories.

Variable rates may be charged to some customers on the basis of different costs of service. Uses that tend to reduce the variation shown by Fig. 8-1, for example, may be subject to special "off-peak" rates because they help the utility generate more of its energy with less expensive fuel.

Nuclear Power Costs

The generating costs associated with nuclear power reactors are divided into the same categories used by the utility industry in general. Capital costs for new facilities tend to be quite high as a result of the sophistication of the technology. Operating and maintenance costs are roughly comparable to those associated with other generating facilities. Although nuclear fuel costs are somewhat complex to compute, they tend to be the lowest among present technologies.

Capital Costs

The capital cost of a typical, large nuclear reactor is now roughly $1 billion. This figure is based on a 10-year lead-time for construction and licensing prior to initial operation.

Capital costs associated with a 1200 Mw(e) pressurized-water reactor are delineated in Fig. 8-2. Dollar amounts are 1977 values based on a scheduled startup in 1987. The largest equipment cost is for the nuclear island, which includes the *nuclear steam supply system* [*NSSS*] (the reactor, coolant, and engineered-safety systems) and all other components designed to assure safe and reliable operation with the fission chain reaction. The turbine island, responsible for the actual generation of electrical power, is the next largest constituent.

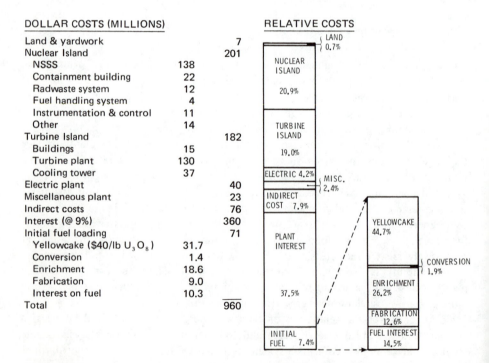

DOLLAR COSTS (MILLIONS)

Land & yardwork	7
Nuclear Island	201
NSSS	138
Containment building	22
Radwaste system	12
Fuel handling system	4
Instrumentation & control	11
Other	14
Turbine Island	182
Buildings	15
Turbine plant	130
Cooling tower	37
Electric plant	40
Miscellaneous plant	23
Indirect costs	76
Interest (@ 9%)	360
Initial fuel loading	71
Yellowcake ($40/lb U_3O_8)	31.7
Conversion	1.4
Enrichment	18.6
Fabrication	9.0
Interest on fuel	10.3
Total	960

RELATIVE COSTS

LAND 0.7%
NUCLEAR ISLAND 20.9%
TURBINE ISLAND 19.0%
ELECTRIC 4.2%
MISC. 2.4%
INDIRECT COST 7.9%
PLANT INTEREST 37.5%
INITIAL FUEL 7.4%

YELLOWCAKE 44.7%
CONVERSION 1.9%
ENRICHMENT 26.2%
FABRICATION 12.6%
FUEL INTEREST 14.5%

FIGURE 8-2

Typical 1977 costs of a 1200-MW(e) PWR system. (Courtesy of *Nuclear Engineering International* with permission of the editor.)

The largest capital cost component shown by Fig. 8-2 is interest (or actually the general carrying charges) that accrues during the 10-year time span from the start of the project until the first revenue is received from the sale of electricity. Any delays which extend this nonoperating period increase the carrying charges and add to the capital-cost component of the rates that must be charged to customers. At the present time, a 2-year delay can produce about a 20 percent increase in the ultimate cost of electricity from a nuclear power plant.

The total of the costs in all Fig. 8-1 categories except those associated with the initial fuel loading are the capital charges to the time of plant startup. These, plus the additional carrying charges that accrue during plant operation, constitute the base for the capital-cost component of the rates that must be charged for the electrical energy.

Operating and Maintenance Costs

The expenses for operating and maintaining a nuclear facility are fairly straight-forward. They include payrolls, supply costs, and administrative charges.

One unique administrative cost is the nuclear liability insurance required by the Federal government. Under the terms of the *Price–Anderson Act*, the owners of a nuclear power plant have a limited liability of $560 million for *off-site* damages caused by a reactor accident. Insurance for this amount is now held jointly by the American Nuclear Insurers—a private insurance consortium—and the federal government. Based on the premium-payment and claim histories (even after the Three Mile Island accident), sufficient private coverage is expected to be available by the mid- to late-1980s to relieve all governmental responsibility.

An increasingly large portion of O & M expenditures at many facilities relates to physical security requirements. Guard forces and equipment are major contributors. (The requirements for security are addressed in some detail in Chap. 20).

Fuel Costs

Fossil fuels used in conventional power plants may be considered as inventory items to be used and replaced as dictated by operations. Nuclear fuel assemblies, by contrast, are characterized by lengthy processing and fabrication times followed by multiyear utilization in a reactor core. On this basis nuclear fuel cycle costs must be treated in somewhat the same manner as capital costs, i.e., carrying charges are a significant factor.

A conceptual economic history of uranium fuel for a light-water reactor may be traced from Fig. 8-3. The upper block represents the major fuel cycle events on a time axis which shows chronology (but which is *not* scaled to show the relationship among the time increments). Some important general features of the events include:

1. acquisition of U_3O_8 [yellow cake] usually through contracts negotiated tens of years ahead of expected use
2. conversion of U_3O_8 to UF_6 with requisite transportation and processing times
3. enrichment of UF_6 over a period of several months (where current policy requires that such services be contracted at least 6 months in advance)
4. fuel element fabrication lasting several months
5. reactor use for 3 to 4 years

6. on-site spent-fuel storage for about 6 months
7. reprocessing to separate uranium, plutonium, and waste products
8. recycle of uranium and plutonium to enrichment and fabrication, respectively, following reprocessing and any necessary conversion

The lower blocks on Fig. 8-3 depict the economic flow for the conceptual closed fuel cycle. The expenses at each step accumulate and cause the outstanding

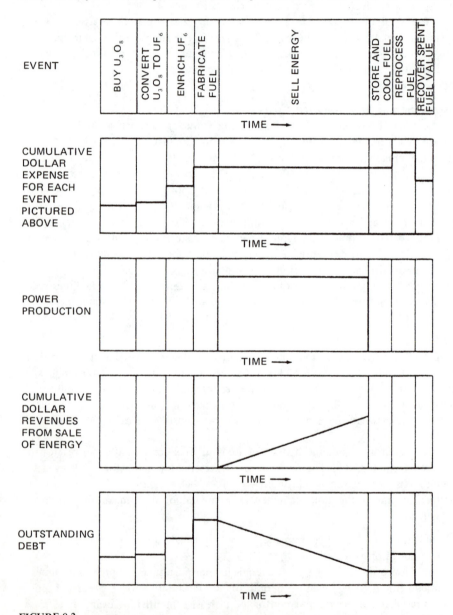

FIGURE 8-3

Economic picture of the nuclear fuel cycle. (From A. Sesonske, *Nuclear Power Plant Design Analysis*, TID-26241, 1973.)

debt to rise until power production begins. The sale of energy (at an assumed constant power level and rate) allows the debt to be reduced uniformly until shutdown. The debt then remains unchanged through storage before increasing with reprocessing and waste handling expenses. The value of the recycled material ultimately reduces the debt to zero as the cycle comes to an end. In principle, the recycled fuel can be used to reduce U_3O_8 requirements for the next cycle or can be sold, either of which have a direct economic increment.

Under current circumstances in the United States, the absence of reprocessing and recycle results in zero value for the residual fissile content of spent fuel. Energy revenues must be adjusted upward so that the debt at the end of the cycle is sufficiently negative to cover charges for long-term storage and waste management with or without reprocessing.

The simple economic picture in Fig. 8-3 is modified for practical applications. The carrying charges associated with the length of the cycle must be superimposed. Since such charges depend on both the size and time history of the outstanding debt, they are not a simple, linear addition to expenses. Revenues, of course, must be increased sufficiently to offset these extra charges.

Comparative Costs

As current generating plants outlive their usefulness and as demand increases, new generating capacity must be built. The decision of which type of plant to build depends heavily on a comparison of the costs for all alternative systems. Economic analyses must account for all possible contingencies, including construction cost escalation, fuel cost and supply problems, costs of appropriate environmental protection, and possible added expenses of delays from regulatory or other actions.

A recent study conducted by the Electric Power Research Institute [EPRI] and reported by Rudasill (1977) compared then-current alternatives for large, base-load generating capacity. Natural gas and oil were quickly excluded from consideration because of cost, supply uncertainties, and impending government regulations. The "new technologies" including solar, geothermal, and fusion were not sufficiently developed to warrant consideration for the study's proposed 1986 startup date.

The final comparison was made between coal and nuclear generating capacity. Regionwise cost estimates, including an indication of uncertainties, are shown in Fig. 8-4. All costs are based on 1976 dollars, but are *levelized* to account for 30 years of operation beginning in 1986. Estimates of inflation and other cost escalations have been factored into the underlying analyses.

The cost comparisons are based on 1976 state-of-the-art power plants. The reference nuclear plant is a 1000 Mw(e) PWR. The reference for coal is two 500-Mw(e) units with flue gas desulfurization and 99.5-percent-efficient electrostatic precipitators. Each meets all 1976 environmental and safety standards and is judged to have comparable reliability, i.e., is expected to generate the same amount of electrical energy over its lifetime.

Coal and uranium price escalations are viewed to be equally uncertain. The regional differences are based largely on the characteristics of the labor force, cost of construction materials, transportation system, and geography. This latter consideration relates to building requirements tailored to meteorologic, seismic, and other site-specific features (as discussed in Chap. 19).

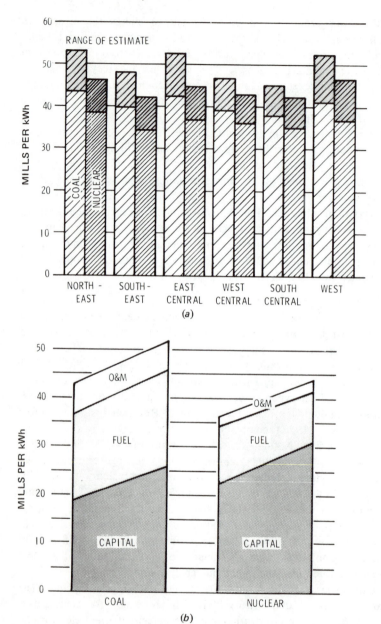

FIGURE 8-4

Estimated levelized costs in 1976 dollars for coal and nuclear plants to be installed in 1986: (*a*) regionwise cost comparison and (*b*) cost breakdown for a typical region. [Adapted courtesy of Electric Power Research Institute (Rudisill, 1977).]

Figure 8-4*a* implies that nuclear plants have a general cost advantage over coal in each region. However, the range of estimate is large enough that either could be favored for any specific site. Nearby coal supplies, for example, can make a very important difference since transportation tends to be a large contributor to fuel costs for coal plants.

Nuclear and coal costs are compared for a representative region in Fig. 8-4*b*. It includes the three basic cost categories and their uncertainties. Capital costs are higher for the nuclear plant as a result of many safety-related considerations. These costs for coal plants, however, are only about 25 percent lower (largely through the expense of antipollution systems).

Fuel costs are relatively high for coal plants because of the large transportation requirements. A typical coal plant must transport 6000 metric tons of fuel for each *day's* operation (equivalent to 60 100-t railroad cars). By contrast, nuclear plants require about a dozen truck shipments of fresh fuel assemblies for each *year* of operation. A typical breakdown of nuclear-fuel-cycle cost contributions may be inferred from the data displayed in Fig. 8-2 for the initial fuel loading of a PWR. Additional components must account for spent-fuel storage, reprocessing, recycle, and waste management (as appropriate).

O & M costs for coal plants tend to be quite high because the maintenance requirements for the antipollution systems are extensive. Historically, maintenance costs for nuclear plants have been more reasonable, as indicated in Fig. 8-4*b*.

The rapid changes that occur in the energy-production field tend to make any cost comparisons somewhat obsolete within a short period of time after they are released. Although the example considered in Fig. 8-4 is no exception, it does serve to illustrate the same general trends which have been observed in both earlier and more recent studies. Current information is generally available in government or EPRI reports or in the magazines described at the end of Chap. 1.

URANIUM

All electrical energy that has been generated from nuclear fission has come directly or indirectly from uranium. The majority of reactors use natural, slightly enriched, or highly enriched uranium. Even the plutonium-fueled systems contain depleted uranium and have had to rely on uranium at some earlier stage for generation of initial plutonium inventories. Thorium-^{233}U reactors employ an isotope of uranium which does not occur in nature, but which has also in some way evolved from natural uranium.

The availability of uranium ore is an important consideration for the continued deployment of the currently operating reactors. It will also have an impact on the potential for expansion, including the phasing-in of advanced-reactor concepts. Processing operations—milling, conversion, and enrichment—are also fundamental if the uranium is to be available in a useable form.

Resources

Exploration for and mining of uranium is often somewhat difficult. The low ore concentrations of less than 0.25 wt% uranium metal are one major concern.

Another problem is that ore bodies are generally "spotty," i.e., limited in volume and widely separated. Typical deposits in one area range from several centimeters to about 6 m thick, and from 1 to about 30 m in length and width. The deposits tend to be separated from each other by regions of very low-grade ore or barren sands.

Exploration

Preliminary exploration activities are based on identification of areas with geologic features similar to those of known uranium deposits. Chemical and radiological testing procedures are employed to verify the actual presence of the resource. Both hand-carried and airplane-mounted radiation detectors can be used to good advantage for early site evaluation.

More detailed analyses require drilling and careful logging of the drill hole and/or evaluation of the resulting core samples. Natural radioactivity and electrical properties can both be useful for this purpose. At the outset, widely spaced bore holes are employed to determine the general features of an area. More extensive local drilling and detailed mapping precedes the onset of actual mining operations.

Resource Evaluation

Since the use of commercial nuclear power is predicated on the availability of uranium at reasonable prices, an accurate assessment of uranium resources is necessary. The National Uranium Resource Evaluation [NURE] program was initiated in 1973 with this in mind. Under the initial direction of the U.S. Atomic Energy Commission [AEC] and currently the Department of Energy [DOE], the program has major objectives to:

- Achieve a preliminary evaluation of domestic uranium resources and favorable exploration areas for the entire country by 1976.
- Conduct more comprehensive evaluations by 1980 and update the preliminary report annually through that year.
- Conduct research and development on new and improved procedures, equipment, and technology for uranium search and assessment.
- Identify new areas favorable for uranium exploration.

Information on ore grade [assay] and extent of deposits is employed to estimate the amount of uranium available at various costs.

Recent NURE estimates are contained in Table 8-2. The "production costs" (also called "forward costs") are exclusive of exploration expenses and capital charges that may have already been incurred; i.e., they include only the projected expenditures on the mining and milling. The values in the table represent the equivalent U_3O_8 content of proven reserves and three categories of potential reserves—probable, possible, and speculative—with decreasing degrees of certainty. By-product reserves are associated with secondary uranium recovery from phosphate and copper operations.

Data like those in Table 8-2 serve as one important basis for determining how much electrical energy can be obtained from domestic uranium supplies. Similar information for other parts of the (non-Communist) world is contained in Table 8-3. It may be noted that the United States accounts for nearly one-third of the world's reasonably assured uranium resources and about half of the estimated

TABLE 8-2
United States Uranium Resource Estimates in Tons U_3O_8 from
"Statistical Data of the Uranium Industry," U.S. Department of Energy[†]

$/lb U_3O_8 production cost	Reserves, tons U_3O_8	Potential resources, tons U_3O_8		
		Probable	Possible	Speculative
$15	290,000	415,000	210,000	75,000
$15–30 Increment	400,000	590,000	465,000	225,000
$30–50 Increment	230,000	500,000	495,000	250,000
By-product, 1977–2000[‡]	120,000	–	–	–
Total	1,040,000	1,505,000	1,170,000	550,000

[†]As of January 1979 [GJO-100(79), 1979].
[‡]By-product of phosphate and copper production.

additional resources. As a point of reference, both the United States and the remainder of the free world are expected to have the same demand for 1980, about 28,000 tons of U_3O_8 equivalent.

Mining

Nearly 90 percent of U.S. uranium resources and current production are centered in the western states of New Mexico, Wyoming, Colorado, and Utah. The nature of each deposit determines the appropriate mining method.

Open-pit or surface mining is implemented whenever uranium ore is within a few hundred meters of the surface and when the overburden can be removed without an excessive amount of blasting. Roughly 40 large operations account for 60 percent of the uranium production in the United States.

Underground mining is employed for uranium deposits which are located at depths greater than a hundred meters or are covered by strata of hard rock. About 40 percent of current U.S. production comes from roughly 150 underground mines.

Because of the large variations in assay and extent of deposits, very little of the material extracted by surface and underground methods is actually uranium. The proposed use of *in situ* solution mining could eliminate this and other problems. In principle the uranium metal would be dissolved, brought to the surface, and then recovered, thereby eliminating the large volume of unwanted material. The radiological hazards associated with radium and radon gas (as described in Chap. 10) could also be reduced to the extent that the process is selective for only the uranium. At the present time the leaching method has not been demonstrated on a large scale, although commercial demonstration projects are being planned for New Mexico and elsewhere.

A major limitation to the development of uranium (and any other energy resource) is the environmental impact associated with extraction. These include both health effects to mine workers and more general effects on the local ecology.

Uranium mining is the most hazardous step in the nuclear fuel cycle, largely due to the prospects for physical injury. Radiological hazards are also present, especially for underground operations where the working environment is somewhat

TABLE 8-3
Free-World Uranium Resource Estimates as of January 1977[†]

	Reasonably assured, 10^3 tons			Estimated additional, 10^3 tons		
Country	<$30/lb U_3O_8	$30–50/lb	Total	<$30/lb U_3O_8	$30–50/lb	Total
North America	913	227	1140	1633	684	2317
United States	690	200	890	1120	330	1450
Canada	217	19	236	510	343	853
Mexico	6	0	6	3	0	3
Denmark	0	8	8	0	11	11
Africa	680	63	743	196	62	258
South Africa	398	55	453	44	49	93
Niger	208	0	208	69	0	69
Algeria	36	0	36	65	0	65
Gabon	26	0	26	6	6	12
Central African Empire	10	0	10	10	0	10
Zaire	2	0	2	2	0	2
Somalia	0	8	8	0	4	4
Madagascar	0	0	0	0	3	3
Australia	376	9	385	57	6	63
Europe	81	417	498	58	57	115
France	48	19	67	31	26	57
Spain	9	0	9	11	0	11
Portugal	9	2	11	1	0	1
Yugoslavia	6	3	9	6	20	26
United Kingdom	0	0	0	0	10	10
Germany, Fed. Rep.	2	1	3	4	1	5
Italy	2	0	2	1	0	1
Austria	2	0	2	0	0	0
Sweden	1	390	391	4	0	4
Finland	2	2	4	0	0	0
Asia	54	4	58	31	0	31
India	39	0	39	31	0	31
Japan	10	0	10	0	0	0
Turkey	5	0	5	0	0	0
Korea	0	4	4	0	0	0
Central & South America	47	31	78	18	1	19
Brazil	24	0	24	11	0	11
Argentina	23	31	54	0	0	0
Chile	0	0	0	7	0	7
Bolivia	0	0	0	0	1	1
Grand total (rounded)	2200	700	2900	2000	800	2800

[†]Data from Atomic Industrial Forum [AIF].

difficult to control. Extensive studies have been conducted on the health effects of direct radiation, radon gas, and mine dust. The observed incidence of excess lung cancers in miners has led to the establishment of lifetime exposure limits that are about a factor of 10 below the defined hazard level.

Potential impacts of uranium mining on the more general environment are somewhat dependent on specific geological features. In New Mexico, for example, uranium is found in sandstone and limestone formations which contain potable groundwater. This and other factors give rise to concerns that:

- High dewatering rates (e.g., estimated at 40,000 acre-feet for the Grants, New Mexico area in 1975) may contaminate surface water and cause changes in groundwater aquifers.
- Erosion effects could change drainage patterns and lead to loss of land productivity.
- By-product radionuclides (^{230}Th, uranium, ^{226}Ra, radon gas, and their daughters) and chemical species (selenium and nitrates) discharged from mines and mills may contaminate air and drinking water supplies.
- Mining operations with associated power generation, transportation, and population changes may cause a deterioration of air quality and "quality of life."

Expansion of mining activities appears to be proceeding with caution pending resolution of these and related issues.

Milling

It has been noted previously that ore from typical U.S. mines assays at less than 0.25 wt% uranium metal. This low value precludes simple separation by conventional metallurgical techniques. Instead, it is necessary to employ somewhat more complex chemical processing in the milling or refining step of the nuclear fuel cycle.

There are currently 14 uranium mills in the United States. Each is located in proximity to mining operations so that uranium ore can be transported over relatively short distances in open trucks or rail cars. If it has not already been done at the mine, the ore is blended to assure uniformity for the chemical processing.

A schematic flowsheet for a typical solvent-extraction milling operation is shown in Fig. 8-5. The crushing and grinding operation reduces the raw ore to a relatively uniform particle size that enhances later removal of U_3O_8 and UO_2 complexes. Leaching is based on dissolution of uranium (and often other trace metals) into either acid or alkaline solution to form a "pregnant leach liquor." The undissolved, nonmetallic constituents, which constitute nearly all of the mass and volume of the ore, remain behind and form the *mill tailings* (for which waste management is described in Chap. 10).

The extraction and stripping circuits in Fig. 8-5 are components of a *solvent-extraction* method for separating chemical elements. As applied in the milling operation, the process uses aqueous and organic liquids which are immiscible, i.e., which will not remain in a mixture with each other. The first aqueous or acid solution dissolves uranium and other metals. It is then mixed physically with an organic solvent that has a large, selective affinity for uranium. Because the organic is lighter than the aqueous, it will float on top along with the

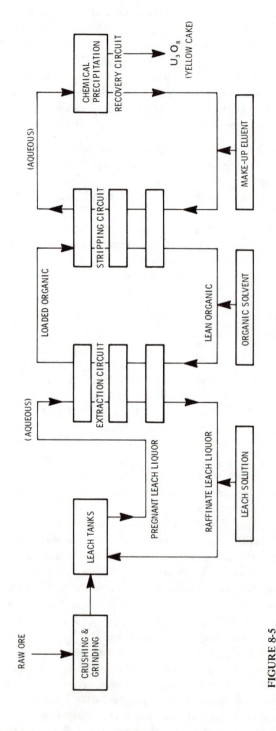

FIGURE 8-5

Typical solvent-extraction flowsheet for a uranium mill (Adapted from "Hydrometallurgy Is Key in Winning U_3O_8," *Mining Engineering,* Aug. 1974, p. 26.)

uranium it removes from the aqueous solution. In this manner the uranium is said to be *extracted* by the organic liquid. The uranium may be *stripped* or removed from the organic by mixing with a second aqueous solution where the latter now has the higher affinity for uranium.

In the flow shown by Fig. 8-5, the pregnant leach liquor enters the extraction circuit where the uranium is extracted by the organic. The depleted or raffinate leach liquor is recycled for further use. The uranium-loaded organic is drawn off and transferred to the stripping circuit where the uranium is stripped to another aqueous solution called an eluent. The lean organic is also recycled. Multiple extraction and stripping operations allow 95 percent or more of the uranium content to be removed.

The final step in the milling process shown in Fig. 8-5 is recovery of the uranium by chemical precipitation, filtering, and drying. The final form is usually ammonium diuranate $[(NH_4)_2U_2O_7$ or ADU] whose yellow color gives rise to the name *yellow cake*. This term is typically applied to mill output independent of the actual chemical composition of the product, although weights and prices are always based on the equivalent content of U_3O_8. Because of its low radioactivity, yellow cake can be transported in standard 200-l [55-gal] drums with 350–450 kg [800–1000 lb] of U_3O_8 each.

Actual milling processes may vary from that just described, often depending on the characteristics of the uranium ore. About 90 percent of current milling operations employ a solvent-extraction process while the remainder use an alkaline-leach method. The two major variants for solvent extraction are use of:

1. mixer-settlers—aqueous and organic liquids are mechanically mixed to enhance physical contact (and thus extraction of uranium) before they are allowed to settle with the organic floating to the top, where it can be skimmed off
2. ion exchange—aqueous liquids contact solid organic spheres contained in wire mesh baskets

The alkaline-leach method employs sodium carbonate [$NaCO_3$] as a highly selective leaching agent. This eliminates the need for solvent-extraction processing (i.e., the extraction and stripping circuits would not be present on the flowsheet in Fig. 8-5).

Each milling process has advantages and disadvantages. The alkaline-leach method, for example, is more efficient but releases selenium and is not very economical for gypsum- or sulfide-rich ores. The major problem associated with solvent extraction is that it releases radium, radon, and their daughters (which causes a waste management concern discussed in Chap. 10).

Conversion

The conversion step of the nuclear fuel cycle usually consists of two operations conducted at the same site—purification of yellow cake from the mill and production of uranium hexafluoride [UF_6] for later use in enrichment. When enriched uranium is not required, as for the CANDU reactor, the purification could take place at the uranium mill, a separate facility, or a fuel fabrication plant.

Very high purity uranium is required for reactor applications because even small amounts of materials like boron, cadmium and the rare-earth elements have sizeable neutron-absorption cross sections. Solvent-extraction and other chemical processes can be used in effective purification methods.

Treatment of purified uranium concentrates with hydrofluoric acid [HF] is employed to produce UF_4, which in turn receives an application of elemental fluorine to become UF_6. Impurities whose fluoride compounds are more volatile than UF_4 can be driven off by heating. Those that are less volatile than UF_6 are left behind when the latter is sublimed to its gaseous form. These two methods coupled with a final fractional distillation process are the bases for generating very high purity uranium hexafluoride.

Enrichment

The LWR, HTGR, and many other reactor designs require fissile ^{235}U at an isotopic concentration greater than the 0.711 wt% [0.72 at%] that exists in natural uranium. Isotope-separation or *enrichment* technologies have been implemented to separate these chemically identical ^{235}U and ^{238}U isotopes.

Of the variety of enrichment procedures that have potential application in commercial nuclear fuel cycles, the following five are representative and are considered further in this chapter:

1. gaseous diffusion
2. gas centrifuge
3. gas nozzle
4. chemical exchange
5. laser excitation

The first three (as implied by their names) are based on the use of the gaseous UF_6. Chemical exchange is implemented between two liquid phases or between a gaseous and a solid phase. The laser process depends on selective excitation of ^{235}U atoms or molecules.

All enrichment processes may be characterized by the amount of separation that occurs in a single unit. This is usually quantified through a *separation factor* α defined as

$$\alpha = \frac{e/(1-e)}{d/(1-d)}$$

where e and d are the final ^{235}U isotopic fractions in the material streams designated as enriched and depleted, respectively. The total separation capability of an integrated system depends on coupling together a number of individual units or stages. The two output streams contain the enriched *product* and depleted *tails*, respectively.

The amount of energy expended in the enrichment process is usually quantified in terms of *separative work*; the energy-per-unit-mass-of-enriched-material is in terms of the *separative work unit* [SWU]. A precise definition of separative work includes factors related to the material quantities and enrichments of the input, product, and tails. For the purposes of this discussion, it suffices to recognize that 1 SWU represents a specific increment of system operation (e.g., energy consumption or time), the net effect of which depends on the desired parameters for the material streams. One kgSWU (i.e., separative work unit based on kg mass) will, for example, take 2.3 kg of natural uranium to 1 kg of 1.37 wt% product and 1.3 kg of 0.2 wt% tails. It will also take 0.73 kg of natural uranium to 3.9 g of 93 wt% product with the remainder 0.2 wt% tails.

Enrichment-facility capacities are rated in kgSWU/y (or tSWU/y, where 1 tSWU = 1000 kgSWU). Total costs divided by capacity are the basis for the enrichment charge in $/SWU or equivalent.

Gaseous Diffusion

The *gaseous diffusion* process is based on differential diffusion of the isotopic constituents in UF_6 gas. According to the kinetic theory of gases, all molecules have the same average kinetic energy $\frac{1}{2}mv^2$ (for mass m and speed v) when they are in thermal equilibrium with each other. The lighter $^{235}UF_6$ molecules then have a slightly greater average speed than the heavier $^{238}UF_6$ molecules.

When UF_6 gas is forced against a barrier of precisely controlled porosity, the faster $^{235}UF_6$ molecules tend to penetrate preferentially. Figure 8-6 represents a cross section of a cylindrical stage. Gas from the high-pressure feed stream diffuses through the barrier into the low pressure outer region. The diffusion occurs in such a manner that the outer region is enriched in ^{235}U while the inner region is depleted in it. In typical operations, the two streams are discharged when about half of the entering gas has passed through the barrier. The enriched stream is pumped to the next "higher" stage, while the depleted stream goes to the next "lower" stage, as shown in Fig. 8-7.

The theoretical or ideal separation factor for a single gaseous diffusion stage is $\alpha = 1.0043$. An actual plant requires about 1200 stages formed into a *cascade* to produce 3 wt% ^{235}U for LWR applications. A cascade of over 3000 stages is needed for the 93 wt% ^{235}U used in the HTGR design.

Uranium hexafluoride is the only known compound of uranium that is suitable for the gaseous diffusion process. The properties of UF_6 dictate that systems employing it be designed:

- to maintain temperatures and pressures above its sublimation level (56°C [134°F] at atmospheric pressure)
- with such metals as nickel and aluminum because of its corrosiveness to the other, more common metals
- to be leak tight and clean, since it is highly reactive with water and is incompatible with organic substances
- with diffusion barriers capable of withstanding the corrosive environment while maintaining separative quality over long periods of time

Despite these severe requirements, diffusion plants have achieved on-line efficiency exceeding 99.5 percent over a period of many years.

The long cascades used for gaseous-diffusion enrichment require up to a year to reach equilibrium and, thus, exclude processing of individual batches of material. Instead, continuous operation occurs on a *toll enrichment* or *SWU-bank* basis with natural UF_6 "deposited" and enriched UF_6 "withdrawn" at a later time. Natural-

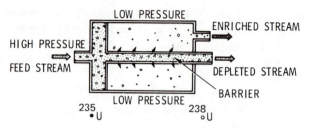

FIGURE 8-6
Gaseous-diffusion enrichment stage schematic. (Courtesy of U.S. Department of Energy.)

uranium feed requirements and SWU charges for given output enrichments and (arbitrary) 0.2 wt% tails can be inferred from Table 8-4. Each kg of 3 wt% ^{235}U output, for example, requires a deposit of 5.479 kg of natural uranium and payment for 4.306 kgSWU. The data for enrichments between the 0.2 wt% tails and 0.711 wt% natural uranium are included in Table 8-4 to allow calculations for operation with tails higher than the arbitrary reference. Operation with a tails assay of 0.3 wt%, for example, would require an *extra* 0.196 kg of natural uranium for each 1 kg of output, but would also require 0.158 kgSWU *less* separative work (i.e., since the tails are being discarded with a higher ^{235}U content, more feed is needed, but less "work" is required to obtain the desired enrichment). For the 3 wt% ^{235}U output in the previous example, 5.675 kg of natural uranium and 4.148 kgSWU would be necessary for each 1 kg output with 0.3 wt% tails.

The data in Table 8-4 shows 0.2-wt% tails assay material as having "zero value" since no feed component and separative work charges would be associated with its withdrawal. In principle, this assay should be set such that the uranium content of the tails has *no economic value* (i.e., so that enrichment of the tails to

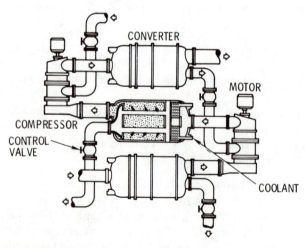

FIGURE 8-7
Section of a gaseous diffusion enrichment cascade. (Courtesy of U.S. Department of Energy.)

TABLE 8-4
Table of Toll Enriching Services[†]

Assay, wt% ^{235}U	Feed component (natural), kg U feed/kg U product	Separative work component, kgSWU/kg product
	Standard table of enriching services[‡]	
0.20	0	0
0.30	0.196	−0.158
0.40	0.391	−0.198
0.50	0.587	−0.173
0.60	0.783	−0.107
0.70	0.978	−0.012
0.711 (natural)	1.000	0.000
0.80	1.174	0.104
0.90	1.370	0.236
1.00	1.566	0.380
1.20	1.957	0.698
1.40	2.348	1.045
1.60	2.740	1.413
1.80	3.131	1.797
2.00	3.523	2.194
2.20	3.914	2.602
2.40	4.305	3.018
2.60	4.697	3.441
2.80	5.088	3.871
3.00	5.479	4.306
3.40	6.262	5.191
3.80	7.045	6.090
4.00	7.436	6.544
5.00	9.393	8.851
10.00	19.178	20.863
90.00	175.734	227.341
98.00	191.389	269.982

[†]From A. Sesonske, *Nuclear Power Plant Design Analysis*, TID-26241, 1973.
[‡]The kg of feed and separative-work components for assays not shown can be determined by linear interpolation between the nearest assays listed.

the level of natural uranium would incur work charges in an amount equal to the cost of buying the same quantity of natural uranium). Table 8-5 shows the influence of the ratio of feed to separative work costs on the optimum tails composition. If uranium were "free," the ratio would be zero and the tails could be very near natural uranium; "free" enrichment would imply essentially zero tails. The tails assay of 0.2 wt% on which Table 8-4 is based is appropriate for a feed-to-separative-work ratio of 1.3. If enrichment capacity is limited, however, a higher-than-optimum tails assay is likely to be favored because it allows higher throughput (at the expense of increased feed requirements and lower utilization of the ^{235}U in natural uranium).

TABLE 8-5
Optimum Enrichment Tails Composition as a
Function of Feed and Separative-Work Costs[†]

Ratio of feed cost to cost of separative work	Optimum tails composition, wt%
0.0	0.711
0.2	0.402
0.4	0.327
0.6	0.282
0.8	0.251
0.9	0.238
1.0	0.227
1.1	0.217
1.2	0.208
1.3	0.200
1.4	0.192
1.5	0.185
2.0	0.158
2.5	0.139
3.0	0.124
4.0	0.103
5.0	0.088
∞	0.0

[†]From A. Sesonske, *Nuclear Power Plant Design Analysis,* TID-26241, 1973.

Gas Centrifuge

When UF_6 gas is contained in a centrifuge rotating at a high speed, the centrifugal forces cause the heavier $^{238}UF_6$ molecules to be driven preferentially toward the outside, while the $^{235}UF_6$ molecules stay more toward the central axis. This is the underlying principle of the *gas-centrifuge* enrichment method sketched in Fig. 8-8. The feed gas enters along the axis, moves downward, and eventually establishes a longitudinal countercurrent flow pattern in the high-speed rotor section. The centrifugal forces and the flow pattern both contribute to a preferential increase in the $^{235}UF_6$ concentration along the centrifuge axis at the bottom of the unit. Product and waste lines draw off the enriched and depleted material, respectively, for further enrichment in a cascade arrangement of centrifuge units. Individual-stage separation factors of 1.1 or greater are possible with this method.

Since the gas centrifuge uses UF_6, it has restrictions on construction materials and operations similar to those identified for gaseous diffusion. The large forces, high speeds, and potential for resonant vibrations which are present with these units can be expected to cause some amount of destructive failure. Thus, cascade design must attempt to minimize propagation of damage to neighboring units. Provision must also be made for rapid changes in cascade connections and replacement of damaged units to assure continuity of production.

Gas Nozzle

The *gas nozzle* (or "Becker nozzle" after the German scientist credited with its development) is one of several separation processes based on the aerodynamic behavior of UF_6 gas. As shown by Fig. 8-9, UF_6 mixed with a light auxiliary gas like hydrogen or helium is caused to flow at high velocity along a fixed curved wall. Centrifugal forces and a skimmer split the flow into light and heavy fractions which are enriched and depleted in ^{235}U, respectively.

The gas nozzle method is capable of achieving separation factors of about 1.01–1.02 for UF_6. Again, this gaseous working fluid results in the limitations described earlier for gaseous diffusion facilities.

Chemical Exchange

A general isotopic exchange reaction can be written in the form

$$AB + A'C \rightleftharpoons A'B + AC$$

for isotopes A and A′ and other chemical species B and C. If A and A′ are ^{235}U and ^{238}U, respectively, such a reaction can serve as the basis for *chemical exchange* enrichment when the equilibrium isotopic content of the compounds differs by some amount.

Research has identified systems with two liquid phases and systems with one solid and one gaseous phase that can implement chemical exchange enrichment. Separation factors of up to 1.002 are anticipated for the process.

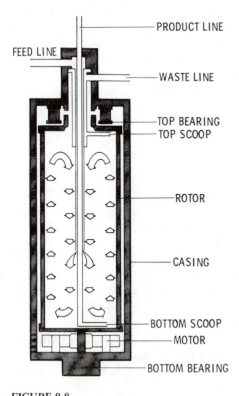

FIGURE 8-8
Gas-centrifuge enrichment schematic. (Courtesy of *Nuclear Engineering International* with permission of the editor.)

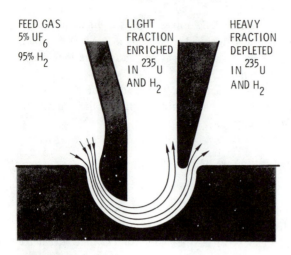

FEED GAS
5% UF$_6$
95% H$_2$

LIGHT
FRACTION
ENRICHED
IN ^{235}U
AND H$_2$

HEAVY
FRACTION
DEPLETED
IN ^{235}U
AND H$_2$

FIGURE 8-9
Gas-nozzle enrichment schematic. (Adapted from "Present State and Development Potential of Separation Nozzle Process," Gesellschaft für Kernforschung M.B.H., Karlsruhe, West Germany, KfK-2067, September 1974.)

Laser Excitation

The laser excitation method for isotope separations is fundamentally different from those considered previously in that it is based not on mass difference but rather on subtle differences in electronic structure of atoms and/or molecules. The quantum nature of such electronic states presents the possibility for a specific laser radiation to affect one isotope while having no effect on another.

The laser methods reported to date rely on bombardment of a beam of UF$_6$ or uranium vapor with multiple laser photons. As a result, the ^{235}U constituent may be removed selectively from the beam through ionization, radiation pressure, or molecular decomposition. Separation factors for a single pass have been reported to be as high as $\alpha = 70$ (i.e., 50 wt% ^{235}U enrichment in one step!).

General Features and Comparisons

Natural UF$_6$ is usually received at enrichment facilities in 13-t steel cylinders. Since its radioactivity is very low, the primary health concern in shipment and initial processing is its reaction with moisture in the air to form uranyl fluoride [UO$_2$F$_2$] and the highly reactive hydrofluoric acid [HF].

Enrichment also adds the potential for criticality. Since the slightly enriched [<5 wt% ^{235}U] uranium used for LWR systems requires moderation for criticality, exclusion of large quantities of water and other low-mass materials from facilities provides an effective control. It is shipped off-site in 2.5-t, water-tight cylinders. The highly enriched [93 wt% ^{235}U] uranium used for the HTGR requires both moderation control and the use of controlled geometry (i.e., high-leakage configurations) to prevent criticality. Its shipment is made in small-capacity, water-tight containers that have oversized exterior dimensions to prevent their packing at high effective material densities.

Information on the enrichment technologies is very highly restricted because

TABLE 8-6
Current Status of Enrichment Technologies

	Gaseous diffusion	Gas centrifuge	Gas nozzle	Chemical exchange	Laser excitation
Separation factor α [†][‡]	1.0043	1.10	1.01–02	1.002	70
Existing capacity [†][‡] (10^6 kgSWU/y)	US 17 France 0.5 UK 0.5 USSR 7–10	UK 0.2 Netherlands 0.2	Germany 0.002 S. Africa 0.006	—	—
New planned capacity [†][‡] (10^6 kgSWU/y)	France 10.3	US 8.75 Germany 1 Netherlands 4 UK 0.2–0.4 Japan 4	Brazil 0.25 S. Africa 5	France 0.05–0.1	—
Energy consumption for new units (kWh/kgSWU) [‡]	2260–2520 (new units)	250	3500–4000	≈800	≈2–3
Cost ($/kgSWU/y) [‡]	200–305	250–370	180–340	≈300	—
Economic optimum plant size [‡] (10^6 kgSWU/y)	9	3 (>0.3 viable)	Flexible	0.2	Very flexible
Material holdup	Very large	Low	Moderate	Very large	≈0
Equilibrium time	≈1 y	≈1 d	<1 y	≈1½ y (3–4 wt%)	≈0
"Proliferability"	Moderate	High	High	Very low	Very high
Advantages	Reliable, established technology	Low energy use, operating flexibility	Flexible plant size	Passivity, low proliferability	Low energy use, ≈0 tails
Disadvantages	Large size and holdup	Maintenance uncertainties	Lack of experience	Low capacity, large holdup	Still in R&D stage, proliferability

[†] Data on gaseous diffusion, centrifuge, and nozzle from K. M. Harmon, "Summary of National and International Radioactive Waste Management Programs 1979," Battelle Pacific Northwest Laboratories, PNL-2941, March 1979.

[‡] Data from D. Kovan, "Enrichment Plants—A Survey of Major New Uranium Enriching Projects," *Nucl. Eng. Int.*, Nov. 1976, pp. 49–51.

of its potential use in nuclear weapons programs. National and industrial interests also limit interchange of any process details that may constitute a "competitive edge" in cost and/or supply capability. Some important features that are known are summarized in Table 8-6 and discussed in the following paragraphs.

Existing enrichment capacity is based largely on gaseous diffusion. The U.S. facilities in Oak Ridge, Tennessee, Paducah, Kentucky, and Portsmouth, Ohio were developed for military purposes but now produce commercial product. The same is likely true of the facilities in the U.S.S.R. Additional capacity of this type is scheduled for both the United States and Europe.

Current production with gas centrifuge technology is centered in the United Kingdom and Holland under the auspices of the Urenco–Centec consortium, of which Germany is also a partner. Both near-term and long-range expansion is planned by this organization. The United States and Japan have also planned gas-centrifuge capacity.

Gas nozzle separation has been implemented only on a small pilot-plant scale. However, South Africa has plans for a large facility and Germany has contracted to build a somewhat smaller one for Brazil. Chemical exchange and laser excitation are still in early research and development stages.

Enrichment costs appear to be somewhat comparable among the alternative technologies. From an energy standpoint, however, the gas centrifuge method has a large immediate advantage, while laser excitation holds the prospect for even more sizeable future savings.

The size of an economically optimized plant, material holdup, and equilibrium time vary among the concepts, as noted in Table 8-6. A final category entitled "proliferability" represents a general evaluation of the ease with which nuclear-weapon material could be obtained either by building a clandestine facility or by diversion from a commercial fuel-cycle facility (as considered further in Chap. 20).

The major advantage of the gaseous diffusion process is its well-established record of reliability. Disadvantages include large system size and related material holdup. Gas centrifuges offer low energy use and flexible operation, but have maintenance uncertainties. The gas nozzle concept may allow for small systems to be economical, but there is still much development work to be completed before reliability can be assured. Chemical exchange is material-intensive and quite slow ($\approx 1\frac{1}{2}$ years to produce 3–4 wt% ^{235}U for LWR use), but for this reason it is most resistant to proliferation. Laser-excitation offers the potential for the lowest energy use and economical separation of nearly all ^{235}U from natural uranium. However, it is still in the R & D stage and could make weapons proliferation much easier.

EXERCISES

Questions

8-1.   Explain each of the three cost components for an electric power plant. Why is plant reliability such an important consideration in nuclear power costs?

8-2.   Identify the fundamental components of the milling, processing, and gaseous-diffusion enrichment operations.

8-3.   Explain the role of UF_6 in the nuclear fuel cycle.

8-4. Describe two potentially harmful environmental impacts of uranium mining.

8-5. List the five basic methods for enriching uranium. Identify the method characterized by:
  a. lowest "proliferability"
  b. highest "proliferability"
  c. greatest current capacity
  d. new technology for near-future U.S. capacity additions
  e. commercial development in Africa and South America

8-6. Describe the general features of a solvent-extraction process.

Numerical Problems

8-7. The fractional cost break-down for the initial fuel loading of a PWR (Fig. 8-2) may be assumed to be somewhat representative for fuel-cycle costs. Using the lowest value for nuclear costs from Fig. 8-4b, estimate the percentage change in fuel cost *and* in the total energy cost associated with:
  a. doubling only enrichment cost
  b. doubling only uranium cost
  c. both of the above

8-8. A 1200-MW(e) LWR may be assumed to have an average yearly requirement for 35 tons of 3.5 wt% ^{235}U enriched uranium over a 30-year lifetime. Using uranium resource estimates for material up to $50/lb and assuming enrichment at 0.2 wt% ^{235}U tails, estimate the number of reactors that can be supported by:
  a. reserves only
  b. reserves plus probable
  c. reserves plus all potential resources
  d. reserves plus potential resources and byproduct

8-9. From the assumptions in Prob. 8-8, estimate the number of reactors that could be supported by the reasonably assured plus estimated additional resources of:
  a. Canada
  b. South Africa
  c. Australia
  d. France

8-10. Calculate the fraction of uranium in U_3O_8, ADU, UF_6, and UO_2.

8-11. What concentration in parts per billion [ppb] would result in 0.025-eV-neutron absorption 1 percent as great as that in *natural* uranium for
  a. boron ($\sigma_a = 760$ b)
  b. cadmium ($\sigma_a = 2450$ b)
  c. samarium ($\sigma_a = 5900$ b)
  d. chlorine ($\sigma_a = 33$ b)

8-12. In an ideal system, the fractional enrichment f for a given process might be expressed by

$$f = \alpha^n$$

for separation factor α and number of stages n. Estimate the minimum number of stages required to enrich natural uranium to 3 wt% and 93 wt%, respectively, for each of the five enrichment technologies. Compare the results for gaseous diffusion to the numbers in the text.

8-13. Estimate the optimum enrichment tails composition for the base case and three options in Prob. 8-7.

8-14. Repeat the calculations in Prob. 1-8 for 0.2 wt% ^{235}U tails assay. Compare the results in terms of fractional savings in uranium ore requirements.

8-15. Repeat the calculations in Prob. 1-8 for 93-wt% ^{235}U product used in the HTGR design.

8-16. Estimate the number of LWRs that could be supplied with low enriched uranium by each country that has commercial enrichment capacity. Compare these results to the projected electrical-generating capacity in 1980 for the same countries. (Assume that 35 t of 3.5 wt% enriched uranium supplies a 1200-MW(e) reactor and that the enrichment tails are 0.3 wt% ^{235}U.)

8-17. Repeat the calculations in Prob. 8-16 for the countries of Western Europe taken as a group.

SELECTED BIBLIOGRAPHY†

Economics

 Elliott & Weaver, 1972 (Chap. 4)
 Gibbs & Hill, 1980
 Graves, 1979 (Chap. 11)
 IEEE, 1979*d*
 Nucl. Eng. Int., Mar. 1975, Nov. 1977, Nov. 1978, Dec. 1979
 Rossin & Rieck, 1978
 Rudisill, 1977
 Sesonske, 1973 (Chaps. 2, 3, 7)

Uranium

 Barnes, 1977
 Benedict, 1981
 Benedict & Pigford, 1957 (Chaps. 4, 6)
 CONAES, 1978a
 Deffeyes & MacGregor, 1980
 Elliott & Weaver, 1972 (Chaps. 5, 6)
 Erdmann, 1979
 Krause, 1977
 Mining Eng., 1974
 Nucl. Eng. Int., Nov. 1978, Dec. 1978, April 1980
 Smith & Valentine, 1979
 WASH-1248, 1974

Enrichment

 Benedict, 1981
 Benedict & Pigford, 1957 (Chaps. 10, 12)
 Davidovitz, 1979
 de la Garza, 1977
 DOE/EDP-0061, 1979
 Elliott & Weaver, 1972 (Chap. 6)
 Mihalka, 1979
 Nucl. Eng. Int., Nov. 1976, Oct. 1980
 Olander, 1978
 Sesonske, 1973 (Chap. 7)

†Full citations are contained in the General Bibliography at the back of the book.

Zaleski, 1976
Zare, 1977

Current Information
  Nucl. Eng. Int.
  Nucl. Ind.
  Nucl. News
  Science
  Sci. Am.

9

REACTOR FUELS

Objectives

After studying this chapter, the reader should be able to:

1. Explain the selection of UO_2 over uranium metal as a reactor fuel.
2. Describe three effects of irradiation which are important to fuel-element design.
3. State the differences in fuel fabrication for mixed-oxide and UO_2 elements.
4. Describe the basic features of the HTGR fuel assembly.
5. Explain the basic principles for multibatch fuel management.
6. Differentiate between SGR and OMR for plutonium utilization.
7. Sketch the basic features of the "full-recycle" mode for an HTGR.
8. Identify the basic features of the design and required testing of an LWR spent-fuel shipping cask.

Prerequisite Concepts

The fuel assemblies for five reference reactors were described in Chap. 1. Each represents the results of a design process which considers economy, reliability, safety, and other important attributes for nuclear reactor systems.

Fuel management procedures are employed to assure high energy production and power capability for a given quantity and composition of fissionable material. Even when spent fuel assemblies are no longer useful for direct energy production, it is possible to recover the residual fissile content for recycle and future use. The large inventory of radioactivity, however, requires special precautions for storage and transportation of spent fuel assemblies.

FUEL–ASSEMBLY DESIGN

The general objectives in the design and engineering of reactor fuel assemblies are to provide

- a geometric arrangement of fissile, fertile, and other materials that can sustain a nuclear chain reaction over a period of several years
- adequate heat transfer and fluid flow characteristics
- failure-free fuel pins that will contain radioactive products over the desired burnup lifetime, through normal and expected transient operations, and under postulated accident conditions
- economy that will help nuclear power be competitive with other energy sources
- fabrication processes for efficient production of standardized, quality-controlled units

Each different type of reactor system meets these and other criteria by somewhat different means.

This section addresses the basic features of fuel-pin and fuel-assembly designs as if they were somewhat isolated from the reactor as a whole. Chapter 11 considers their role as component parts of an integrated system concept.

The earliest reactor fuel elements were made from uranium metal. However, they tended to suffer from unacceptably large dimensional changes as a result of thermal cycling and/or radiation damage. The possible extent of radiation-induced changes is well illustrated by the before-and-after comparison in Fig. 9-1. Alloying materials, which can limit the dimensional instability, are often found to have unacceptably large tendencies toward parasitic neutron absorption (i.e., they have sizeable absorption cross sections), especially for thermal neutrons.

The ceramic uranium dioxide and uranium carbide fuels have emerged as the current favorites for reactor systems. They have satisfactory radiation-damage and fission-product-retention properties up to high burnup levels, are essentially inert to high temperature coolants, and have little poisoning effect in the core.

Oxide Fuels

Although the LWR, CANDU-PHWR, and LMFBR designs differ from each other in many respects, they are similar in their use of small oxide fuel pellets stacked in metal cladding tubes. Their fuel assemblies are also designed under somewhat comparable performance criteria. Typical (interactive) limitations on fuel assemblies include:

- temperatures—preclude melting (e.g., below $2,850 \pm 15°C$ for UO_2), excessive component expansion, fission-product release, and damaging chemical reactions among fuel, clad, and/or coolant

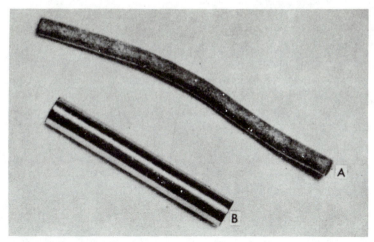

FIGURE 9-1
Irradiation-induced growth of a uranium rod: (*a*) after, and (*b*) before. (From A. N. Holden, *Physical Metallurgy of Uranium,* © 1958, Addison-Wesley, Reading, Massachusetts. Fig. 11-1. Reprinted with permission.)

- cladding stress and strain—balance coolant pressure, differential pellet–clad expansion, pellet swelling, and fission-gas pressure at all times, including provision for the cyclic effects that follow power-level changes and which can cause metal fatigue
- pressure—assure that fission gases do not cause either clad rupture or an excessive increase in the thickness of the gap between the pellet and the clad during normal operations or transients
- clad corrosion—account for thickness reductions and oxide-film formation that occur with burnup and may cause changes that have impacts on the ability to meet other limiting conditions
- fuel densification—balance pellet grain size and porosity to limit formation of gaps in the pellet stack.

Fuel-element design analyses are complicated by the several changes which occur in oxide pellets as a result of neutron irradiation and fissile burnup. Radiation damage produced by the fission fragments changes the pellet microstructure, which in turn modifies the physical and mechanical properties. Noble-gas and oxygen release leads to variations in volume, stoichiometry, and chemical reactivity. A few representative concerns are considered below.

General swelling of fuel pellets tends to occur as a result of the accumulation of both gaseous and solid fission products. The magnitude of the effect appears to be a function of the operating temperature and the initial void content of the fuel. It has also been observed that swelling varies with burnup in a nearly linear manner up to a certain "critical" level, above which sharp increases may occur.

An extreme example of the restructuring that can occur to pellets as a result of high-power operation is shown by Fig. 9-2. These experimental results represent irradiation of LMFBR mixed-oxide (Pu + depleted uranium) fuel in a cosine-shaped flux. The attendant variation in linear heat rates produces different amounts of restructuring. Comparison of the 3.84 kW/m [12.6 kW/ft] sample in Fig. 9-2 with Fig. 9-3

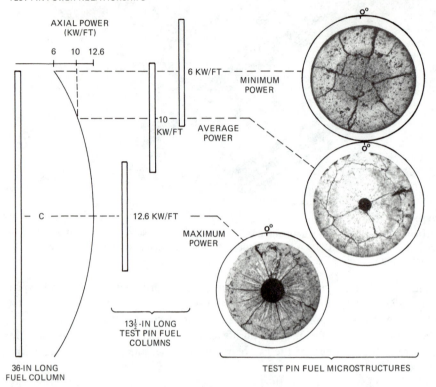

TRANSIENT OVERPOWER TEST SERIES
TEST PIN POWER RELATIONSHIPS

FIGURE 9-2

Mixed-oxide fuel restructuring versus linear heat rate. (Photograph courtesy of the Hanford Engineering Development Laboratory, operated by Westinghouse Hanford Company for the U.S. Department of Energy.)

highlights the most important features of the restructuring. The central void is not a result of center-line melting but is instead caused by the migration of individual pellet voids in the direction of increasing radial fuel temperature. The void is surrounded by a full-density columnar-grain region. The growth of equiaxed grains is a common effect in ceramic materials operated at high temperatures. In the low-temperature region near the clad interface, the original microstructure appears to be maintained. The structures shown in the figures complicate temperature-distribution calculations (especially as compared to the very simple models employed in Chap. 7).

During routine reactor operation, fission product gases are released slowly from the fuel matrix to the central void and the pellet-clad gap. The latter along with any fuel swelling affect the heat transfer between the fuel pellet and the clad. The fission gases that are retained by the fuel are also of concern, since they could be released rapidly by a large transient temperature change, e.g., as produced by a reactivity-insertion accident. If such release compromises the integrity of the fuel pins, it could provide a mechanism for core-wide propagation of damage.

The complexity of the process of fuel-element design is highlighted by the flow chart in Fig. 9-4. Using input data on linear heat rate, coolant temperature, and fuel-pin characteristics, the interactions among various mechanical, metallurgical, and chemical processes must be evaluated. Irradiation-induced effects add a time-dependent component and require substantial repetition of the full range of analyses.

LWR Fuel

Fuel pins for both light-water reactor types consist of short, small-diameter UO_2 fuel pellets stacked into free-standing cladding tubes of a zirconium alloy. The features of a typical pin are shown in Fig. 9-5. The pellets are pressed from powder and loaded into the long cladding tubes. The end plugs are welded in place to seal the fuel pin. The dished ends in the pellets and use of the plenum spring are designed to minimize axial expansion of the pellet column at operating temperatures. The plenum space at the top of the pin provides a volume to accommodate fission-product gases.

The early LWR designs employed stainless-steel cladding primarily because of its relatively low cost and good structural properties. However, it is also a thermal-neutron poison. The development of zircaloy (primarily zirconium alloyed with a small amount of iron and traces of other elements) reduced the problem substantially. It has been found that zirconium clad allows use of about 1 wt % [235]U less enrichment than stainless steel for comparable reactivity. However, the potential for exothermal

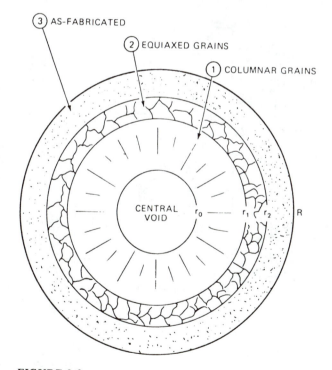

FIGURE 9-3
Features of mixed-oxide fuel restructuring. (From D. R. Olander, *Fundamental Aspects of Nuclear Reactor Fuel Elements*, TID-26711-P1, 1976.)

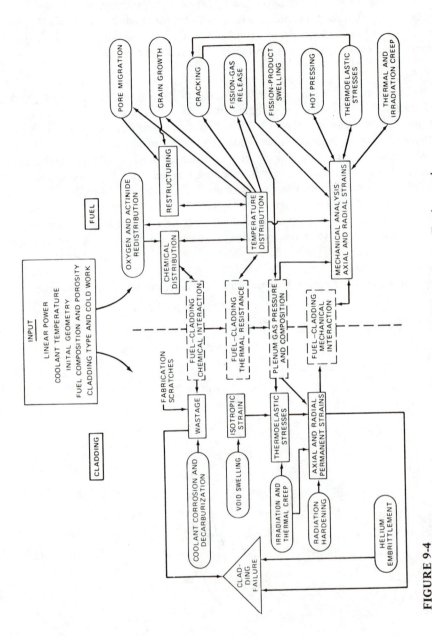

FIGURE 9-4

Flow chart for representative fuel-rod design interactions. (From D. R. Olander, *Fundamental Aspects of Nuclear Reactor Fuel Elements*, TID-26711-P1, 1976.)

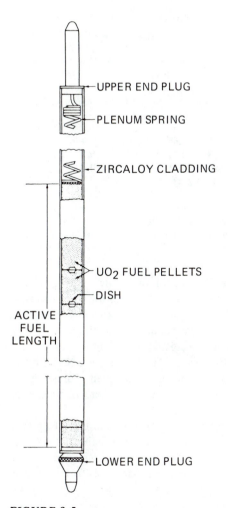

UPPER END PLUG

PLENUM SPRING

ZIRCALOY CLADDING

UO$_2$ FUEL PELLETS

DISH

ACTIVE
FUEL
LENGTH

LOWER END PLUG

FIGURE 9-5
Typical LWR fuel pin. (H. E. Williamson and D. C. Ditmore, *Reactor Technology*, vol. 14, no. 1, 1971.)

zirconium–water reactions is a drawback from a reactor safety standpoint (as considered in Chap. 16).

Although the PWR and BWR employ conceptually similar fuel-pin designs, there are variations related to overall reactor system differences. The second and fourth columns of Table 9-1 show, for example, that rod diameter and clad thickness are greater and power density is lower for current BWR designs compared to their PWR counterparts. The table also contains data on the most recent design revisions in which the array number has been increased for each system. Since the overall outer dimensions of the fuel assemblies remain unchanged, the result is lower average linear heat rates at comparable or larger average power densities. The reduced linear heat rates generally enhance overall power capability (as inferred from the discussions in Chap. 7).

TABLE 9-1
Typical Comparative Data for Recent Fuel Design Modifications in PWR, BWR, and CANDU Systems

	PWR		BWR		CANDU	
Design parameter	15 × 15	17 × 17	7 × 7	8 × 8	28-pin	37-pin
Rod diameter (mm)	10.7	9.50	14.3	12.5	15.2	13.1
Active fuel height (m)	3.66	3.66	3.66	3.66	0.475	0.475
Clad thickness (mm)	0.61	0.58	0.81	0.86	0.38	0.38
Pellet-clad diametrical gap (mm)	0.19	0.17	0.28	0.23	0.089	0.089
Average linear heat rate (kW/m)[†]	23.1	17.8	23.3	19.8	26.5	25.7
Average power density (kW/1)[‡]	106	105	51	56	85.2	109

[†]Calculated from core thermal power and total linear meters of fuel.
[‡]Calculated from core thermal power and active core volume (for CANDU, volume is that for pressure tubes only).

One interesting example of the complex interaction of fuel design parameters is related to fuel densification and "clad creepdown" problems identified in some early zircaloy-clad PWR fuel pins. The pellets, which had been pressed to about 93 percent of their theoretical density, were found to densify as initial irradiation caused their pores to collapse. As the volume of the pellets decreased, gaps formed in the pellet stack. The cladding then crept down into the gaps under the driving force provided by the high pressure ($\approx$ 15.5 MPa or 2250 psi) coolant. Subsequent failure of the clad led to release into the primary coolant of gaseous and other volatile fission products.

The densification-creepdown phenomenon has had a profound impact on fuel-rod design. Pellets are now pressed to higher initial density and with carefully controlled porosity. An inert-gas backfill is employed in the clad tube to resist the external coolant pressure. Since this backfill pressure will later be enhanced as fission product gases are released, a careful balance of design parameters is required to minimize the likelihood of clad failure by overpressurization.

The next major challenge in LWR fuel design appears to be related to the desire to extend burnup. Especially when recycle is precluded (as is now the case in the United States and elsewhere), there is a strong incentive to enhance energy output per unit of uranium resource. There will be a place for many innovations in this area.

CANDU-PHWR Fuel

Fuel for the CANDU pressurized-heavy-water reactor is similar to that of the LWR in its use of UO_2 pellets and zircaloy clad. However, there are also notable differences, some of which are identified in Table 9-1.

The CANDU assemblies are slightly less than 0.5 m in length to be consistent with on-line refueling. This short length does not require that the cladding tubes be free-standing like those for the LWR. In fact, the clad is quite thin and is allowed to creep down onto the pellets. Zircaloy bearing pads are employed to separate fuel pins from each other and from the pressure tube wall (as shown by Fig. 1-9). A combination of low burnup for the natural uranium fuel and the short length of the fuel pins limits fission gas production and the need for the plenum space designed into LWR pins (e.g., as shown in Fig. 9-5).

Like its LWR counterparts, the CANDU fuel assembly has evolved over a period of time. The most recent change noted in Table 9-1 is the use of 37 small-diameter pins instead of 28 larger ones. By this means, the system can be operated at a substantially greater average power density while the average linear heat rate (and power peaking) is slightly lower.

LMFBR Fuel

Current liquid-metal-fast-breeder reactor designs are predicated on the use of 15–20 wt % PuO_2 mixed with depleted UO_2. Proposed operating conditions are much more severe for the LMFBR than for LWR systems, as shown in Table 9-2. Burnup, specific power, fast flux and fluence, operating temperatures, and fission gas release all contribute to tougher fuel-pin design requirements.

The fuel pin for the Fast Flux Test Facility [FFTF] shown in Fig. 9-6 is typical of LMFBR designs. Small-diameter mixed-oxide pellets are loaded into stainless steel cladding tubes. The pellets are about 90 percent of theoretical density to accommodate large fission-product volumes without excessive swelling. (The densification problem with the PWR is not a concern here because the sodium coolant is maintained at low pressure.) Stainless steel has good strength and compatibility with sodium coolant. It also has a minimal neutron poisoning effect for the fast neutrons in the LMFBR design.

Figure 9-6 shows two other interesting features of the FFTF fuel pin. The wrap wire is used to separate the pins by a sufficient distance to allow for sodium coolant flow without providing for excessive moderation of fast neutrons. The tag-gas capsule contains xenon and krypton in an isotopic ratio that is the same for each fuel pin in a given assembly but which differs for each fuel assembly. After a pin is welded shut, the capsule is broken to distribute its contents throughout the pin. Fuel assemblies which develop leaking pins during operation can often be identified by analyzing the Kr–Xe isotopic ratio in the cover gas above the sodium coolant level.

Although the use of mixed-oxide fuel pellets is common to nearly all LMFBR designs, one exception is worthy of note. The Experimental Breeder Reactor [EBR-2] used a uranium–plutonium–fission alloy clad in stainless steel during the early 1960s.

TABLE 9-2
Comparison of LWR and LMFBR Fuel-Design Targets[†]

Parameter	LWR	LMFBR
Maximum burnup, MWD/T	30,000	100,000
Specific power, kW(th)/kg	25–35	170–200
Fast (> 0.1 MeV) flux, $n/cm^2 \cdot s$	2×10^{14}	5×10^{15}
Clad fast (> 0.1 MeV) fluence, n/cm^2	10^{22}	3×10^{23}
Coolant outlet temperature, °C	320	570
Maximum clad temperature, °C	340	660
Fission gas release during operation, %	5	80

[†]Adapted from J. Weisman and L. Eckart, "Fuel Rod Design," NFCEC-1, CES-S1, 1978.

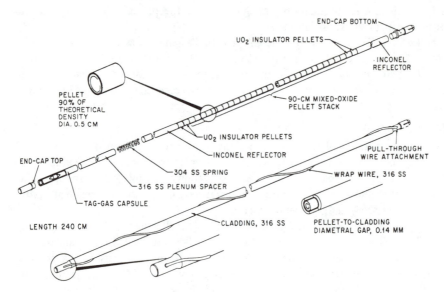

FIGURE 9-6

Fuel pin employed in the Fast Flux Test Facility [FFTF]. (Courtesy of the Hanford Engineering Development Laboratory, operated by Westinghouse Hanford Company for the U.S. Department of Energy.)

The fissium, or fission-product, content of the fuel resulted from incomplete reprocessing of spent fuel. At the end of each burnup period, fuel assemblies were transferred to a nearby fuel-cycle facility where pyrometallurgical techniques were used remotely to remove some of the fission products (leaving behind some of the Zr, Mo, Tc, Ru, Rh, and Pa) and to reconstitute the alloy for use in a new fuel assembly. Closed-cycle operations like this are of particular current interest since a material safeguards advantage is often perceived when pure, separated plutonium is not present during any processing step (see Chap. 20).

Carbide Fuels

Uranium carbide has received some consideration as an alternative to uranium dioxide for LWR and LMFBR fuels. However, most development activities have been related to the microspheres employed in the high-temperature gas-cooled reactor.

The HTGR fuel particles have been developed for high thermal and radiation stability. Two basic particles are shown in Fig. 9-7. The fissile particle—often called TRISO because it has three different coating types—is characterized by:

1. a highly enriched [93 wt % ^{235}U] uranium carbide center
2. a buffer layer of pyrolytic graphite to stop fission fragments, provide fission gas volume, and limit swelling
3. an inner pyrolytic graphite layer to limit fission product migration
4. a silicon carbide barrier coating for overall particle strength, for stopping migration of certain solid fission products like barium and strontium, and to implement physical separation of particle types during reprocessing (as considered in the next chapter)

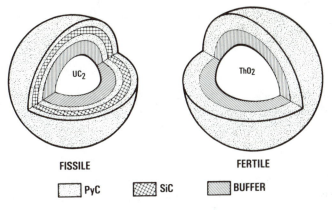

FIGURE 9-7
Fissile and fertile microsphere fuel particles for the high-temperature gas-cooled reactor [HTGR].
(Courtesy of the General Atomic Company.)

5. an outer pyrolytic graphite layer to protect the somewhat brittle SiC coating and to
   hold the overall microsphere together via radiation-induced dimensional changes

The fertile BISO particle consists of a thorium oxide center plus buffer and pyrolytic
graphite layers. Its simplicity may be attributed largely to the low fission rate expected
in the bred ^{233} U.

The effects of irradiation on TRISO-coated particles (similar to those of carbide
used in the HTGR) are shown in Fig. 9-8. Although density and shape changes are

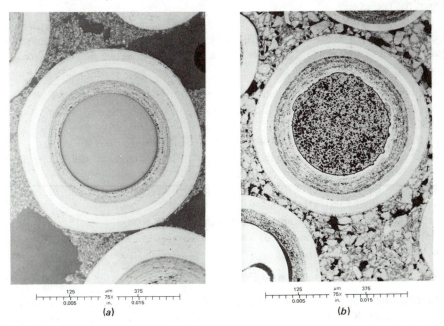

FIGURE 9-8
TRISO-coated $(4Th, O)_2$ fuel particles: (*a*) unirradiated, and (*b*) irradiated to 12 percent burnup
at $1000°C$. Magnification 75×. (Courtesy of Oak Ridge National Laboratory, operated by the
Union Carbide Corporation for the U.S. Department of Energy.)

noted, the microspheres are found to be effective at preventing release of fuel and fission products under a substantial range of thermal stress and radiation damage.

FABRICATION

An important basis for both the pellet-clad and microsphere fuel designs is ease of fabrication. All of the reference reactors, thus, have fuel assemblies that can be highly standardized and quality controlled at a reasonable cost.

Oxide Fuels

All of the reference reactors which use UO_2 or mixed-oxide [sometimes called MO_2 or MOX] pellets employ similar fuel fabrication techniques. The most substantial difference occurs when plutonium is present and many operations must be conducted in glove boxes to protect personnel from its chemical and radiological toxicity. (Since plutonium and its daughter products do not emit highly penetrating radiations, heavy shielding and/or remote operations are generally *not* required.)

The first step of the fabrication process is production of UO_2 powder. This may be done at a separate facility or at the fabrication plant itself. In either case, slightly enriched uranium for LWR and depleted uranium for LMFBR use generally enter in the form of UF_6 from gaseous-diffusion or gas-centrifuge enrichment. Through a series of operations, the solid UF_6 is sublimed to its gaseous form, reacted with a variety of chemical agents, and ultimately converted to dry UO_2 powder. The natural UO_2 employed in CANDU fuel is received in this form directly from the milling-purification step of the fuel cycle.

The flow diagram in Fig. 9-9 is characteristic of mixed-oxide fuel fabrication operations. With deletion of the initial PuO_2 stream, it also has many of the same features as all of the standard UO_2 operations. Important steps in a generic pelletizing process include:

- calcining [or "baking"] plutonium nitrate solution to form PuO_2 powder
- blending of the depleted UO_2 and PuO_2 to a uniform mixture and milling it to a uniform consistency
- adding binder materials which serve to control the final porosity of the pellets
- drying, screening, and dry-pressing into green pellets (named because UO_2 pellets have this color and are also "uncured")
- heating at low temperature to remove binder additives and at high temperature to sinter the pellets to their final density
- centerless grinding to final dimensions (typically ± 0.01 mm)
- inspecting by visual, dimensional, and chemical techniques to eliminate unacceptable pellets

The assembly of fuel pins then proceeds by:

- forming the pellets into stacks, weighing and recording their content, and vacuum outgassing at elevated temperatures to remove moisture and organic contaminants
- loading of the dry pellet stack plus internal hardware (if any) into a clean, dry cladding tube which has one end cap welded in place

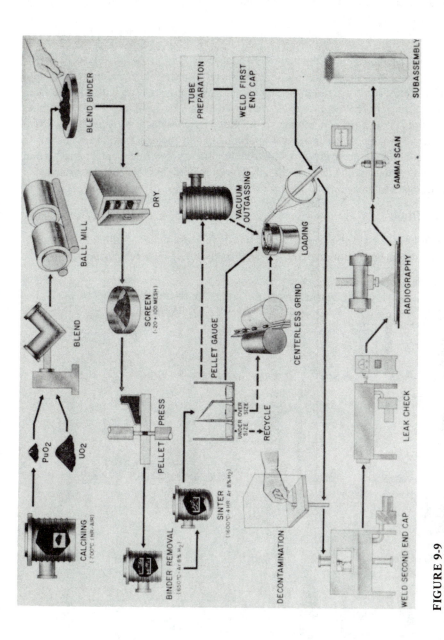

FIGURE 9-9

Mixed-oxide fuel fabrication steps (Adapted from Education and Research in the Nuclear Fuel Cycle, edited by David M. Elliot and Lynn E. Weaver. Copyright 1972 by the University of Oklahoma Press.)

- inspecting and decontaminating the loaded tubes
- welding second end cap after evacuation and inert-gas backfill (at atmospheric or higher pressure, as desired)
- leak testing, radiography, nondestructive assay of fissile loading, and inspection of welds

Following some combination of dimensional inspections, cleaning, etching, and autoclaving (to form a corrosion-resistant oxide film by treatment with high-pressure steam) the pins are formed into their fuel assemblies. The BWR and PWR assemblies have fuel-pin arrays formed by use of retention grids (which are sometimes called "egg crates" because of their general appearance) and the other hardware shown by Figs. 1-7 and 1-8, respectively. CANDU fuel bundles employ interelement spacers and end support plates, as shown by Fig. 1-9. The wrap wires on LMFBR pins provide for the tightly packed hexagonal geometry of Fig. 1-11.

Carbide Fuels

As would be expected, the fabrication of HTGR fuel assemblies is quite different from that considered above for the conventional oxide fuels. Figure 9-10 shows the basic

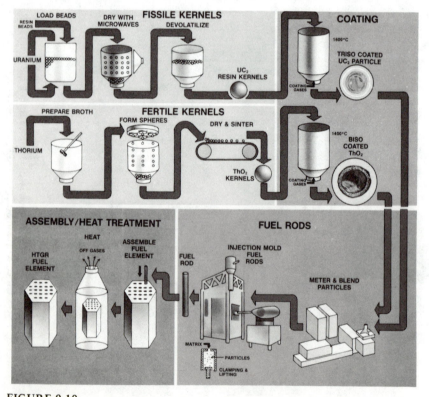

FIGURE 9-10
Fuel fabrication process for high-temperature gas-cooled reactor [HTGR] fuel. (Courtesy of General Atomic Company.)

features of one manufacturing process. A number of steps depend on the use of *fluidized bed* technology, where droplets or particles are suspended for relatively long periods of time by carefully adjusted upflow of gases. During such suspension, drying and/or coating is facilitated.

The first step of the manufacturing process for the fissile TRISO particles is loading uranium nitrate solution into weak-acid resin spheres, as shown by Fig. 9-10. The resulting microspheres are then dried and sintered to produce high-density UC_2 kernels. The various carbon and silicon carbide coatings are applied by controlled fluidized-bed treatment with hydrocarbon and silane gases, respectively.

The fertile BISO particles are produced by preparing a thorium hydrosol [sol] broth from ThO_2 powder and dilute nitric acid. Sol droplets from a vibrating nozzle are converted to gelatin [gel] spheres by extracting the water content in a column which contains ammonia gas. After a screening process is employed to select spheres of the proper size, drying and sintering in a high-temperature furnace produce the final ThO_2 kernels. The two graphite coatings are applied by a fluidized-bed process.

Finished BISO and TRISO particles are metered at the desired ratio and blended with a graphite shim. The mixture is then formed into a packed bed for injection molding with a carbonaceous binder, as shown in Fig. 9-10. The rods are loaded into machined hexagonal graphite blocks of the type depicted by Fig. 1-10. The fuel assembly is heated in a high-temperature furnace to remove volatile components of the fuel rod binder and to convert the binder residue to a carbon char matrix. The cured fuel element is complete after it has been fitted with graphite positioning dowels.

General Considerations

The high economic value of fuel assemblies and their constituent parts leads to requirements for careful handling during all processing steps. Finished assemblies are cleaned and packaged with protective covers for temporary on-site storage. Shipment is accomplished by using specially-designed, shock-mounted casks, each of which holds one or more fuel assemblies. Individual or multiple casks are generally transported by truck to the appropriate reactor site.

Unirradiated uranium fuel constitutes minimal radiological and chemical hazards. Plutonium, however, must be isolated from contact with operating personnel. This is most readily accomplished by using sealed glove boxes, which, as the name implies, allow workers to handle the material with specially designed gloves. Devices and procedures for material entry and removal are also provided. All mixed-oxide operations up to and including final fuel-pin welding require the use of glove boxes. Once the pins are sealed and have been certified as free of contamination, they may be handled in the open like their uranium counterparts.

If the fuel contains highly active radionuclides that emit penetrating radiations, glove boxes are inadequate. Remote operations behind heavy shielding may be necessary. Although procedures and equipment are well developed for such operations, their cost and inconvenience suggest that remote activities be a last resort. The HTGR fuel fabrication processes have been developed to be fairly ammenable to remote operations at such time as ^{233}U recycle (with the attendant ^{232}U contamination noted in Chap. 6) becomes desirable. Use of ^{233}U or fission-product-contaminated plutonium in any of the other reference reactor designs would require substantial modification of existing processes.

All facilities which handle plutonium or other highly radioactive materials must take steps to limit radiation exposures and radionuclide releases to as-low-as-reasonably-achievable [ALARA] levels. Careful design of ventilation and filtration systems provides the basis for one important control mechanism. A valuable adjunct is routine monitoring by radiation-safety personnel and by remote sensors.

Criticality safety is applied to all fabrication operations except those for the natural-uranium CANDU fuel. In slightly enriched uranium processes, exclusion of moderator assures subcriticality. Plutonium and highly enriched uranium pose more difficult problems that are dealt with by employing geometrically favorable equipment and stringent administrative procedures.

UTILIZATION

The basic goals of fuel utilization in a nuclear power reactor are:

1. a "flat" power distribution to maximize power capability
2. maximum burnup from a minimum amount of fuel
3. minimum fuel cycle costs

Although the goals cannot be met completely and simultaneously, judicious in-core fuel management provides viable compromises. Recycle of the residual fissile content of spent fuel may also enter into the balance.

In-Core Fuel Management

If all fuel for a given reactor were of the same enrichment, and loaded and unloaded as a single batch, power peaking would be relatively high and the system would suffer from uneven, relatively low burnup. These factors would result in an excessively low power capability and great diseconomy in fuel utilization, respectively.

Current operating practice is based on *multiple-batch fuel management* with several (typically three or four) initial or effective enrichments. Careful loading by enrichment results in substantial flattening of the power distribution. By discharging one irradiated batch and charging a fresh batch, the fuel that remains in the core can be driven to relatively higher burnup. Fuel cycle cost, fuel element performance, and time-dependent power capability all contribute to determination of the enrichment and target burnup for the individual fuel batches. Use of a limited number of fuel enrichments reduces both costs and quality-assurance requirements.

Fuel management in PWR systems employ three enrichment batches. A typical initial core loading is shown in Fig. 9-11. The highest enrichment at the core periphery and the remaining batches checkerboarded in the central region maintain power peaking at a relatively low level. At the end of the first cycle (when the core runs out of reactivity), the lowest enrichment batch is discharged, the remaining two "effective enrichment" batches are arranged in the center, and a fresh batch is loaded at the core periphery. Similar procedures are conducted on a roughly annual basis. After some period of time, it may be possible to establish an "equilibrium" loading pattern like that shown in Fig. 9-11. This situation would be characterized by use of the same fresh-fuel enrichment and the same overall pattern for all successive cycles. In practice, an equilibrium cycle is somewhat unlikely as plant operating conditions tend to change from year to year.

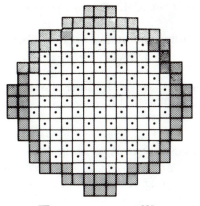

☐ BATCH A (2.0 wt % ^{235}U)

☐ BATCH B (2.5 wt % ^{235}U)

■ BATCH C (3.0 wt % ^{235}U)

(a)

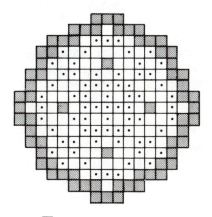

☐ TWICE-BURNED FUEL

☐ ONCE-BURNED FUEL

■ FRESH FUEL

(b)

FIGURE 9-11
Representative fuel management patterns for (a) initial and (b) equilibrium PWR cores.

PWR fuel management employs both "shuffling" and rotation of irradiated assemblies to minimize power peaks. In addition, control-rod patterns and temperature coefficients of reactivity must be reanalyzed for each cycle.

The BWR uses a four-batch scheme with the fresh fuel loaded at the periphery of the core and the other three batches located inside in a "scatter" pattern. A typical fuel batch starts at the outside, replaces an interior batch after its first cycle, remains in the same location for three cycles, and then is discharged. These procedures are somewhat simpler than those for the PWR since the BWR has smaller fuel assemblies and more flexibility for controlling power shape with control rods and recirculation flow rates (the latter as described in Chap. 12).

Fuel management for the HTGR is substantially more flexible than for LWRs. Effective batch enrichments can be changed merely by adjusting the ratio of fissile TRISO and fertile BISO particles in fuel assemblies. Four-batch fuel management in the radial direction has similarities to that for the LWR designs. However, the HTGR has up to eight assemblies stacked vertically which allow axial fuel zones and essentially three-dimensional power-shaping capability.

The principles of LMFBR core fuel management are the same as for other systems. However, since the plutonium production rate offsets much of the depletion effect, the power shape remains relatively stable without a great deal of assembly shuffling. Maximizing burnup and minimizing power peaks at the core–blanket interface are major objectives of the fuel management.

Management of the LMFBR blanket elements is more challenging. Since the axial blanket material is an integral part of the core fuel assemblies, no real flexibility is available. The radial blanket, however, consists of separate assemblies which usually contain large-diameter rods of depleted uranium in a tightly packed lattice. Shuffling and rotation of elements are used to minimize uneven plutonium production associated with the steep neutron-flux gradients in the blanket region. Since the fission rates are quite low in these blanket assemblies, their residence time in the system may be substantially longer than that of the core fuel.

The CANDU reactor has ultimate fuel management flexibility through on-line refueling. Although all bundles are of the same initial (natural U) composition, use with a computer inventory system allows power shapes and bundle burnups to be optimized at the same time excess core reactivity is maintained at a low level.

Refueling

Implementation of fuel management is tied to fuel loading and reloading procedures in all of the reference reactors. Fresh-fuel and spent-fuel storage areas are usually located in a separate fuel handling building. One or more remotely operated transfer systems carry fuel assemblies from the storage areas, through the containment building, and to the reactor core, and vice versa.

The basic features of the CANDU fuel-handling facilities and sequences are shown by Fig. 9-12. Fresh fuel is transferred from its storage area through an equipment lock in the reactor containment building. It is then loaded into the "charge" portion of the fueling machine. The two portions of the machine attach to opposite ends of the same pressure tube without disrupting the flow of heavy-water coolant or reactor energy production. As the charge-machine inserts a fresh fuel bundle into the channel, a spent bundle is pushed through in the direction of the coolant flow to the accept-machine. The two portions of the fueling machine are identical, allowing the procedure to be implemented in either direction (since the coolant is *oppositely* directed in *adjacent* channels).

Spent CANDU fuel from the fueling machine is transferred into a discharge room through a specially designed port as shown in Fig. 9-12. After failed or leaking bundles are packaged into overpack cans, the spent fuel travels under water through the transfer canal to the service-building storage bays.

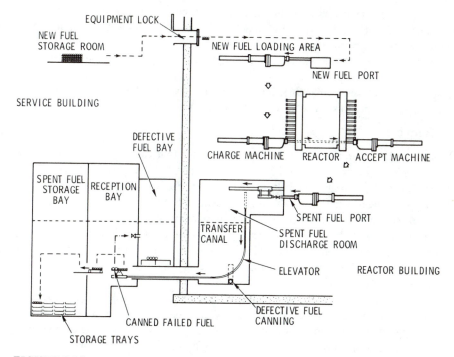

FIGURE 9-12

CANDU fuel-handling sequence. (Courtesy of Atomic Energy of Canada, Limited.)

Refueling of the other reference reactors requires that the systems be shut down and allowed to cool (i.e., let fission products decay) for some period of time. Then the reactor vessel must be opened to change fuel assemblies (e.g., see Chaps. 12-15).

The two LWR designs require a few days of fuel cooling time before the reactor vessel can be opened by loosening the head assembly. The reactor cavity is flooded with water to a height of 10 m or more above the core before the vessel head and then fuel assemblies are removed. The depth of water allows the assemblies to be fully withdrawn from the vessel while still having enough water above them to provide radiation shielding for operating personnel. A spent assembly is extracted from the core and moved to the edge of the cavity by an overhead crane, moved to a more horizontal position, removed from the containment building through the water-filled transport tube, returned to a vertical position, and placed into a spent-fuel storage rack. Other core assemblies are shuffled to new locations. Fresh assemblies are brought from the fuel-handling building by reversing the procedures used for removal of spent fuel. Typical LWR refueling outages are 15-20 days in length, of which only 20-30 percent of the time is actually involved in fuel movement operations. The remaining time is tied up with cooldown and advance preparations, maintenance, reassembly, and preoperational testing.

HTGR refueling is accomplished remotely from the control room after the vessel has been depressurized. A fuel-handling machine enters the reactor vessel through a refueling penetration and lifts one fuel block into a transfer cask. The cask then transfers the spent-fuel assembly to a storage well in the fuel building. The process is repeated until each block in an axial stack has been removed. Introduction of fresh fuel reverses the procedures.

Refueling of LMFBR systems is accomplished remotely in a sodium environment. After initial cooldown, an in-vessel transfer machine lifts a spent assembly from the core to an ex-vessel machine. The latter then carries the assembly through a sodium-filled channel to storage in the fuel-handling building.

Fuel Recycle

All discharged reactor fuel has some fissile inventory that can be recycled to reduce overall uranium resource and enrichment requirements. It has been estimated that full uranium and plutonium recycle in LWR's, for example, could cut their yellow cake and enrichment demands by about 25 percent and 18 percent, respectively.

The economy of recycle depends on cost comparisons between current fuel cycle operations and those for reprocessing and recycle fabrication. Large required expenditures for safety- and safeguards-related systems tend to make recycle less attractive. Rising yellow cake and enrichment charges, by contrast, lead to more favorable economics.

Since federal policy precludes commercial reprocessing, the only current recycle operations in the United States are experimental in nature. France, however, is pursuing an aggressive plutonium recycle program for both LWR and LMFBR systems. Other countries also view recycle favorably (as noted by their commitment to reprocessing considered in the next chapter), although their long-term plans are less developed.

LWR Considerations

Fissile mass flows for typical BWR and PWR initial and replacement (nonrecycle) cores are summarized in Table 9-3. The ^{235}U content of the discharge fuel may be

TABLE 9-3
LWR Fuel Utilization Parameters[†]

Parameter	Boiling-water reactor	Pressurized-water reactor
Thermal efficiency, %	32.5	32.5
Capacity factor, %	85	85
Specific power, MW(th)/t U	22	34
Initial core, average		
Irradiation level, MW(th)·d/t U	21,000	26,000
Fresh-fuel assay, wt % ^{235}U	2.2	2.8
Spent-fuel assay, wt % ^{235}U	0.8	0.9
Fissile plutonium discharged, kg/t U	4.9	5.7
Replacement loadings, typical		
Irradiation level, MW(th)·d/t U	28,000	32,000
Fresh-fuel assay, wt % ^{235}U	2.6	3.4
Spent-fuel assay, wt % ^{235}U	0.8	0.9
Fissile plutonium discharged, kg/t U	5.4	7.4

[†]From A. Sesonske, *Nuclear Power Plant Design Analysis*, TID-26241, 1973.

noted to exceed that of natural uranium for all cases. The plutonium content, equivalent to 0.5–0.7 wt % of the heavy metal, is an isotopic mixture as described in Chap. 6.

Uranium recycle still has an unresolved problem related to the buildup of ^{236}U due to nonfission capture by ^{235}U. Since reenrichment on a continuous-process basis would cause some contamination of all material in the system, it may be necessary to establish a "fissile-equivalent" algorithm for assigning value to material of off-normal composition. Another possible option is use of a dedicated enrichment plant (perhaps gas-centrifuge or, eventually, laser) for all uranium recycle.

Plutonium may be employed in LWR's in two general ways. In the *self-generated recycle* [SGR] mode, a reactor uses only the mass of plutonium it produces. The *open-market recycle* [OMR] mode considers any amount of plutonium from a small quantity to a full core. The SGR option allows reactor fuel management based on a fixed-fraction plutonium inventory. For OMR operations, the system must be designed with enough flexibility to accommodate anywhere from zero to 100 percent plutonium as the fissile content of the core.

The fissile content of recycle fuel batches is most likely to include both plutonium and slightly enriched uranium. One fabrication option is to mix the two homogeneously so that each fuel pin has the same composition. This has an advantage in terms of power peaking, but would, of course, mean that glove-box operations would be necessary for each pin. The other option—to produce separate uranium and mixed-oxide [PuO_2 plus depleted UO_2] pins—reduces the need for glove-box operations. However, it also leads to some burnup-dependent mismatches between the pin types that can result in power peaking problems. Such problems are potentially greatest for separate uranium and plutonium assemblies, but also exist in the "small-island" concept where each assembly contains some plutonium pins. Overall economic considerations—mainly fabrication costs versus power capability—ultimately serve as the basis for selecting among the procedures.

HTGR Recycle

The basic HTGR design has many recycle options for uranium and thorium, some of which are represented on Fig. 9-13. Four possible (arbitrarily-named) cycles are:

1. Nonrecycle—Highly enriched uranium is used in TRISO fuel for each cycle with discharged ^{235}U and ^{233}U stored or sold.
2. Full recycle—All uranium is recycled in the BISO particle along with the thorium, and makeup ^{235}U is fabricated into TRISO particles.
3. Type I segregation—Bred ^{233}U is recycled with makeup ^{235}U in TRISO particles, but discharged ^{235}U is not recycled.
4. Type II segregation—Makeup ^{235}U is in TRISO particles with thorium and recycled ^{233}U in BISO (or other separate) particles.

The highly enriched uranium is recycled once or not at all, based on considerations shown by Table 9-4. The initial ^{235}U content is depleted by 92 percent and 99 percent during first and second cycles, respectively. The very large ^{236}U content (with its reactivity penalty) precludes reenrichment or use past the second cycle.

Thorium is not likely to be recycled directly because it contains the short

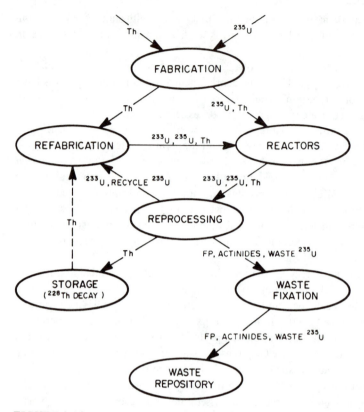

FIGURE 9-13
HTGR recycle options. (Courtesy of Oak Ridge National Laboratory, operated by the Union Carbide Corporation for the U.S. Department of Energy.)

half-lived ^{228}Th isotope from the ^{232}U decay chain (of Fig. 6-4). Instead, it may be stored for later use as indicated on Fig. 9-13.

There is very little plutonium produced from the highly enriched uranium feed in the reference HTGR design. However, it is quite feasible to use a lower enrichment so that both plutonium and ^{233}U would be generated. This, of course, would compli-

TABLE 9-4
Composition of Discharged HTGR Uranium Feed Material[†]

Isotope	Original composition		Composition after one 4-year cycle		Composition after two 4-year cycles	
	g/kg U	%	g/kg U	%	g/kg U	%
U-234	10	1.0	5	1.9	2	1.2
U-235	930	93.0	75	28.6	7	4.3
U-236	2	0.2	133	50.8	118	72.4
U-238	58	5.8	49	18.7	36	22.1

[†]Adapted from R. C. Dahlberg et al., "HTGR Fuel and Fuel Cycle Summary Description," GA-A12801 (Rev.), January 1974.

cate fuel-cycle operations (not to mention Fig. 9-13) by introducing a new material stream. Whether such options will be employed is dependent on the viability of both the recycle operations and the reactor concept itself.

Other Systems

The plutonium inventory in spent natural-uranium CANDU fuel bundles is relatively low and near-term recycle is unlikely. However, utilization of either plutonium or ^{233}U could be implemented at almost any time because of the great flexibility afforded by on-line refueling.

The reference LMFBR design depends entirely on plutonium recycle. It is, thus, viable only with commercial reprocessing of spent fuel. Further, wide-spread utilization may also require initial-core loadings from LWR sources.

SPENT FUEL

Storage and transportation of spent fuel assemblies are important components of the nuclear fuel cycle, whether or not reprocessing and recycle occur. Interim storage at the reactor site allows the assemblies to cool prior to reprocessing or disposal. As on-site storage space is filled, spent fuel must be shipped to away-from-reactor-storage or disposal sites. Reprocessing and recycle would also require spent-fuel shipment.

Storage

Most of the world's spent fuel is currently stored in water basins at LWR or CANDU sites. Water is a particularly convenient medium because it is inexpensive, can cool by natural convection, and provides shielding and visibility at the same time. LWR fuel assemblies are stored vertically in a fixed-lattice structure. CANDU fuel is placed horizontally in trays which reside in the high-capacity storage bay shown in Fig. 9-12.

Dry storage wells with external water cooling have been designed for HTGR fuel blocks. It is intended that LMFBR fuel assemblies will be stored in liquid-soidum-filled basins.

High-Density Storage

Since CANDU fuel bundles have relatively low burnup and cannot be critical in ordinary water, high-density storage poses no special problems. Substantial storage capacity (up to enough for the entire lifetime) has been designed into each reactor because reprocessing and recycle have not been considered to be near-term options.

Spent LWR assemblies have traditionally been stored in lattice structures that maintain adequate spacing to allow for natural-convection cooling and criticality prevention. The pools were generally designed to accommodate a limited number of discharge batches in anticipation of impending recycle. Since this is apparently not to occur in the foreseeable future, many of the operating reactors in the United States are finding it necessary to increase storage capacity. Tight-packed storage is most readily achieved by employing neutron poisons which are integral to the structures. Some amount of forced cooling may also be necessary for fuel assemblies with short decay times.

Away-From-Reactor Storage

An alternative or adjunct to increasing on-site spent-fuel storage capacity is the use of away-from-reactor [AFR] facilities. When intended for interim storage, existing and proposed AFR operations have many similarities to the present at-reactor water basins. Long-term and/or waste-disposal concepts may differ substantially (as considered in the next chapter).

One away-from-reactor storage facility is currently operated by the General Electric Company at its Morris, Illinois site. The facility has a capacity of about 750 t or 20–25 discharge batches. All fuel assemblies are received in specially designed rail casks (as described shortly). The cask is moved from the rail car into an air-locked room where it is decontaminated as necessary. The "clean" cask is transferred to an underwater storage area where the fuel assemblies are removed and placed into a specially-designed "basket" (one type holds nine BWR assemblies, the other four PWR assemblies). The baskets are then transferred into the fuel-storage area where they are latched into place in a manner that retains the fuel in a vertical position.

Additional AFR storage would be provided by a proposed expansion of 1100 t at the Morris facility. It is also possible that the AGNS/Barnwell reprocessing plant may be used in a similar manner. A federally built facility is also possible as an adjunct to some of the waste-management policies described in the next chapter.

Transportation

Until they have been irradiated in a reactor, fuel cycle materials do not pose any serious transportation hazards. This is not true for the spent fuel, which generally contains sizeable quantities of plutonium and is "hot" both thermally and radiologically. Shipments of spent fuel are, thus, subject to very stringent domestic and/or international regulations.

Regulations

Domestic responsibility for the safe shipment of all commercial nuclear materials is shared by the Nuclear Regulatory Commission [NRC] and the Department of Transportation [DOT]. DOT establishes general regulations for packaging, loading, and operation of vehicles, as well as for inspection, shipping documents, and accident reporting. The NRC sets packaging standards for highly-radioactive materials. It also recommends performance standards related to the potential for adverse effects on operating personnel and/or the general public due to radiation or release of radioactive material. Currently accepted international regulations have many similarities to their domestic counterparts.

Packages are generally divided into two categories from a radiological standpoint:

1. Class A—"small" quantities which if released directly to the environment would have minimal impact
2. Class B—non-class A package which must be designed for containment under (defined) normal and accident conditions

As provided by U.S. regulations, class B packages, including spent-fuel transport containers, must be able to withstand all normally anticipated transport contingencies with:

- no leakage of material
- radiation levels < 200 mr/h at the surface and 10 mr/h at 1 m
- no criticality
- surface temperature < 80°C

Each container type must also be shown to be capable of withstanding the following accident sequence with most of the shielding still intact and with only limited escape of coolant and/or inert radioactive gases:

- free fall of 10 m on a flat, unyielding horizontal surface (an approximation to a 95 km/h [60 mi/h] vehicle crash)
- puncture test with a 100-cm drop onto a 15-cm-diameter bar located to maximize damage
- thermal test with a temperature of ≈ 800°C for 30 min (an approximation to a gasoline or kerosene fire)
- water immersion at a depth of > 1 m for 8 h

Spent-Fuel Casks

Casks for transporting spent-fuel assemblies must be designed to accommodate both high radiation levels and large decay-heat loads while meeting all necessary regulatory requirements. One such container for LWR fuel is the General Electric IF-300, shown separately and rail-car mounted in Fig. 9-14. Its important components are:

- a fuel basket holding seven PWR assemblies or 18 BWR assemblies
- a stainless-steel cylinder sealed with a closure head
- depleted-uranium or lead shielding material
- an outer stainless-steel shell
- a corrugated stainless-steel outer jacket holding neutron shielding (usually water)
- fins to aid forced cooling (if required) and minimize impact damage

The basic features of a simpler shipping cask for six HTGR fuel assemblies are shown in Fig. 9-15. Since the assemblies are expected to be stored for at least 6 months between discharge and transport, and since they have a large inherent heat capacity, forced cooling should not be required. It is anticipated that at least initially the same container would be used for both fresh and spent fuel assemblies.

With slight modifications of the fuel basket, the cask in Fig. 9-14 could readily accommodate either CANDU or LMFBR fuel assemblies as necessary. The latter situation, however, might be somewhat limiting as a small number of these high-power-density, short-cooled assemblies would have the same heat and radiation characteristics as a full load of LWR assemblies. Thus, special casks are being designed for LMFBR use.

Crash Tests

In 1977–1978 Sandia Laboratories prepared and conducted a series of full-scale crash tests for spent-fuel casks of a type similar to that in Fig. 9-14. The tests were not intended to replace the series required for regulatory certification. Instead, one purpose was to evaluate current capabilities for predicting crash results. The other major purpose was to demonstrate for the general public the overall safety of the transport method even under highly unlikely accident conditions.

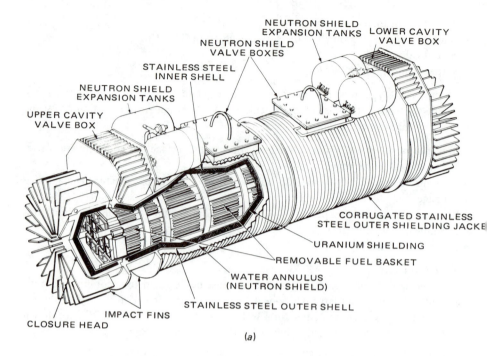

(a)

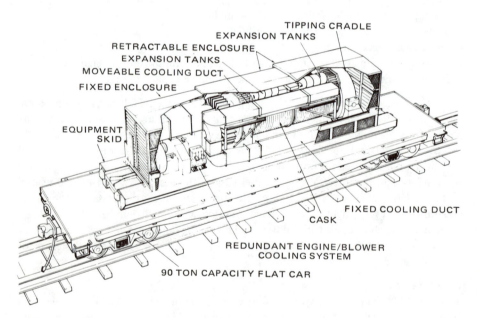

(b)

FIGURE 9-14

General Electric IF-300 spent-fuel shipping cask (*a*) separately, and (*b*) in normal rail-transport configuration. (Courtesy General Electric Company.)

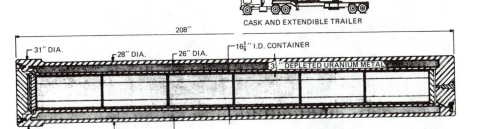

FIGURE 9-15
HTGR spent-fuel shipping cask. (Courtesy Oak Ridge National Laboratory, operated by the Union Carbide Corporation for the U.S. Department of Energy.)

The series consisted of:

1. 97 and 130 km/h crashes of tractor-trailer rigs with a spent-fuel cask into a massive, stationary concrete barrier
2. a 130 km/h locomotive crash on a stationary, cask-loaded tractor-trailer rig at a simulated grade crossing
3. a 130 km/h impact of a special railcar-mounted cask into the concrete barrier with a subsequent 125-min burn in JP-4 fuel at 980–1150°C

The results of the tests demonstrated that the predictive methods are very accurate. Since all casks survived with minimal damage and no postulated leaks of radiation or radioactive material, public concerns should be reduced. The photographs in Fig. 9-16 show some representative features of the test series.

EXERCISES

Questions

9-1. Explain why UO_2 is more widely used as a reactor fuel than is uranium metal.

9-2. If plutonium is used in LWR fuel assemblies, what differences occur in the fabrication process?

9-3. Describe the basic design features of the HTGR fuel assembly. How does this give the HTGR a more flexible fuel management program than the LWR designs?

9-4. Explain the basic concepts of multiple-batch fuel management. What feature makes the LMFBR unique in this regard?

9-5. Differentiate between SGR and OMR for plutonium utilization. What are their impacts on LWR design and operation?

9-6. Sketch the basic features of the "full-recycle" mode of HTGR operation. Identify the limitations of [235]U and thorium recycle.

9-7. Explain how standard and recycle fuel-management procedures in the CANDU are different from those of the other reference reactors.

9-8. Describe the most basic features of an LWR spent-fuel shipping cask and of its testing program required by the NRC. What is the role of depleted uranium in the cask? Why?

FIGURE 9-16
Spent-fuel shipping cask full-scale accident test sequences: (a) and (b) 130 km/h tractor-trailer crash; (c) and (d) 130 km/h locomotive crash.

254

(e)

(g)

(f)

FIGURE 9-16

Spent-fuel shipping cask full-scale accident test sequences (*Continued*): (*e*)–(*g*) 130 km/h railcar impact, burn, and post-test status of cask. (Photographs Courtesy of Sandia National Laboratories.)

Numerical Problems

9-9. One of the more severe design limitations for LMFBR fuel is based on the volume of fission gas released from its fuel pellets. Using data in Table 9-2, estimate the ratio of LMFBR-to-LWR lifetime gas release per unit mass of fuel.

9-10. From the data in Table 9-4, estimate the average capture-to-fission ratio for highly enriched uranium spending 4 years in an HTGR. Recalculate this parameter using 0.025-eV cross sections from Fig. 2-16.

9-11. Assume that the fission-product decay power (Eq. 2-10) must fall to less than 1 percent of the steady-state reactor operating level before LWR refueling can begin. What is the minimum time in days required for this "cooling" process?

9-12. The use of stainless-steel clad in an LWR is "expensive" to neutron economy.
  a. Calculate the ratio of the 0.025-eV microscopic cross sections for iron to zirconium.
  b. From Table 8-4, estimate the fractional increase in both uranium resource and SWU requirements for the extra 1 wt % ^{235}U needed with stainless clad.

9-13. Calculate the core-average enrichment for the initial loading pattern in Fig. 9-11.

9-14. Estimate the masses of each plutonium isotope in the equilibrium discharge batches (Table 9-3) from the BWR and PWR, respectively. Assume that the isotopic fractions inferred from the highest burnup on Fig. 6-2 are valid in both cases.

9-15. If HTGR thorium can be recycled only after its ^{228}Th content has decayed to 0.1 percent of its initial level, how long must it be kept in storage?

9-16. Assuming that the ^{238}U mass differences in Table 9-4 are as a result of plutonium production, estimate the potential energy contribution of each cycle's production as a fraction of that of the initial ^{235}U loading.

SELECTED BIBLIOGRAPHY[†]

Fuel Design
    Elliot & Weaver, 1972 (Chaps. 2, 3, 9, 10, 11)
    Graves, 1979 (Chaps. 8, 9)
    Hanson, 1978
    Holden, 1958
    Nucl. Eng. Des., Feb. 1980
    Olander, 1976
    Thompson & Beckerley, 1973
    Weisman & Eckart, 1978
    Zebroski & Levenson, 1976

Fabrication and Utilization
    Dahlberg, 1974
    Elliot & Weaver, 1972 (Chaps. 7, 12–15)
    Graves, 1979 (Chaps. 12, 13)
    Lotts & Coob, 1976
    Notz, 1976
    Nucl. Eng. Int., Aug. 1978, April 1979
    Sesonske, 1973 (Chap. 7)

Spent-Fuel Storage and Transportation
    Astrom & Eger, 1978
    Dukert, 1975
    Elliot & Weaver, 1972 (Chap. 13)

[†]Full citations are contained in the General Bibliography at the back of the book.

General Electric, 1979
Grella, 1977
Jefferson & Yoshimura, 1977
Nucl. Eng. Int., Dec. 1980
NUREG-0404, 1978
Symposium, 1978

Other Sources with Appropriate Sections or Chapters
APS, 1978
Connolly, 1978
Erdmann, 1979
Foster & Wright, 1977
Glasstone & Jordan, 1980
Glasstone & Sesonske, 1967
Knief, 1981
Lamarsh, 1975
Murray, 1975
WASH-1250, 1973

10

REPROCESSING
AND WASTE MANAGEMENT

Objectives

After studying this chapter, the reader should be able to:

1. Identify the basic steps in spent-fuel reprocessing.
2. Describe at least two unique engineering problems of spent-fuel reprocessing.
3. Identify the three basic methods of radioactive waste handling.
4. Explain the unique environmental role of radon gas in the uranium fuel cycle.
5. Differentiate between low-level, contact-handled, remote-handled and high-level radioactive wastes.
6. Identify the major options for interim storage of spent fuel.
7. List at least five basic waste disposal alternatives.
8. Describe the general characteristics of the Waste Isolation Pilot Plant [WIPP] conceptual design.
9. Identify the potential beneficial uses for radioactive waste constituents.
10. Explain the significance of the "Oklo phenomenon" to radioactive waste disposal.

Prerequisite Concepts

Nuclear Fuel Cycle	Chapter 1
Reprocessing	
Waste Management	
Decay Heat	Chapter 2
Dose Limits and MPC	Chapter 3
Neutron Balance Controls	Chapter 4
Wastes and Hazard Index	Chapter 6
Solvent Extraction	Chapter 8

Waste management is the final step in a closed nuclear fuel cycle. If recycle is employed, reprocessing operations separate fission-product and actinide wastes from fuel material; otherwise, the spent fuel assemblies themselves may be the major waste form.

REPROCESSING

Chemical processing of spent reactor fuel, i.e., reprocessing, is a well-developed technology which dates back to 1943 and the Manhattan project. The first operations were somewhat crude but quite effective. They also provided a basis for development of the current Purex process.

Although the chemical principles of reprocessing are relatively simple, large-scale implementation has presented some interesting problems and drawn forth ingenious solutions. Current limits to use are largely sociopolitical rather than technological.

Purex Process

Since its inception in the early 1950s at the Savannah River Plant, the Purex process has been the world's workhorse for both military and commercial reprocessing of uranium fuel materials. The simplified flowsheet in Fig. 10-1 identifies the important generic components.

Spent fuel is received and placed into buffer storage in a manner equivalent to that described in the last chapter. Its residence time will normally be short since the storage is intended to enhance continuity of operation rather than to allow for additional decay time.

The *head-end* operations shown on Fig. 10-1 begin with the mechanical disassembly of fuel bundles. This is usually accomplished by the "brute-force" method of chopping them into small pieces. When the segments are placed in nitric acid, the fuel is dissolved while the *cladding hulls* form the first waste stream.

The extraction and partition operations in Fig. 10-1 are both based on solvent-extraction processes of the same nature as those used in uranium milling (Chap. 8). The nitric acid solution forms the aqueous phase, while tributyl phosphate [TBP] in a kerosene-like carrier is the organic phase. The basis of the separation process is most readily described in terms of *distribution coefficients* D_i defined by

$$D_i = \frac{\text{concentration of } i \text{ in organic phase}}{\text{concentration of } i \text{ in aqueous phase}}$$

for chemical species i. In general, such coefficients vary with the quantity of material present and with the composition and concentration of the two phases. The curves in Fig. 10-2 are representative of the variation of distribution coefficients with the aqueous-phase acid concentration. The top curves are for uranium and plutonium in their usual chemical charge states of $+6$ and $+4$, respectively. The other curves show

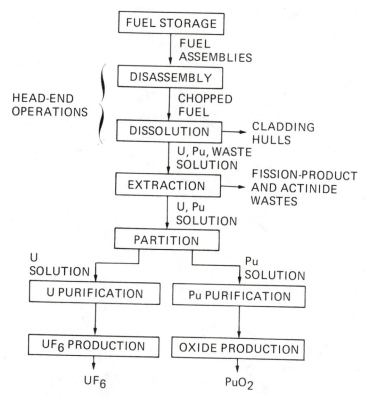

FIGURE 10-1
Simplified flowsheet of the Purex process for spent reactor fuel.

the behavior of zirconium (clad residues and fission products), ruthenium [Ru], niobium [Nb], the rare-earth elements, and the "gross beta" (i.e., general beta-emitting fission products). Solvent-extraction is most useful when the coefficients for materials of interest differ greatly. A separation factor α is defined as the ratio of product and impurity distribution coefficients, or

$$\alpha = \frac{D_{\text{product}}}{D_{\text{impurity}}}$$

The larger the value of α, the more complete the separation in a single step.

The fission product removal step in Fig. 10-1 is implemented by solvent-extraction operations. The initial separation relies on maintaining nitric acid and organic concentrations that provide a high separation factor between the heavy-metal product (U^{6+} and Pu^{4+}) and the general clad (Zr), fission product, and transuranic inventories. Figure 10-2 suggests, for example, that this could be accomplished by maintaining a high acid concentration. A second solvent-extraction reduces the fission-product content further.

When plutonium is converted from Pu^{4+} to Pu^{3+} by chemical or electrostatic

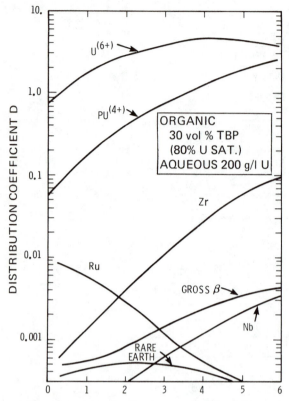

FIGURE 10-2

Effect of nitric acid concentration on the distribution of plutonium and fission products in the presence of uranium. [Adapted from S. M. Stoller and R. B. Richards (eds.), *Reactor Handbook—Volume II: Fuel Reprocessing*, 2nd ed., copyright © 1961 by Interscience Publishers. Reprinted by permission of John Wiley & Sons, Inc.]

means, it is much more readily separated from uranium by solvent extraction. This is the basis for the partition step in Fig. 10-1. Final purification of the uranium and plutonium streams is accomplished by a combination of solvent extraction and ion exchange operations. For recycle operations, it is most convenient if the reprocessing plant converts the uranium product to UF_6 for reenrichment and the plutonium product to plutonium nitrate for later use in fuel fabrication (e.g., see Fig. 1-2).

Technology Base

The Purex process involves relatively standard chemical engineering operations. However, the unique features of spent reactor fuel create technological challenges related to:

- chemical separation requirements
- radiological health
- radiation damage

- maintenance operations
- criticality control

The final product of the design and fabrication procedures described in the previous chapter is chemically inert reactor fuel with a high retention for fission-product and actinide impurities. Reprocessing, of course, seeks to convert spent fuel to a form with the *opposite* characteristics. The latter can be accomplished only with very highly corrosive chemical reagents (and the requisite corrosion-resistent process equipment, which tends to be quite expensive).

Reprocessing separations must be substantially more complete than is common in the chemical industry. If fuel material has impurity levels in excess of parts-per-billion [ppb] to parts-per-million [ppm] of certain species, radiation levels can complicate handling and fabrication or parasitic neutron absorption can degrade the ultimate energy value. The high economic value of uranium and plutonium also dictates that as little as possible of these products be allowed to escape the system with the waste streams. (The plutonium content also determines the long-term toxicity of the wastes.)

The high activity of spent fuel poses potential radiological hazards to both operating personnel and the general public. Occupational concerns are minimized by conducting all appropriate operations remotely in heavily shielded "canyons" and by controlling airborne radionuclide concentrations in working areas. Public exposures are readily limited by proper treatment of gaseous and liquid effluents and by careful storage of fission-product and actinide wastes.

The radiation environment of the spent fuel has the potential for damaging components and instrumentation, enhancing corrosion, and causing breakdown of organic solvents. The extreme importance of the latter effect has prompted the development of radiation-resistant extraction agents like TBP. It has also spurred the design of special process equipment which assures high separation with minimum contact time between the organic solvent and the waste-rich solutions. Since the extent of the radiation damage is proportional to the time, solvent lifetimes and, thus, process economy are enhanced.

The processing operations themselves are complicated by the high radiation levels and the related shielding. Extensive instrumentation and remote sampling systems monitor the various processes. Equipment is designed to be as reliable and maintenance-free as possible. Because no facility is ever completely trouble-free, maintenance concepts are crucial to the design of reprocessing plants. *Direct-maintenance* operations employ equipment with a minimum number of moving parts and have provisions for remote decontamination prior to hands-on activities. *Remote-maintenance* facilities employ equipment modules and piping runs that can be removed and replaced remotely by operators in heavily shielded cranes. Both concepts encourage ingenuity on the part of design engineers.

Nuclear criticality safety is a particularly challenging problem in a reprocessing plant because of the wide variety of fissile-material forms. The most positive control is based on the use of high-leakage geometries ["geometrically favorable" configurations], e.g., long, small-diameter solvent-extraction columns and thin, flat "slab" storage tanks as shown in Fig. 10-3. Administrative controls on such parameters as fissile mass, enrichment, moderation, reflection, and neutron poisons must be employed when geometric control is not feasible.

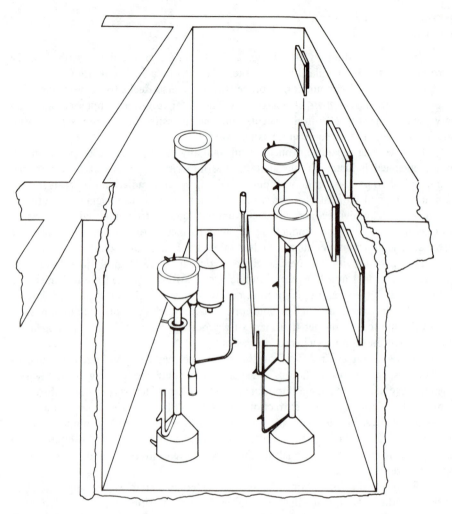

FIGURE 10-3
Arrangement of the plutonium processing cell at the Barnwell Nuclear Fuel Plant. (Courtesy Allied General Nuclear Services.)

Since reprocessing operations are so well shielded, criticality is of less concern as a radiation hazard than as an indication of insufficient design safety. The major "environmental" impact of such an accident might actually be the personnel layoffs during plant shutdown and the ensuing investigation.

Thorium Fuels

The *Thorex process* can be used to separate thorium from uranium in a manner similar to that by which the Purex process separates plutonium from uranium. Although it has yet to be implemented on a large scale, the viability of the process is well established.

A most interesting application of reprocessing in a thorium fuel cycle relates to the HTGR fuel of Figs. 1-10 and 9-7. The microsphere design facilitates physical separation of the ^{235}U and ^{233}U isotopes. A proposed head-end process calls for:

- crushing the entire irradiated graphite fuel block into small fragments
- burning graphite from the fragments in a fluidized bed to remove the outer layer from the TRISO particles and reduce the BISO to uranium–thorium ash
- separating the ash from the remaining SiC-clad TRISO particles by centrifuge to form two process streams
- dissolving the U–Th ash to facilitate Thorex separations of waste products, uranium (^{233}U plus a small amount of the other isotopes from broken TRISO particles), and thorium
- grinding the TRISO stream to crack the SiC coating from the residual uranium and allow for burning, dissolution, and ultimate fission-product removal

Each product stream is likely to be converted to a uranyl nitrate solution or another form consistent with the fabrication process described in the previous chapter. If the ^{233}U stream is stored, its radiation level will build with time as shown, for example, by Fig. 6-5.

Status

The technology for fuel reprocessing is well established in the United States. The experience base is summarized by Table 10-1. Extensive military applications have been augmented by 6 years of commercial operation at the Nuclear Fuel Services Plant in New York state. The Midwest Fuel Reprocessing Plant at Morris, Illinois proved to be inoperable because of difficulties with its hydrofluor process. It is currently being used for spent-fuel storage as noted in the previous chapter. Much of the Barnwell Nuclear Fuel Plant is ready for operation, but has yet to process any spent fuel pending federal policy changes.

The status of worldwide reprocessing activities is summarized in Table 10-2. France has an aggressive program consistent with its commitment to LWR plutonium recycle and LMFBR development. Other major activities are centered in Germany, Japan, and the United Kingdom.

FUEL–CYCLE WASTES

Each step in the nuclear fuel cycle, including waste management, produces some radioactive wastes. Although compositions, quantities, and activities vary greatly, all streams are amenable to one of three waste handling principles.

The first method considered for any waste disposal is called *dilute and disperse*. Whenever concentrations can be reduced to well below MPC standards, release to the general environment is allowed (within the constraints of the as-low-as-reasonably-achievable [ALARA] criterion). Low-activity gaseous wastes are generally diluted and dispersed by mixing with the atmosphere at the top of a tall stack. Liquid effluents are usually diluted partially in-plant and further at the point of discharge.

Wastes that have high activity but short half-lives may be handled by the *delay and decay* method. Short-term holdup allows radiation levels to decrease of their

TABLE 10-1
Summary of United States Reprocessing Experience

Facility	Date(s) operational	Capacity (tons/year)	Process	Maintenance	Status
Hanford (WA)	1944–1972	Military	Various	Remote	May be reactivated in 1980s
Savannah River (SC)	1954	Military	Purex	Remote	Operating
Idaho	1953	Naval reactors	Purex	Direct	Operating
Nuclear Fuel Services (NY)	1966–1972	300	Purex	Direct	Shutdown for expansion; will not reopen due to inability for backfit to new regulations
Midwest (IL)	–	–	Hydrofluor	Remote	Process will not work on production scale; plant modification not planned at present
Barnwell (SC)	?	1500	Purex	Direct	Ready to operate pending federal decision for commercial reprocessing

TABLE 10-2
Commercial Reprocessing Activities Outside of the United States

Nation	Plant	Capacity[†] (t/y)	Date operational	Status[†]
Belgium	Eurochemic (Mol, international sponsorship)	60	1966	Shut down by Eurochemic in 1974; scheduled for Belgium take-over in 1982 and restart in 1984
Brazil	Pilot plant	3	1984	Planned with West German technology
France	SAP (Marcoule)	7.5		Operational pilot plant for mixed oxide (supporting LMFBR program)
	UP (Marcoule)	900–1200	1958	Early military plant now operational for natural U metal fuel (from gas-cooled, graphite-moderated reactors)
	PURR	100	1989–1990	Planned for mixed oxide for LMFBR's
	UP_2 (La Hague)	1000	1966	Operational for natural U metal fuel; 400 t/y head-end for LWR fuel to be expanded to 800 t/y; intended ultimate use for natural U metal fuel only
	UP_3 (La Hague)	800 + 800	1986 1989– 1990	Planned for LWR oxide fuel; first line to service foreign customers (contracts as of 3/79 with Austria, Belgium, Japan, the Netherlands, Sweden, Switzerland and West Germany; second line for domestic use)
Germany (West)	WAK (Leopoldshafen)	40	1970	Operational for reprocessing and as a test facility
India	Trombay	60	1965	Pilot plant for military use and for U metal or oxide fuels
	Tarapur	100	1978	Operational for oxide fuels from LWR and PHWR
	Madras	"Industrial"		For PHWR and thorium FBR fuels
Italy	EUREX (Saluggia)		1969	Operational for test reactor and oxide fuels
	ITREC (Trisaia)	4.5		Operational for U–Th and Pu–U fuels
	Commercial	1200		Under evaluation
Japan	Tokai	210	1977	Operational for oxide fuels
	Commercial	1800		Planned by private industry
Pakistan	Commercial	100		Planned with French technology[‡]

(See footnotes on page 268.)

TABLE 10-2
Commercial Reprocessing Activities Outside of the United States (*Continued*)

Nation	Plant	Capacity[†] (t/y)	Date operational	Status[†]
Spain		2		Design state
United Kingdom	Windscale	1500–2500	1964	Operating for gas-cooled reactor fuels; has new head-end for 400 t/y of oxide fuel
	THORP (Windscale)	1200	1987	Planned for oxide fuels
	Dounreay	9–10	1978	Operating for mixed oxide fuels (for Prototype Fast Reactor [PFR])
USSR	Klopin Radium Institute	≈ 1	1973	Operating experimental facility
		1500		Reported under construction

[†]Data from K. M. Harmon, "Summary of National and International Radioactive Waste Management Programs 1979," Battelle Pacific Northwest Laboratories, PNL-2941, March 1979.

[‡]Has been cancelled following French insistence on use of co-processing as a nonproliferation measure (see Chap. 20).

own accord. Delay times of from hours to days often reduce concentrations (and requisite dilution volumes) to the point where dispersal is feasible.

The final method is to *concentrate and contain* those wastes which are not amenable to the other two procedures. The volumes of gaseous and liquid wastes may often be reduced to minimize storage volumes. Solidification enhances the ability to contain the wastes over long periods of time.

Front-End Wastes

The most hazardous wastes in the front end of the uranium fuel cycle are those containing uranium daughter products and/or toxic chemicals. If there is recycle, small amounts of plutonium are present in fuel-fabrication wastes.

Mining and Milling

Although the natural uranium isotopes decay very slowly, ore bodies have been in place long enough to accumulate sizeable quantities of radioactive daughter products. The stability of the ore precludes large-scale migration of these materials under normal conditions. However, mining and milling operations make many of the constituents much more accessible for environmental release.

Radon gas is the most troublesome of the daughter products because it has a potential mobility not found among the other elements. Although the uranium oxides in the ore are fairly effective at radon retention (recall that UO_2 is used as a fuel form because it retains fission gases!), some release does occur. Mining operations then allow general entry into the working areas. Milling procedures result in additional releases.

Most of the initial radon content escapes during milling. However, some of its parent or precursor nuclides generally remain in the mill tailings to provide a long-term source of the gas. The major concerns are the ^{226}Ra and ^{222}Rn products in the ^{238}U decay chain. The radium isotope alpha decays with a half-life of 1600 years (a relatively short time compared to the uranium). The ^{222}Rn daughter has a half-life of 3.8 days, which allows it ample opportunity to migrate from the tailings and deposit elsewhere. The several short-lived daughters of ^{222}Rn increase the effective activity.

The mill tailings have a mass comparable to that of the mined ore. Because the activity is relatively low, they are stored in the open, often in or around a "tailings pond." Limiting radon releases during and after the lifetime of the mill is an important environmental concern. Secondary recovery of radium during the milling operations is an effective, if somewhat expensive, means of reducing the ultimate radon content. The tailing piles may also be chemically stabilized to inhibit release of radon and other daughter products. In the case of radon, the latter method is a variant of delay and decay, since retention for a month or two (i.e., ≈ 10 half-lives) is adequate to reduce the potential near-term hazard.

Fabrication

Uranium wastes from conversion, enrichment, and fabrication are of minor concern from a chemical or radiological standpoint. They consist of contaminated clothing, lubricants, and other maintenance-related items which can be compacted and stored in steel drums for an indefinite period of time.

Mixed-oxide fabrication adds small amounts of plutonium to the waste streams. Since it would not be designated as a waste if the plutonium were readily recoverable, accidental releases are unlikely. Packaging and handling of these wastes can be accomplished in a manner similar to that used for the uranium wastes.

Reactor Operations

Radioactive wastes which appear directly during reactor operation are generally of two types. Fission products enter the coolant as a result of cladding leaks or failures. They consist primarily of noble gases, halogens, and tritium. The second waste constituent is activation products generated from neutron irradiation of the coolant and its additives. These wastes vary with the coolant.

Noble Gases

The chemically inert noble gases separate readily from liquid coolant (and any other medium). Since many of the constituent nuclides have short half-lives, the delay and decay principle is appropriate. Relatively large gas volumes, however, preclude long-term storage.

The most limiting problems with the noble gases are experienced in the BWR. Here, the single-loop, direct-cycle coolant releases the gases on a continuous basis as they enter the turbine. Limited storage-tank volumes require release after only about one day of decay time. Gaseous-effluent MPC restrictions require that the noble gases be discharged from a very tall stack.

The multiple-loop reactors employ the primary system as a "delay tank" for an initial period of time. Then the gases are "bled off" to storage tanks on a predetermined schedule. Such procedures allow the pressurized-water reactor [PWR], for example, to hold back noble-gas release by up to 60 days. The resulting 100-fold reduction in activity allows building-level dispersal.

Tritium and Soluble Boron

Tritium, the radioactive isotope of hydrogen, is produced by several mechanisms in operating reactors. It is readily incorporated with water molecules as HTO and very rarely as $T_2 O$. Since isotopic concentration or separation is not feasible, this relatively long-lived $[T_{1/2} = 12.3$ years$]$ radionuclide is subjected to the dilute and disperse procedure.

A primary production mechanism for tritium is ternary [three-fragment] fission. Even the small amounts released through the clad to the coolant constitute an important waste management problem for all reactors. Additional tritium sources result from neutron absorption in deuterium or boron. The heavy-water CANDU system experiences large tritium production from the ^2D (n, γ) reaction. However, since most tritium will reside in the stationary moderator volume, essentially in-place storage is facilitated (see Chap. 13).

The use of soluble boron for reactivity control in the PWR has a double-edged effect on system wastes. Negative reactivity is readily inserted by adding a small amount of high-concentration boric acid to the coolant. A positive insertion, however, requires a general dilution of the coolant with a large volume of boron-free water. This latter situation, of course, produces a comparable amount of liquid waste that is contaminated with fission products, tritium, and other radionuclides. Special liquid waste tanks are built into PWR's just to handle boron dillution. The second complication is that the ^{10}B (n, 2α) reaction in the soluble boron is responsible for the production of a substantial amount of extra tritium. Ultimately, the tritiated water must be diluted and discharged.

Solid Wastes

Solid radioactive wastes from reactor operations are generally associated with cleanup of gaseous and liquid waste streams or with testing and maintenance activities. The chemically reactive gases like the halogens are readily immobilized on filters. Elemental impurities (as opposed to isotopic impurities, e.g., HTO in H_2O) can often be removed to demineralizer resins. Other liquid wastes are concentrated or reduced to solids in an evaporator.

Testing and maintenance operations generate contaminated clothing, gloves, wipes, and tools. These and the filters, resins, and concentrates are packaged into steel drums like the fabrication wastes. Some liquids are mixed in concrete binder and placed in the same or different drums.

Storage

Interim spent-fuel storage produces a minimum amount of additional radioactive wastes (assuming leaking assemblies are packaged). From the time of reactor shutdown, the absence of new fission-product generation and lower thermal gradients limit general release from the fuel assemblies.

The spent fuel itself is the fuel cycle's final waste form if reprocessing is not to be implemented. Since the assemblies were not designed for an indefinite lifetime, extended water-basin storage may be expected to compromise their integrity.

Reprocessing

The fundamental purpose of reprocessing is separation of the fuel materials from fission-product and actinide wastes (Fig. 10-1). With this separation come concerns for:

- release of gaseous products following mechanical disassembly and dissolution
- treatment of particulates and nitrogen oxides from processing operations
- storage of the separated radioactive wastes

Off-gassing of large quantities of tritium, iodine (mainly ^{129}I and ^{131}I), and retained noble gases (primarily ^{85}Kr) necessitates strict effluent control. Although much of the tritium is incorporated into water solution, some must be dispersed along with the inert noble gases. Release of the iodine radionuclides is minimized by filtration. Although in-place decay occurs during storage, the gas inventories in the fuel are very large at the time of reprocessing. Their release constitutes the major environmental constraint for the operations.

The particulate and oxide wastes are controlled with proper use of adsorption beds, particulate filters, and scrubbers. These and the other testing and maintenance operations ultimately produce drummed solid waste similar to that from fabrication and reactor applications.

Liquid Wastes

The separated fission-product and actinide wastes leave the active processing in nitric acid solution. In their as-generated form, stainless-steel tanks are required for storage. Less expensive, mild-steel tanks may be employed if the solutions are made alkaline.

Following leakage problems with early (i.e., Manhattan Project vintage) tanks, current designs incorporate full double walls. They provide leak detection capability and opportunity for timely transfer of material to other tanks. The heat-load from the radioactive products necessitates use of forced cooling. Most tanks incorporate an extensive network of water piping for this purpose.

Solidification

Since liquids are subject to leakage and spills, it is desirable to convert spent-fuel wastes to a solid form. Current federal regulations call for:

- storage of commercial wastes in liquid form for no longer than 5 years from the time of separation
- conversion to a solid form for on-site storage past 5 years
- transfer of solidified wastes to a federal repository no later than 10 years after separation

In the absence of commercial reprocessing and a federal repository, the regulations are academic. They do, however, serve as guidelines for technology development related to solidification, management, and ultimate disposal of wastes.

High-level liquid wastes from reprocessed LWR spent fuel contain fission products, residual uranium and plutonium, and other actinides. If separation follows 180 days of cooling, the wastes from each metric ton of uranium generate heat at a rate of 15–20 kW and occupy a volume of about 1200 l. Prior to solidification (nominally at 5 years), the heat rate would be down to a few kilowatts and the volume could be reduced to about 600 l.

The high radiation levels and heat loads for the wastes suggest that any solid product must have good radiation and thermal stability. It is also desirable for the material to be chemically stable, so that constituent radionuclides will not be readily leached or volatilized out. Meeting these and other constraints with a small volume and an inexpensive product is a major design challenge.

At the present time, there are two solidification processes that are reasonably well developed. *Drying and calcination* is based on high-temperature "baking" of the liquid wastes. (The term "calcination" derives from a similar method for producing calcium oxide or, as it is more commonly known, lime.) Fluidized-bed, spray, and rotary-kiln methods have been developed to produce granular or powdered solids like those on the left side of Fig. 10-4. The pot calcination procedure generates porous caked solids similar to the samples on the right. (All calcined solids would likely require further treatment before being suitable disposal forms.)

The *vitrification* process produces glass or glass-like solids. The black, borosilicate glass solid in the top center portion of Fig. 10-4 is typical of the product material. Design-stage operations have been developed around in-can melting and rotary-kiln continuous-melting concepts. The latter is the basis for the pilot plant built in France on the site of the Marcoule reprocessing plant.

These solidification methods, along with others that are in earlier developmental stages, are being evaluated for compatibility with the various waste management options. Final selections are to be keyed to the ultimate waste isolation and disposal.

There would be 70–80 l of solidified high-level waste from the reprocessing of each metric ton of LWR spent fuel. This corresponds to roughly 2.5–3.0 m^3 for each

FIGURE 10-4
Solidified waste forms from calcination (white powder and porous solids) and vitrification (black glass) processes. (Courtesy of U.S. Department of Energy.)

GW(e)·year of electricity generated. It has been noted that the volume would readily fit into two to three standard, four-drawer file cabinets. Viewed another way, the average U.S. family of five deriving *all* their electricity from nuclear power would have an annual waste contribution that would occupy the volume of a standard aspirin bottle.

WASTE MANAGEMENT

Nuclear fuel cycle wastes may be classified according to their radiation and heat-generation levels, as well as their composition. The following working definitions[†] serve as bases for further discussions:

- low-level wastes [LLW] —actinide content low enough ($<$ 10 nCi/g) to permit disposal by surface burial
- *contact-handled transuranic [TRU] wastes*[‡] —actinide content $>$ 10 nCi/g, minimal heat generation, surface dose rate low enough ($<$ 200 mrem/h) to permit handling by contact (as opposed to remote) methods
- *remote-handled transuranic [TRU] wastes*[‡] —actinide content $>$ 10 nCi/g, some heat generation, surface dose rates high enough ($>$ 200 mrem/h) to require remote handling and/or shielding
- *high-level wastes [HLW]* —waste products from the extraction step of the reprocessing operation (e.g., Fig. 10-1) or, currently, spent-fuel assemblies

The drum-packaged wastes from enrichment, fabrication, reactor operation, and spent fuel storage are low-level wastes. Testing and maintenance activities in reprocessing and waste management facilities also produce wastes of this type. They are generally stored on-site for a period of time before being shipped to designated sites for shallow burial.

The contact and remote-handled TRU wastes originate from reprocessing operations other than the extraction step. They are solidified as necessary and packaged into drums or steel canisters. The contact-handled TRU wastes require no cooling and minimal shielding. Depending on their actual heat and radiation levels, the remote-handled TRU wastes may require some cooling and from light to heavy shielding.

The high-level wastes are a mixture of fission products and actinides. The HLW from reprocessing would be stored initially was liquids and then solidified as noted previously. Spent fuel assemblies, of course, are already in a solid form.

General waste management strategies consider both *storage*—interim, retrievable emplacement—and *disposal*—permanent, nonretrievable storage. Either may be applied to spent fuel or solidified reprocessing wastes. Fuel assemblies are likely to remain intact for interim storage and, if reprocessing is not permitted at a future date, may be the form for final disposal. With reprocessing, fissile and other constituents can be removed to limit the total amount of disposed material.

[†]That is, practical definitions applied to the Waste Isolation Pilot Plant [WIPP] (described later in this chapter).
[‡]These two categories are occasionally lumped together as intermediate-level waste [ILW]. However, the continuing need for subcategories limits the usefulness of the ILW designation.

Storage

In the absence of reprocessing, a *throw-away ["once-through"] fuel cycle* is said to exist. The spent fuel assemblies are first placed into interim storage, perhaps sequentially at different facilities. Ultimately, the intact assembly becomes the final disposal form.

The prospect of a throw-away cycle prompted the U.S. Department of Energy [DOE] to formulate a long-term policy calling for:

- the government to take ownership of spent fuel at approved storage sites
- the utility to pay the cost of transportation to the site and pay a one-time fee to cover both the interim storage and eventual permanent disposal (early estimates are $3 million/year for a 1000-Mw(e) plant—an amount which is about $2\frac{1}{2}$ percent of projected yearly costs)
- return of the spent fuel (or an equivalent credit) plus the unused portion of the storage/disposal fee if reprocessing becomes a viable option at a later time
- acceptance of foreign spent fuel on a case-by-case basis when space is available and (especially) when nonproliferation objectives may be enhanced

Although set forth in the fall of 1977, nearly 3 years later it has yet to be implemented by Congressional action. The policy has, however, served to encourage the development of spent-fuel isolation strategies and systems.

Interim fuel storage options may be divided into two categories. Unpackaged assemblies are to be handled in essentially the manner now employed in at-reactor and away-from-reactor spent-fuel storage pools. Packaging of the assemblies allows for their isolation in one of several engineered or geologic facilities.

Packaged Fuel

Prolonged storage has not typically been an important fuel-assembly design criterion. Thus, viable long-term isolation is dependent on packaging in an appropriate container.

A facility for storing packaged spent fuel must provide for heat removal by active or passive means, criticality control by spacing and/or poisoning, monitoring, and equipment for handling operating wastes generated by corrosion and/or leakage. The ability to overpack and store leaking canisters is also important to the overall facility design.

Concepts for packaged spent-fuel storage include

- water basins
- air-cooled vaults
- surface silos
- near-surface heat sinks or storage wells
- geological formations

The basin differs from current interim storage procedures only by its use of a fuel container. The air-cooled vault is similar to the basin concept except for the use of forced-circulation air cooling. The surface silo and storage well are being developed as noted below. Geologic formations may be equally applicable to storage and disposal as considered later in the chapter.

It is important that the container for packaging spent fuel be compatible with the storage concept. Sealed metal cylinders appear to be logical choices. If radiant heat transfer from the fuel to the container is not adequate, a gas, powdered metal, or glass might be introduced as a filler. A common configuration that can be used with several storage concepts and/or for final disposal is also highly desirable.

Packaging and Handling

The Commercial Spent Fuel Packaging and Handling Program has been charged by DOE with engineering development and testing of a variety of packaging, handling, and storage concepts for spent reactor fuel. One specific responsibility is preparation of packages for its own and other storage demonstration programs.

The programs are located at the Nevada Test Site [NTS] where they employ the Engine Maintenance and Disassembly [EMAD] Facility for remote-operation packaging of the high-activity spent fuel. Since early 1979, two PWR assemblies have been in place in storage wells and one has been in a surface silo. Future plans included similar storage of BWR assemblies. Underground storage in granite of 13 spent-fuel packages is also planned (based on a concept similar to that described for salt later in this chapter) as part of the Climax Project.

The *surface-silo* concept relies on vertical container storage in a concrete housing that provides for natural-convection air cooling, as shown by Fig. 10-5. This is some-

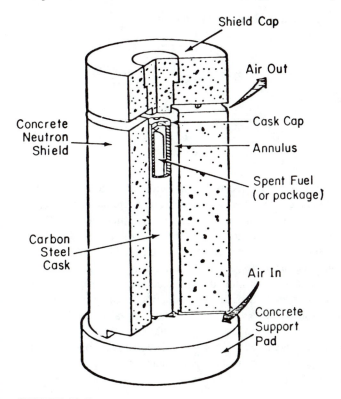

FIGURE 10-5
Surface silo concept for storage of packaged spent-fuel assemblies. (Courtesy of U.S. Department of Energy.)

times referred to as the "Stonehenge" concept since an array of units will bear some resemblance to the famous English site of the same name. The primary advantages of the surface silo are its total passivity and low maintenance requirements. Large land use and material requirements are the dominant drawbacks.

The *near-surface heat sink* or *storage well* employs vertical concrete-and-steel-lined holes with a covering concrete plug. Heat transfer is by conduction to the ground and air convection from the surface of the plug. Although construction may be somewhat easier, this concept appears to have the same general advantages and disadvantages as the surface silo.

Disposal

Few portions of the nuclear fuel cycle are more subject to changing policies than is waste management. The most definitive statement on the subject at the time of this writing is the *Draft Environmental Impact Statement: Management of Commercially Generated Radioactive Waste* (DOE/EIS-0046-D, 1979). It "describes ten alternative methods for disposal of nuclear wastes and evaluates their anticipated environmental impacts." Since the statement is based entirely on the existing data base, the extent of each evaluation is limited by the state of knowledge on the specific concept.

The ten alternatives identified in the impact statement are:

1. geologic disposal using conventional mining techniques
2. subseabed geologic disposal
3. very deep hole concept
4. rock melting concept
5. reverse-well disposal
6. chemical resynthesis
7. island disposal
8. ice sheet disposal
9. space disposal
10. partitioning and transmutation

Among the 10, only the first three are receiving substantial attention from DOE at the present time. The remainder are much more highly conceptual in nature. The final two continue to receive some attention by DOE and/or other agencies. Coordination of the current DOE activities is centered at the Office of Nuclear Waste Isolation [ONWI], operated by the Battelle Memorial Institute in Columbus, Ohio.

Each of the 10 concepts has as its goal near-permanent removal of high-level radioactive wastes from the human environment. As a practical matter, the goal may be modified to suggest isolation until a particular reduction in the potential hazard associated with the wastes is achieved, e.g., a hazard index equivalent to that of uranium ore as shown by Fig. 6-9 may be considered appropriate.

Radioactive waste management relies on a multiple-barrier approach similar in concept to that used in designing nuclear reactors. Although there are differences for each of the methods, the four typical barriers are:

1. waste form
2. canister

3. disposal medium
4. human institutions

The first two taken together constitute the *engineered barriers* to radionuclide release. Initial retention is provided by the form of the solid spent-fuel assemblies or solidified reprocessing wastes. Metal or ceramic canisters with "getter"-material fillers and/or overpacks provide the next barrier level. Recent policy of the U.S. Nuclear Regulatory Commission [NRC] has emphasized these engineered features as the primary barriers for radionuclide retention.

The disposal medium is intended to provide massive *natural barriers* that will prevent intrusion of the waste material into the biosphere. These range from stable geologic formations to outer space. Human *institutional barriers* like adequate repository markers may also be of value.

Environmental impact statements must be prepared for every "major federal action" and must include a thorough evaluation of alternative concepts (as discussed further in Chap. 19). Current DOE policy for waste includes rigorous consideration of a limited set of alternatives. Since it has been recognized that low expenditure levels produce only "paper studies," the major alternative concepts have sufficient funding to allow feasibility studies which include experiments, core sampling, etc. The principle programs cover several geologic disposal concepts thoroughly, subseabed disposal at a lower level, and the very deep hole concept at a still lower level. It is proposed that a number of candidate sites for the geologic and subseabed alternatives be evaluated and identified in the 1981–1986 time window.

Geologic Disposal

In the common usage, the terms *geologic disposal*, "conventional geologic disposal," and "geologic disposal using conventional mining techniques" are often employed interchangeably. Current DOE programs in this area are divided into two major categories:

1. the *Waste Isolation Pilot Plant* [WIPP] proposed for a bedded-salt deposit near Carlsbad, New Mexico as a repository for military TRU wastes and as a site for a limited experimental program with high-level wastes
2. alternative-site evaluation and selection for possible disposal of commercial wastes in domed-salt, bedded-salt, basalt, granite, tuff, and shale deposits

The WIPP project has progressed through site selection and conceptual design. The basalt program ["B-WIPP"] at Hanford, Washington is also well advanced. The commercial-waste programs anticipate site selections on the 1981–1986 time scale.

The long-term safety of geologic storage and disposal depends on the characteristics of the site and of the overall design of the related systems. Figure 10-6 summarizes important inputs for the development of a site-specific system concept. Waste, biosphere, and geosphere characteristics coupled with logistic considerations provide a basis for selection of a site and a system. System qualification depends on assessment of operational reliability, nuclide release and consequences, and geologic, chemical, and mechanical stability.

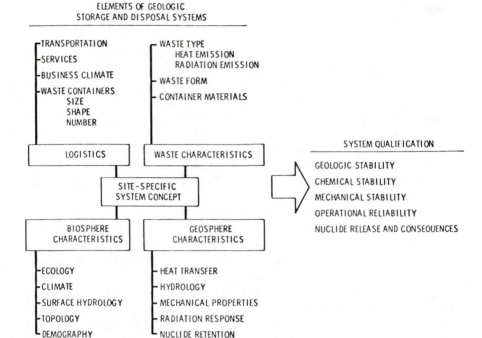

FIGURE 10-6
Decision components for qualification of a waste disposal site. (Courtesy of U.S. Department of Energy.)

Waste Isolation Pilot Project

Although it is intended primarily for military wastes, the Waste Isolation Pilot Plant [WIPP] has similarities to all other geological disposal concepts. The advanced development of the project suggests that the "WIPP Conceptual Design Report" (SAND 77-0274, 1977)[†] can serve as an appropriate basis for further consideration of geological disposal.

The WIPP project is intended to demonstrate the feasibility of storage and disposal of solid radioactive wastes in the mined chambers of a bedded-salt deposit. The three facets of the operation are:

1. disposal of contact-handled TRU waste
2. disposal of remote-handled TRU waste
3. experimentation with typical and "spiked" military HLW

As a pilot operation, it is intended to handle limited quantities of waste in a recoverable manner. A lengthy period of operation on a trial basis would precede its possible conversion to a permanent disposal facility.

[†]The WIPP program is in a state of flux that is even greater than that for waste disposal as a whole. Since policy changes have little real effect on the more fundamental aspects of the system design, this readily available report is employed here.

The major reasons for selecting bedded salt as the WIPP storage medium were well identified by a 1957 study prepared by the National Academy of Sciences. It was recognized that salt has good physical properties in the form of high thermal conductivity, plasticity, and strength. Salt deposits are also geologically stable with low deformation forces and a general absence of groundwater. This latter property is well demonstrated by the mere existence of the deposit (i.e., if there had ever been much groundwater, the salt would have been dissolved and transported elsewhere). The wide national distribution of salt deposits shown in Fig. 10-7 provides substantial flexibility for site selection.

The WIPP project also benefits from experience gained in two other operations. During 1963-1966, Oak Ridge National Laboratories conducted a series of waste storage experiments called Project Salt Vault in an unused salt mine near Lyons, Kansas. Valuable data was gathered despite the final decision that the particular mine was not suitable as a repository. The Asse salt mine in West Germany is unique as the world's only operating underground waste storage facility. Operating experience at Asse with low-level wastes since 1964 and with intermediate-level wastes more recently has been credited as very helpful to the WIPP design effort.

The site in the Delaware Basin in southeastern New Mexico was selected for the WIPP project based on the following favorable characteristics of the deposit:

- depth, thickness, and lateral extent
- tectonics (low deformation forces)
- good hydrology with few deep drill holes
- low mining activity and resource potential
- low population density
- good land availability and accessibility

FIGURE 10-7

Distribution of salt deposits in the United States. (Courtesy of Oak Ridge National Laboratory, operated by the Union Carbide Corporation for the U.S. Department of Energy.)

Figure 10-8 shows a representative geologic section through the proposed site. The upper formations effectively isolate the salt from both surface and ground waters. The major resource, potash, is particularly low-grade in the vicinity of the site. The relative locations of the low- and high-level waste storage areas are also shown on the figure.

The bi-level, conceptual design of the WIPP project is shown by the cut-away drawing in Fig. 10-9. Mined corridors are provided for waste storage at two different levels. (There is *no large cavern*, as is sometimes inferred from this cut-away drawing!) The upper level is intended for the contact-handled TRU wastes contained in steel drums. The lower level is designed for canisters of solidified remote-handled TRU wastes and experiments. The two levels are independent with separate surface facilities and ventilation, transport, and storage shafts.

In one surface facility, the contact-handled TRU wastes are received and moved to storage locations by conventional means (e.g., worker-operated fork lifts). All remote-handled TRU and high-level wastes arrive at a separate facility in heavily shielded railroad shipping casks (similar in design to the spent-fuel shipping casks described in the previous chapter). As may be inferred from Fig. 10-10, the cask is removed and decontaminated before the canisters are moved remotely to the transfer shaft, and then to a transfer cell on the lower storage level. Each canister is moved from the cell to a final storage location in a drilled hole in the corridor floor. The

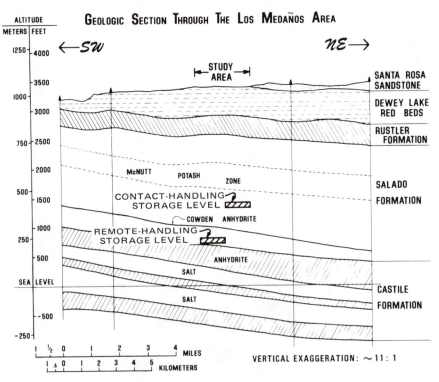

FIGURE 10-8

Geologic section through the proposed site for the Waste Isolation Pilot Plant [WIPP]. [Courtesy of Sandia National Laboratories (SAND 77-1401, 1978).]

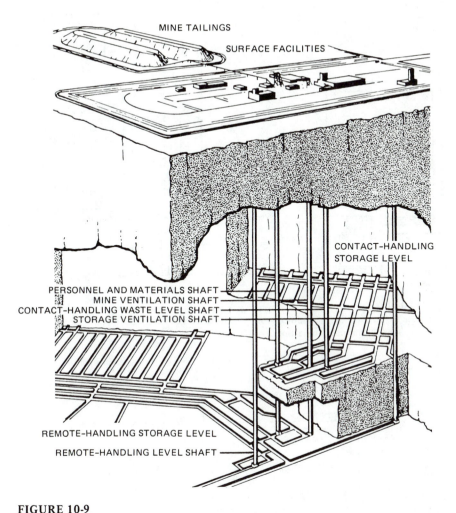

MINE TAILINGS

SURFACE FACILITIES

CONTACT-HANDLING STORAGE LEVEL

PERSONNEL AND MATERIALS SHAFT
MINE VENTILATION SHAFT
CONTACT-HANDLING WASTE LEVEL SHAFT
STORAGE VENTILATION SHAFT

REMOTE-HANDLING STORAGE LEVEL

REMOTE-HANDLING LEVEL SHAFT

FIGURE 10-9
Artist's conception of the Waste Isolation Pilot Plant [WIPP]. (Courtesy of Sandia National Laboratories.)

latter is accomplished with a special remote-controlled vehicle in the sequence shown in Fig. 10-11. The canisters may be retrieved at a later time by reversing the procedures with a vehicle designed to be able to remove both the backfill material and the canister.

The final WIPP design is likely to differ somewhat from the conceptual version. One suggested modification calls for reducing mining requirements by employing a single storage level with canister placement in the two corridor walls instead of the floor. Contact-handled TRU waste drums and boxes could be placed in the remaining corridor space following canister emplacement.

Supporting studies scheduled prior to initial operation of the WIPP include:

- rock mechanics and thermal properties
- computer analyses of mine designs

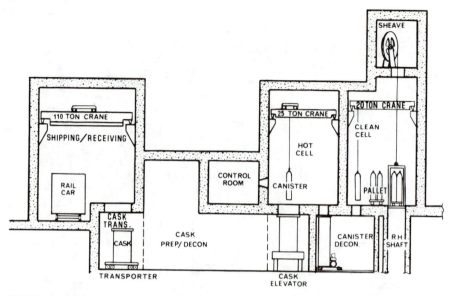

FIGURE 10-10
Flow diagram for surface remote-handled-waste facility at the Waste Isolation Pilot Plant [WIPP].
(Courtesy of Sandia National Laboratories.)

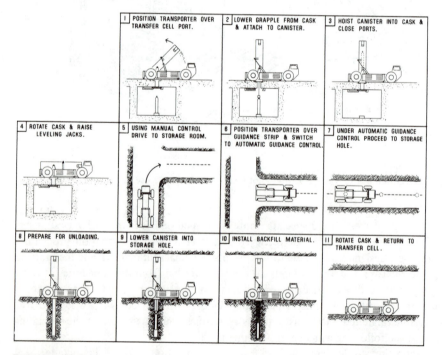

FIGURE 10-11
High-level waste canister storage sequence for the Waste Isolation Pilot Plant [WIPP]. (Courtesy
of Sandia National Laboratories.)

- waste characterization
- geochemical analyses of waste-canister–salt interactions

Equally extensive analyses are to be conducted during operation. Selection of the WIPP as an actual disposal site will require favorable results from both sets of studies.

Subseabed Geologic Disposal

The *subseabed geologic disposal* concept is the most advanced of the alternatives to conventional (land-based) geological disposal. It is being pursued through a five-phase program with decision gates which call for:

1. identification of natural barriers, which if undisturbed by the wastes or human activities, would contain the radionuclides for the time period of interest
2. development of the methodology that would allow the use of the barriers
3. determination by field testing of whether the capability to store wastes exists
4. engineering field testing of a total concept
5. if all previous steps are positive, initiation of full-scale operations

Initial evaluations have been based on a classical multiple-barrier approach which relies on the waste form, a canister, sediment or rock, and the ocean. The sediment barrier appears particularly promising, since the waste diffusion time alone is estimated at 10^6 y per 100 m. Sorption processes would be expected to add many orders of magnitude to the potential containment time.

The most favorable oceanic regimes for radioactive waste disposal are those with:

- high geologic stability
- low resource content (mineral and organic)
- large area
- low accessibility to other human activities
- deep sediment thickness
- low biological productivity
- international waters

Studies by various domestic and international groups working in unison have determined that the criteria seem to be met best by "mid-plate/mid-gyre" regions, i.e., those toward the center of both the earth's tectonic plates (discussed further in Chap. 19) and the major ocean currents (gyres).

Several concepts for subseabed disposal are shown in Fig. 10-12. The free-fall emplacement methods that rely on self-sealing of the sediment appear to be the simplest. However, the trenching and drilled-hole concepts should also be reasonable alternatives.

The subseabed has many promising characteristics for radioactive-waste disposal. However, one of the most important—inaccessibility to human activities—also tends to make experimental evaluations more difficult. Additionally, the necessary international treaties are likely to pose severe political problems. Current DOE plans do call for selection of experimental sites by the mid-1980s.

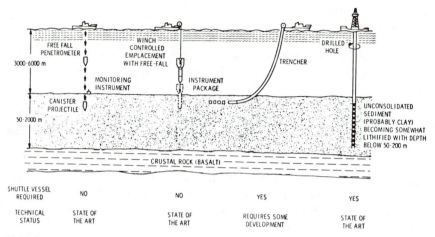

FIGURE 10-12
Engineering concepts for seabed disposal. (Courtesy of U.S. Department of Energy.)

Very Deep Hole Concept

Another potential waste management alternative is to drill or sink a shaft to isolate high-level wastes in a *very deep hole*. The concept relies on the surrounding rock to contain the wastes, and on the great depths to delay their release and reentry into the biosphere. As a point of reference, current technology allows excavation of narrow holes to 10 km and of a wide shaft to about 4 km (as compared to less than 1 km for the proposed WIPP project).

Although DOE funding has been provided for this approach, many serious questions remain unanswered. Until geologic characteristics are evaluated more thoroughly, the fundamental question of "how deep is deep enough" cannot be answered. It is also not known if the hole would need to be cased. Engineering problems associated with emplacement are a third concern.

Partitioning and Transmutation

In the context of waste disposal, the term *partitioning* implies separation of actinide and fission-product wastes from each other. This has the general effect of separating the waste hazards according to their relative lifetimes (e.g., as shown by the curves in Fig. 6-9). The relatively short-lived fission products, then, need only be stored for a few hundred years before they decay out. On the other hand, the actinides may be placed in a repository or handled by a different procedure.

It has been proposed that the separated actinides could be reintroduced into a nuclear reactor. Continued *transmutation* would eventually produce species that fission spontaneously and, thereby, convert actinides to shorter-lived fission products. Since this material would be a neutron poison, fuel-cycle costs would be adversely affected. The production of unrecoverable actinide wastes during refabrication operations suggests a limited usefulness for the transmutation approach to waste management.

Space Disposal

Although it is not part of current DOE programs, extraterrestrial or *space disposal* of radioactive wastes is receiving some attention from the National Aeronautics and Space Administration [NASA]. The cost per unit mass strongly suggests limiting this alternative to separated actinide wastes.

The basic components of a space disposal system are waste containers and a space shuttle, as shown in Fig. 10-13. The final destination could be a solar-escape trajectory, a solar orbit, or impact with the sun itself. Evaluation of the ability to maintain stable orbits for long periods of time, shuttle and container reliability, cost, safety, and international policies will contribute to the ultimate feasibility of this concept.

Other Concepts

Of the 10 alternatives identified in the "Draft Environmental Statement," half have been described above. The remaining entries are interesting (if unfunded) concepts.

Chemical resynthesis would process the wastes into the form of synthetic minerals designed to be compatible and, perhaps, in thermodynamic equilibrium with repository rock. *Rock melting* would be facilitated by emplacing waste into a deep underground hole or cavity. Decay heat would melt the rock that would, in turn, dissolve the waste. When the solution ultimately refreezes, a relatively stable storage medium results. *Reverse-well disposal* relies on injection of waste in liquid or slurry form into deep wells in a manner common to the oil and gas industry.

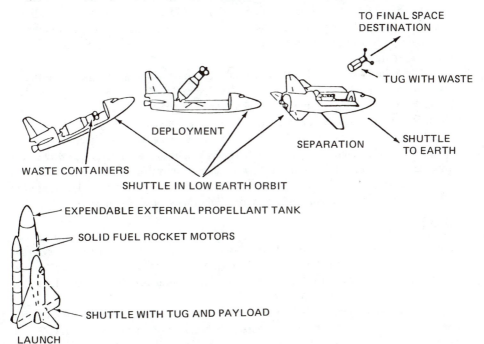

FIGURE 10-13
Space disposal concept for nuclear wastes. (Courtesy of U.S. Department of Energy.)

Island disposal would emplace solidified wastes into a geologic formation located below an island. *Ice sheet disposal* would rely on decay heat to allow a container to melt through a thick sheet of ice. The container would lodge itself on the underlying bedrock, with refreezing providing the necessary isolation.

Beneficial Uses

A number of the constituents in high-level radioactive wastes have the potential for sterilization, heat-source, and/or other industrial applications. Although such *beneficial uses* were not addressed in the "Draft Environmental Impact Statement," DOE does have current programs related to selectively partitioned defense wastes. Future applications to commercial wastes should be straightforward (assuming that reprocessing is facilitated).

Gamma radiation from certain radionuclides has been used for many years to sterilize or otherwise preserve food products. It is also equally possible to sterilize sewage and other waste materials to allow their use as livestock feed. Sandia Laboratories has successfully operated a large sewage-sludge disinfection facility since 1978 on ^{137}Cs extracted from defense wastes.

Isotopic heat sources (e.g., ^{90}Sr, ^{137}Cs, and ^{238}Pu) are readily available. However, the separation costs (in dollars and energy) tend to exceed the potential output and preclude use for other than very specialized applications, e.g., heart pacemakers or spacecraft power supplies.

An energy system based on bulk high-level wastes might produce saturated steam at 180°C over a 10-year period. Security considerations, however, would likely limit siting to a well-controlled site like a military base. Transportation and potential steam damage to containers must also be considered in assessing the overall economy of such a program.

A number of elements which are rare or nonexistent in nature are found in relatively large quantities in high-level wastes. Examples include the valuable catalyst rhodium [Rh], and technitium [Tc], which has potential applications as a corrosion inhibitor, alloying agent, and semiconductor constituent. It has yet to be demonstrated that the fission-product nuclides could be employed as effective and safe replacements. However, since the two materials would be available in quantities near or in excess of expected demands, further consideration is worthwhile.

Radioisotope applications in medicine and industry are also possible. Separation and transportation costs, however, may preclude many or all of these potential uses.

Oklo Phenomenon

It is not possible to perform definitive, real-time experiments on the behavior of long-lived waste constituents prior to implementation of a nuclear waste disposal policy. However, studies of a 1.7-billion-year-old natural reactor—referred to as the *Oklo phenomenon*—provide some reassurance.

In the Oklo region of Gabon in West Africa, it is believed that nature operated one or more fission reactors in some very rich uranium ore deposits (> 10 wt % uranium metal). A combination of thick ore seams, high enrichment,[†] and the presence

---

[†] ^{235}U has a shorter half life than ^{238}U and, thus, was relatively more plentiful at that earlier time. In fact, the isotopic ratio was comparable to that of the slightly enriched uranium now employed in LWR systems.

of groundwater allowed a sustained neutron chain reaction to continue for over 100,000 years. The system is thought to have operated at a low power density in a self-controlling mode based on the water-temperature-feedback mechanism.

Scientific investigations have shown that the plutonium produced by the Oklo phenomenon experienced minimal movement from the site of its initial formation. This demonstrated retention ability provides some assurance that plutonium and the other long-lived actinide elements can be isolated and contained by geologic formations.

EXERCISES

Questions

10-1. Identify the basic steps in spent-fuel reprocessing. Describe at least two major problems that are not present in more conventional chemical engineering operations.

10-2. Describe the three basic principles for handling radioactive wastes. Compare the BWR and PWR in terms of their major operating waste problems as related to the principles.

10-3. Differentiate between low-level, contact-handled, remote-handled, and high-level radioactive wastes on the basis of heat content, radiation level, and packaging for storage and disposal.

10-4. Identify the present and near-future options for interim storage of spent fuel.

10-5. List at least five different waste disposal alternatives. Identify those which are receiving funding support from DOE.

10-6. Identify two similarities between the Oklo reactor and a BWR. Explain the major significance of the phenomenon to radioactive waste management.

10-7. Describe the basic characteristics of the Waste Isolation Pilot Plant [WIPP].

10-8. Identify the beneficial use for the ^{137}Cs in radioactive wastes.

Numerical Problems

10-9. Estimate the separation factors for uranium and
   a. Pu
   b. Zr
   c. gross β
   d. rare earths
   for a reprocessing operation characterized by the distribution coefficients in Fig. 10-2. Assume a nitric acid concentration of $4M/1$.

10-10. Draw the ^{238}U decay chain, including half-lives, from data in the "Chart of the Nuclides." Which radon isotope has the most important environmental role in the nuclear fuel cycle? Why?

10-11. Calculate the enrichment for natural uranium at the time of the Oklo phenomenon. Would a natural reactor be possible now?

10-12. Sandia Laboratories' facility for sewage sludge disinfection uses 15 capsules, each of which contains approximately 70,000 Ci of ^{137}Cs ($E_\gamma = 0.662$ MeV).
   a. Calculate the mass of this radionuclide which has a half-life of 30 years.
   b. A dose of 2×10^5 rad can reduce the population of dangerous salmonella bacteria by a factor of a million. Using the expression in Prob. 3-7, estimate the time a sludge sample would need to spend at 1 m from the total ^{137}Cs source to reduce the bacteria level by this factor.

10-13. A new PWR core has about 800 ppm of soluble boron. Toward the end of the cycle this drops to 20 ppm. In each case, a 10 ppm dilution corresponds to a reactivity change of about 0.1 percent.

    a. Assuming the reactor to be a container that may be filled to any volume, calculate the amount of water that must be added to each initial liter to produce the 10 ppm dilution for fresh and end-of-cycle cores.

    b. What is the significance of the result of (a) to liquid-waste management?

    c. In which case would the tritium level be highest?

10-14. Neutron irradiation of boron produces tritium through three different reactions. Complete the following and write the reaction equations:

    a. $^{10}_{5}B$ (n, 2_____) $^{3}_{1}T$

    b. $^{10}_{5}B$ (n, α) _____ (n, nα) $^{3}_{1}T$

    c. $^{11}_{5}B$ (n, T) _____

The last reaction has a threshold energy of 14 MeV and a cross section of 5 mb. Is it likely to be important in PWR systems? Why?

10-15. In a CANDU reactor, tritium is produced by both ternary fission and radiative capture in deuterium.

    a. Calculate the ^{3}T production rate per unit volume of fuel, assuming a ^{235}U atom density of $2 \times 10^{20}/cm^3$, a thermal flux of $5 \times 10^{13}/cm^2 \cdot s$, and $\sigma_f = 577$ b. Assume also that one ternary fission occurs for every 10,000 fission events.

    b. Calculate the ^{3}T production rate per unit volume of D_2O coolant assuming a deuterium density of $6 \times 10^{22}/cm^3$, $\sigma_a = 0.52$ mb, and the same flux as (a).

    c. How long would the rate in (b) need to be maintained to reach the MPC for ^{3}T in water? (Use $T_{1/2} = 12.3$ years for ^{3}T.)

10-16. It has been projected that the United States might have 625 GW(e) of nuclear installed capacity by the year 2000.

    a. Calculate the volume of solidified high-level waste associated with 1 year's operation.

    b. To what depth would this cover a standard U.S. football field (100 yd $\times$ 50 yd) or soccer field (120 yd $\times$ 80 yd)?

    c. Repeat the above calculations assuming 4500 GW(e)·years of nuclear electricity has been generated through the year 2000.

SELECTED BIBLIOGRAPHY[†]

Reprocessing
    Bebbington, 1976
    Benedict, 1981
    Benedict & Pigford, 1957 (Chaps. 7, 8)
    Erdmann, 1979
    Glasstone & Sesonske, 1967 (Chap. 8)
    Knief, 1981
    Long, 1978
    Nucl. Eng. Int., Aug. 1978
    Sesonske, 1973 (Chap. 7)
    Stoller & Richards, 1961
    WASH-1250, 1973 (Chaps. 1, 2)

[†]Full citations are contained in the General Bibliography at the back of the book.

Wastes and Waste Management
    APS, 1978
    Cohen, 1977
    Cowan, 1976
    DOE/EIS-0046-D, 1979
    ERDA-76-43, 1976
    ERDA-76-162, 1976
    Erdmann, 1979
    Fisher, 1978
    Glasstone & Jordan, 1980 (Chaps. 8–10)
    Haley, 1980
    Hammond, 1979
    Harmon, 1979
    Henry & Turner, 1978
    IRG, 1979
    Koplik, 1979
    Lapp, 1977
    McElroy & Burns, 1979
    Nevada, 1980
    Nucl. Eng. Int., Nov. 1976, Jan. 1978
    Nucl. Tech., Dec. 1974
    NVO-210, 1980
    Oceanus, 1977
    Olds, 1980*b*
    Sagan, 1974
    SAND77-0274, 1977
    SAND77-1401, 1978
    SAND79-0182, 1979
    Strauss, 1979
    WASH-1250, 1973 (Chaps. 1, 4, 7)
    WASH-1297, 1974
    Weart, 1977
    Westinghouse, 1979*b*

IV

NUCLEAR REACTOR SYSTEMS

Goals

1. To identify representative features of the design process for nuclear reactor systems.
2. To describe and compare the basic components and operation of the five reference reactors identified in Chap. 1 and of several other interesting systems
3. To explain the origins and applications of various reactor features in terms of the principles outlined in Part II.

Chapters in Part IV

Reactor Design Process
Light-Water Reactors
Heavy-Water-Moderated and Graphite-Moderated Reactors
Thermal-Breeder Reactors
Fast-Breeder Reactors

11

REACTOR DESIGN PROCESS

Objectives

After studying this chapter, the reader should be able to:

1. Identify the five major categories of the reactor design process.
2. Explain why slightly enriched uranium is a preferred fuel material for LWR systems.
3. Differentiate between water- and nonwater-cooled reactors in terms of their thermal efficiency.
4. Perform calculations of reactor thermal efficiencies.

Prerequisite Concepts

Reactor Steam Cycles	Chapter 1
Reactor Fuel	Chapter 1
Four-Factor Formula	Chapter 4
Reactor Control	Chapters 5–7
Breeding/Conversion	Chapter 6
Nuclear Limits	Chapter 7
Nuclear Power Costs	Chapter 8
Fuel-Pin Designs	Chapter 9

Many of the unique features of nuclear fission have been introduced in the previous chapters. Each of these can have important impacts on the design and operation of nuclear reactors. All specific requirements and/or restrictions must be integrated to assure overall viability of each reactor concept.

PRINCIPLES

The design process may be divided roughly (and arbitrarily) into the following five categories:

1. nuclear design
2. materials
3. thermal-hydraulics
4. economics
5. control and safety

Each is somewhat complex by itself, and much more so when considered in relation to the others. The often-conflicting goals ultimately lead to design compromises.

Nuclear Design

The most explicitly nuclear considerations in reactor design relate to neutron economy. Utilization of the fissile, fertile, and other constituents of the core are the major contributors.

The composition and geometry of fuel assemblies plus their interspersed coolant and/or moderator determine the energy spectrum of the neutrons. This, in turn, determines the relative probability for fission, fertile-to-fissile conversion, and parasitic capture.

Materials

One of the most perplexing problems that faced the reactor pioneers was that their best nuclear designs could not be built and operated because of materials limitations. A number of the related problems are described in Chap. 9.

It is of primary importance to select materials for fuel, cladding, moderator, coolant, and structural components on the basis of compatibility with each other (e.g., minimum corrosion and chemical reaction). Thermal and radiation stability are also necessary characteristics. High strength and other favorable mechanical properties are equally desirable.

Thermal Hydraulics

Proper thermal-hydraulic design is necessary to assure efficient removal of fission energy for ultimate electric power generation. The primary concern is to have temperature distributions and coolant flow characteristics that preclude exceeding either linear-heat-rate or DNB limits.

These requirements tend to favor high thermal conductivities and large heat capacities. High melting and boiling points are also generally desirable characteristics for structural materials and coolants, respectively. (The low melting point for the sodium coolant in the LMFBR is, of course, advantageous).

A large surface-to-volume ratio for the fuel enhances heat transfer. Uniform power density (i.e., low power peaking factors) increases core power capability.

A high steam temperature and a large temperature differential across the turbine both enhance the efficiency of electric energy production. This, in turn, suggests that the reactor core should have a high coolant outlet temperature as well as a large temperature differential (or enthalpy rise).

Economics

The basic economic premise of reactor design is to minimize *overall* energy costs. Since income depends on both the charge rate and the quantity of energy produced, plant reliability is an extremely important economic consideration (as described in Chap. 8). The tendency to favor inexpensive materials is, then, tempered by the necessity for long-term stability.

Capital costs depend on material, fabrication, and construction requirements. Simple, proven technologies may have advantages in terms of cost-outlay and/or reliability. For example, reactors like the HTGR and LMFBR that produce "modern," dry-steam conditions [$\approx$ 540°C or 1000°F] can employ turbine-generators of the type developed for fossil-fuel applications. The water reactors, by contrast, must use specially designed wet-steam units which are more expensive. The latter also have a reduced efficiency for converting thermal energy to electricity, another economic drawback. (Efficiency is considered again later in this chapter.)

Since the net electrical output determines income, low in-plant energy consumption is desirable. Low power requirements for coolant pumping in appropriate primary and secondary loops are a major economic advantage.

Low operating and maintenance [O&M] costs are enhanced by system reliability. Simplified and/or minimized waste handling requirements can also limit those charges. A compact, low-mass—i.e., high-power-density—core tends to have relatively low fuel costs. High burnup can also be an advantage.

An integrated fuel cycle tends to have minimum costs when each step is optimized. Low yellow cake and separative work requirements, coupled with efficient fuel fabrication, are always favored. Convenient and inexpensive spent-fuel storage, reprocessing, and waste management are also desirable.

Control and Safety

Control strategies are developed to facilitate steady-power operations and to limit the severity and/or mitigate the consequences of potential accidents. The goal of safety design is to minimize radiation exposures for operating personnel and the general public.

An integrated control system generally employs control rods to maintain or adjust the power level, compensate long-term depletion effects, and provide scram or trip capability. Soluble and/or burnable poisons may be traded off with control rods as appropriate.

Automatic monitoring and protective systems are employed to identify abnormal operating conditions and shut down the neutron chain reaction as necessary. Negative temperature feedbacks have been noted to have a general stabilizing effect on routine operations

Uniform core power density precludes large temperature gradients and stresses. Lower power densities reduce the potential heat source from fission-product decay.

Reactors and fuel cycle operations designed for multiple-barrier containment of fission-product and actinide materials tend to limit radiation exposures. Low inherent radioactive waste quantities can also reduce relative hazards.

INTERACTIONS

Development of an integrated reactor design requires a substantial amount of interaction among the often-conflicting design principles considered above. Economic and safety considerations provide some of the most stringent constraints. Only those designs that are cost-competitive can be expected to penetrate the electric utility market. On the other hand, only demonstrated safety will allow a system to be licensed (e.g., by the U.S. Nuclear Regulatory Commission according to the process described in Chap. 19) and, therefore, operated. In this sense, then, safety is also an overriding economic consideration.

With economics and safety providing many constraining conditions, the design process may be viewed according to the representative interactions shown by Fig. 11-1. Nuclear, thermal-hydraulic, and materials-design components result in the composition, geometry, and operational specifications that characterize the particular reactor system.

Nuclear design begins with preliminary cross section data. Since the calculational models rely on flux-averaged cross sections (as described in Chap. 4), iterations are required as the design concept evolves. The flux, power distribution, and burnup contribute to a final specification of fuel loading, i.e., enrichments and fuel management pattern.

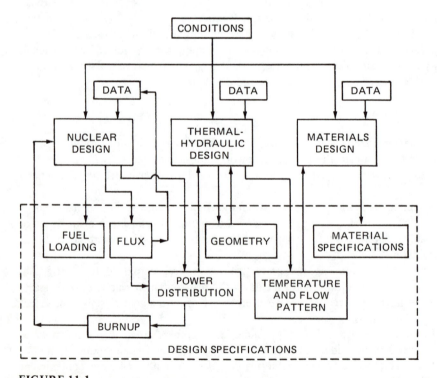

FIGURE 11-1
Reactor design interactions. (From A. Sesonske, *Nuclear Power Plant Design Analysis*, TID-26241, 1973.)

The thermal-hydraulic design block in Fig. 11-1 interacts with the power distribution to develop a geometry and its related temperature and flow distributions. These also contribute importantly to material specifications.

In the interest of simplicity, Fig. 11-1 excludes a number of the important iterative interactions among the components. Economic and safety evaluations for each preliminary design provide the conditions [constraints] appropriate to the next generation of specifications.

The scope of this book precludes a more detailed discussion of actual reactor design practices. However, two examples are provided to illustrate typical procedures.

Fuel-Assembly Design

The design process detailed in Chap. 9 considers oxide fuel pins somewhat in isolation from the reactor as a whole. However, the composition and geometry of the fuel assemblies are very closely coupled to the overall system concept.

A typical core-design approach for an LMFBR outlined by Sesonske[†] is based on nine somewhat arbitrary stages. A negative or otherwise unacceptable result anywhere suggests return to a previous stage with respecification of appropriate parameters. The sequence may be summarized as:

1. Thermal analysis
   - set primary inlet and bulk outlet temperatures
   - assume radial peaking factors to calculate nominal central channel outlet temperature
   - assume axial flux profile and engineering factor to calculate hot-channel coolant temperature
   - calculate clad surface temperature profile for hot-channel assuming a clad surface heat flux and an empirical heat transfer coefficient
   - set clad and gap materials and dimensions
   - calculate fuel-surface temperature profile
2. Fuel pin composition and diameter selection
   - for given fuel material use thermal conductivity and peak temperature to determine limiting linear heat rate [kW/m or kW/ft]
   - set pellet diameter based on consideration of fabrication and fuel inventory costs
   - recalculate heat flux and fuel temperatures
3. Core sizing
   - calculate number of fuel pins from core power and length
   - choose array geometry and spacing
   - compare to acceptible values based on economics (conversion ratio) and safety (reactivity coefficients)
   - calculate axial and radial power profiles
   - calculate required coolant velocity
4. Fuel-cycle economic analysis—calculate fuel-cycle costs based on burnup analysis

[†]A. Sesonske, *Nuclear Power Plant Design Analysis*, TID-26241, 1973, pp. 396–402. This reference contains four detailed flow charts which are not reproduced here. A more comprehensive design study is presented on pp. 404–445.

5. Fuel-pin structural analysis
   - determine backfill pressure and plenum volume based on fission gas release
   - calculate pressure, radiation, and thermal-stress effects to verify suitability of clad selection
6. Hydraulic analysis
   - calculate pressure drop through core and flow distribution for structural design
   - determine total pin length (fuel, gas plenum, and blanket as appropriate)
   - perform pumping power calculation
7. Safety analysis
   - calculate reactivity coefficients and apply them to accident analyses
   - evaluate fuel expansion, slumping, and bowing effects
8. Fuel element reliability analysis—evaluate design on the basis of available test data for similar configurations
9. Post-irradiation handling considerations
   - establish cooling requirements
   - evaluate reprocessing solution compatibility

The depth of coverage involved in each stage is highly variable. The more novel the concept, the more extensive the required analyses. Entirely different pathways and sequences could, of course, also produce the same desired results.

Lattice Effects

Typical fuel assemblies consist of a *lattice* [geometric array] of clad fuel pins. Such arrangements have been developed for high retention of radioactive constituents according to the multiple-barrier safety concept. They also facilitate energy removal as coolant flows lengthwise between the pins.

The use of a lattice arrangement in thermal reactors is crucial to achieving economic goals in both a neutronic and cost sense. Neutron economy tends to be enhanced by an increased resonance escape probability [p in the four-factor formula]. Since fast neutrons from fission are likely to leave the fuel and achieve thermal energies in the external moderator, they tend to escape absorption in the resonance-energy region and experience a net increase in multiplication. Costs are reduced by the ability to achieve high burnup with fuel characterized by relatively low yellow cake and enrichment requirements.

The economic considerations are found to be quite sensitive to both the size of the fuel pins and the spacing between them. The latter effect is important because it controls the fuel-to-moderator ratio and, thus, the extent of neutron thermalization. The following example illustrates the interplay among lattice parameters, burnup, enrichment, and fertile-to-fissile conversion for fuel characteristic of LWR systems.

The LWR designs are characterized by use of water moderator and slightly enriched uranium fuel. For fuel pins of fixed geometry, moderation effects are determined by pin spacing and water density or, equivalently by the hydrogen-to-uranium ratio [H/U] of the lattice. Figures 11-1 and 11-2 illustrate the effects of enrichment and H/U on the initial conversion ratio and fuel burnup for 1.0-cm-diameter rods in large, critical arrays.

The initial (i.e., early-in-burnup-lifetime) conversion ratio is seen from Fig. 11-1

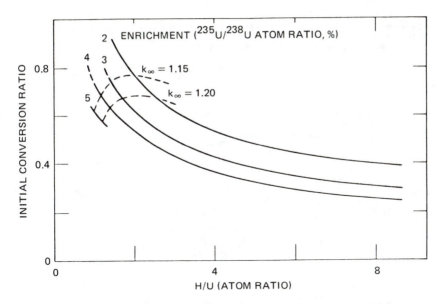

FIGURE 11-2
Initial conversion ratio for 1.02-cm [0.400-in] fuel pins in a large, critical reactor. (From A. Sesonske, *Nuclear Power Plant Design Analysis*, TID-26241, 1973.)

to be enhanced by low H/U and low enrichment. In such a "dry," tightly packed lattice, fission neutrons are not well moderated and, thus, are readily absorbed in the resonances. Low enrichment leads to the highest conversion as most neutrons are absorbed in ^{238}U resonances, ultimately producing ^{239}Pu. At increased enrichment, ^{235}U fission resonances compete for neutrons and, thus, reduce conversion.

As might be expected, the lower the fuel enrichment, the shorter the burnup lifetime for any given H/U, as shown by Fig. 11-3. Burnup falls rapidly at low H/U as the "dry" lattice does not thermalize neutrons sufficiently to take advantage of the large thermal-fission cross section for ^{235}U. At higher H/U the lattice becomes too "wet" and results in excessive leakage and hydrogen absorption. The shape of the curves in Fig. 11-3 suggests that an H/U of about 3 is optimum. For a typical LWR enrichment of 3 wt % [$\approx$ 3 percent atom ratio], this corresponds to a burnup of 25,000 MWD/T. A conversion ratio of about 0.6 would be expected according to Fig. 11-3.

Although the example illustrates qualitative effects well, it must be noted that the diameter of the fuel pin also has an important impact. The conversion and burnup behavior must ultimately be balanced with fuel-cost, thermal-hydraulic, and safety considerations for a viable design to be obtained.

REACTOR SYSTEMS

Each nuclear reactor has some common features based on its utilization of the fission process. Differences also exist due to the wide range of materials and geometries that can sustain a neutron chain reaction.

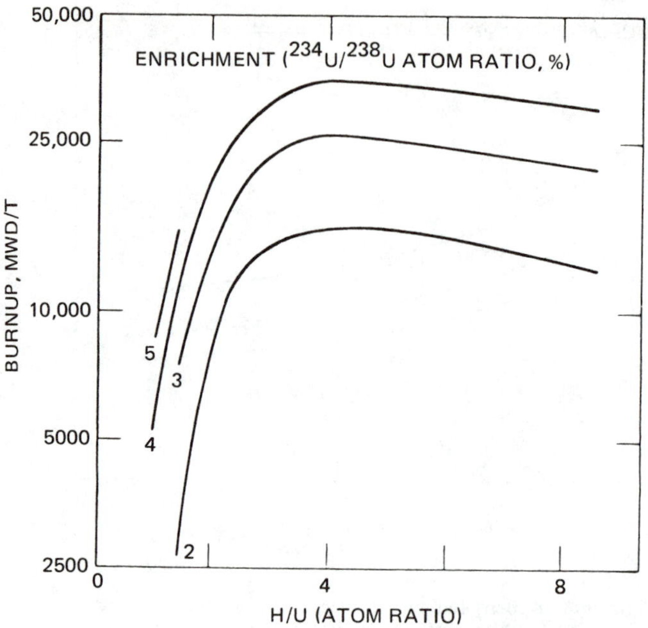

FIGURE 11-3

Burnup lifetime for 1.02-cm [0.400-in] fuel pins in a large, critical reactor. (From A. Sesonske, *Nuclear Power Plant Design Analysis*, TID-26241, 1973.)

The remaining four chapters in this part describe some basic features of the five reference reactors and several other systems. Coverage is generally limited to the areas of:

- steam cycle
- fuel assemblies
- reactivity control
- protective system
- power monitoring

Consideration of important features related to safety design is deferred to Chaps. 16 and 17.

Steam Cycle

Current application of commercial nuclear power is almost entirely for production of electricity by use of a steam cycle. The one-, two-, and three-loop steam-cycle concepts were described briefly in Chap. 1. All reactors have a reactor vessel and primary-coolant loops. Heat exchangers may provide an interface to one or more secondary loops.

Despite the actual arrangement of the energy-removal system, the primary objective of the reactor is to convert the heat energy from fission into electrical energy. The efficiency of the process ultimately depends on the steam conditions that exist at the turbine-generator.

According to the fundamental laws of thermodynamics, it is not possible to convert heat completely to another energy form. The theoretical maximum conversion is limited by the operating temperatures of the conversion system according to the expression

$$\eta = \frac{T_{in} - T_{out}}{T_{in}}$$

for efficiency η, and *absolute* [kelvin K or rankine R] inlet and outlet temperatures T_{in} and T_{out}, respectively. The quantity η—also known as the *Carnot efficiency*—is enhanced by a large temperature differential and, thus, a low outlet temperature.

Steam inlet temperatures at the turbine are ultimately limited by operating characteristics of the reactor and any associated heat-transport systems. Turbine outlet temperatures are controlled by the availability of condenser cooling water. Once-through cooling from a river or lake generally provides the lowest outlet temperature and the best efficiency. Artificial lakes and cooling towers tend to result in lower efficiencies, but they do have some advantages in terms of minimizing potential environmental impacts (as described in Chap. 19).

The *thermal efficiency* of a power plant—the ratio of the electrical energy output to the thermal energy generated, e.g., MW(e)/MW(th)—is an important cost consideration. The higher the efficiency, the better the utilization of the fuel. Since condenser temperatures are not readily subject to reduction,[†] maximum steam temperatures are favored to enhance efficiency.

All light- and heavy-water reactors require the presence of liquid-phase coolant to provide the moderation necessary for criticality. If the *critical temperature* of 375°C [706°F] for water is exceeded, liquid cannot exist even at arbitrarily large pressures. A practical operating limit of about 340°C [650°F] is needed to assure both criticality and heat-removal capacity (since steam is a very poor coolant compared to liquid water). The critical temperature must also be avoided to preclude the large pressure change that would accompany a general liquid-to-vapor transition in the coolant. A change of such magnitude would certainly compromise the integrity of the reactor vessel and the entire primary coolant system.

These concerns have the effect of limiting steam temperatures and thermal efficiencies. Since it is possible for liquid water and steam to coexist, moisture separators and "wet-steam" turbines must be employed. Typical thermal efficiencies are held to 34 percent or less. The special turbines and relatively low efficiencies each constitute economic penalties.

Gaseous, liquid-metal, and molten-salt coolants can be used to produce "dry" steam at well above the critical temperature. "Modern" steam temperatures of about 540°C [1000°F] allow the use of "conventional" turbines characteristic of fossil-fueled plants. Thermal efficiencies of 40 percent or greater result.

Since even the best steam-cycle efficiencies are relatively low, alternative uses are being considered for the fission energy. A direct, gas-turbine cycle, for example, is feasible for the HTGR. A number of process-heat applications may also be appropriate, especially for the high-temperature systems.

[†]Seasonal changes do produce condenser temperature changes and variations in thermal efficiency.

Other Considerations

The structure of fuel assemblies may be coupled to the mechanisms and procedures for reactivity control, protection, and flux monitoring. Examples are provided in Chaps. 12–15 for the following reactor systems:

Light-water reactors [LWR]
- Boiling-water reactors [BWR]
- Pressurized-water reactors [PWR]

Pressurized-heavy-water reactors [PHWR]
High-temperature gas-cooled reactors [HTGR]
Pressure-tube graphite reactors [PTGR]
Thermal-breeder reactors
- Light-water breeder reactor [LWBR]
- Molten-salt breeder reactor [MSBR]

Fast-breeder reactors
- Liquid-metal fast-breeder reactor [LMFBR]
- Gas-cooled fast reactor [GCFR]

EXERCISES

Questions

11-1. Identify the five major areas of reactor design interaction. Describe at least two components of each.

11-2. Describe the nuclear design basis for use of ≈ 3 wt % enriched uranium fuel for the LWR designs.

11-3. Define Carnot efficiency. Why do LWRs operate at lower thermal efficiency than LMFBRs?

Numerical Problems

11-4. Consider as a reference the H/U ratio, which maximizes the reactivity of a particular LWR fuel-pin lattice. Will operation with a lower ratio [under-moderated system] or a higher ratio [overmoderated system] result in a negative moderator-temperature coefficient? Why?

11-5. The condenser is designed as a heat exchanger between the cooling water and the turbine outlet. Assuming a theoretical efficiency of 32 percent and a steam inlet temperature of $290°C$, calculate the effective outlet temperature.

11-6. The temperature of a river used for power plant cooling varies from a minimum of $5°C$ in the winter to $30°C$ in the summer. Assume that the result of Prob. 11-5 is based on the minimum temperature and that half of any increase must be added to the turbine outlet temperature. Calculate the thermal efficiency for the maximum river temperature. What fractional reduction in electrical output would this represent?

SELECTED BIBLIOGRAPHY[†]

Reactor Design Analyses
    Etherington, 1958 (Chap. 13)
    Glasstone & Sesonske, 1967 (Chaps. 12, 13)
    Sesonske, 1973 (Chaps. 7–9)

[†]Full citations are contained in the General Bibliography at the back of the book.

12

LIGHT-WATER REACTORS

Objectives

After studying this chapter, the reader should be able to:

1. Sketch the steam cycles for the two different LWR designs.
2. Identify the major differences between BWR and PWR fuel assemblies.
3. Differentiate between the BWR and PWR systems in terms of routine reactivity control and trip shutdown.
4. Identify and explain four trip parameters in a typical protective system.
5. Describe the basic operating principles for boron-ion-chamber, fission-chamber, and rhodium detectors.

Prerequisite Concepts

LWR Steam Cycles and Fuels	Chapter 1
Nuclear Reactions	Chapter 2
Reactivity Control Methods	Chapters 4–7
Fuel and Moderator Feedbacks	Chapter 5
LWR Fuel Design	Chapter 9
In-Core Fuel Management	Chapter 9
Reactor Design	Chapter 11
LWR Fuel Lattices	Chapter 11

The light-water reactors [LWR] have many similarities based on the dependence on ordinary water for cooling and moderation. Fuel pins, for example, have comparable geometries and composition.

The boiling-water reactors [BWR] employ a direct cycle where steam is produced in the reactor core. The fuel assemblies and control systems are tailored to the presence of the boiling coolant.

The pressurized-water reactors [PWR] maintain the coolant in its liquid form under the influence of high primary-system pressure. Steam is produced in heat exchangers known as steam generators. Fuel assemblies and control methods are again consistent with the nature of the coolant.

BOILING-WATER REACTORS

Major evolution of the boiling-water reactor design has been centered with the General Electric Company [GE]. They are the sole vendor in the United States and currently dominant elsewhere in the world (often through licensing agreements with local manu-facturing concerns). Large-scale systems have also been developed by AEG and KWU in the Federal Republic of Germany, ASEA-Atom in Sweden, and Hitachi and Toshiba in Japan. Since all are quite similar conceptually, the GE BWR-6 system is considered as representative for the purpose of the discussions which follow.

Steam Cycle

The basic features of the steam system for the direct-cycle BWR are shown in Fig. 12-1. Feedwater enters the reactor vessel, has its flow adjusted by the recirculation system, and leaves as steam. The high-pressure [HP] turbine stage receives steam at about 290°C [550°F] and 7.2 MPa [1000 psi]. By use of successive low-pressure [LP] stages and a condenser loop in a standard regenerative cycle, a maximum thermal efficiency of about 34 percent is obtained.

Details of the BWR reactor vessel are shown by Fig. 12-2. The active core height of 3.8 m may be noted to be a relatively small fraction of the 22-m interior height of the vessel. Control rods and drives occupy the space below the core. The upper portion of the vessel is occupied by an extensive separator-and-dryer network (required be-

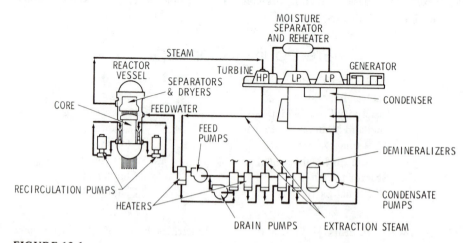

FIGURE 12-1

Boiling-water reactor [BWR] steam cycle schematic diagram. (Courtesy of General Electric Company.)

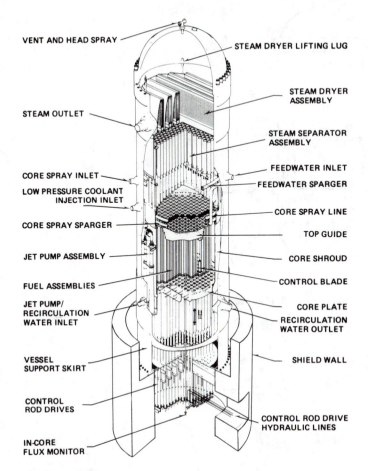

VENT AND HEAD SPRAY

STEAM DRYER LIFTING LUG

STEAM DRYER ASSEMBLY

STEAM OUTLET

STEAM SEPARATOR ASSEMBLY

CORE SPRAY INLET

FEEDWATER INLET

LOW PRESSURE COOLANT INJECTION INLET

FEEDWATER SPARGER

CORE SPRAY LINE

CORE SPRAY SPARGER

TOP GUIDE

JET PUMP ASSEMBLY

CORE SHROUD

FUEL ASSEMBLIES

CONTROL BLADE

JET PUMP/ RECIRCULATION WATER INLET

CORE PLATE

RECIRCULATION WATER OUTLET

VESSEL SUPPORT SKIRT

SHIELD WALL

CONTROL ROD DRIVES

CONTROL ROD DRIVE HYDRAULIC LINES

IN-CORE FLUX MONITOR

FIGURE 12-2
Boiling-water reactor [BWR] vessel. (Courtesy of the General Electric Company.)

cause the system is operated under two-phase [saturation] conditions). The vessel has an inner diameter of about 6.4 m and is constructed of 15-cm-thick stainless-steel-clad carbon steel.

The recirculation and jet pumps shown on Figs. 12-1 and 12-2, respectively, are especially important to the BWR concept. They allow fine control of core flow rate and, through moderator void feedback, of power level. Water enters the feedwater inlet, flows downward between the vessel wall and the shroud, is distributed by the core plate, flows upward through the core and upper structure, and leaves by the steam outlet. About 30 percent of the feedwater flow, however, is diverted to the two recirculation loops. The water is then circulated at high pressure and velocity to the jet-pump nozzle as shown by Fig. 12-3. This forced flow causes suction which, in turn, draws feedwater into the jet pump as shown by Fig. 12-4. The resulting mixed flow in the jet pump has the net effect of increasing the total coolant flow rate above that based on the system's pressure differential alone.

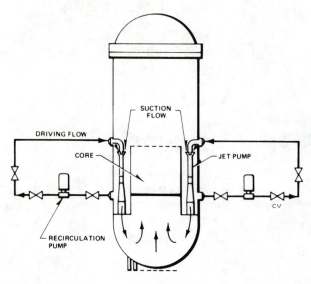

FIGURE 12-3
Jet pump recirculation system for a BWR. (Courtesy of General Electric Company.)

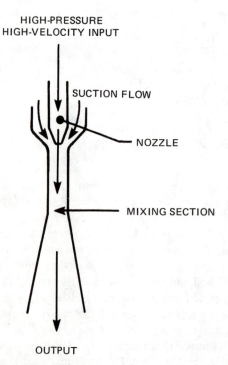

FIGURE 12-4
Principle of BWR jet-pump operation.

Fuel Assemblies

The basic features of the BWR fuel assembly are shown in Fig. 1-7 from Chap. 1. A cross-section of a four-assembly *fuel module* and a core arrangement are provided in Fig. 12-5. Each module includes a cruciform-shaped control blade and the adjacent fuel assemblies.

As described in Chaps. 1 and 9, BWR fuel assemblies consist of zircaloy-clad fuel rods containing slightly enriched UO_2 pellets. The 8×8 fuel pin array is surrounded by a zircaloy fuel channel to prevent cross-flow of coolant between bundles. (In the absence of the channel, general boiling could void large regions of the core and result in general DNB or dryout conditions.)

Fuel rods of up to four different enrichments are loaded into each assembly. Whenever the control rod is not fully inserted in the fuel module of Fig. 12-5, the space is filled with water. This results in a substantial thermal-flux peak which would tend to produce high power density in neighboring pins. By loading fuel pins of lower enrichment next to the water channel and the water gaps between modules, power peaking can be reduced and power capability enhanced. The corner pins have the lowest enrichment because there is water on two sides. The other pins are arranged to flatten the power distribution for the assembly as a whole. The current BWR-6 design also employs two unfueled rods to which water can be added as an internal moderator.

Reactivity Control

Short-term control of reactivity is provided by recirculation flow and poison control rods. Whenever feasible, flow adjustment is the favored procedure.

The recirculation system in Fig. 12-3 employs the jet pumps to control the general flow of water through the reactor core. Because of the negative moderator void feedback, core power varies with coolant/moderator flow rate. Increased flow rates tend to reduce coolant temperatures and the amount of boiling. The resultingly

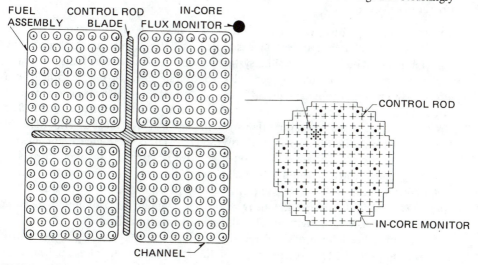

FIGURE 12-5
Fuel module and core arrangement for a BWR. (Adapted from General Electric Company.)

higher water density makes neutron moderation more effective and, thus, increases the reactivity. Decreased flow rates enhance boiling and lower the reactivity. Under normal operating conditions, a critical system may be made supercritical (or subcritical) by increasing (decreasing) flow. The continuing feedback interactions, however, cause criticality to be reestablished at a higher (lower) power level. Power changes up to 25 percent are readily implemented by this procedure.

Larger power level changes, as well as long-term depletion compensation, are accomplished by using the boron-carbide [$B_4 C$] loaded cruciform control rods. As shown by Fig. 12-5, one such four-bladed rod is associated with each of the fuel modules in the core. The rods have a poison length comparable to the active fuel height. They are driven from the bottom of the core because their reactivity worth is greater in the liquid region than in the steam-water region toward the top. During normal operation, incremental movements are implemented electrically with a locking-piston design that prevents inadvertent removal. Each of the 137 control rods in the BWR-6 design may be moved independently to facilitate power-level changes, power shaping, and/or depletion compensation.

Long-term reactivity control is also provided by burnable poisons. A currently favored procedure employs gadolinia [$Gd_2 O_3$] mixed uniformly into the UO_2 pellets in several fuel pins in each fuel assembly. The gadolinia-loaded rods may also be positioned to reduce power peaking in the bundle. A disadvantage is the residual poisoning from the unburned gadolinia that stays with the fuel through its depletion lifetime.

Another viable burnable poison is the boron-loaded curtains which have been employed in earlier BWR systems. They are designed for placement in the gaps between fuel modules. The curtains cause a potentially undesirable flux depression, but they have the advantage of being fully removable at any time fuel is loaded or unloaded.

Protective System

The BWR reactor protective system is designed to insert all of the control rods (i.e., "trip" or "scram" them) if certain designated operating limits are exceeded. The insertion is facilitated by a hydraulic system employing high-pressure nitrogen. Under routine conditions, a normally open relay controlling nitrogen release is held in a closed position by an electric current. Loss of current from a trip signal or from a power failure allows the relay to open and cause the rods to be fully inserted. The BWR protective system has features that are similar in *concept* to those described later in this chapter for the PWR design (see GE/BWR-6 (1978) or other references for details).

It is important that there be enough negative reactivity in the control rods to assure complete neutronic shutdown at all times during the core lifetime. In practice, a *shutdown margin* of several percent in reactivity [$\% \Delta\rho$ or $\% \Delta k/k$] is desirable. The *stuck rod criterion* calls for the margin to be met under the assumption that the control rod of highest reactivity worth fails to insert into the core.

A typical breakdown of BWR reactivities near beginning-of-core-lifetime [BOL] is shown in Table 12-1. Several percent of the reactivity of the clean, unirradiated core is offset at operating temperatures by the void-feedback effect. Control reactivity is provided by a combination of control rods and burnable poisons. Since cool-

TABLE 12-1
Typical Reactivity Inventory for a Boiling-Water Reactor [BWR] System[†]

		Reactivity, % $\Delta k/k$
Clean, 20°C		+ 25
Control rods (highest worth rod stuck)	− 17	
Burnable poisons	− 12	
Total control		− 29
Shutdown margin		− 4

[†]From A. Sesonske, *Nuclear Power Plant Design Analysis*, TID-26241, 1973.

down of the core returns the void reactivity, the shutdown margin is computed from the cold, clean condition. A similar margin must be assured at all times in core life as the fuel and burnable poisons deplete.

Power Monitors

The BWR uses several types of in-core monitors in locations shown by Fig. 12-5 to map power and flux levels and distributions. Boron-lined ionization chambers employ the ^{10}B(n, α) reaction to allow the neutron level to be measured in terms of the electric current from the charged alpha particles. In a similar manner, a fission-chamber detector generates a current from charged fragments as neutrons cause fission in a thin foil. Neutron levels are correlated to power for control and protection operations.

Fission chambers give accurate flux readings during startup and low-power operations. The boron-lined chambers are employed to monitor activities from inter-mediate- to full-power operating levels. Detailed flux mapping may be accomplished through the use of axially movable ionization chambers called traversing in-core probes [TIP].

PRESSURIZED-WATER REACTORS

A majority of the world's pressurized-water reactors [PWR] have been manufactured by the Westinghouse Electric Company. Although most of these systems have been built in or exported from the United States, some are the result of licensing agreements with other international corporations.

The other domestic vendors of PWR systems are the Babcock & Wilcox Company and Combustion Engineering, Inc. Major international operations are conducted in the Federal Republic of Germany (BBR, KWU, and Siemens), France (Framatome), and Japan (Mitsubishi). The U.S.S.R. has large domestic operations centered in Atom-mash and exports through Atomenergoexport.

The state-of-the-art technology of the three U.S. vendors is the basis for discussions in this section. All other PWR systems have features similar to the ones embodied in these designs.

Steam Cycle

The principle features of the steam system for the two-loop PWR are shown in Fig. 12-6. The primary loop contains liquid water at a pressure of 15.5 MPa [2250 psi].

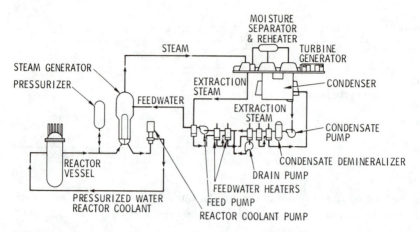

FIGURE 12-6

Pressurized-water reactor [PWR] steam cycle schematic diagram. (Adapted from K. C. Lish, *Nuclear Power Plant Systems and Equipment.* Published and copyright © 1972 by Industrial Press, Inc. Reprinted with permission.)

The core outlet temperature is about 340°C [650°F]. The coolant pump regulates flow in the loop and the reactor vessel. The pressurizer is designed to maintain the system within a specified range of pressures.

The secondary loop provides feedwater to the steam generator which, in turn, produces steam at about 290°C [550°F] and 7.2 MPa [1000 psi]. Since the PWR steam conditions are essentially the same as those for the BWR, the regenerative turbine cycles in Figs. 12-1 and 12-6, respectively, are identical, as is the typical 34 percent thermal efficiency. The two-loop PWR arrangement, however, confines fission-product and other activity to the primary loop and does not allow its general entry to the turbine. (Some of the differences between the PWR and BWR systems in terms of their related waste handling problems are discussed in Chap. 10.)

A typical PWR reactor vessel is shown in Fig. 12-7. It is about 13 m high by 6.2 m in diameter. The vessel is constructed of low-alloy carbon steel with a wall thickness of about 23 cm, including a 3-mm stainless-steel clad on the inner surface. Coolant enters the vessel inlet nozzle, flows downward between the vessel and the core barrel, is distributed at the lower core plate, flows upward through the core, and leaves by the outlet nozzle.

The pressure of the primary loop is maintained by use of a pressurizer like that of Fig. 12-8. The device is designed to contain steam in the upper portion and liquid water below. A positive pressure surge is compensated by introducing water through the spray nozzle to condense some of the steam. A negative surge activates an electrical heater which converts some of the water to steam with a resulting pressure increase.

Steam generators are employed to transfer heat energy from the primary loop to the secondary loop. One Westinghouse design uses four separate steam generators in the configuration shown in Fig. 12-9. Each is nearly 21 m high with a 4.5-m diameter in the upper portion. Other PWR systems employ two, three, or four steam generators of somewhat different sizes.

Two different types of steam generators are employed in current PWR systems. In each case, the primary coolant flows through tubes (the "tube side" of the heat exchanger) while the secondary water occupies the space between the tubes and the outer shell (the "shell side" of the heat exchanger). The Westinghouse and Combustion systems use *U-tube steam generators* similar to that shown in Fig. 12-10a. Primary coolant enters at the bottom left, flows inside a bundle of tubes (of which only two

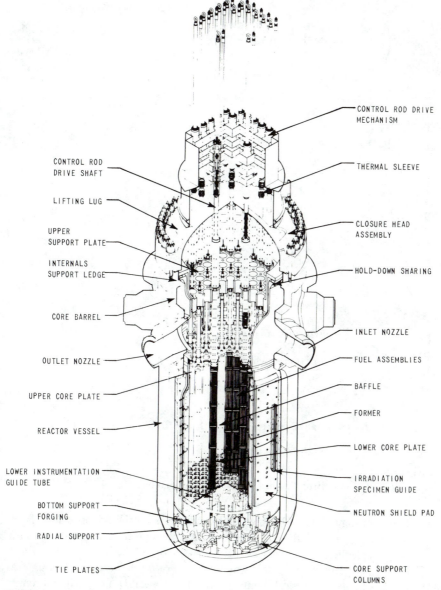

FIGURE 12-7

Pressurized-water reactor [PWR] vessel. (Courtesy of Westinghouse Electric Corporation.)

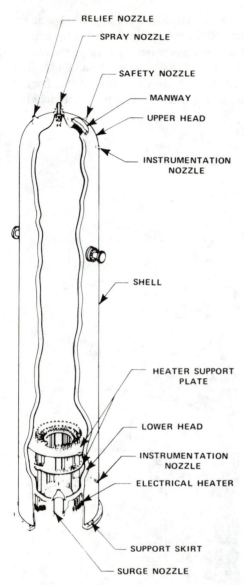

RELIEF NOZZLE

SPRAY NOZZLE

SAFETY NOZZLE

MANWAY

UPPER HEAD

INSTRUMENTATION
NOZZLE

SHELL

HEATER SUPPORT
PLATE

LOWER HEAD

INSTRUMENTATION
NOZZLE

ELECTRICAL HEATER

SUPPORT SKIRT

SURGE NOZZLE

FIGURE 12-8
Pressurizer for PWR. (Courtesy of Westinghouse Electric Corporation.)

are shown by the figure) in a U-shaped path, and leaves on the opposite side. Feed-water is introduced toward the bottom, flows counter-current among the U-tubes, and travels through the steam-separator region at the top of the unit prior to leaving.

The Babcock & Wilcox system employs a *once-through steam generator* as shown by Fig. 12-10*b* (in a to-scale comparison with a typical U-tube design). Primary coolant is introduced at the top, flows downward through straight tubes, and exits through two outlets. Feedwater enters the steam generator, flows downward near the

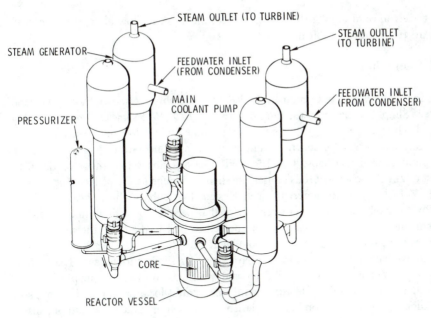

FIGURE 12-9
Arrangement of a PWR nuclear steam supply system. (Courtesy of Westinghouse Electric Corporation.)

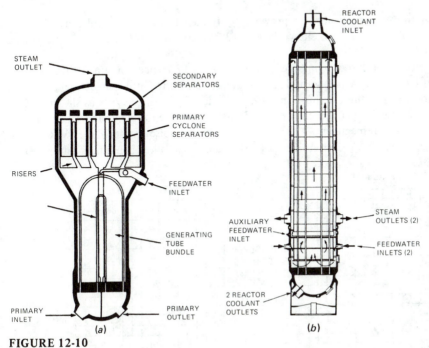

FIGURE 12-10
Comparison of features and sizes of (*a*) U-tube and (*b*) once-through steam generators for PWR systems. (Adapted courtesy Babcock & Wilcox Company.)

wall, travels upward among the tubes, reverses direction to flow along the outer wall, and exits from one of the steam outlets.

Fuel Assemblies

The basic features of the Combustion PWR fuel assembly are shown by Fig. 1-8 from Chap. 1. Simplified patterns for the 16 × 16 and 17 × 17 designs are provided in Fig. 12-11. Each square in the grid represents a possible fuel-pin location. The circles indicate locations reserved for control-rod insertion or instrumentation.

As described in Chaps. 1 and 9, PWR fuel assemblies like those of the BWR consist of zircaloy-clad fuel rods containing slightly enriched UO_2 pellets. The 16 × 16 and 17 × 17 arrays each represent recent designs that enhance linear-heat-rate limits. The PWR assemblies employ an open lattice that permits some flow mixing between adjacent units. Since the coolant is not allowed to boil in the core, the possibility of general voiding is not the problem it could be in a BWR.

Individual PWR fuel assemblies have pins of a single enrichment. The 17 × 17 lattice employed by Babcock & Wilcox and Westinghouse has 24 one-pin locations occupied by control-rod guide tubes with one central location containing an instrument tube. The 16 × 16 Combustion design employs four guide tubes and one instrument tube, each of which displaces four pin locations in the lattice.

Reactivity Control

Short-term reactivity control is provided by full-length poison control rods. The control rod drive mechanisms shown in Fig. 12-7 cause the rods to enter from the top of the core.

The control rods employed with the 17 × 17 lattice contain 24 fingers connected by a spider as shown in Fig. 12-12. Each is inserted into a single fuel assembly. Control rods for the Combustion 16 × 16 lattice are of three types as shown in Fig. 12-13. The four-finger rod enters a single assembly while the eight- and 12-finger span three and five, respectively. Although the traditional poison material for PWR control rods has been B_4C, current designs favor use of an alloy containing silver (80%),

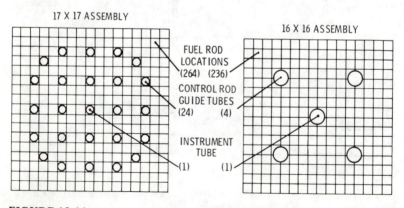

FIGURE 12-11
Fuel-assembly grid for current PWR designs.

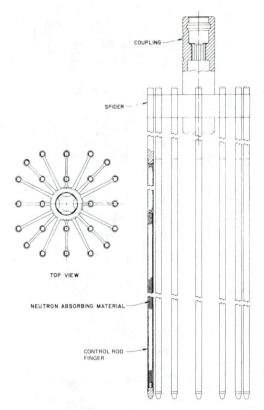

FIGURE 12-12
Control rod with 24 poison fingers for PWR use. (Courtesy of Babcock & Wilcox Company.)

indium (15%), and cadmium (5%). The latter are slightly weaker absorbers and tend to produce somewhat less severe flux gradients than their B_4C counterparts.

Most PWR control rods have a poison length equivalent to the active fuel height. However, some *part-length rods* [PLR] have a partial poison loading. Typical configurations (modifications of Figs. 12-12 and 12-13a) have the lower 25 percent of each finger occupied by standard poison material, while the remainder contains stainless steel. These rods have a unique capability for shaping the axial flux that is not available from the full-length rods. The PLR can be particularly useful for control of axial xenon oscillations, should they occur.

A representative control rod pattern for the Combustion PWR design is shown in Fig. 12-14. Routine control operations are accomplished using the regulating rods. Those remaining are employed for shutdown.

Groups of from four to nine regulating rods (identified in Fig. 12-14b) are moved as a unit. They are arranged in a symmetric pattern that suppresses the core flux in a relatively uniform manner with low power peaking.[†] Only the first regulating

---

[†]Previous Figs. 7-9 and 7-10 are generally representative of the kind of power peaking that could be expected for insertion of the first control group identified in Fig. 12-14. The axial shapes are based on core-averaged values while the radial shapes would be characteristic for one of the rows of fuel assemblies into which the rods are inserted (e.g., the vertical row containing the central control rod and the two other Group-1 rods).

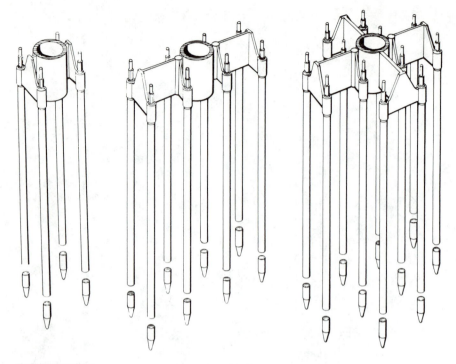

FIGURE 12-13
Control rods with four, eight, and 12 poison fingers for PWR use. (Courtesy of Combustion
Engineering, Inc.)

group is employed at full power, with the others entering the core sequentially with
power reduction. In a typical planned shutdown from full power, for example, the
first regulating group is driven fully [100 percent] into the core. The second group
begins its travel when the first reaches the 60 percent mark. The other regulating
groups also start into the core when their predecessors reach 60 percent of full insertion. Shutdown-rod groups are driven into the core to assure a margin of subcriticality.
Restart of the system employs the reverse of this procedure.

The eight part-length rods [PLR] in Fig. 12-14 are generally operated in two
symmetric groups of four each. As noted previously, their role is related predominately to axial-flux shaping rather than reactivity control.

Intermediate- to long-term reactivity control is provided largely through the use
of soluble poison (also called "chemical shim") in the form of boric acid [*soluble
boron* or simply *solbor*]. As has been discussed in Chaps. 6 and 7, its use minimizes
the need for routine control-rod insertion and thereby provides for a more uniform
power distribution, i.e., low peaking and enhanced power capability. Under normal,
full-power operating conditions, it is typical for *only* the first regulating group (e.g.,
the nine four-finger rods of Fig. 12-14) to be inserted, and to less than 25 percent of
the core length.

The critical soluble boron concentration decreases with fuel burnup as shown in
Fig. 12-15. The poisoning effect of the initial buildup of ^{135}Xe and ^{149}Sm requires

dilution of the boron as shown by the sharp drop of the curve at low burnup. Continued steady-state operation allows gradual dilution and/or natural boron-depletion processes to match fuel burnup effects. When it becomes necessary to shut down and restart the system during the cycle, active solbor addition and dilution programs may be implemented as necessary to minimize control rod use.

Since soluble poisons tend to make the moderator temperature coefficient more positive (by the mechanism described in Chap. 5), fixed burnable poisons are often employed as well. Their presence is seen from Fig. 12-15 to reduce the solbor

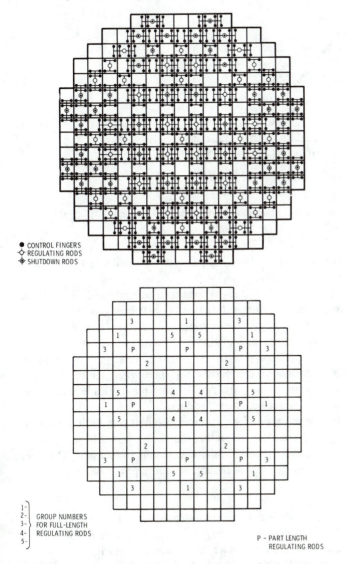

● CONTROL FINGERS
✧ REGULATING RODS
✦ SHUTDOWN RODS

1-
2- GROUP NUMBERS
3- FOR FULL-LENGTH
4- REGULATING RODS
5-

P - PART LENGTH
     REGULATING RODS

FIGURE 12-14

Representative control-element pattern and regulating group designation for the Combustion Engineering System 80 PWR. (Adapted courtesy of Combustion Engineering, Inc.)

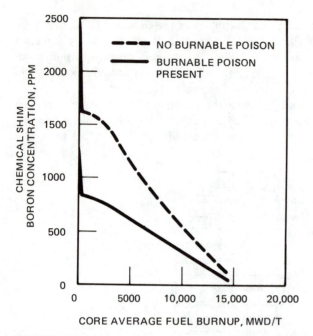

FIGURE 12-15
Typical soluble boron concentration as a function of burnup for a fresh core. (Courtesy of Westinghouse Electric Corporation.)

requirements. The differential, however, becomes successively smaller as the burnable poisons deplete throughout core lifetime.

The burnable poisons for the PWR designs consist of separate rods (sometimes called "shim rods") placed into the fuel-rod lattice. The Babcock & Wilcox and Westinghouse designs employ 9 to 20 poison rods attached to a spider (i.e., an arrangement similar to that of the control rod in Fig. 12-12) and inserted into fuel-assembly guide tubes. Current Babcock & Wilcox designs use pellets of relatively dilute boron carbide [B_4C] in a matrix of aluminum oxide [Al_2O_3]. The pellets are contained in zircaloy cladding tubes (analogous to the structure of fuel pins). The Westinghouse designs use borosilicate glass rod in stainless-steel cladding.

The individual burnable poison rods for the current Combustion PWR are similar to those used by Babcock & Wilcox. However, rather than being placed in guide tubes, they occupy fixed lattice locations that would otherwise belong to fuel pins. The shim rods are generally positioned in a manner that reduces assembly power peaking.

Protective System

All full-length control rods in the PWR systems are mounted to their drives by electromagnets. Any interruption of the magnet current, by loss of power or through a trip signal, causes the rods to drop into the core under the influence of gravity.

A typical breakdown of PWR reactivities near beginning-of-core-lifetime [BOL] is provided in Table 12-2. The reactivity of the cold, unirradiated core is reduced by 5 percent upon achieving full power due to the fuel- and moderator-temperature defects. Equilibrium xenon and samarium produce an additional decrease of 3 percent.

Consistent with the general PWR control philosophy, the worths of the soluble and burnable poisons are sufficient to compensate the excess core reactivity at full-power operating conditions. The control rods are available to counteract the temperature-defect and xenon worths that reappear with or following shutdown. It is especially important that the solbor not be diluted and replaced by control rods at full power. If the system is critical at a given large insertion, the reactivity "investment" is, of course, not available for shutdown purposes. Control rod use must, then, be matched by power-level reductions, i.e., power-dependent insertion limits [PDIL] are imposed on reactor operations. As noted previously, a typical full-power limit is about 25 percent for the first regulating group.

The PWR protective system (and that of the BWR as well) is designed to initiate a scram or trip of all full-length control rods if predetermined parameter limits are or appear to be exceeded. Figure 12-16 indicates the fundamental inputs to the Westinghouse system. The direct trip signals are based on:

- high flux
- high pressurizer pressure
- manual, operator-initiated action
- safety injection (automatic activation of one or more of the emergency core cooling systems [ECCS] described in Chap. 17)
- low–low steam generator water level (a certain range of low levels provide warning signals to the operators; at the low end of the range a trip is induced)
- steam/feedwater flow mismatch
- temperature differential indicating an overpower condition
- temperature differential for overtemperature

The system is highly redundant since any signal (other than, perhaps, the manual one) is likely to be closely followed by a related signal for a different parameter (e.g., high flux and overpower ΔT). With a control-rod trip, the protective system also initiates appropriate operating changes in both the primary and secondary coolant loops.

The interlock block in Fig. 12-16 provides another series of trip signals with limits predicated on the turbine load and average neutron flux. These are of primary use during various power level changes where flux and coolant conditions are expected to vary with time. The interlocks also serve to enable or inhibit, as appropriate, control changes (e.g., implementing power-dependent insertion limits).

TABLE 12-2
Typical Reactivity Inventory for a Pressurized-Water Reactor [PWR] System

	Reactivity, % $\Delta k/k$	
Clean, 20°C		+ 22
Clean, hot full power	+ 17	
Clean, hot full power, equilibrium xenon and samarium	+ 14	
Control rods (highest worth rod stuck)	− 12	
Burnable poison	− 7	
Soluble boron	− 7	
Total control worth		− 26
Shutdown margin		− 4

Programmed control-rod insertion is the preferred shutdown method because it allows time for heat-balance adjustments between the primary and secondary loops. A trip is readily accommodated, but is somewhat more disruptive of operations. Thus, there is a large incentive to develop a system that trips under all appropriate actual conditions, while having a minimum number of spurious trips.

Reliability is enhanced by using four sensors for each input parameter (e.g., in Fig. 12-16) along with a two-out-of-four logic block. The latter requires signals from at least two sensors (for the *same or different* parameters) before a trip is initiated. Each circuit employs a *fail-safe* design where a malfunction indicates a trip condition. The two-of-four logic allows operations to continue with a single failed circuit (and during the repair thereof) while providing assurance that actual trip conditions can be detected by the remaining active sensors.

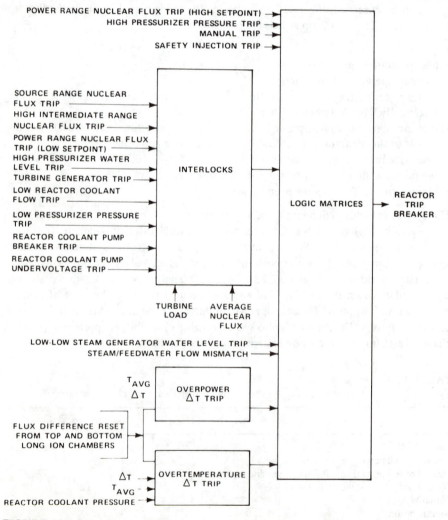

FIGURE 12-16

Typical protective system inputs for a PWR. (Courtesy of Westinghouse Electric Corporation.)

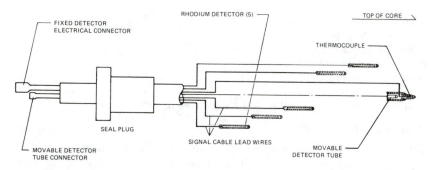

FIGURE 12-17

In-core instrument string for the Combustion Engineering System 80 PWR. (Courtesy of Combustion Engineering, Inc.)

Power Monitors

Flux signals for the PWR protective system are provided by boron-lined ion chambers located just outside of the reactor vessel. In-core instrumentation is used primarily for assessing overall core performance.

The Combustion in-core instrument string shown in Fig. 12-17 contains three separate measurement devices. The large instrument tube in the 16 × 16 fuel assembly (Fig. 12-11) allows this string to incorporate greater capability in a single unit than can be obtained with the smaller tubes in the Babcock & Wilcox and Westinghouse 17 × 17 assembly. However, the individual functions of the Combustion design are readily available separately in the other systems.

The segmented rhodium detectors in Fig. 12-17 provide a continuous readout of neutron flux at five axial positions for steady-power operation. The rhodium is activated by neutron exposure with the subsequent beta-decay providing a direct electric current.

Movable detectors are available for detailed axial flux mapping in individual assemblies. Each of two such detectors are shifted among the instrument strings and axial traverses are implemented in the moveable-detector tubes.

Since the in-core instruments are inserted from the bottom of the core (Fig. 12-7), the thermocouple in Fig. 12-17 monitors the core outlet temperature. This reading plus the flux values and other operating parameters are fed to a dedicated computer which generates linear-heat-rate [kW/ft], departure-from-nucleate-boiling-ratio [DNBR], and burnup data.

EXERCISES

Questions

12-1. Sketch the steam cycles for the two different LWR designs. Include typical temperatures and pressures.

12-2. Identify the major differences between BWR and PWR fuel assemblies in terms of pin enrichments, lattice pattern, and external hardware.

12-3. Differentiate between BWR and PWR systems on the basis of reactivity control with:

a. control rods
b. bunable poisons
c. soluble poisons
d. flow adjustment
Explain the basis for shutdown by reactor trip in each design.

12-4. Identify and explain four of the trip parameters in a typical protective system. What is two-of-four logic and how is it employed in a PWR system?

Numerical Problems

12-5. In the actual Maine-Yankee reactor core, 80 of the fuel assemblies contain 16 boron shims and 68 of them contain 12 boron shims. Recalculate the core-averaged linear heat rate (Prob. 7-5) in kW/ft and kW/m for this case. What effect does the introduction of the shims have on the design-target heat-flux factor?

12-6. Write equations for the four most likely reactions of neutrons with boron in an ionization chamber. Using cross-section and natural-abundance data from the Chart of the Nuclides, calculate the relative probability for each.

12-7. Consider the rhodium neutron detectors employed in PWR systems.
 a. Write equations for neutron capture in naturally occurring ^{103}Rh and the subsequent beta decay of the product.
 b. When a radionuclide is produced by irradiation in a constant neutron flux, its activity $A(t)$ approaches an equilibrium level A_∞ according to the expression

$$A(t) = A_\infty(1 - e^{-\lambda t})$$

 for decay constant λ. Calculate the time (in minutes and number of half-lives) required for the beta-signal from the rhodium detector to reach 90 percent, 99 percent, and 99.9 percent of its equilibrium level at a constant power level.
 c. Based on the result in (b), is it feasible to use this type of detector to initiate a reactor trip? Why?

12-8. A fission chamber contains a thin foil of ^{235}U with a total of 10^{20} atoms.
 a. Calculate the fission rate it would experience in a 0.025-eV neutron flux of 10^{14} n/cm$^2\cdot$s.
 b. Assuming that each fission produces an average of four collectible positive charges, calculate the current in the system associated with the flux in (a).

12-9. Identify three similarities between the Oklo natural reactor and a current BWR.

SELECTED BIBLIOGRAPHY[†]

BWR

    GE/BWR-6, 1978
    GESSAR
    Lish, 1972 (Chap. 3)

PWR

    B-SAR-241
    Babcock & Wilcox, 1975
    Babcock & Wilcox, 1978
    Babcock & Wilcox, 1980

[†]Full citations are contained in the General Bibliography at the back of the book.

CESSAR
Combustion Engineering, 1978
Lish, 1972 (Chap. 2)
RESSAR
Westinghouse, 1975
Westinghouse, 1979a

Individual Reactors
Safety Analysis Reports [SAR] from the U.S. Nuclear Regulatory Commission docket (for every U.S. reactor)
Nucl. Eng. Int. (description plus wallchart)
Douglas Point (General Electric BWR), Nov. 1973
Forsmark 3 (ASEA-Atom BWR), Sept. 1976
Cherokee-Perkins (Combustion PWR), Dec. 1977
Fessenheim (Framatome PWR), Sept. 1975
Oconee (Babcock & Wilcox PWR), April 1970
Snupps (Westinghouse PWR), Nov. 1975
Trillo (KWU PWR), Sept. 1978

Current Status Sources
Nucl. News—"World List of Nuclear Power Plants" updated semiannually and published in February and August issues; summarizes status of currently operable and planned power reactors.
Nucl. Eng. Int.—"Power Reactors 19 (XX)" updated annually and published as the July Supplement issue; contains a summary of design parameters and operating history for past, present, and planned power reactors.

Other Sources with Appropriate Sections or Chapters
Connolly, 1978
Foster & Wright, 1977
Lamarsh, 1975
Murray, 1975
Nero, 1979
Sesonske, 1973
WASH-1250, 1973
WASH-1400, 1975 (esp. App. IX)

13

HEAVY-WATER-MODERATED AND GRAPHITE-MODERATED REACTORS

Objectives

After studying this chapter, the reader should be able to:

1. Identify three types of heavy-water-moderated and three types of graphite-moderated reactors.
2. Describe the features of the CANDU system that allow for on-line refueling.
3. Identify two reactivity control methods for the CANDU system that are not available in LWRs.
4. Explain the basis for a positive reactivity feedback due to coolant boiling in light-water-cooled, heavy-water-moderated reactors.
5. Describe the reserve shutdown system for the HTGR.
6. Identify the major similarity and the major difference between the HTGR and pebble-bed reactor fuel cycles.
7. Describe the basic features of fuel assemblies for the PTGR designs.

Prerequisite Concepts

Although the light-water reactors represent a majority of the world's present electrical generating capacity, heavy-water-moderated and graphite-moderated systems are also significant contributors. The pressurized-heavy-water reactor [PHWR] developed by Canada is the dominant representative in the former category. The major variants are systems that employ boiling light-water coolant.

A large number of gas-cooled, graphite-moderated reactors are in operation. However, potential future expansion appears to be concentrated on the high-temperature gas-cooled reactor [HTGR] and the German thorium high-temperature reactor [THTR] which is also known as the pebble-bed reactor.

Another viable graphite reactor is the Soviet pressure-tube graphite reactor [PTGR]. This system employs boiling-water coolant in a series of pressure tubes.

HEAVY–WATER–MODERATED REACTORS

Heavy-water moderator may be employed in reactor systems cooled by heavy water, light water, an organic fluid, or a gas. A few of the more prominant designs are described in this section.

Deuterium $[^2_1D]$ exists in nature in a ratio of 1:6500 with ordinary hydrogen $[^1_1H]$. By contrast, reactor applications generally employ heavy water $[D_2O]$ with a 400:1 isotopic ratio. The well-developed process for separation is quite energy-intensive and results in an expensive product ($\approx$\$140/kg in 1978–1979). Thus, there is a substantial incentive to design and operate reactor systems with minimal heavy-water losses. It is also important to minimize the likelihood of contamination with ordinary water.

Pressurized-Heavy-Water Reactors

Systems classified as pressurized-heavy-water reactors [PHWR] employ D_2O as both the moderator and the coolant. One such design is similar in concept to the PWR in its use of a reactor vessel and a single coolant/moderator volume. Use of heavy water does require lattice changes as would be inferred, for example, by data like that in Figs. 11-2 and 11-3. The pressure-vessel PHWRs also employ natural uranium fuel and on-line refueling schemes.

The other major PHWR employs one moderator volume and separate coolant contained in pressure tubes. State-of-the-art development of such systems is embodied in the Canadian Deuterium Uranium–Pressurized Heavy Water [CANDU–PHW] reactors manufactured by Atomic Energy of Canada Limited [AECL]. Since the latter strongly dominate the current world market for heavy-water reactors, the CANDU[†] design is the basis for the descriptions below.

Steam Cycle
A simplified sketch of the CANDU steam cycle is provided by Fig. 13-1. Primary heavy-water coolant is pumped through an array of pressure tubes that contain the

---

[†]According to the usual convention, the term CANDU is essentially synonymous with CANDU–PHW. Design variations are specified more completely (e.g., CANDU–BLW for the system employing boiling light-water coolant).

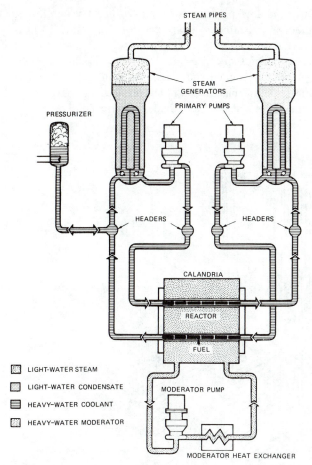

FIGURE 13-1
Simplified steam-cycle schematic for the CANDU–PHW system. (Courtesy of Atomic Energy of Canada Limited.)

fuel elements. The pressurizer maintains coolant pressure as in the PWR design (Fig. 12-7). The heated coolant flows through the tube side of the steam generator. The secondary loop is comparable to that of the PWR (Fig. 12-6).

The need to avoid excessive heavy water losses limits the primary loop to a temperature of about 310°C [590°F] and a pressure of 10 MPa [1450 psi]. As a result, steam reaches the turbine at about 260°C [500°F] and 4.7 MPa [680 psi], leading to a thermal efficiency of 28–29 percent.

Reactor Concept

The basic features of the CANDU reactor system are shown in Fig. 13-2. Heavy-water moderator is contained in a reactor vessel [*calandria*] which is about 7.6 m in diameter by 4 m deep. The vessel is penetrated by 380 calandria tubes which are fastened securely to the tubesheets at each end of the vessel, as shown in cross-section by Fig. 13-3. This contiguous moderator volume is maintained at a temperature of 70°C [158°F] and low pressure to minimize heavy-water losses. Since fission gammas cause moderator heating, a heat exchange loop (Fig. 13-1) is required.

Each calandria tube accommodates one pressure tube. Spacers and an intervening gas annulus (Fig. 13-3) minimize contact and, thus, heat transfer between the two. Twelve fuel bundles (Fig. 1-9) are placed end-to-end in each of the pressure tubes. Coolant flows through the tubes and the fuel bundles simultaneously.

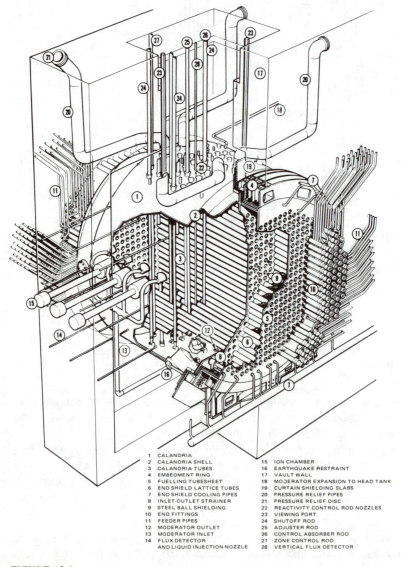

1	CALANDRIA		
2	CALANDRIA SHELL	15	ION CHAMBER
3	CALANDRIA TUBES	16	EARTHQUAKE RESTRAINT
4	EMBEDMENT RING	17	VAULT WALL
5	FUELLING TUBESHEET	18	MODERATOR EXPANSION TO HEAD TANK
6	END SHIELD LATTICE TUBES	19	CURTAIN SHIELDING SLABS
7	END SHIELD COOLING PIPES	20	PRESSURE RELIEF PIPES
8	INLET-OUTLET STRAINER	21	PRESSURE RELIEF DISC
9	STEEL BALL SHIELDING	22	REACTIVITY CONTROL ROD NOZZLES
10	END FITTINGS	23	VIEWING PORT
11	FEEDER PIPES	24	ADJUSTER ROD
12	MODERATOR OUTLET	25	SHUTOFF ROD
13	MODERATOR INLET	26	CONTROL ABSORBER ROD
14	FLUX DETECTOR	27	ZONE CONTROL ROD
	AND LIQUID INJECTION NOZZLE	28	VERTICAL FLUX DETECTOR

FIGURE 13-2
Reactor arrangement for the CANDU–PHW system. (Courtesy of Atomic Energy of Canada Limited.)

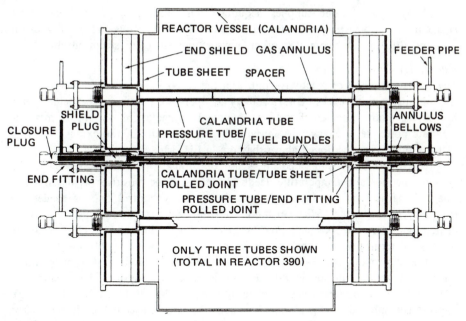

FIGURE 13-3
Simplified reactor cross section for the CANDU–PHW reactor. (Reproduced from *Nuclear Engineering International* with permission of the editor.)

The heavy-water coolant enters and leaves the pressure tubes through feeder pipes positioned at right angles to the main flow (Figs. 13-2 and 13-3). The refueling machines (Fig. 9-12) may, thus, make firm contact with the end fittings of the tubes without stopping the coolant flow. Removal of the end and closure plugs to the machine allows refueling operations to proceed with some redistribution, but no general interruption, of coolant flow.

The CANDU system employs two completely separate coolant loops with oppositely directed flow in adjacent channels. This allows all refueling to be in the direction of coolant flow (Chap. 9) while minimizing potential flux-shape concerns. It also provides a reactor-safety advantage as considered in Chap. 17.

Circumferential radiation shielding of the vessel is provided by a concrete vault containing light water. The end faces are shielded by placement of small steel balls in the space between the pressure tubes (Figs. 13-2 and 13-3). Heat removal is implemented by water flow through end-shield cooling lines.

The CANDU fuel bundles (Fig. 1-9 and Table 9-1) consist of short zirconium-clad fuel pins containing natural UO_2 pellets. All bundles for a given reactor are identical.

The use of natural uranium, heavy water, and on-line refueling are fundamental to the integrated CANDU design. Lacking enrichment capability, the Canadian government opted to develop heavy water production capacity. Viability of the CANDU concept depends on the availability of D_2O for initial calandria loading as well as smaller amounts for primary-coolant makeup. On at least one occasion, it has been necessary to drain a small unit to provide enough heavy water to start up a new, larger reactor.

Early heavy water supplies acquired from the United States have been augmented by domestic production. The newest D_2O facility is co-located with the four-unit Bruce reactor complex and employs 12 percent of its steam output.

Reactivity Control

The primary method of reactivity control in the CANDU system is on-line refueling. By this mechanism, excess core reactivity at operating power can be held to very low levels. Four other methods are employed to make minor adjustments and facilitate power level changes. A few member components of the extensive control systems are shown positioned above the core in Fig. 13-2.

Since the heavy-water coolant is not the primary moderator in the CANDU system, the coolant temperature feedback tends to be positive (as described in Chap. 5). This requires much more precise reactivity control than is needed for reactor systems like the LWRs. Such control is implemented through use of a sophisticated computer network which monitors neutron flux and initiates reactivity changes.

Short-term reactivity balance is maintained by zone control absorbers positioned as shown in Fig. 13-4a. Each of 14 chambers can contain light water which serves as a neutron poison in this heavy-water-moderated system. The quantity of water in each compartment is controlled by manipulation of inlet valves. Helium gas pressure is employed to expell water at a constant rate. Adjacent flux detectors feed their signals to the computer for positive monitoring of the effects of the zone-control absorbers.

Motor-driven, stainless-steel adjuster rods are employed for flux shaping and minor reactivity corrections. Figure 13-4b shows the relative positioning of 18 such

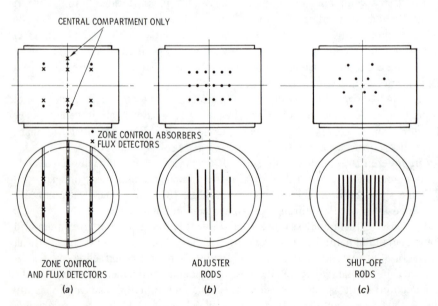

FIGURE 13-4

Reactivity control systems for the CANDU–PHW reactor: (a) zone-control absorbers and flux detectors; (b) adjuster rods; and (c) shut-off rods. (Reproduced from *Nuclear Engineering International* with permission of the editor.)

rods in an early CANDU design. Current practice employs 21 adjuster rods moved in unison in groups of from two to five.

Four mechanical control absorber rods of stainless-steel-clad cadmium are normally positioned above the core. They may be driven in to supplement the zone-control system or dropped to affect a rapid power reduction.

Longer-term reactivity balance can be maintained by the addition of small amounts of soluble poison to the moderator. Since boron burns out slowly, it is used when the core is first loaded with fresh fuel or when it has an inordinate amount of fresh fuel. The faster burnout of gadolinium can be employed to match xenon buildup (e.g., after a prolonged shutdown). The low poison levels employed in the CANDU (e.g., as opposed to those of the PWR) allow effective removal by an ion exchange system. Routine addition and removal of poison is controlled by the reactor operators rather than by the computer system.

Protective System

Reactor trip is facilitated in current designs by insertion of 28 stainless-steel-clad cadmium shutoff rods. Figure 13-4c shows the arrangement of a comparable 11-rod set from an earlier CANDU design. The rods are mounted to their drives by a direct-current clutch. A trip signal releases them for a spring-assisted gravity drop into the core.

The shutoff rods are designed for a reactivity worth of about -8% $\Delta k/k$ with the two rods of highest worth failing to insert. This assures an adequate shutdown margin under all credible reactor operating conditions.

A backup shutdown system relies on rapid injection of concentrated gadolinium nitrate solution into the bulk moderator through six horizontally distributed nozzles. Fast-acting helium-pressure valves facilitate the poison injection. Prior to the advent of this backup shutdown procedure, an equally effective method applied in several earlier CANDU systems relied on a gravity dumping of the heavy-water moderator into a special tank below the level of the reactor.

Each shutdown system employs an independent triplicated logic system which senses the requirement for reactor trip. Apart from a few concept-related differences, the parameters are similar to those for the PWR displayed by Fig. 12-16.

Power Monitoring

Extensive in-core flux monitoring is employed in the CANDU system. Since signal processing is an absolute requirement for both routine reactivity adjustment and reactor trip, comparable primary and backup computer systems are necessary.

Fission chambers and ionization counters are employed for flux level monitoring. Self-powered detectors of platinum and vanadium, respectively, provide more detailed information on flux distribution.

The platinum detectors are sensitive to both neutrons and gamma radiation. Based on a fast response time, they are employed for both the regulating and shutdown functions. The relatively long equilibrium time for vanadium restricts these detectors to use in the regulating system.

CANDU Modifications

The basic CANDU-PHW design offers a number of possibilities for alternative operation. One of these (noted in Chap. 9) is to employ fuel bundles containing mixed-oxide or ^{233}U–Th pins. As long as reactivities are roughly comparable, fuel bundles may be handled interchangeably.

The possibility of breeding on a ^{233}U–Th cycle (Fig. 6-1) had led to consideration of an organic-cooled reactor [CANDU-OCR]. With the exception of the coolant, it is conceptually identical to the system described above. Organic coolants can provide low neutron absorption as well as higher operating temperatures than are feasible with heavy water. The latter enhances thorium-to-^{233}U conversion as well as thermal efficiency. Development efforts on the CANDU-OCR are continuing.

Light-Water Cooled Reactors

The pressure and temperature limitations placed on the CANDU-PHW by its use of heavy-water coolant may be relaxed somewhat by employing light-water coolant. Designs based on boiling light-water coolant include the CANDU-BLW, the British *steam-generating heavy-water reactor* [SGHWR], and a prototypical Japanese system. Although at one time the British had planned for major use of the SGHWR, the long-term implementation of this and the other two reactor designs is somewhat in doubt.

Many features of the CANDU-BLW have been adapted from its heavy-water-cooled counterpart. Although the vessel is similar to that of Fig. 13-2, it is reoriented so that the pressure tubes are vertical. The short fuel bundles and the on-line refueling system are comparable. The British and Japanese reactors employ a similar pressure-tube design but with full-length fuel assemblies with off-line and on-line refueling, respectively.

Water enters at the bottom of the pressure tubes, flows upward to remove heat from the fuel, and leaves as a steam–water mixture. The output from the tubes is collected in one or more steam drums where moisture separation occurs prior to the steam's transport to the turbine.

Since the separate heavy-water volume is responsible for most of the moderation, the net reactivity effect of the light-water coolant is that of a neutron poison. Boiling, voiding, or a general density reduction is equivalent to a reactivity increase. This, of course, results in a positive coolant feedback mechanism which, in turn, necessitates sophisticated computer control.

It has been found that the magnitude of the coolant feedback may be reduced somewhat by employing slightly enriched UO_2 or mixed-oxide fuel rather than natural uranium. Since this requires enrichment and/or reprocessing in addition to heavy-water production, the fuel cycle would be more complicated than that of the basic CANDU-PHW.

GRAPHITE-MODERATED REACTORS

The earliest of the gas-cooled, graphite-moderated reactors employ carbon dioxide [CO_2] as a coolant. Over 40 such systems are currently used for commercial generation of electricity. The seven units in France have power levels of 40 MW(e) to 540 MW(e) and have come on line between 1959 and 1972. The 31 operating units in the United Kingdom range in power from 32 MW(e) to 625 MW(e), with startup dates from 1962 to 1977. The United Kingdom has six additional reactors of this type scheduled to begin operation between late 1980 and 1982.

Both France and the United Kingdom have ceased further development of gas-cooled reactors. This is, at least in part, due to limitations on high-temperature use of carbon dioxide coolant. At elevated temperatures the CO_2 tends to cause excessive corrosion in piping and steam generator components.

The state-of-the-art gas reactor concepts are predicated on use of helium coolant. This essentially eliminates the corrosion concerns, but at the expense of developing new technology for handling large quantities of this light gas at high temperatures and pressures. The two concepts which are in development stages are the high-temperature gas-cooled reactor [HTGR] and the thorium high-temperature reactor [THTR]. Although neither is expected to have much short-term impact on electric power generation, certain inherent safety features suggest that additional future development may be highly desirable.

High-Temperature Gas-Cooled Reactors

The basic HTGR design has been developed in the United States by the General Atomic Company, a joint venture of Gulf Oil and Royal Dutch Shell. The prototype 40 MW(e) Peach Bottom I reactor demonstrated the concept during operation from 1967–1974. The 330 MW(e) Fort St. Vrain reactor located near Platteville, CO began commercial operation in April 1979.

The basic design for a larger 1160 MW(e) HTGR is well advanced. At one time several utilities had placed orders for such a system. General economic problems, however, dictated against their incurring the somewhat higher risk of a new concept. As of late 1979 the system is not being actively marketed by General Atomic. The unique aspects of the system, including some inherent safety features (described in Chaps. 16–17), suggest the value of expanding on the previous discussions (Chaps. 1 and 9) of the HTGR.

Steam Cycle

The basic features of the steam system for the two-loop HTGR are shown in Fig. 13-5. The primary loop is contained entirely within the prestressed concrete reactor vessel [PCRV]. The helium circulators pump the coolant through the core. Helium at a temperature of about 740°C [1370°F] and a pressure of 4.9 MPa [710 psi] enters the steam generator.

The steam generators receive feedwater and convert it to superheated steam at about 510°C [960°F] and 17.2 MPa [2500 psi]. It is said to be superheated because the temperature exceeds that at which the steam evolved from the liquid water. Above the critical temperature of 375°C [710°F], all steam is superheated since it cannot coexist with water at any pressure. This steam feeds the high-pressure [HP] turbine section.

Steam released from the HP stage serves to drive the primary helium circulators before it reenters the steam generator as shown in Fig. 13-5. In the lower section of the steam generator, a reheat cycle provides a steam output at about 540°C [1000°F] and 4.0 MPa [585 psi]. The reheat product feeds the intermediate-pressure [IP] and then the low-pressure [LP] turbine sections. Overall thermal efficiency is roughly 40 percent.

A cut-away drawing of the prestressed-concrete reactor vessel [PCRV] is shown by Fig. 13-6. It is about 28 m high by 30 m in diameter, compared to core dimensions of 6.3 m and 8.5 m, respectively. The vessel is constructed by pouring concrete on site (as opposed to the off-site fabrication and transportation required of the steel LWR vessels). Strength is provided by vertical and circumferential prestressing cables.

The PCRV encloses six sets of helium circulators and steam generators in a manner shown by Fig. 13-6. Helium enters above the core, flows downward through the fuel blocks, and enters the steam generator. Three sets of auxilliary circulators and steam generators are also included. Provisions are also made for control-rod drives, refueling penetrations, and helium purification.

The basic features of the HTGR steam generator are shown in Fig. 13-7. The tube side of the device consists of helical coils wrapped circumferentially with respect to the central axis. Feedwater traverses the upper series of coils and leaves through an axial tube as superheated steam. The reheat cycle employs the coils below the helium inlet. Helium from the reactor core enters the shell side of the steam generator, flows downward through the reheat section, upward near the axis, downward through the superheater/evaporator/economizer sections, and finally upward between the shroud and PCRV liner to leave from the top of the unit.

It has been proposed that the high temperature of the gaseous coolant may allow steam-cycle applications to be bypassed. The use of a gas turbine for generation of electricity is one possibility that is receiving a good deal of attention. Process-heat applications may also be practical either by themselves or in concert with cogeneration of electricity.

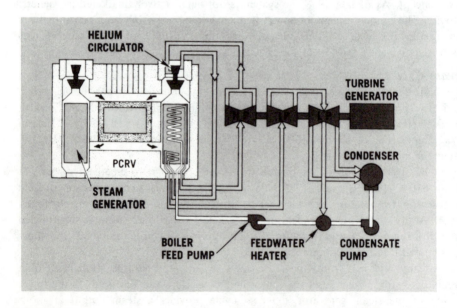

FIGURE 13-5
High-temperature gas-cooled reactor [HTGR] steam-cycle schematic diagram. (Courtesy of General Atomic Company.)

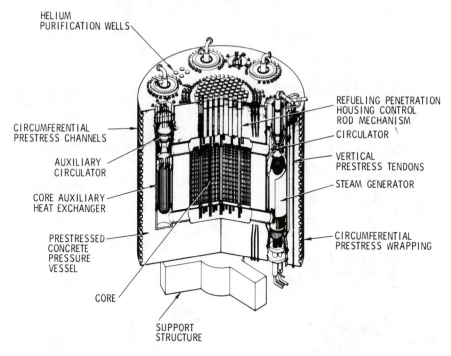

HELIUM
PURIFICATION WELLS

CIRCUMFERENTIAL
PRESTRESS CHANNELS

AUXILIARY
CIRCULATOR

CORE AUXILIARY
HEAT EXCHANGER

PRESTRESSED
CONCRETE
PRESSURE
VESSEL

CORE

SUPPORT
STRUCTURE

REFUELING PENETRATION
HOUSING CONTROL
ROD MECHANISM
CIRCULATOR

VERTICAL
PRESTRESS TENDONS

STEAM GENERATOR

CIRCUMFERENTIAL
PRESTRESS WRAPPING

FIGURE 13-6
Prestressed concrete reactor vessel for the HTGR system. (Courtesy of Oak Ridge National
Laboratory, operated by the Union Carbide Corporation for the U.S. Department of Energy.)

Fuel Assemblies

The HTGR fuel has been described in some detail (Figs. 1-10 and 9-7). Two basic
fuel assembly types—designated as standard and control—are shown in Fig. 13-8.
Each is a hexagonal graphite block containing small holes that accommodate coolant
flow and hold stacks of fuel rods, respectively. It may be recalled that the rods
consist of a mixture of TRISO and BISO fuel particles like those in Fig. 9-7.
Separate locations for burnable poisons are also present.

The standard assembly in Fig. 13-8a has a symmetric spatial arrangement of
fuel. The control assemblies in Fig. 13-8b contain two control-rod channels and a
reserve shutdown hole. A third assembly type (which is not pictured) is a modified
control assembly where the lower portion is not penetrated by the three large holes.
This latter block is intended for use at the bottom of the core as a positive stop for
the control rods and the reserve shutdown system.

Reactivity Control

Short-term control of reactivity is provided by pairs of control rods which operate
in the two larger penetrations in the control assemblies. Figure 13-9 shows the
arrangement of the HTGR into 73 fuel regions, each consisting of six standard
assemblies surrounding a single control assembly. Each rod is suspended from a
flexible steel cable and positioned by the motor drives located above the core.

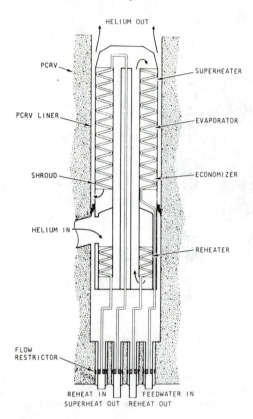

FIGURE 13-7
Steam generator for a large HTGR. (Courtesy of General Atomic Company.)

The seven-assembly regions are also the fundamental units for fuel-management operations. Removal of a refueling penetration and the control-rod drive mechanism (Fig. 13-6) allows all fuel blocks in a region to be removed from the core remotely as described in Chap. 9.

Long-term reactivity control is provided by burnable poisons of B_4C particles loaded into carbon rods. Up to six rod stacks may be inserted into the corner locations of each standard assembly (Fig. 13-8a) and four into each control assembly (Fig. 13-8b). A basic HTGR design goal is to employ enough burnable poisons so that seven or fewer control-rod pairs (roughly 2.5% $\Delta\rho$) will be sufficient for routine control during sustained full-power operations.

Protective System

During normal operations, the pulley for the control-rod cables maintains contact with the drive motor by a dc holding voltage. Removal of the voltage under reactor trip conditions allows the rods to drop into the core under the influence of gravity.

A rough estimate of the reactivity inventory of a large HTGR at beginning-of-core-lifetime [BOL] is provided in Table 13-1. In a manner similar to that found for the pressurized-water reactor, fuel- and moderator-temperature defects reduce

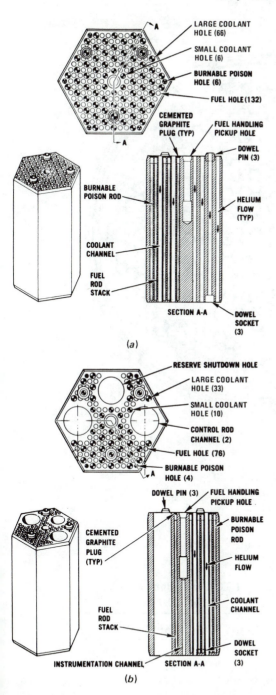

FIGURE 13-8
Fuel assemblies for a large HTGR: (*a*) standard assembly; and (*b*) control assembly. (Courtesy of General Atomic Company.)

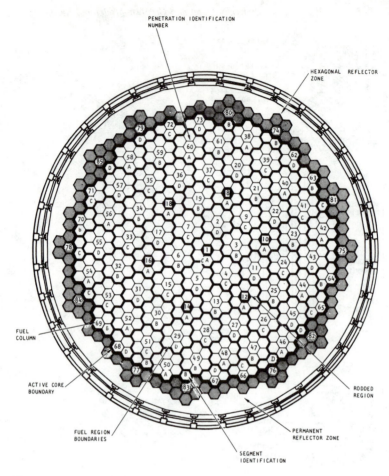

FIGURE 13-9
Reactor core configuration for the HTGR. (Courtesy of General Atomic Company.)

TABLE 13-1
Estimated Reactivity Inventory for a Large HTGR at Beginning-of-Core Lifetime

	Reactivity, $\% \Delta k/k$	
Clean, 27°C		+22
Clean, hot full power	+16	
Clean, hot full power, equilibrium xenon and samarium	+13	
Burnable poisons	−10	
Control rods (less stuck rod)	−18	
Total control		−28
Shutdown margin		−6

the cold core reactivity by about 6 percent, with xenon and samarium poisoning causing another 3 percent decrease.

The HTGR also has another poisoning effect which appears later in core lifetime. It results from the presence of ^{233}Pa in the ^{232}Th-to-^{233}U conversion chain (Fig. 2-16). During operation, the ^{233}Pa acts as a poison. Shutdown allows it to decay to fissile ^{233}U which, of course, results in a positive reactivity insertion. The reactivity swing related to ^{233}Pa decay would be as much as about $+5\%$ $\Delta k/k$ during the middle of a cycle.

Control reactivity may be divided between the burnable poisons and the control rods as shown in Table 13-1. As has been noted to be standard practice, the reference control rod worth is that with the individual rod pair of highest worth assumed to be "stuck" and having failed to insert into the core.

The HTGR protective system employs a two-out-of-three trip logic. Trip parameters have similarities to those shown for the PWR in Fig. 12-16. Differences are related to the nature of the respective primary coolants (e.g., high moisture content in the helium produces a trip in the HTGR).

The HTGR incorporates a reserve shutdown system to assure subcriticality should the rods fail to insert after a trip signal. The basic constituent is a hopper of 14-mm-diameter graphite spheres with about a 40 percent loading of B_4C. The hopper is contained in the refueling penetration (Fig. 13-6) and connects to the smaller of the control-assembly penetrations (Fig. 13-8*b*). As shown by Fig. 13-10, a

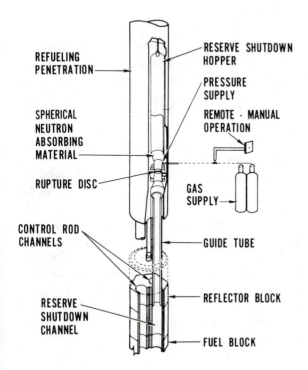

FIGURE 13-10
Reserve shutdown system for the HTGR. (Courtesy of General Atomic Company.)

rupture disc holds the spheres above the core during normal operations. A gas supply is provided to maintain the pressure in the hopper at a level which balances the coolant pressure in the core. Upon a signal from the operator (at the reactor console or at a remote location outside of the control room), the gas supply line is closed to cause the disc to rupture. The spheres travel through a guide tube and the upper reflector blocks to enter the reserve shutdown channel in the fuel blocks.

The system is designed with sufficient negative reactivity worth (nominally -15% $\Delta k/k$) to affect a full shutdown in the absence of any control-rod insertion. The hoppers are interconnected in groups of six or seven with a relatively uniform distribution across the core. The control system allows insertion of individual groups or the entire array. There is a vacuum system available to remove the spheres following a planned or inadvertent insertion.

Power Monitoring

Routine control and trip functions are based on ex-core flux detectors. Fission chambers are employed at low power levels. Boron-lined ionization chambers provide the needed capability for higher power conditions.

In-core instrumentation is used for informational purposes. Moveable flux detectors housed in the refueling penetrations (Fig. 13-6) are designed to make an axial traverse of the control assembly stack through the instrumentation channel (Fig. 13-8*b*).

Pebble-Bed Reactor

The thorium-high-temperature reactor [THTR], generally referred to as the *pebble-bed reactor*, is being developed by Hochtemperataur-Reaktorbau GmbH [HRB] in the Federal Republic of Germany. More than a decade's worth of experience has been accumulated from the 15-MW(e) AVR unit, which achieved initial criticality in 1966 and produced its first commercial electricity in 1967. This is the basis for a demonstration 296-MW(e) plant targeted for 1982 completion and conceptual designs of 1120-MW(e) commercial-scale units.

The basic pebble-bed reactor has many similarities to the HTGR (whose development it has closely paralleled). Figure 13-11 shows a cut-away of each. Both employ comparable steam cycles and prestressed concrete reactor vessels [PCRV]. The use of U–Th carbide fuels, graphite moderator, and helium coolant is also a common denominator.

The pebble-bed fuel elements are the unique feature of the concept. As shown in Fig. 13-12 (and inferred from Fig. 13-11), the fuel is contained in 6-cm-diameter graphite spheres. Coated fuel particles of similar design to those used for the HTGR (Fig. 9-11) are contained in a graphite matrix which is covered by an outer graphite shell.

Reactivity control is provided by addition and/or removal of the fuel pebbles from the core hopper. Based on the geometrical arrangement in Fig. 13-11, gravity facilitates the exchange. Tapered control rods mounted at the top of the core can be inserted for routine control and reactor trip functions.

Pressure-Tube Graphite Reactor

Commercial development of the *pressure-tube graphite reactor* [PTGR] (sometimes referred to as a *light-water-cooled graphite-moderated reactor* [LWGR] in the West)

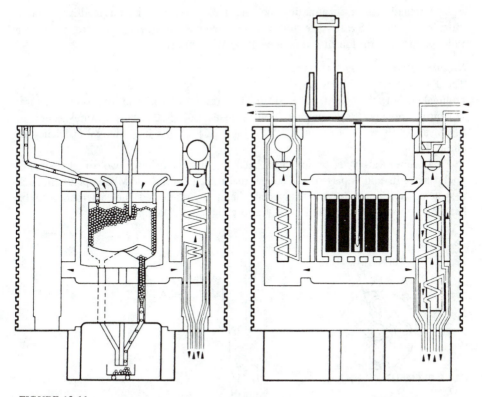

FIGURE 13-11
Comparison of the pebble-bed and HTHR gas-cooled reactor concepts. (From H. Oehme in "Gas-Cooled Reactors: HTGR and GCFR," CONF-740501, May 7–10, 1974.)

has been centered in the U.S.S.R. The World's first commercial electricity was generated in 1954 by a 5-MW(e) unit of this type located at Obninsk near Moscow. Fourteen years of successful operation encouraged development of various 100- and 200-MW(e) units which entered service between 1958 and 1968. A pair of 1000-MW(e) plants have been in operation near Leningrad since 1973. The pseudomodular design of the PTGR has led the Soviets to consider future development of 1500- and 2000-MW(e) units.

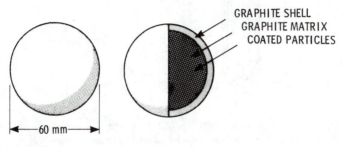

FIGURE 13-12
Fuel elements for the pebble-bed reactor.

Detailed design information on the PTGR reactors is not readily available outside of the U.S.S.R. The following paragraphs, however, serve to provide a rudimentary description of a reference 1000-MW(e) unit.

Steam Cycle

The PTGR has similarities to the CANDU–BLW in its use of pressure tubes and boiling-light-water coolant. The steam is drawn off to one of four moisture-separating boilers. Each pair of boilers feeds one 500-MW(e) turbine-generator. Steam at 280°C [540°F] and 6.5 MPa [940 psi] results in a thermal efficiency of about 31 percent.

A superheat-cycle option has been incorporated into two of the smaller operating units and is part of the conceptual design for a 2000-MW(e) plant. In such a system, steam is returned to special channels in selected fuel assemblies to be heated to about 450°C [840°F] (well above the critical temperature).

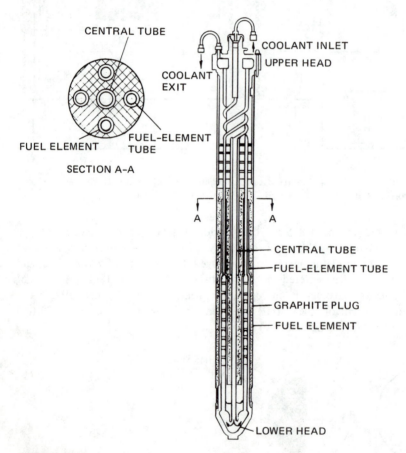

FIGURE 13-13
Fuel assembly for the pressure-tube graphite reactor [PTGR]. (Adapted from L. Konstantinov, "Commercial Reactors of the U.S.S.R.," Argonne Center for Educational Affairs, November 2, 1976.)

Reactor System

A cylindrical pile of graphite blocks roughly 10 m in diameter and height serves as the neutron moderator. It is enclosed in a gas-tight shell of carbon steel. The central radial portion of the pile is pierced with the pressure-tube fuel assemblies.

A schematic diagram of a fuel assembly from the Obninsk reactor is provided in Fig. 13-13. A cylindrical graphite plug contains five thin-walled coolant tubes. Clad fuel elements are fixed around four tubes in the assembly. The central tube is used only for coolant flow. Water enters the coolant inlet, flows downward through the central tube, distributes itself at the bottom of the assembly, travels upward through the fuel element tubes, and leaves as a steam–water mixture.

The current fuel-assembly design is similar to that of Fig. 13-13, but with 18 fuel elements. The fuel material itself is UO_2 enriched to 1.8 wt% in ^{235}U. The elements are clad in zircaloy tubes 13.5 mm in diameter and 0.9 mm thick. The active fuel length of each element is 7.0 m, compared to an assembly length of about 13 m.

The reactor core consists of nearly 1700 pressure-tube assemblies emplaced vertically to form a uniform array in the graphite pile. The active fuel core has an effective diameter of 11.8 m and a height of 7.0 m.

Control

The PTGR is designed to allow on-line replacement of fuel assemblies. In contrast to the CANDU systems and the pebble-bed reactor, however, one unit must be fully removed before another can be added.

The reactivity control afforded by refueling is augmented by the use of 180 neutron-poison control rods. These same rods are inserted by the protective system to produce a shutdown following a reactor trip.

Neutron flux monitoring is accomplished by using fission-chamber and rhodium detectors as described in the last chapter. The in-core units are installed in holes bored into the graphite blocks.

EXERCISES

Questions

13-1. Identify three types of heavy-water-moderated and three types of graphite-moderated reactors.

13-2. Describe the features of the CANDU system that allow for on-line refueling.

13-3. Identify two reactivity control methods for the CANDU system that are not available in LWRs.

13-4. Explain the basis for a positive reactivity feedback due to coolant boiling in light-water-cooled, heavy-water-moderated reactors. What fuel change can reduce this effect?

13-5. Describe the reserve shutdown system for the HTGR.

13-6. Identify the major similarity and the major difference between the HTGR and pebble-bed reactor fuel cycles.

13-7. Describe the basic features of fuel assemblies for the PTGR designs.

Numerical Problems

13-8. Draw to-scale rectangles whose sides represent the approximate heights and diameters of the vessels of the BWR, PWR, CANDU, HTGR, and PTGR.

13-9. Estimate the fraction of ^{233}Pa that will not beta-decay to ^{233}U in an HTGR thermal neutron flux of 1×10^{14} n/cm^2·s.

13-10. A 638-MW(e) CANDU system requires 263 Mg of heavy-water moderator and 199 Mg of coolant.

a. Calculate the value of this inventory at $140/kg.

b. Assuming a capital cost of $1000/kW(e), calculate the fractional cost of the heavy-water inventory.

13-11. CANDU generating costs are divided into the three usual categories plus a fourth for heavy-water losses. A typical system loses an annual average of 1.3 kg/h. Assuming it operates at 80 percent of capacity (i.e., generates this fraction of the maximum possible electrical energy at rated power), calculate:

a. the cost per kW·h of the heavy-water losses

b. the fraction of total generating costs represented by the previous result (use the lowest cost on Fig. 8-4 for a rough-estimate base).

13-12. Typical reactor-grade heavy water is 99.75 wt% D_2O, with the remainder being H_2O. Considering thermal neutron absorption by the two hydrogen isotopes, estimate the fraction that occurs in the deuterium.

13-13. Repeat Prob. 12-7 for the vanadium neutron detector employed in the CANDU reactor.

13-14. It has been proposed that $^3_2$He ($\sigma_a = 5330$ b) be introduced into the pebble-bed reactor as a reserve shutdown medium. Recalling from basic chemistry that 3 g of ^{3}He contains 6.02×10^{23} atoms and occupies 22.4 l at standard temperature and pressure [STP]:

a. Calculate its macroscopic cross section at STP.

b. If 12 kg is required at a cost of $80/l, calculate the total cost.

c. Assuming a plant cost of $1000/kW(e) for the 296-MW(e) demonstration unit, calculate the fractional cost of the ^{3}He inventory.

SELECTED BIBLIOGRAPHY†

CANDU
>   AECL, 1976
>   Haywood, 1976
>   Hinchley, 1975
>   McIntyre, 1975
>   McNelly & Williamson, 1977
>   *Nucl. Eng. Int.,* June 1970, June 1974

HTGR
>   CONF-740501, 1974
>   GASSAR
>   GFHT, 1976
>   Lish, 1972 (Chap. 4)
>   *Nucl. Eng. & Des.,* 1974

Pebble-Bed Reactor
>   CONF-740501, 1974
>   COO-4057-6, 1978

†Full citations are contained in the General Bibliography at the back of the book.

PTGR
> Emel'yanov, 1977
> Konstantinov, 1976
> Lewin, 1977

Individual Reactors
> Safety Analysis Reports [SAR] from the U.S. Nuclear Regulatory Commission docket (for every U.S. reactor)
> *Nucl. Eng. Int.* (description plus wall chart)
>> Point Lepreau (CANDU–PHW), June 1977
>> Gentilly 1 (CANDU–BLW), Nov. 1972
>> Winfrith (U.K. SGHWR), June 1968
>> Fugen (Japan HWR), Aug. 1979
>> Fulton (General Atomic HTGR), Aug. 1974
>> Heysham (U.K. Advanced Gas-Cooled Reactor), Nov. 1971
> Current Status (see Chap. 12 Selected Bibliography)

Other Sources with Appropriate Sections or Chapters
> Connolly, 1978
> Foster & Wright, 1977
> Lamarsh, 1975
> Murray, 1975
> Nero, 1979
> Sesonske, 1973
> WASH-1535, 1974

14

THERMAL BREEDER REACTORS

Objectives

After studying this chapter, the reader should be able to:

1. Explain the requirement for thorium-based fuel in thermal breeder reactors.
2. Identify at least two reasons why the light-water breeder reactor [LWBR] was not considered feasible at the time the early LWR systems were developed.
3. Describe the seed-blanket concept of the LWBR fuel assembly.
4. Identify the three important features of on-line reprocessing in the molten-salt breeder reactor [MSBR].
5. Explain how the same salt fluid can be used for both the reactor core and the breeding blanket in the MSBR concept.

Prerequisite Concepts

The primary resource for the current thermal reactor designs is the 0.711 wt% ^{235}U content of natural uranium. This is augmented somewhat by conversion of fertile

^{238}U to plutonium or ^{232}Th to ^{233}U. Whether the systems operate in a once-through or recycle mode, fresh ^{235}U is required with each new fuel batch.

A breeder reactor, by definition, produces more fissile material than it burns. An external supply of fissile ^{235}U, ^{233}U, and/or plutonium is thus required only for the initial core and, perhaps, the first few reloads. Eventually enough material is produced to sustain the system. Continuing operation at this point is dependent only on externally supplied depleted uranium or thorium. Since this has the net effect of shifting the reactor fuel from the scarce ^{235}U to the abundant ^{238}U and thorium, a breeder system accesses a substantially greater energy resource. A current LWR, for example, is able to employ 2 percent or less of the energy available in natural uranium where, by contrast, an LMFBR may utilize 60 percent or more. Thorium, which is somewhat more abundant than uranium, is useful as a reactor fuel only when converted to ^{233}U. Its use has the potential for extending the ^{235}U resource by a hundred-fold or more.

As noted from Fig. 6-1, the average number of neutrons produced per neutron absorbed in fuel η is greatest for thermal and intermediate-energy neutrons in ^{233}U and for fast neutrons in ^{239}Pu. Thermal-breeder reactor concepts employ a ^{233}U-Th fuel cycle on this basis. Fast-breeder reactor designs favor a plutonium-^{238}U cycle because it has the largest η values for fast neutrons.

One of the best developed concepts for a thermal-breeder reactor is the heavy-water-moderated, thorium-based CANDU-OCR described briefly in the previous chapter. The light-water breeder reactor [LWBR] is another design that is receiving important attention. The molten-salt breeder reactor [MSBR] is a novel concept for future development.

LIGHT–WATER BREEDER REACTORS

The light-water breeder reactor [LWBR] concept employs ^{233}U-Th fuel with water coolant/moderator. It is the only known approach for which breeding appears possible with ordinary water.

An LWBR was not considered feasible when the LWR systems were developed in the 1950s. However, the following factors have suggested that the initial evaluation is no longer valid:

- Recent experimental data (Fig. 6-1) suggests that η for ^{233}U is larger than originally thought, especially for a slightly harder (i.e., less thermal) neutron spectrum than that employed in the LWR.
- Development of zircaloy cladding has decreased fuel-pin parasitic absorption in comparison with that of stainless steel.
- Neutron leakage is reduced substantially by the large core sizes of current designs.
- A novel seed-blanket design eliminates the need for neutron poisons to control reactivity.

Development

The U.S. Department of Energy is sponsoring research and development activities for the LWBR concept. One major goal of the program is to confirm that breeding can be achieved using ^{233}U–Th fuel in a light-water reactor. An equally important goal is to demonstrate that an LWBR thorium core can be installed and operated with a standard PWR vessel and steam system.

Program implementation has centered around the 60-MW(e) Shippingport nuclear power station—the first U.S. commercial nuclear reactor and the first commercial unit of its size in the world. The PWR core has been replaced by an LWBR thorium design. Commercial operation began December 2, 1977 (which happened to be the 20th anniversary of the initial criticality of the Shippingport reactor). A 3- to 4-year run is planned before the thorium core is removed and subjected to "proof of breeding" evaluations. It is expected that the expended core will contain about 1 percent more fissile material than its initial loading.

Although UO$_2$ and mixed-oxide fuel lattices have been studied extensively (Chap. 9), uncertainties exist concerning the nuclear and materials characteristics of ^{233}U–Th oxide fuel with zirconium clad and the water coolant/moderator. Continuing research and development activities in these areas, and others related to fuel assembly design and reactivity control, are important adjuncts to the operational portion of the program.

Fuel Assemblies and Reactivity Control

The Shippingport LWBR core consists of 12 fuel modules surrounded by a reflector blanket as shown by Fig. 14-1. It is installed with the vessel, primary coolant loop, and balance of plant essentially intact from the reactor's previous operation as a PWR. Figure 14-2 is a cut-away drawing of the new core and the vessel.

Each of the fuel modules consists of the movable seed and fixed blanket regions in Fig. 14-3. The central *seed* is a closely spaced hexagonal array of fuel pins. The fuel pellets are designed to contain up to about 6 wt% ^{233}UO$_2$ in ThO$_2$.

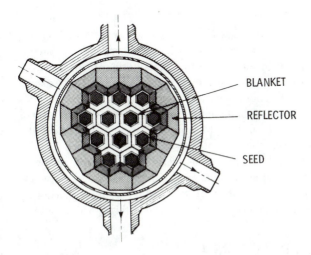

BLANKET

REFLECTOR

SEED

FIGURE 14-1
Radial cross section of the light-water breeder reactor [LWBR] core. (Courtesy of U.S. Department of Energy.)

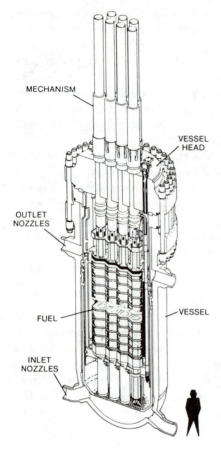

MECHANISM

VESSEL
HEAD

OUTLET
NOZZLES

FUEL

VESSEL

INLET
NOZZLES

FIGURE 14-2
Axial cross section of the Shippingport reactor
vessel with the LWBR core installed. (Courtesy of
U.S. Department of Energy.)

The cladding is zirconium alloy. The tight lattice spacing results in a "drier" lattice and a "harder" neutron spectrum than are common in LWR applications. Such a spectrum enhances the prospects for breeding with the ^{233}U–thorium fuel.

The *blanket* region consists of larger diameter rods in a less closely spaced lattice. They contain pellets of up to about 3 wt% $^{233}UO_2$ in ThO_2.

The fuel pins for both the seed and blanket contain ThO_2 pellets over a portion of their length, as indicated on Fig. 14-3. The arrangement provides the basis for reactivity control without the use of poisons. When the seed is fully inserted in the blanket, a relatively low-leakage, high-reactivity operating geometry results. Shutdown is facilitated by withdrawal of the seed to the high-leakage configuration shown by Fig. 14-3. The absence of neutron poisons and the resulting increase in neutron economy are fundamental for breeding to be feasible in this ^{233}U–Th system.

MOLTEN–SALT BREEDER REACTORS

The molten-salt breeder reactor [MSBR] concept is a potential alternative to the LWBR as a thermal breeder reactor. Although not currently subject to extensive research and development activities, the MSBR design has a number of attractive features.

The basic MSBR design employs a molten-salt fluid which acts as both the reactor core and the primary-loop coolant. The fuel is ^{233}U-Th. Moderation is provided by graphite assemblies that contain channels for flow of the core/coolant. On-line reprocessing of the fuel is a necessary component of the MSBR concept.

Current conceptual designs are based in part on the molten-salt reactor experiment [MSRE] program conducted by Oak Ridge National Laboratory between 1965 and 1969. Although the breeding ratio appears to be limited to about 1.06, the fuel management economy afforded by on-line refueling may provide a doubling time of 22 years.

Steam Cycle

The basic features of a conceptual 1000-MW(e) MSBR are identified by Robertson (1971). A steam-cycle schematic is shown in Fig. 14-4. The three-loop arrangement, which resembles that of the LMFBR, includes an intermediate salt loop to preclude contact of the radioactive primary loop with any of the steam/water mixture in the steam generator.

The primary loop contains all of the fissile and fertile material. An outlet temperature of 700°C [1300°F] at 0.3 MPa [50 psi] would be considered typical for such a system.

The moving core contains all fission-product and actinide nuclides. Since this also includes the delayed neutron precursors, neutron activation can occur *anywhere* in the primary loop.

The intermediate heat exchanger provides communication between the circulating fuel and a loop containing a "clean" working fluid—a salt mixture of

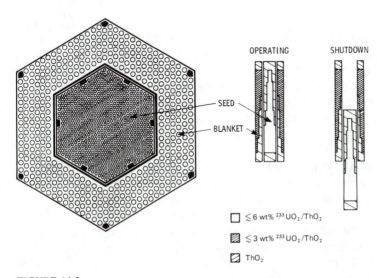

OPERATING   SHUTDOWN

SEED

BLANKET

□ ≤ 6 wt% ^{233}UO$_2$/ThO$_2$

▨ ≤ 3 wt% ^{233}UO$_2$/ThO$_2$

▨ ThO$_2$

FIGURE 14-3
Detailed radial cross section and configurations for operation and shutdown in an LWBR fuel module. (Courtesy of U.S. Department of Energy.)

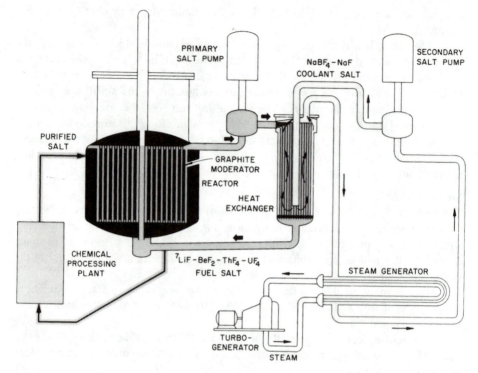

FIGURE 14-4
Steam-cycle schematic for the molten-salt breeder reactor [MSBR] concept. (Courtesy of Oak Ridge National Laboratory, operated by the Union Carbide Corporation for the U.S. Department of Energy.)

sodium fluoride [NaF] and sodium fluoroborate [$NaBF_4$]. The pressure in the intermediate loop is greater than that of the core to assure that any leakage would be to the fuel loop (i.e., to prevent radioactivity from entering the clean loop).

The steam generator exchanges heat between the intermediate coolant loop and the turbine steam loop. Supercritical steam at about 540°C [1000°F] and 24 MPa [3500 psi] can lead to a thermal efficiency approaching 44 percent.

Fuel

The conceptual MSBR would employ a single fluid for both the core and the breeding blanket. A mixture containing fluoride salts of lithium, beryllium, thorium, and uranium can provide a clear, nonviscous solution with very low parasitic neutron absorption as well as excellent thermal and radiation stability.

The neutron multiplication in the salt varies with the amount of graphite moderator present. At a high moderator-to-fuel ratio, for example, resonance absorption in ^{232}Th tends to be reduced while thermalization increases ^{233}U fission and, thus, multiplication (as is comparable to the situation for slightly-enriched uranium described by Fig. 11-2). A smaller ratio reduces multiplication by enhancing resonance absorption in the fertile ^{232}Th—a net favorable effect on breeding ^{233}U. By varying the moderator and fuel fractions, the same fluid is employed for both the reactor core and breeding blanket. A typical molten-salt composition of LiF(72%)–BeF$_2$(16%)–ThF$_4$(12%)–^{233}UF$_4$(0.3%), for example, has a multiplication factor of $k_\infty \approx 1.03$ in the small-diameter flow channels of the core and $k_\infty \approx 0.4$ in the large channels of the blanket. The corresponding moderator-to-fuel ratios are about 6.7 and 1.5, respectively.

Good breeding characteristics depend entirely upon on-line reprocessing. The following features are of special importance:

- general removal of fission products and transuranic poisons
- reduction of the gaseous ^{135}Xe concentration by at least a factor of 10 through sparging of the fuel salt with helium and by controlling the porosity of the graphite moderator to limit gas retention
- early isolation of ^{233}Pa so that it can decay to ^{233}U rather than be a parasitic neutron absorber (i.e., undergo a (n, γ)-reaction to become nonfissile ^{234}Pa.)

In the case of the latter, ^{233}U would be separated from the ^{233}Pa at regular intervals for return to the primary system and/or for sale (as the *net* production due to the breeding process).

Reactivity Control

Adjustment of the fissionable, fission-product, and actinide-poison contents of the fuel salt provides a core which has a very low excess reactivity at all times. Thermal expansion of both the fuel and the graphite moderator provide negative temperature feedback mechanisms. Poison control rods are used both for routine reactivity control and for reactor-trip shutdown.

The delayed-neutron-precursor fragments circulate with the fuel salt in the primary loop. This is an additional time-dependent variation which must be considered in the neutron-kinetics evaluation (e.g., added to Eqs. 5-8) for the MSBR. Although the net effect of the moving source is a reduction in the effective fractions and lifetimes of the delayed neutrons, control problems are not anticipated.

Other Considerations

The MSBR concept has a number of highly favorable characteristics. On-line reprocessing enhances neutron economy and resource utilization. It also eliminates the need for fuel fabrication and limits out-of-core inventory requirements.

On-line reprocessing is also responsible for what may be the major drawback of the MSBR. In the conventional, solid-fuel reactors, much of the waste-product inventory decays in place during both operation and the interim storage period which follows. The MSBR, by contrast, separates such wastes while they are still relatively "fresh." The absence of the automatic "delay and decay" provision is expected to result in stringent operating constraints for fission-gas and tritium releases.

EXERCISES

Questions

14-1. Explain the requirement for thorium-based fuel in thermal breeder reactors.

14-2. Identify at least two reasons why the light-water breeder reactor [LWBR] was not considered feasible at the time the early LWR systems were developed.

14-3. Describe the seed-blanket concept of the LWBR fuel assembly.

14-4. Identify the three important features of on-line reprocessing in the molten-salt breeder reactor. Explain why its environmental impact may be potentially greater than that of the standard reprocessing.

14-5. Explain how the same salt fluid can be used for both the reactor core and the breeding blanket in the MSBR concept.

Numerical Problems

14-6. Estimate the time required for 50 percent and 99 percent, respectively, of a fixed quantity of ^{233}Pa to decay to fissile ^{233}U.

14-7. Of the thermal neutrons absorbed by the reference MSBR fuel composition, estimate the fraction which are:

a. absorbed in the heavy metal

b. absorbed in ^{233}U

c. responsible for ^{233}U fission

(Assume that the percentage composition of the fuel is on a molar basis.)

SELECTED BIBLIOGRAPHY[†]

Light-Water Breeder Reactor
    DOE/ET-0089, 1979 (Sec. IV-C)
    ERDA-1541, 1975

Molten-Salt Breeder Reactor
    Engel, 1979
    ORNL-4782, 1972
    Robertson, 1971
    Simnad, 1971 (Sec. 70)
    WASH-1222, 1972

Other Sources with Appropriate Sections or Chapters
    Foster & Wright, 1977
    Lamarsh, 1975
    Nero, 1979
    WASH-1535, 1974

[†]Full citations are contained in the General Bibliography at the back of the book.

15

FAST BREEDER REACTORS

Objectives

After studying this chapter, the reader should be able to:

1. Explain the preference for plutonium as a fuel in fast breeder reactors.
2. List five countries with major LMFBR programs.
3. Identify the two different primary-system arrangements for an LMFBR.
4. Describe the major differences between fast-reactor and LWR fuel assemblies.
5. Identify two major similarities and differences between the GCFR and HTGR designs.
6. Describe the redundant shutdown systems for the SNR-300 LMFBR.
7. Identify the relative advantages of the LMFBR and GCFR with respect to each other.

Prerequisite Concepts

Breeding is achieved with some difficulty with the CANDU-OCR, LWBR, and MSBR concepts described in the previous chapters. Even the most promising ^{233}U-thorium fuel system is characterized by relatively low η values (Fig. 6-1) for neutrons in the thermal [$\lesssim$0.1 eV] energy range. With the unavoidable losses that accompany thermalization, breeding ratios only slightly greater than unity and long doubling times are anticipated.

If the energy spectrum of fission neutrons (Fig. 2-8) is not modified greatly, corresponding η values (Fig. 6-1) may be noted to be substantially higher than is possible in thermal systems. The fissile plutonium isotopes [^{239}Pu and ^{241}Pu] are especially favorable for breeding at neutron energies above about 0.1 MeV. Thus, the *fast breeder reactor* [*FBR*] concepts favor a plutonium-^{238}U fuel cycle.[†]

A fast neutron spectrum is maintained only if moderators are effectively excluded from the reactor core. Liquid-sodium coolant is an integral part of the liquid-metal fast-breeder reactor [LMFBR] concept because it is a relatively poor moderator while having good heat transport properties. The gas-cooled fast reactor [GCFR], or gas-cooled fast-breeder reactor [GCFBR], employs helium coolant in part for its low tendency to scatter fast neutrons.

LIQUID-METAL FAST-BREEDER REACTORS

The liquid-metal fast-breeder reactor [LMFBR] has been one of the world's most studied concepts. Figure 15-1 traces the development timetable for the seven most active countries. Table 15-1[‡] provides a summary of the major LMFBR projects in the figure as well as other long-range plans.

The world's first nuclear electricity was generated by the Experimental Breeder Reactor [EBR-1] in 1952 (two years before the Soviet Obninsk PTGR gained the distinction as the first *commercial* unit). Since that time about 150 reactor-years of experience have been accumulated on the liquid-sodium-cooled LMFBR systems. Operable units as of the beginning of 1980 represent roughly 3000 MW(th), with nearly double that capacity scheduled to come on line within about 5 years.

The United States led the early development of the LMFBR concept with EBR-1 and several other small experimental units. Commercial electrical generation began with the Experimental Breeder Reactor-2 [EBR-2] and the Enrico Fermi-1 reactor. EBR-2 has exhibited an excellent on-line record for power production despite the demands of its basic experimental programs. The prototypical Fermi-1 unit, developed by a utility consortium for full commercial application, operated for only a brief period of time before it was decommissioned.

Current U.S. efforts are centered around the imminent completion of the Fast Flux Test Facility [FFTF], continuing research and development for the Clinch River Breeder Reactor Project [CRBRP], and conceptual design of the Prototype

---

[†]India does plan to employ a thorium cycle, but primarily because it has such extensive reserves of this latter material.

[‡]Differences in dates between the table and Fig. 15-1 are based on the distinction between initial operation and full-power or commercial operation, respectively.

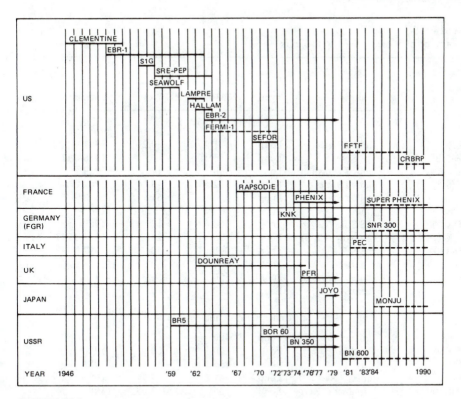

FIGURE 15-1
World breeder reactor development timetable. Horizontal lines show service years for experimental units, or commercial operating years for power reactors. Arrows piercing 1979 indicate units currently operating. (Adapted from *Power Engineering*. © Copyright by Technical Publishing, a Division of Dun-Donnelley Publishing Corp., a Company of the Dun & Bradstreet Corporation, 1979–all rights reserved.)

Large Breeder Reactor [PLBR]. The ultimate fate of both the Clinch River and PLBR projects is uncertain pending resolution of national policy concerns.

The experimental BR-5, BOR-60, Dounreay, Rapsodie, KNK, and Joyo units have provided valuable LMFBR experience for their respective countries. The operating prototypical BN-350, Phenix, and PFR units are to be supplemented by the BN-600, SNR-300, and Monju systems. The 1200-MW(e) Super Phenix, the first near-commercial-sized LMFBR, is progressing well on a construction schedule geared to 1983 operation. The other commercial-scale units identified in Table 15-1 (e.g., CDFR, SNR-2, and BN-1600) are still in their design phases, generally awaiting additional operating experience with their predecessor units. The French are already planning a second-generation version of Super Phenix for operation in the 1990s at 1450–1500 MW(e).

International cooperation on LMFBR development has been substantial. Most notable is the multinational nature of the SNR-300, SNR-2, and Super Phenix projects. The Federal Republic of Germany, Belgium, and the Netherlands on a 70-15-15 percent basis formed the RWE consortium to finance and construct the

TABLE 15-1
Summary of World LMFBR Projects[†]

Name	Country	Type	Power MW(th)	Power MW(e)	Initial operation
		Operable			
EBR-1	United States	Loop	1.2	0.2	1951
BR-5/BR-10[‡]	U.S.S.R.	Loop	5/10	–	1958/1973
Dounreay	United Kingdom	Loop	60	14	1959
EBR-2	United States	Pool	62.5	18.5	1963
Fermi-1	United States	Loop	300	61	1963
Rapsodie	France	Loop	40	–	1967
BOR-60	U.S.S.R.	Loop	60	12	1969
KNK-1/KNK-2[‡]	Germany (FRG)	Loop	58	20	1971/1977
BN-350	U.S.S.R.	Loop	1000	350[§]	1972
Phenix	France	Pool	567	250	1973
PFR	United Kingdom	Pool	600	250	1974
Joyo	Japan	Loop	100	–	1974
		Under construction			
FFTF	United States	Loop	400	–	1980
BN-600	U.S.S.R.	Pool	1420	600	1980
PEC	Italy	Loop	135	–	1981
Madras[¶]	India	Loop	42.5	17	1981
SNR-300	Germany (FRG)	Loop	736	312	1981
Super Phenix	France	Pool	2900	1200	1983
CRBRP	United States	Loop	975	350	1985
Monju	Japan	Loop	714	300	1986
		In design			
CDFR	United Kingdom	Pool	3250	1320	–
SNR-2	Germany (FRG)	Loop	3750	1300–1500	–
PLBR	United States	–	2980	1000	–
BN-1600	U.S.S.R.	–	–	1600	–
Super Phenix-II	France	Pool	–	1450–1500	–

[†]Data assembled from: J. S. McDonald and S. Golan, "Engineering Aspects of the Pool Type LMFBR, 1000 MW(e)," EPRI NP-645-SY, April 1978; J. Moore, "International Progress with Liquid-Metal Fast Reactors," *Nucl. Eng. Int.*, July 1979; F. C. Olds, "The Impressive Super Phenix," *Power Eng.*, August 1979; and E. H. Hill, "U.S. and Foreign Breeder Reactors," CRBRP-PMC 78-03, Summer 1978.
[‡]Upgraded or otherwise modified.
[§]Bn-350 produces 150 MW(e) plus 200-MW(e)-equivalent process heat for desalination.
[¶]Will employ a thorium cycle.

SNR-300 project. The United Kingdom later acquired a 1.65 percent financial interest in the project to lead to the formation of a new group called SBK.

Shares in the Super Phenix project are currently held by France (51%), Italy (33%), and SBK (16%). In a similar manner, the SNR-2 shares are held by SBK (51%), Italy (33%), and France (16%).

System-to-system differences seem to be greater for the LMFBRs than for most other reactor designs. A number of the more basic options are identified in the remainder of this section. For the purpose of comparison with the other reactor types, characteristics of the Super Phenix or the PLBR concepts are assumed to be representative.

Steam Cycles

All current LMFBR designs employ liquid-sodium coolant for its good heat transfer properties and low neutron moderation characteristics. However, the sodium is highly reactive with water and is also subject to neutron activation. The desire to prevent "hot" [activated] sodium from contacting water due to possible leaks in the steam generator is responsible for the use of an intermediate loop of "clean" sodium (Fig. 1-5).

The use of the primary and intermediate sodium loops in the LMFBR design has been implemented in two conceptually different arrangements. Figure 15-2*a* shows the basic features of the *loop-type* system where the intermediate heat exchanger is located outside of the reactor. Primary sodium coolant enters toward the bottom of the vessel, flows upward through the core, and exits to the intermediate heat exchanger. The intermediate loop then carries the heat energy to the steam generator.

The *pool-type* LMFBR in Fig. 15-2*b* includes the reactor core, primary pumps, and intermediate heat exchangers in a large-volume pool of sodium in the reactor vessel. Sodium is discharged from the intermediate heat exchanger to the pool, is eventually drawn to the pump inlet, is forced upward through the core, and reenters the heat exchanger. The flow baffle serves to prevent the heated core-outlet stream from mixing with the general pool volume.

Both the loop- and pool-type designs employ multiple primary and intermediate coolant circuits to divide the heat load and provide redundant backup capability in the event of component failures. They are also configured to enhance natural-convection circulation if forced cooling ceases. As shown by Fig. 15-2, the operating temperatures may be essentially the same for the two types of systems. Each steam-cycle concept is employed in several of the current-generation LMFBR systems as identified in Table 15-1.

Typical reactor vessel configurations for the commercial-size loop- and pool-type LMFBRs are shown in Figs. 15-3 and 15-4, respectively. Representative vessel dimensions are 14 m diameter by 28 m height (SNR-2 loop-type), and 21 m diameter by 19 m height (Super Phenix pool-type). The vessels are generally 5- to 8-cm-thick stainless steel.

Several steam generator designs are employed for LMFBR systems. Once-through and U-tube units (e.g., similar in basic features to their PWR counterparts of Fig. 12-10) are employed in SNR-300 and the PFR and CDFR systems, respectively. The Phenix and Super Phenix systems use once-through steam generators with helical (rather than straight) tubes.

The steam generator design for Clinch River and PLBR is shown in Fig. 15-5. This "hockey stick" design uses bent tubes to reduce the stresses associated with thermal expansion. As in all of the LMFBR steam generators, the water–steam flow is through the tubes while the sodium is on the shell side.

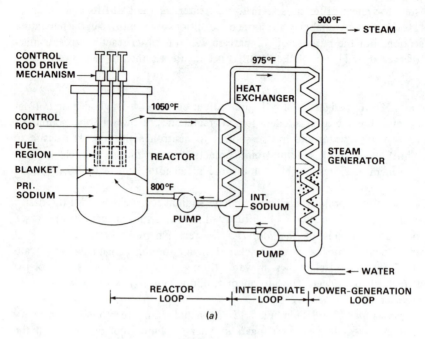

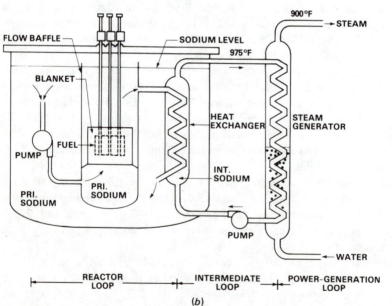

FIGURE 15-2
Conceptual steam-cycle arrangements for LMFBR systems: (a) loop design; and (b) pool design. (Courtesy of Clinch River Breeder Reactor Plant.)

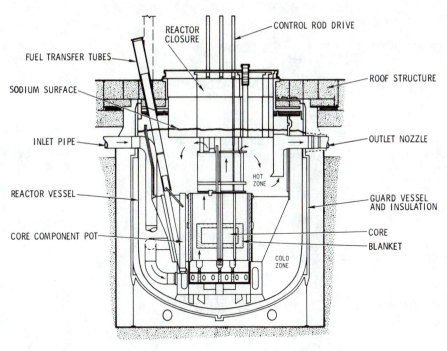

FIGURE 15-3
Basic features of a loop-type LMFBR system. (Courtesy of S. H. Fistedis, Argonne National Laboratory.)

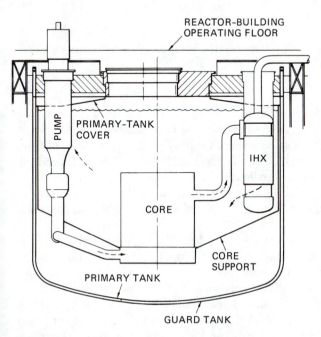

FIGURE 15-4
Basic features of a pool-type LMFBR system. (Courtesy of General Electric Company.)

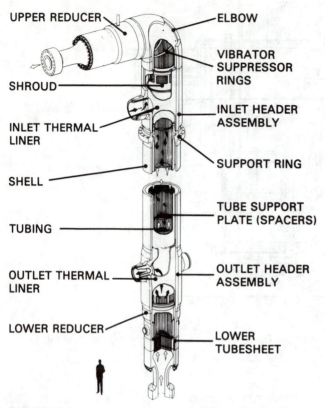

UPPER REDUCER

ELBOW

VIBRATOR
SUPPRESSOR
RINGS

SHROUD

INLET HEADER
ASSEMBLY

INLET THERMAL
LINER

SUPPORT RING

SHELL

TUBE SUPPORT
PLATE (SPACERS)

TUBING

OUTLET THERMAL
LINER

OUTLET HEADER
ASSEMBLY

LOWER REDUCER

LOWER
TUBESHEET

FIGURE 15-5
"Hockey stick" steam generator for an LMFBR. (Courtesy of Clinch River Breeder Reactor Plant.)

Fuel Assemblies

The design, fabrication, and utilization of LMFBR mixed-oxide fuel assemblies have been described in Chaps. 1 and 9. Core and blanket assemblies for the Soviet BN-600 system shown in Fig. 15-6 are typical of many systems. The small-diameter driver fuel pins contain a central stack of mixed-oxide fuel pellets with depleted-uranium pellets above and below forming the axial blanket. All are contained in a single stainless-steel cladding tube. Spacing is maintained by the wire wrap (e.g., as in Fig. 9-6). The channel around the array directs the flow of the liquid-sodium coolant.

The radial blanket assemblies in Fig. 15-6 have the same outer dimensions as the driver assemblies. However, they consist of larger-diameter pins that contain only depleted-uranium pellets. This dimensional change is allowable because the fission and heat generation rates are relatively lower in the blanket assemblies.

The BN-600 driver assemblies have a total length of 3.5 m, a fuel height of 0.75 m, and upper and lower axial blanket heights of 0.4 m each. The higher-power LMFBRs call for longer assemblies and an active fuel height of about one meter.

One variation on the arrangement in Fig. 15-6 is to shorten the driver-fuel pins and employ separate pins for one or both of the axial blanket regions. The Phenix and Super Phenix systems each use similar large-diameter pins for the radial fuel assemblies and the upper axial blanket. The EBR-2 driver fuel assembly contains blanket pins for both the upper and lower axial blankets. (The latter had also employed the metallic fuel described briefly in Chap. 9.)

Fuel utilization patterns for LMFBRs are generally of two types. Examples for the Clinch River plant are shown in Fig. 15-7. Configurations with driver-fuel assemblies in the center and blanket assemblies located only around the periphery are referred to as *homogeneous*. *Heterogeneous* patterns intersperse some number of blanket assemblies to more interior locations. Because of their general appearance, the latter are often called "bull's-eye" configurations. Both patterns in Fig. 15-7 also accommodate primary and secondary control assemblies, plus radial shield assemblies.

Reactivity Control and Protective System

Since by definition breeder reactors produce more new fissile material than they expend, reactivity changes tend to be smaller than for their converter counterparts. The fast-neutron spectrum also reduces the poisoning effects of ^{135}Xe and ^{149}Sm (which are so important in thermal reactor systems) and other fission products.

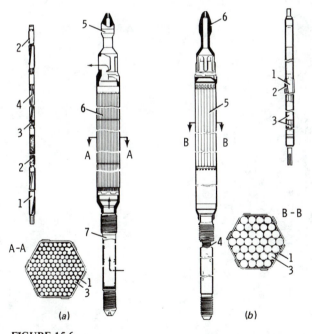

FIGURE 15-6
Core and blanket assemblies for the Soviet BN-600 LMFBR. (*a*) BN-600 fuel assembly: (1) pin cladding, (2) slugs of depleted uranium, (3) fuel pellets, (4) wire wrapped fin, (5) fuel assembly head, (6) fuel pin assembly, (7) stem. (*b*) BN-600 radial blanket assembly: (1) pin cladding, (2) wire wrapped fin, (3) depleted uranium, (4) blanket assembly stem, (5) blanket pin assembly, (6) blanket assembly head. (Reproduced from *Nuclear Engineering International* with permission of the editor.)

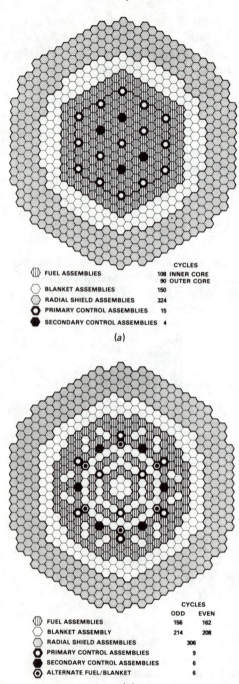

FIGURE 15-7
Clinch River LMFBR core in (*a*) a homogeneous and (*b*) a heterogeneous configuration. (Courtesy of Clinch River Breeder Reactor Plant.)

Neutron-poison control rods are used for all control and trip functions. Reactivity requirements for a typical plutonium-fueled fast reactor system are provided in Table 15-2. The total control requirement of 7.4% $\Delta k/k$ is predicated on being able to override the core's maximum reactivity worth (the excess for burnup lifetime plus the defect values) with an adequate shutdown margin. Since the LMFBR does not employ burnable poisons (as are common to the thermal reactors), the maximum reactivity inventory occurs at the beginning of each depletion cycle when there is no fission-product inventory.

The standard LMFBR control method employs rods which occupy fuel-assembly locations (e.g., as indicated on Fig. 15-7). In general the rods are divided into two groups, each of which can produce full neutronic shutdown even when one "stuck rod" fails to insert.

One Clinch River design (Fig. 15-7a), for example, calls for 19 core locations to be allocated to control. Power and flux-distribution control are provided by 15 primary rods, each of which contains 37 B_4C pins in a hexagonal array. An electromagnetic coupling between these rods and their drive motors provides the basis for a gravity-initiated reactor trip, as is employed in the PWR systems. Four secondary rods with a latched-disconnect mechanism provide a fully independent shutdown system. Insertion of any 14 primary rods *or* three secondary rods results in complete shutdown.

The SNR-300 employs a nine-rod primary shutdown system which is similar to that considered above. The secondary system, however, differs as shown in Fig. 15-8. These three rods each consist of a series of linked-segment absorbers which normally reside below the core. Under trip conditions, the rods are pulled into the active region by a spring-loaded system. This secondary system is designed to implement shutdown even if the core becomes deformed and would inhibit gravity insertion of the long, rigid primary rods.

GAS–COOLED FAST–BREEDER REACTORS

The gas-cooled fast-breeder reactor [GCFBR], or more commonly, the gas-cooled fast reactor [GCFR], is an alternative to the LMFBR for breeding plutonium in a fast neutron spectrum. However, it is still in the research and development stage, as no such units are currently in operation.

TABLE 15-2
Control Requirements for a Typical Fast Reactor System[†]

Control characteristics	% $\Delta k/k$
Burnup (equilibrium cycle)	3.7
Temperature defect: cold to hot, zero power	0.5
Temperature defect: hot, zero power to hot, full power	0.8
Shutdown margin	2.4
Total requirement	7.4

[†]From A. Sesonske, *Nuclear Power Plant Design Analysis*, TID-26241, 1973).

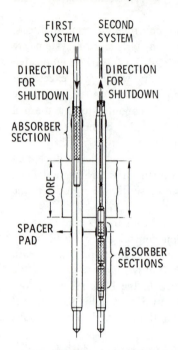

FIRST   SECOND
SYSTEM   SYSTEM

DIRECTION   DIRECTION
FOR   FOR
SHUTDOWN   SHUTDOWN

ABSORBER
SECTION

CORE

SPACER
PAD

ABSORBER
SECTIONS

FIGURE 15-8
Primary and secondary shutdown rods for the SNR-300
LMFBR. (Reprinted from *Nuclear Engineering International* with permission of the editor.)

Of three major coolant options for the GCFR, helium has received the most attention based on extension of existing HTGR and THTR technology. The United Kingdom and the U.S.S.R. have lower-level research efforts related to CO_2 and N_2O_4 [nitrogen tetroxide] coolants, respectively.

The General Atomic Company and a consortium of U.S. utilities have been the primary proponents of the helium-based GCFR. The U.S. Department of Energy has provided limited funding for further development of the concept as an alternative to the LMFBR. International agreements are also in effect with the Federal Republic of Germany and Switzerland. Current activities are geared to preliminary design of a 300-MW(e) demonstration plant and conceptual design of a larger commercial unit.

Reactor System

Many of the principle features of the 300-MW(e) GCFR system are shown in Fig. 15-9. The primary similarities to the HTGR (Fig. 13-6) are related to the use of a prestressed concrete reactor vessel [PCRV] containing the entire primary coolant system. The common use of helium coolant results in comparable steam conditions and thermal efficiencies for the two gas-reactor concepts.

The GCFR core is designed much like that of the LMFBR. Fuel assemblies consist of stainless-steel-clad mixed-oxide fuel pins. It may also be noted that the active core region is very compact like that of the LMFBR.

Helium has several advantages over sodium as a fast-breeder reactor coolant. Since it is relatively transparent (i.e., does little scattering) to neutrons, the fission-neutron spectrum remains "hard" and allows for a high breeding ratio (Chap. 6). This same feature also limits the potential reactivity effect of coolant voiding

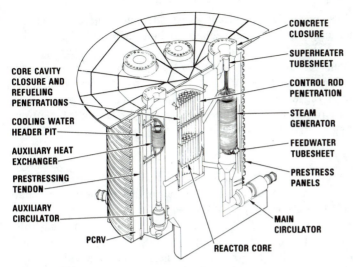

FIGURE 15-9
Reactor arrangement for a 300-MW(e) GCFR. (Courtesy of General Atomic Company.)

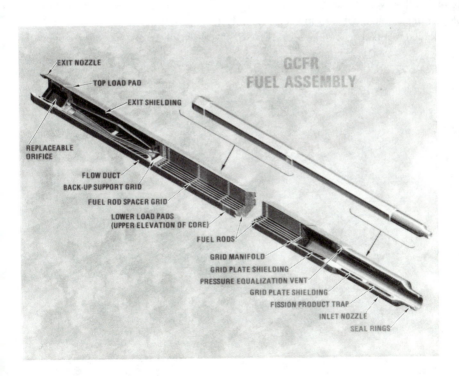

FIGURE 15-10
Fuel assembly for a GCFR. (Courtesy of General Atomic Company.)

(Chap. 5). Finally, since helium is chemically inert and not subject to neutron activation, there is no need for an intermediate coolant loop like that of the LMFBR.

By contrast, helium is at a major disadvantage to sodium in terms of natural-convection cooling afforded by large, passive coolant volumes (Figs. 15-3 and 15-4). The potential for water to mix with helium in the steam generator is another important concern as it provides a mechanism for enhanced neutron moderation and resulting reactivity insertion.

Fuel Assemblies

The GCFR fuel assembly in Fig. 15-10 consists of a lattice of small-diameter fuel pins enclosed in a flow duct. The design may be noted to bear a general resemblance to the LMFBR assembly (Fig. 15-6).

The GCFR fuel pin in Fig. 15-11 appears to be similar to its LMFBR counterpart (Fig. 9-6). Pellet and clad composition and sizes are generally comparable. The fuel and axial blanket lengths are also nearly the same for the two systems. To enhance heat transfer to the helium coolant, however, the GCFR clad is ribbed and the lattice spacing is roughly twice as great.

Another unique feature of the GCFR fuel assembly is the system that equalizes the pressure between the inside and outside of the cladding. The fuel rods are vented through a helium purification system to the inlet of the primary system circulator. The venting system includes a charcoal trap in each rod (Fig. 15-11) and a large integral charcoal trap in each assembly (Fig. 15-10).

Radial blanket assemblies for the GCFR consist of an array of larger diameter

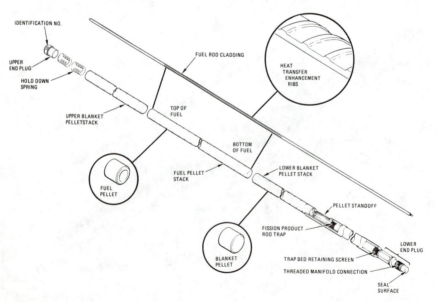

FIGURE 15-11
Fuel pin for a GCFR. (Courtesy of General Atomic Company.)

pins in the standard flow duct as for their LMFBR counterparts (Fig. 15-6). Hexagonal control rods are another similarity between the two fast reactor systems.

EXERCISES

Questions

15-1. List five countries with major LMFBR programs.

15-2. Explain the preference for plutonium as a fuel in fast breeder reactors.

15-3. Identify the two different primary-system arrangements for an LMFBR. Sketch the basic features of each.

15-4. Describe the major differences between fast-reactor and LWR fuel assemblies.

15-5. Identify the two major similarities and differences between the GCFR and HTGR designs.

15-6. Describe the redundant shutdown systems for the SNR-300 LMFBR.

15-7. Identify the relative advantages of the LMFBR and GCFR with respect to each other.

Numerical Problems

15-8. Ten nations have substantial commitments to LMFBR development. Considering the units that are currently operating (Fig. 15-1) and firmly committed (Table 15-1 "under construction" and "in design" units with specified type and thermal power), calculate the total thermal capacity for each nation. Rank the nations from highest to lowest commitment. For multinational efforts, divide the capacity in proportion to the financial support.

15-9. Extend Prob. 13-8 to include loop- and pool-type LMFBRs. Using data from the text or App. III, sketch the core sizes on the same figures.

SELECTED BIBLIOGRAPHY †

LMFBR

      Barthold, 1979
      CRBRP-ARD-0230, 1978
      CRBRP/PSAR
      Davies, 1978
      ERDA-1, 1975
      ERDA-1535, 1975
      French Atomic Energy Commission, 1980
      Gray, 1978
      McDonald, 1978
      Nucl. Eng. Int., June/July 1975, Dec. 1976, May 1977, Jan. 1979, July 1979
      Olds, 1979*a, b*
      PMC-XX-YY, 19XX
      Rippon, 1980
      Seaborg & Bloom, 1970
      Vendryes, 1977
      Veziroglu, 1978
      WASH-1101, 1973
      Wett, 1978

†Full citations are contained in the General Bibliography at the back of the book.

GCFR
>CONF-740501, 1974
>GCFR/PSAR

GCFR
>CONF-740501, 1974
>GCFR/PSAR
>*Nucl. Eng. & Des.,* 1977
>*Nucl. Eng. Int.,* Dec. 1978
>Snyder, 1978

Individual LMFBRs
>*Nucl. Eng. Int.* [description with (*) or without wall chart]
>>BN-600 (U.S.S.R.), June/July 1975
>>Clinch River (U.S.), Oct. 1974*
>>Creys–Malville/Super Phenix (France), May 1977, June 1978*
>>FFTF (U.S.), Aug. 1972*
>>Joyo (Japan), Nov. 1978
>>Phenix (France), July 1971,* July 1974
>>PFR/Dounreay (United Kingdom), Aug. 1971*
>>SNR-300/Kalkar (Fed. Rep. Germany), July 1976*
>Current Status (see Chap. 12 Selected Bibliography)

Other Sources with Appropriate Sections or Chapters
>Connolly, 1978
>Foster & Wright, 1977
>Lamarsh, 1975
>Murray, 1975
>Nero, 1979
>Sesonske, 1973

V

REACTOR SAFETY AND SAFEGUARDS

Goals

1. To introduce the physical phenomena that may contribute to reactor accidents.
2. To describe sequences of events for unmitigated reactor accidents.
3. To extend the reactor descriptions in Part IV to include engineered safety systems.
4. To describe the basic features of risk assessment methodologies applied to reactor safety.
5. To consider the accident at the Three Mile Island [TMI] Reactor as a case study on safety.
6. To outline the role of governmental regulation on nuclear fuel-cycle and reactor design and operation.
7. To describe the fundamental principles of nuclear material safeguards and to differentiate between domestic and international applications.

Chapters in Part V

Reactor Safety Fundamentals
Assessment of Reactor Accident Risk
The Three Mile Island Reactor Accident
The Regulatory Process
Nuclear Safeguards

16

REACTOR SAFETY FUNDAMENTALS

Objectives

After studying this chapter, the reader should be able to:

1. Describe the five sources of energy available in reactor accidents.
2. Explain the concept of design-basis accidents and identify the major categories into which they may be classified.
3. Identify the three ways to initiate a loss-of-cooling accident in a reactor.
4. Describe the design-basis loss-of-coolant accident [LOCA] for a light-water reactor.
5. Identify the major differences in the "beyond-design-basis" accidents for LWR, CANDU, HTGR, and LMFBR systems.
6. Describe the four principle events in an LWR meltdown that lead to substantial increases in fission-product release rates.
7. Identify the four categories of fission products which would reside in the containment after an accident and the two ways they could be released to the general environment.

Prerequisite Concepts

An operating nuclear power reactor generates a sizeable inventory of radioactive products (e.g., Table 6-3). Prevention of their release to the general environment is the basis for multiple-barrier design of fuel assemblies and overall reactor systems.

Nearly 98 percent of all radioactive products are retained by the fuel assemblies as long as sufficient cooling is provided to prevent fuel melting. Thus, a major objective of nuclear reactor safety is control of the energy released in the system to the extent required to preclude such melting.

The safety of each reactor facility has come to be based on design for the following three types of accidents:

1. anticipated operational occurrences which should result in no abnormal releases of radioactivity
2. low-probability events which may cause small releases
3. potentially severe accidents of extremely low probability

Safety at the first level is enhanced by a highly reliable system with minimum failure of equipment. The second category recognizes that the design must accommodate some component failures and operating errors during the lifetime of the facility. Analyses of the final type of accidents establish performance criteria for engineered safety systems and serve as bases for site-suitability evaluations.

Discussions in this and the following chapter are generally limited to the five designated reference reactor systems—BWR, PWR, CANDU-PHW, HTGR, and LMFBR. Systems of these types represent the bulk of both current electrical generating capacity and safety evaluation efforts. An understanding of their basic responses and features is also readily applicable to most other reactor designs.

ENERGY SOURCES

The probability of fuel melting as a result of a reactor accident depends directly on the amount of energy that is present. Typical energy sources have the following origins:

- stored energy
- nuclear transients
- decay heat
- chemical reactions
- external events

Both the magnitude and timing of contributions from each category are important to accident evaluation.

Stored Energy

The fuel, coolant, and structures store some amount of heat energy at all times during reactor power operation. The amounts depend on the material properties like the heat capacity and the temperature distributions (e.g., as in Fig. 7-4). The major concern is that a redistribution of this heat energy and the resulting temperature profile from the fuel to the clad and coolant may result in direct damage and/or an extended heat-transfer mismatch.

A very important form of stored energy in the water-cooled reactors is the latent heat associated with the liquid–steam phase transformation. Under normal operating conditions, high pressure allows all-liquid (**PWR** and **PHWR**) or saturation (**BWR**) conditions to be maintained at elevated temperatures. If the pressure is suddenly relieved, however, the coolant vaporizes rapidly under the influence of this stored energy.

The nature of the initiating event determines the extent and timing of stored energy release during an accident. The overall potential magnitude of this source, however, is well defined by the geometry and temperature distribution of the core. The design process is geared to assuring that feasible energy redistribution and/or depressurization does not lead to component damage.

Nuclear Transients

A positive insertion of reactivity causes a transient power increase which may result in a higher stable power level or a large power pulse (shown, for example, by Figs. 5-4 and 5-5, respectively). Although the former may add more total energy to the system, the rapid changes that occur during a pulse can sometimes lead to more serious effects in the system, including fission-gas release, vaporization, and/or fuel fragmentation.

Serious transients in thermal systems which lead to core damage tend to be self-terminating because of the delicate balance of geometry and moderation required for criticality (e.g., as described for **LWR** lattices in Chap. 11). By contrast, fast systems may be more reactive in a modified (e.g., coolant-free or compacted) geometry and achieve recriticality following initial neutronic shutdown.

The total energy associated with one or more reactivity transients depends on the magnitude and time dependence of the insertion as well as the nuclear and physical feedback response of the fuel. Quantification of this energy source is generally difficult and quite uncertain, especially where recriticality is possible.

Decay Heat

Stored energy with or without a nuclear-transient source tends to dominate the initiation of reactor accidents. However, as soon as heat transport can occur to the surrounding environment, the fission-product decay heating becomes a relatively more important energy source. It may be recalled that this source is as high as about 7 percent of operating power at the time of shutdown from a lengthy run.

The long-term nature of the decay heating may be inferred from Eq. 2-10. This energy source is a large contributor for a substantial period of time after shutdown of the neutron chain reaction. In the absence of some type of forced cooling, it is generally sufficient to cause the ultimate melting and/or destruction of reactor fuel assemblies.

Chemical Reactions

Fuel, clad, and coolant materials are selected to be essentially unreactive with each other under normal operating conditions (Chaps. 9 and 11). However, at the elevated temperatures that may be produced by nuclear transients and/or decay heat during an accident, energetic chemical reactions may be facilitated. Table 16-1 presents data related to a number of potentially energetic reactions that are possible in various nuclear reactor systems.

The reactions of zirconium and stainless steel are important to the water-cooled reactors since these materials are employed for clad and structures, respectively. Reactions with oxygen or water at elevated temperatures are highly exoergic (i.e., release energy), as indicated by the examples in Table 16-1. Under accident conditions, the reactions are likely to occur before clad melting and at rates that increase with temperature. Substantial oxidation reduces clad integrity and may result in fragmentation with a further increase in the reaction rate. Overall, the energy release from metal-water reactions is roughly proportional to the amount of metal involved.

As noted in Chap. 9, zirconium-based cladding is favored over stainless steel from the standpoint of neutron economy. Table 16-1 indicates, however, that the zirconium reactions with either water or oxygen generate substantially more energy than those with stainless steel. Thus, some potential safety disadvantages accrue from the decision to use zircaloy cladding for fuel pins in water-cooled reactors.

TABLE 16-1
Properties of Potentially Energetic Chemical Reactions of Interest in Nuclear Reactor Safety[†]

Reactant R	Temperature (°C)	Oxide(s) formed	Heat of reaction[‡] with: Oxygen (kcal/kg R)	Water (kcal/kg R)	Hydrogen produced with water (l/kg R)
Zr (liq.)	1852[§]	ZrO_2	−2883	−1560	490
SS (liq.)	1370[§]	F_eO, Cr_2O_3, NiO	−1330 to −1430	−144 to −253	440
Na (solid)	25	Na_2O	−2162	−	−
Na (solid)	25	NaOH	−	−1466	490
C (solid)	1000	CO	−2267	+2700	1870
C (solid)	1000	CO_2	−7867	+2067	3740
H_2 (gas)	1000	H_2O	−29,560	−	−

[†]Adapted from T. J. Thompson and J. G. Beckerley, eds., *The Technology of Nuclear Reactor Safety*, Vol. 1, by permission of The MIT Press, Cambridge, Mass. Copyright © 1964 by the Massachusetts Institute of Technology.

[‡]Positive values indicate energy that must be added to initiate an endoergic reaction; negative values indicate energy released by exoergic reactions.

[§]Melting point.

The very rapid, exoergic reaction between sodium and water (e.g., as in Table 16-1 for solid sodium) is the fundamental reason for isolating the LMFBRs radioactive primary sodium coolant loop from the steam generator by use of an intermediate loop (Figs. 1-6 and 15-2). Such reactions are also a major concern for potential core accidents. In addition to energy generation, sodium hydroxide [NaOH] or sodium oxide [Na_2O] are produced from reactions with water or oxygen, respectively. The former, a highly corrosive base, corrodes certain stainless steels. The Na_2O can cause plugging and thereby restrict the flow of unreacted liquid sodium.

If oxygen enters a graphite reactor, exoergic reactions occur with the carbon to generate CO_2 or carbon monoxide [CO]. Even the endoergic graphite-water reactions are significant since some of their products—CO and hydrogen—participate in secondary exoergic reactions.

Hydrogen is evolved from the reactions of water with zirconium, sodium, and graphite. Its reaction with oxygen to produce water is quite energetic (e.g., as indicated by Table 16-1). The reaction may also be explosive under certain circumstances.

When molten fuel, clad, or structure contacts water or liquid sodium, another rapid energy release—a *vapor explosion*—is possible. These complex processes occur only with very large temperature differences between the molten material and the coolant and within a specific range of pressures.

External Energy Sources

Natural and human events external to the reactor system each have the potential to initiate or otherwise contribute to accidents. Natural events include phenomena such as floods, hurricanes, tornadoes, and earthquakes. Aircraft impacts and industrial explosions are among the nonnatural occurrences that must be considered.

The potential effects of external energy sources vary greatly among specific reactor sites. Further considerations of reactor siting requirements are deferred to Chap. 19.

Fuel Failure Propagation

The interaction among the three mechanisms for energy addition—nuclear transients, decay heat, and chemical reactions—will determine the course of an accident from local origin to whole-core involvement. Major factors include the potential for:

- fission gas release, fuel fragmentation, and clad bursting
- fuel pin slumping with physical contact among adjacent fuel pins
- fuel and clad melting

The details of such geometry changes will determine the coolability, likelihood of recriticality, and the overall involvement of the core.

ACCIDENT CONSEQUENCES

A very wide range of potential accidents must be evaluated to assure that nuclear reactor operation does not result in excessive radiation exposures to the general

TABLE 16-2

Representative Reactor Accident Classifications for a PWR System[†]

Class #	Description	Example(s)
1	Trivial incidents	Small spills; small leaks inside containment
2	Misc. small releases outside containment	Spills; leaks and pipe breaks
3	Radwaste system failures	Equipment failure; serious malfunction or human error
4	Events that release radioactivity into the primary system	Fuel defects during normal operation; transients outside expected range of variables
5	Events that release radioactivity into the secondary system	Class 4 and heat-exchanger leak
6	Refueling accidents inside containment	Drop fuel element; drop heavy object onto fuel; mechanical malfunction or loss of cooling in transfer tube
7	Accidents to spent fuel outside containment	Drop fuel element; drop heavy object onto fuel; drop shielding cask—loss of cooling to cask, transportation incident on site
8	Accident initiation events considered in design basis; evaluation in the safety analysis report	Reactivity transient; rupture of primary piping; flow decrease; steamline break
9	Hypothetical sequences of failures more severe than Class 8	Successive failures of multiple barriers normally provided and maintained

[†]WASH-1250, 1973.

public. It has been common practice to divide the spectrum of such accidents into categories according to the severity of the potential consequences. One such classification scheme for PWR systems is contained in Table 16-2. These nine accident classes may be noted to range from trivial incidents with little or no release of radiation to very severe sequences of events which include successive failures of the multiple barriers provided for fission product retention.

Although the class designations in Table 16-2 are no longer used extensively, evaluations of the entire range of accidents will always be appropriate. The final two entries tend to receive the most detailed attention because of their complexity. The *design basis accidents* serve as a standard for evaluating the acceptability of reactor designs and specific sites. The class 9, or "beyond-design-basis," accidents are postulated to explore the effects of failures of engineered safety systems and the possibilities for fuel melting.

Design-Basis Accidents

The design-basis accidents involve postulated failure of one or more important systems. Evaluations of such accidents by employing conservative assumptions must assure that radiological consequences will be within pre-established limits. In this sense, the accidents serve as the *basis* for assessing the overall acceptability of a specific reactor *design*, generally on a specific site.

Design-basis accidents for most reactors may be classified according to the following general initiating mechanisms:

- control-rod withdrawal
- spent-fuel handling
- steam-line break
- loss of cooling
- external events

The first four categories are considered below in terms of cause, progression, analysis assumptions, and/or design-based systems for consequence mitigation. Requirements related to external events are described in Chap. 19.

Control-Rod Withdrawal

Withdrawal of a control rod in all of the reference reactor designs is equivalent to a removal of neutron poison and an insertion of positive reactivity. Accidents of this type may be divided into three categories:

- uncontrolled rod withdrawal from a subcritical condition
- uncontrolled rod withdrawal at power
- control-rod ejection (PWR only)

Withdrawal from a subcritical condition produces a continuous reactivity addition which could produce a power excursion and eventual fuel failure. Protective-system trip circuits (e.g., Fig. 12-16) for excess startup rate, excessively short period, and overpower may be expected to mitigate the accident. In the unlikely event of failure of these safety features, negative temperature feedbacks reduce energy release until coolant-temperature or overpressure trips are activated. If control rods are withdrawn while the reactor is at power without a corresponding increase in the turbine-cycle load, the coolant temperature increases as the core power and heat flux increase. A heat-flux/coolant mismatch results in critical heat flux or DNB. Overpower, overpressure, and high coolant-temperature trip levels are generally set to minimize the likelihood of DNB.

Failure of a control-rod housing in a PWR could facilitate rapid ejection of the rod and its drive under the influence of the high primary-system pressure. Such ejection produces a reactivity insertion as well as potentially large local power peaking. With only the first regulating control-rod group (e.g., as in Fig. 12-14) inserted at full power, ejection of a peripheral rod produces a highly asymmetric power distribution. The core protective system and the power-dependent rod insertion program must be coordinated to assure that a trip occurs before linear heat rate or DNB limits are exceeded.

Spent-Fuel Handling

Spent-fuel handling accidents are the most credible of those related to the design-basis evaluations. Since the reactor-vessel head is open, an accident may allow the volatile fission products to be transported quickly to various parts of the containment. Mechanical damage to the fuel assembly, criticality, or failure to maintain adequate cooling could result in substantial release of radioactivity.

The potential for mechanical damage is minimized by using well-designed handling tools, interlocks, and operating procedures. Criticality prevention and adequate cooling are assured by use of storage arrays and heat-removal systems designed to mitigate credible contingencies.

Steam-Line Break

A major break in a steam line produces a "cold-water" reactivity insertion (Chap. 5) in all multiple-loop systems which have a negative power feedback. (In the single-loop BWR, such a break is equivalent to a loss-of-coolant accident as described next.) The stored energy in the high-temperature, high-pressure steam produces a very rapid "blowdown" with a substantial cooling effect on the steam generator and, hence, on the primary coolant loop. This cooling results in a positive reactivity insertion in systems with negative temperature feedbacks.

While large, negative temperature feedbacks tend to mitigate most other accidents, they can enhance the severity of the steamline-break accident. These "cold-water" accidents, thus, may lead to limitations on the magnitude of negative temperature coefficients of reactivity.

Loss of Cooling

Some of the most limiting of the design-basis accidents for each reactor system are those associated with the loss of ability of the coolant to remove heat from the fuel. Complete loss of the coolant is the most severe accident. Small losses of fluid and/or the loss of coolant flow may also have important consequences.

The *loss-of-coolant accident* [*LOCA*] is most likely in the water-cooled reactors where the stored energy content of the high-pressure, high-temperature coolant may be released to the containment by rupture of an exposed pipe. HTGR systems with their primary coolant loops contained entirely within the reactor vessel (Fig. 13-6) are not as readily susceptible to extensive coolant loss. The low-pressure sodium coolant in LMFBRs is also not subject to such rapid removal.

Design-basis LOCA analysis of LWR systems calls for the following scenario:

- A double-ended, "guillotine" pipe break in a primary coolant line to allow free coolant flow from both ends.
- Coolant flashes to steam under the influence of the stored energy and is discharged rapidly into the containment building.
- Although the coolant loss shuts down the system neutronically, reactor trip is initiated by an under-pressure reading to the protective system to assure continued subcriticality.
- The emergency core cooling systems [ECCS] operate to cool the core and prevent excessive decay-heat-driven damage.
- Radioactivity in the coolant is retained by the containment structure with natural deposition processes and active removal systems eventually reducing overall levels of radioactivity.
- Heat removal systems maintain ECCS effectiveness and reduce containment pressure.

When the engineered safety features operate as designed, the core is cooled with a minimum amount of local fuel failure and radioactivity release. (Such systems are discussed in some detail in the next chapter.)

The pressure-tube design of the CANDU reactor provides for oppositely directed coolant flow in adjacent channels. This separation, in turn, generally allows at least half of the core to be isolated if a major pipe break occurs. Although the positive moderator feedback mechanism (Chap. 5) tends to cause an initial power increase with loss of coolant, the Doppler feedback in the fuel soon produces an overall neutronic shutdown in the system. The remaining features of the LOCA for the CANDU system parallel those noted above for the LWR design.

Loss of core cooling capacity without a general loss of the coolant fluid may be initiated by halting the flow in the primary or secondary coolant loops. A *loss-of-flow accident* [*LOFA*] may occur in any system experiencing pump failure in the primary system. By contrast, a *loss-of-heatsink accident* [*LOHA*] depends on a heat exchanger failure (e.g., loss of feedwater flow that inhibits removal of heat energy from the primary loop). Either mechanism should lead to a scenario with the following features:

- Coolant temperature increases.
- Protective system initiates a reactor trip on over-temperature or over-pressure.
- Stored-energy redistribution and decay heat result in coolant heating with voiding or density reduction.
- Natural convection cooling limits fuel damage until some circulation can be restored by safety systems.

As in the LOCA, design features are expected to mitigate the consequences of the accident to minimize the amount of fuel failure and subsequent release of radioactivity for each of the reactor designs.

Meltdown Scenarios

The "beyond-design-basis," class 9 accidents have the potential for causing core meltdown. Based on hypothetical sequences of events which include successive failures of engineered safety systems, analysis of such accidents provides a basis for design of emergency cooling systems and containment structures.

Meltdown ultimately requires the heat production rate to exceed the removal rate. This may be initiated by either overpower or undercooling conditions. The former are most readily obtained by reactivity transients. The latter are related to reduced cooling through loss of coolant flow or loss of the coolant itself.

The sequences described in the following paragraphs assume substantial to complete failure of engineered safety systems. They should be recognized as extreme *upper-bound* scenarios rather than as anticipated accidents. The design and operation of the safety systems as well as overall risk assessments are considered in the next chapter.

LWR Loss-of-Coolant Accidents

The limiting design-basis accident for light-water reactors is the loss-of-coolant accident [LOCA]. Coolant losses range from small amounts for leaks in minor piping to large amounts for rupture of a major primary-coolant line. The engineered safety systems have been designed primarily to mitigate the large-LOCA scenarios.

The two class-9 loss-of-coolant accidents for the LWR are:

1. complete circumferential rupture of the primary vessel
2. double-ended, "guillotine" rupture of the largest primary coolant pipe

Since massive vessel rupture would essentially eliminate emergency core cooling and produce certain meltdown, great efforts are made to assure that its probability is extremely low. The double-ended pipe break, which allows relatively unimpeded coolant loss, becomes the event of most interest.

Figure 16-1 shows sequences of events in an unmitigated LOCA which would lead to core meltdown. Reactivity insertion [overpower] or double-ended failure of the primary system [undercooling] are the potential initiating events.

A large reactivity insertion could in principle cause failure of the fuel and clad, pressure-pulse generation, and ultimate failure of the primary system. Since the required rates of reactivity insertion do not appear to be obtainable, this scenario is discounted as being nonmechanistic.

The direct primary-system failure is the more likely initiating event on Fig. 16-1. Its most direct meltdown scenario includes:

- Initial blowdown driven by the system's stored energy.
- Damaging system loads, possibly including missile generation (that can damage safety systems or even breach the containment).
- Continued coolant blowdown leads to DNB and begins core heat-up (moderator loss causes neutronic shutdown to remove the potential nuclear-transient energy source).
- Decay heat and metal–water reactions continue heat-up to the eventual point of fuel melting.

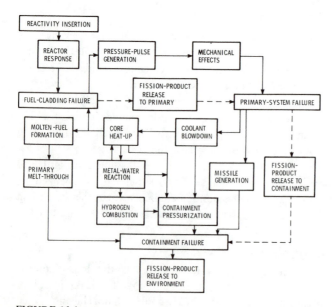

FIGURE 16-1

Loss-of-coolant accident [LOCA] sequences for light-water reactors. (Adapted from A. Sesonske, *Nuclear Power Plant Design Analysis*, TID-26241, 1973.)

- Local fuel melting propagates to full-core involvement with ultimate melt-through of the primary system.
- Containment failure occurs by overpressurization if blowdown and chemical reactions, with or without hydrogen combustion, are sufficient driving forces; or
- containment failure may occur by melt-through otherwise.

With loss of coolant and no accident mitigation by the engineered safety systems, containment failure and the associated release of fission products to the environment appears to be inevitable.

The "Reactor Safety Study" (WASH-1400, 1975), which is described in more detail in the next chapter, identifies the basic features of core meltdown scenarios for both types of light-water reactors. It includes the following appropriate information for an unmitigated LOCA in a PWR system:

- Blowdown causes DNB in about 0.25 s and results in loss of most of the coolant in 10-11 s.
- Loss of the coolant/moderator results in neutronic shutdown.
- Zirconium–water reactions produce hydrogen and add energy to that available from decay heat.
- Clad melting begins and is followed by fuel melting.
- About 80 percent of the core may melt in place before it moves en masse to the lower support plate and hence to the bottom of the vessel.
- Vessel melt-through is expected within about 1 h from the time of molten-fuel-mass contact.
- When the molten core (plus zirconium, iron, and their oxides) contacts the containment floor, spalling and vaporization of free water in the concrete produce a high penetration with a typical melt-through time expected to be about 18 h.
- Water and carbon dioxide generated by fuel–concrete interactions could also cause overpressurization failure of the containment before melt-through.
- If molten UO_2 contacts water in the pressure vessel or on the containment floor, steam explosions are possible which could overpressurize the containment, damage engineered safety systems, and/or enhance fission product release through fuel fragmentation.
- Hydrogen produced by zirconium–water and other reactions (in an amount equivalent to reaction of 75 ± 25 percent of the zirconium in the system) is expected to burn, rather than explode, and add to the containment-overpressurization threat.

An unmitigated LOCA in a BWR would have many similar features to those described above for a PWR. Significant differences, however, include:

- Lower system pressure results in a longer ($\approx$30-s) blowdown which may remove all water from the vessel and lead to a "dry" meltdown excluding both the zirconium–water reaction and the possibility for an in-vessel steam explosion.
- A tight-fitting primary containment with an atmosphere inerted to $<$5 percent oxygen content may prevent hydrogen flammability.

Double-ended LOCAs for both LWR designs would, of course, be mitigated by engineered safety systems. Such systems and representative impacts on accident progression are described in the next chapter.

CANDU Loss-of-Coolant Accident

An unmitigated LOCA in a CANDU pressurized-heavy-water reactor would have certain similarities to those described for the LWR systems. The stored energy in the coolant leads to blowdown and fuel heating. With failure of all safety systems, fuel melting eventually occurs.

Unique design features of the CANDU system, however, produce some major differences as well. The interconnections of the pressure tubes and steam generators allow for possible isolation of the section affected by the coolant-line break. The large moderator volume and the surrounding water shield (Fig. 13-2) serve as major heat sinks to inhibit fuel melting. Only if the accident scenario includes breaching of the calandria and a general loss of the moderator volume would whole-core fuel melting be possible on a relatively short time scale. The more likely circumstance involves at most only localized melting. Prolonged absence of moderator cooling, however, could lead to its removal by boiling and might present the possibility for a relatively long-term core meltdown.

LMFBR Hypothetical Core Disruptive Accidents

The class 9 accidents for a liquid-metal fast-breeder reactor are substantially different from those of the light-water reactors. The low coolant pressure excludes the possibility of a rapid loss-of-coolant accident. The large volume of liquid sodium in the reactor vessel provides natural-convection cooling which may be able to stabilize a partially melted core. On the other hand, the LMFBR is not initially in its most reactive configuration and may become supercritical as a result of coolant expulsion and/or melting of fuel.

Hypothetical core disruptive accidents [HCDA] are employed in the design basis evaluations for the LMFBR. The major sequences shown by Fig. 16-2 are initiated by transient overpower [TOP] or loss of flow [LOF], respectively. The progression of each scenario is also dependent on "loss of protective action," i.e., failure of the core protective system to induce a trip and effect shutdown. For this reason, emphasis is placed on the independent shutdown systems described in Chap. 15. No credit is taken for any engineered safety systems (as was the case for the LWR sequences described above).

The TOP scenario in Fig. 16-2 is initiated by a large, rapid reactivity insertion. Without a reactor trip there is:

- Fuel-pin failure with fuel expelled to coolant channels to block the coolant flow.
- Heat-up accompanied by sodium voiding and boiling, fission-gas release, and fuel-coolant interactions.
- Accident termination if fuel is swept out by the sodium to produce a subcritical geometry which can be cooled in place; or
- transition to gross core disruption if the geometry is either subcritical but not coolable or supercritical to produce an energetic burst.
- If disruption continues, fuel melt-out and boil-out or mechanical disassembly (via

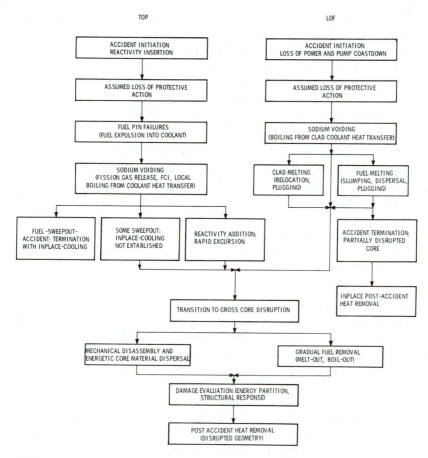

FIGURE 16-2
Possible progressions for hypothetical core disruptive accidents [HCDA] in liquid-metal fast-breeder reactors. (From K. Kleefeldt et al., "LMFBR Post-Accident Heat Removal Testing Needs and Conceptual Design of a Test Facility," KfK-Ext 8/76-4, Karlsruhe, March 1977.)

recriticality) ultimately lead to a subcritical geometry where the energy releases and structural responses will determine the final state of the system.
- The ability of natural processes to remove decay heat in the disrupted geometry determines whether in-place cooling or vessel melt-through occur.

It has been estimated, for example, that in the Clinch River Breeder Reactor 28–46 percent of the core could be cooled on the horizontal surfaces of the upper plenum while only 0.8–8.0 percent of the core material could be cooled on the bottom of the vessel. Thus, a detailed history of the disruption process is necessary for evaluation of the ultimate progression of the accident.

The LOF scenario in Fig. 16-2 begins with a loss of pump power or an extensive flow blockage. Again, assuming failure of the protective system, the coolant heats with:

- sodium voiding and boiling (perhaps with a reactivity insertion if the void feedback is positive as described in Chap. 5)
- clad melting with relocation and plugging of coolant channels
- fuel melting with dispersion, channel plugging, and/or fuel assembly slumping
- accident termination if partial disruption leaves the core in a subcritical geometry which can be cooled by natural convection of the liquid sodium
- gross core disruption if geometry is not subcritical and/or coolable
- ultimate termination of the accident as in the TOP scenario

Both HCDA scenarios have a wide range of possible outcomes. The differences from the LWR accidents are related largely to the respective coolants. Since a rapid loss of coolant is not credible for the LMFBR, natural convection cooling is viable for an extended period of time. The excellent heat transfer properties of the sodium also allow for the possibility of its mixture with molten fuel in a configuration that will not melt through the containment. The various engineered safety systems are designed to enhance the effectiveness of sodium cooling during all stages of the accident. Such systems for the LMFBR and other reactors are described in the next chapter.

HTGR Core Heat-Up Accidents

The most serious potential accidents in high-temperature gas-cooled reactors involve unmitigated core heat-up. Despite the high pressure of the coolant, its rapid expulsion is precluded by containment of the entire primary coolant system in the high-integrity PCRV (Fig. 13-6). A gradual depressurization is the most credible coolant loss scenario.

A core heat-up accident would be driven by a general loss of cooling capacity with or without depressurization. Even with failure of the protective system to initiate one of the redundant trip methods (Chap. 13), negative moderator and fuel feedbacks reduce core reactivity as the coolant increases in temperature. The large heat capacity of the graphite fuel-moderator complex limits the heat-up rate to a few degrees per second, allowing a substantial period of time for external action. If the accident continues unchecked, however, the core eventually vaporizes or, in the presence of oxygen, burns (rather than melting).

As for the other reactor designs, HTGR accidents are readily mitigated by the engineered safety systems. Circulation of either cooled helium or air would be expected to prevent general destruction of the core.

Fission-Product Release

The ultimate concern in the serious reactor accidents is the potential for release of fission products to the general environment. These species leave the fuel on a somewhat continuous basis until stable cooling is established. However, substantial increases in release rates would be noted at times which correspond to specific events in the accident sequence.

The "Reactor Safety Study" (WASH-1400, 1975) identifies the following mechanisms for escape of fission products from LWR fuel:

- gap release—clad rupture facilitates the escape of the volatile products which have migrated to the pellet-clad gap

- meltdown release—fuel melting releases additional volatile fission products
- vaporization release—chemical interactions between molten fuel and concrete produce vapors which facilitate fission-product escape
- oxidation release—steam explosion enhances oxidation and release of fission products

These or comparable categories are also generally applicable to the other reactors which employ metal-clad oxide fuel.

Table 16-3 summarizes the estimated cumulative percentage releases through the successive LWR mechanisms. The oxidation [steam explosion] releases are of necessity based on the fraction of the core involved and the fraction of the fission products which remain in the fuel. The following paragraphs describe each LWR mechanism.

Pellet-Clad Gap

When zircaloy-clad rupture occurs between 800 and 1100°C, the fission products which have migrated to the pellet surface, gas gap, or plenum may be released. Further release from the fuel pellets is quite slow until the melting temperature is reached.

The inert noble gases leave the pin as soon as the clad ruptures. The halogens and alkali metals tends to react chemically with the clad and other materials to reduce their escape fraction to about one-third of what it would be otherwise. Other fission products are characterized by low volatility which limits their release to small fractions of 1 percent.

Meltdown

The first large release of fission products in reactor accident sequences occurs with fuel melting as indicated by Table 16-3. If melting occurs on a pellet-by-pellet basis, the high surface area can lead to large releases. On the other hand, if major sections

TABLE 16-3
Estimated Fission Product Release from an LWR Meltdown Accident[†]

Fission products	Cumulative release percentage			
	Gap	Meltdown	Vaporization[‡]	Oxidation[§]
Noble gases (Kr, Xe)	3.0	90	100	90 $(X)(Y)$
Halogens (I, Br)	1.7	90	100	90 $(X)(Y)$
Alkali metals (Cs, Rb)	5	81	100	–
Te, Se, Rb	10^{-2}	15	100	60 $(X)(Y)$
Alkaline earths (Sr, Ba)	10^{-4}	10	11	–
Noble metals (Ru, Mo)	–	3	8	90 $(X)(Y)$
Rare earths (La, Sm, Pu) & refractories (Zr, Nb)	–	0.3	1.3	–

[†]Adapted from WASH-1400 (1975).
[‡]Exponential loss over 2 h with a half-time of 30 min. If a steam explosion occurs first, only the core fraction not involved in the explosion can experience vaporization.
[§]X = fraction of core involved in steam explosion; Y = fraction of inventory remaining for release by oxidation.

of fuel assemblies collapse into a molten mass before melting themselves, a substantial portion of the fission products may be retained.

Most releases occur early in the melting process. Iron and oxide crusts which form later are expected to reduce the overall escape of fission products. The atmosphere of steam, hydrogen, and fission products would not be likely to cause oxidation or accelerate release.

With melting, all of the noble gases and halogens are available for release, although trapping in large masses may reduce the fraction to about 90 percent. The volatile alkali metals and tellurium-group members have varying degrees of retention based on chemical reactivity. The remaining products also experience small releases.

Vaporization

When the molten mass of fuel and structural material penetrates the reactor vessel, it is exposed to oxygen, steam from the containment atmosphere, and cooling water, respectively. Interaction of the molten fuel with concrete adds CO_2 to the local accident environment. The oxidizing atmosphere may produce dense aerosol clouds [smokes] in which some lighter particles follow the bulk flow while those that are heavier condense out on structures or settle back to the melt.

Sparging (by the concrete-reaction gases) and natural convection produce exponential release of virtually all volatile products still in the fuel. The remaining fission products experience relatively small releases driven by oxidation and aerosol formation.

Steam Explosion

A steam explosion produces and scatters finely divided UO_2. It may also facilitate a breach of the containment to allow general escape of fission products to the environment.

As the UO_2 particles cool, they oxidize exothermically to form U_3O_8 and release volatile fission products. It is expected that 60-90 percent of remaining noble-gas, halide, tellurium, and noble-metal fission products would be released instantaneously under such conditions.

Leakage from Containment

Essentially all of the fission products released during the early stages of the meltdown would move from the primary system into the containment. These, plus the products released after vessel melt-through, form the inventory available for leakage to the general environment.

The overall change in the concentration of a specific fission product in the containment atmosphere may be represented by the balance equation

$$\begin{array}{c} \text{Net increase} \\ \text{in concentration} \end{array} = \begin{array}{c} \text{rate of release} \\ \text{from primary system} \end{array} - \begin{array}{c} \text{removal} \\ \text{rate} \end{array} - \begin{array}{c} \text{leakage} \\ \text{rate} \end{array}$$

$$\frac{dC_i}{dt} = R_i(t) - \sum_j \lambda_{ij} C_i - \alpha_i C_i$$

for component i with concentration C_i, release rate R_i, removal coefficient λ_{ij} for process j, and leakage coefficient α_j. The rate of release for each nuclide is governed by the mechanisms identified in Table 16-3.

The removal mechanisms include both natural processes and those associated with the engineered safety systems. The former are generally based on decay, chemical reactivity, and deposition. The latter tend to enhance natural processes (as considered in the next chapter).

The fission-product source available for release in LWR accidents is often divided into four categories (WASH-1400, 1975):

- noble gases
- elemental iodine
- organic iodides
- particulates and aerosols

Since the first category consists of chemically inert gases, no natural processes provide for effective removal. The organic iodides are similarly unreactive, including being insoluble in water. Reduction in the potential hazard of both categories is dependent on long-term holdup (i.e., the delay and decay principle) in the containment.

Elemental iodine, the other major gaseous fission-product form, is quite reactive chemically. Therefore, substantial removal from the containment atmosphere occurs by this natural process. The solid fission products (including most species that are volatilized at fuel-melt temperatures) appear in particulate and/or aerosol† form. Natural deposition and gravitational settling result in some removal. Agglomeration of small particles or aerosols to form larger ones usually enhances the settling processes.

While the rate at which specific fission products leak to the general environment depends on their concentration, it is also a function of the internal state and integrity of the containment. Leakage from an intact containment is determined by the system pressure. A typical LWR design, for example, has daily leakage of less than 0.1 volume percent as tested at a specified overpressure.

If the containment is breached by overpressurization, leakage is controlled by the generation of the noncondensable gases [H_2 and CO_2] and steam. A typical LWR meltdown (WASH-1400, 1975), for example, might exchange up to 200 volume percent per day through a 9-cm-diameter hole.

A melt-through failure of the containment leads to a rapid pressure equalization. Fission-product release to the outside environment is then controlled by the nature of the ground through which the gases, particulates, and aerosols must pass. A typical LWR melt-through (WASH-1400, 1975) would be expected to occur at 9–12 m below grade. The diffusion of noble gases to the surface could occur as quickly as about 10 h in sandy soil to as slowly as years in hard-packed, fine clay. Soluble and particulate fission products may be filtered by a factor of a thousand or more, while the noble gases and organic iodides are relatively unaffected.

---

†For a rough, working definition, aerosols may be considered to be small particulates (generally ≤ 0.3 mm in diameter) which follow the bulk flow of the gases in which they are suspended.

EXERCISES

Questions

16-1. Describe the five sources of energy available in reactor accidents.

16-2. Explain the concept of design-basis accidents and identify the major categories into which they may be classified.

16-3. Identify the three ways to initiate a loss-of-cooling accident in a reactor.

16-4. Describe the design-basis loss-of-coolant accident [LOCA] for a light-water reactor.

16-5. Identify the major differences in the "beyond-design-basis" accidents for LWR, CANDU, HTGR, and LMFBR systems.

16-6. Describe the four principle events in an LWR meltdown that lead to substantial increases in fission-product release rates.

16-7. Identify the four categories of fission products which would reside in the containment after an accident and the two ways they could be released to the general environment.

Numerical Problems

16-8. Consider a 1000-MW(e) reactor with a thermal efficiency of 32 percent. Calculate its decay-heat load after shutdown at times of
   a. 1 min
   b. 1 day
   c. 1 month
   d. 1 year

16-9. Consider a system where the entire content of fission product θ has been released from the fuel to the containment.
   a. Derive an expression for the time-dependent concentration $C_\theta(t)$ with $C_\theta(t=0) = C_0$, constant leakage coefficient α_θ, radioactive-decay constant λ_θ, and constant removal coefficients $\lambda_{\theta j}$.
   b. Assuming $\lambda_\theta = \alpha_\theta$ and $\Sigma_j \lambda_{\theta j} = 0$, determine the time required for the concentration to fall to half its initial level.
   c. Repeat (b) for $\Sigma_j \lambda_{\theta j} = \alpha_\theta$ and $2\alpha_\theta$.
   d. Use the results of (b) and (c) to determine the amount of θ that will leak from the containment at time α_θ^{-1} for each of the three removal rates.

SELECTED BIBLIOGRAPHY[†]

Reactor Safety (General)
   Farmer, 1977
   Lewis, 1977
   Sesonske, 1973 (Chap. 6)
   Thompson & Beckerley, 1964, 1973

LWR Safety
   WASH-1250, 1973 (Chaps. 2, 5)
   WASH-1400, 1975

CANDU Safety
   van Erp, 1977

[†]Full citations are contained in the General Bibliography at the back of the book.

HTGR Safety
    General Atomic, 1976

LMFBR Safety
    CONF-761001, 1976
    CRBRP-1, 1977
    Fauske, 1976

17

ASSESSMENT
OF REACTOR ACCIDENT RISK

Objectives

1. Identify the five basic categories of engineered safety systems and provide an example of each for the LWR, CANDU, HTGR, and LMFBR designs.
2. Describe three pathways whereby airborne radioactivity can interact with a human population.
3. Explain Starr's concept of risk perception.
4. Describe the difference between fault-tree and event-tree methodologies.
5. Justify the 10^{-9} per-reactor-year probability for the most serious LWR accident calculated in WASH-1400.
6. Identify the major criticisms of WASH-1400 by the Lewis Committee.
7. Explain the role of the LOFT program in LWR safety.
8. Describe the status of probabilistic risk assessment applications to HTGR and LMFBR systems.

Prerequisite Concepts

Radiation Dose & Effects	Chapter 3
Reactor Designs	Chapters 12, 13, 15
Accident Energy Sources	Chapter 16
Design-Basis Accidents	Chapter 16
Meltdown Scenarios	Chapter 16

The previous chapter described the progression of loss-of-cooling accidents under the assumption that there were no safety systems available to prevent or mitigate the ultimate release of radioactive material. In actual practice, extensive sets of redundant engineered safety systems are integral to all reactor designs.

While the previous chapter emphasized reactor-related scenarios, evaluation of overall accident risk must also include an assessment of the environmental consequences and the probable frequency for each event. The extensive "Reactor Safety Study" (WASH-1400, 1975) is the definitive document for risk from LWR systems. The other reference reactors have also been evaluated, but in somewhat less depth.

ENGINEERED SAFETY SYSTEMS

The ability to control accident consequences is extremely limited once the radionuclides escape from the containment building. Thus, the goal of reactor safety reduces to limiting such releases by prevention, or at least mitigation, of accidents through the use of *engineered safety systems.*[†]

Conceptual engineered safety features related to LWR loss-of-coolant accidents [LOCA] are depicted by Fig. 17-1. These include:

- reactor trip [RT] to provide positive and continued shutdown of the nuclear chain reaction
- emergency core cooling [ECC] to limit fuel melting
- post-accident heat removal [PAHR] to prevent containment overpressurization
- post-accident radioactivity removal [PARR] to reduce the radionuclide inventory available for release
- containment integrity [CI] to limit radionuclide release

These concepts are also readily extended to the other reference reactor designs.

For most reactor designs, the functions identified in Fig. 17-1 are interrelated as described in the following pages. It is, therefore, not uncommon for the entire array of features to be referred to simply as the emergency core cooling systems [ECCS].

PWR Safety Features

The basic features of engineered safety systems for a typical pressurized-water reactor are shown in Fig. 17-2. These are intended to mitigate the consequences of design-basis and beyond-design-basis LOCA sequences.

Reactor Trip

The reactor trip function is facilitated by gravity insertion of the control rods mounted above the core (Fig. 12-7). The core protective system is designed to induce the trip when operating limits (e.g., Fig. 12-16) are exceeded. Redundant sensors, fail-safe design, and separated cable runs add reliability to the system.

Normal PWR shutdown requires full control-rod insertion at operating soluble-boron concentrations (e.g., as inferred from Table 12-2). While a LOCA produces its own shutdown through the moderator feedback (i.e., loss of moderation), the negative reactivity worth of the control rods is not sufficient by itself to assure

---

[†]It has also been common to refer to such systems as *engineered safeguards*, or simply *reactor safeguards.* Current practice, however, tends to favor reserving the word "safeguards" to describe procedures and/or systems for protecting nuclear materials from theft or diversion (e.g., as considered in Chap. 20).

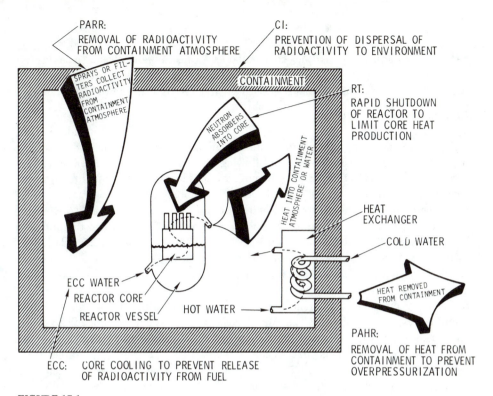

FIGURE 17-1
Conceptual engineered safety systems for LWRs. (From WASH-1400, 1975.)

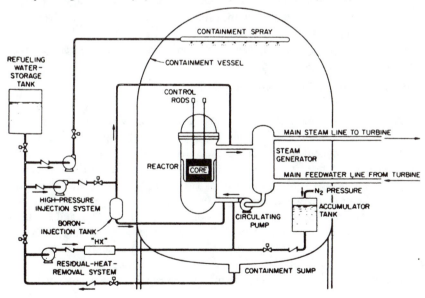

FIGURE 17-2
Engineered safety systems for a PWR. (From W. B. Cottrell, "The ECCS Rule-Making Hearing," *Nuclear Safety*, vol. 15, no. 1, Jan.–Feb. 1974.)

continued subcriticality as the core refloods. Thus, borated water is used for emergency cooling (Fig. 17-2). It is usual practice, in fact, to provide a soluble boron concentration that is sufficient by itself to cause full shutdown.

Emergency Core Cooling

The emergency core cooling systems [ECCS] in Fig. 17-2 consist of independent high-pressure injection, accumulator, and low-pressure injection systems designed to reflood the core with cooling water following a LOCA. Since the accident-initiating break nullifies the effectiveness of certain injection locations, each of the three systems consists of multiple loops with access points distributed around the reactor vessel. Redundant off- and on-site power supplies are also provided for the active systems.

The high-pressure injection systems are designed to operate when the primary-loop pressure falls from its nominal 15.5 MPa [2250 psi] to about 12.5 MPa [1800 psi]. By pumping the borated water from the refueling water-storage tank to the vessel inlet, a reasonable coolant flow can be maintained for a small-break LOCA. The flow also provides cooling for a large, design-basis LOCA.

The accumulators contain borated water under nitrogen pressure. If the primary pressure drops below 12.5 MPa [1800 psi], the system is "armed" to allow coolant injection when the pressure falls to 4–5.5 MPa [600–800 psi]. [It may be noted that the vessel coolant inlet is actually above the core (Fig. 12-7), rather than below as shown by Fig. 17-2.] The accumulators are sized to provide coolant for about 30 s, enough time for blowdown to be complete and for the low-pressure injection system to become effective.

The low-pressure injection systems use the residual-heat-removal [RHR] pumps to provide post-LOCA cooling with water from the refueling tank. When the tank is emptied, the system is converted to use the water which will have accumulated in the buildup sump (largely from condensation of the steam released to the containment by the accident).

Heat and Radioactivity Removal

The post-accident-heat-removal [PAHR] function is facilitated by containment water-sprays which are initially supplied from the refueling water-storage tank and later from the building sump. By condensing the steam, the sprays remove heat energy from the containment and lower the pressure.

A complementary feature of the post-accident-heat-removal function becomes important as the refueling water is exhausted. The alternate supply—i.e., the sump water—is relatively hot and would become more so with recirculation through the low-pressure-injection and spray systems. Its ability to cool the core and containment depends on removal of some of the heat energy. The heat exchanger ["HX"] of the residual-heat-removal system in Fig. 17-2 serves such a cooling function. It employs a heat sink located outside of the containment building.

Post-accident removal of radioactivity may take several forms, including:

- containment spray additives like sodium hydroxide [NaOH] or sodium thiosulfate [$Na_2S_2O_3$] to "wash" reactive fission products from the containment atmosphere
- charcoal adsorbers capable of fission-product removal at high temperatures and humidities

- high-efficiency particulate air [HEPA] filters
- reactive coatings for passive removal of halide fission products

The inert noble gases and organic halides are not readily subject to removal from the containment atmosphere. As with the other products, however, assuring containment holdup (i.e., delay and decay) tends to mitigate their potential hazard.

Containment

The initial requirement for containment integrity is the operation of isolation valves to close off containment penetrations. Such valves are designed to operate automatically under overpressure conditions characteristic of a large-break LOCA. One type of isolation valve for a main steam line is shown in Fig. 17-3. This hydraulic valve is specially designed for quick closure when actuated remotely by an external signal.

A PWR containment arrangement typical of currently operating plants is shown in Fig. 17-4. It consists of an inner leak-tight steel liner surrounded by a reinforced concrete shell. The figure includes the relative positions of the reactor vessel, coolant pumps, and steam generators, as well as the accumulators and containment sprays that are described in the preceding paragraphs.

Current design practice favors a double-containment system which differs from that of Fig. 17-4 by maintaining an annular region between the steel liner and the concrete shell. Additional provision for removal of the fission products which enter

FIGURE 17-3
Hydraulic isolation valve for a PWR main steam line. (Courtesy of Flow Control Division, Rockwell International.)

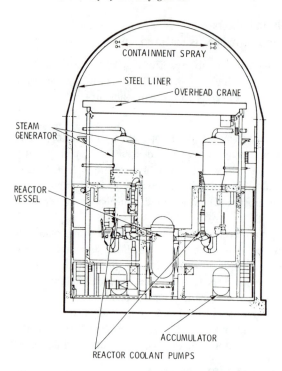

FIGURE 17-4
Typical PWR containment. (From WASH-1400, 1975.)

the annulus can provide a substantial reduction in overall containment leakage. (The general concept is explored later in this chapter as related to the LMFBR design.)

Ultimate long-term containment integrity for any structure depends on prevention of overpressure and melt-through failure (Chap. 16). Proper operation of the other engineered safety systems should assure continued containment. On the other hand, once a meltdown occurs the containment may be breached by one of the two mechanisms.

Integrated Accident Response

The relative roles of the emergency-core-cooling and post-accident-heat-removal systems on containment integrity are emphasized by several hypothetical accident progression examples from WASH-1400 (1975). For each of four selected scenarios, Fig. 17-5 shows containment pressure as a function of the elapsed time (on a logarithmic scale) after a double-ended LOCA.

The design-basis accident shown by the solid line includes rapid coolant blowdown accompanied by containment pressurization within a few seconds. If all engineered safety systems function as intended, pressure reduction begins in minutes and approaches atmospheric levels within about $\frac{1}{2}$ h. Continued operation of the sump-recirculation and· heat-removal systems assures long-term stability.

If the low pressure-injection system fails to draw sump water after the refueling tank is empty, meltdown ensues as soon as the water in the vessel boils off. The containment sprays (assumed to draw adequately from the sump) prevent excessive pressurization, although some buildup occurs (Fig. 17-5, path 2) due to zirconium–steam and fuel–concrete interactions. Containment failure ultimately occurs by melt-through.

If the spray rather than low-pressure injection fails to draw from the sump, there is no effective steam condensation. Containment pressurization then occurs as the hot core boils off coolant (Fig. 17-5, path 3). The pressure eventually builds to breach the containment.

Failure of both the spray and injection systems to draw water from the sump results in the complex behavior shown by Fig. 17-5, path 4. The lack of any core cooling results in rapid boiloff, zirconium–steam reactions, and hydrogen generation. The pressure increase is unimpeded due to the spray failure. As soon as all vessel water has boiled away, natural condensation allows some pressure decrease. Vessel melt-through results in concrete interactions, evolution of steam and CO_2, and renewed pressure increases. Overpressure failure would be possible at this point. Otherwise, as the concrete reactions slow, natural condensation may again reduce the pressure until containment melt-through finally reduces it to an equilibrium level.

Analysis of sequences such as those in Fig. 17-5 provided the basis for the risk assessment in WASH-1400 (1975). Further applications are described later in this chapter.

BWR Safety Features

The major features of the engineered safety systems for the original boiling-water reactor design are shown by Fig. 17-6. Although current systems are somewhat different, the fundamental principles are still valid.

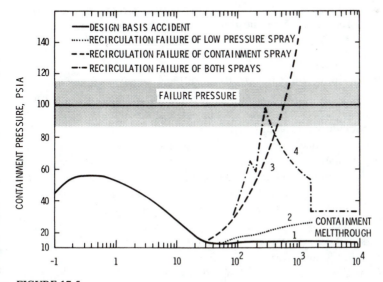

FIGURE 17-5
Containment pressure response for a typical PWR to a design-basis LOCA with assumed safety system failures. (Adapted from WASH-1400, 1975.)

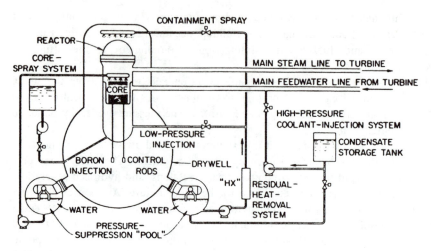

FIGURE 17-6

Engineered safety systems for an early BWR. (From W. B. Cottrell, "The ECCS Rule-Making Hearing," *Nuclear Safety*, vol. 15, no. 1, Jan.–Feb. 1974.)

Reactor Trip

The reactor trip function is readily accomplished by mechanical insertion of the cruciform control rods mounted below the core (Fig. 12-2). The core protection system is designed to facilitate the trip when operating limits (e.g., similar to those for the PWR in Fig. 12-16) are exceeded. Redundant sensors, fail-safe design, and separated cable runs add reliability to the system.

Injection of soluble boron with the emergency cooling water provides a backup to assure subcriticality. The LOCA itself produces neutronic shutdown from lack of moderation. The presence of the rods and/or the soluble boron are of most importance as the core refloods.

Emergency Core Cooling

The BWR system in Fig. 17-6 employs three different emergency core cooling systems [ECCS]. Each system, in turn, consists of multiple loops, all of which need not be operable to provide the needed cooling capacity. Redundant off-site and on-site power sources enhance reliability.

The high-pressure coolant-injection enters the reactor vessel through a feed-water line (actually located *above* the core as shown by Fig. 12-2). It is capable of providing cooling water immediately after LOCA initiation at essentially any system pressure. Diesel-powered pumps from the residual-heat-removal [RHR] systems are used in conjunction initially with the water in the condensate storage tank and later with the water that collects in the pressure-suppression pool.

The low-pressure core spray and low-pressure injection systems in Fig. 17-6 are available after the system pressure is reduced by the LOCA. Each relies on pumping water from the suppression pool for discharge to the core. The spray line is connected to a sparge ring (Fig. 12-2) that distributes the flow among the core fuel assemblies. The low-pressure injection system employs the residual-heat-removal-system components to provide additional coolant (entering *above* the core, e.g., as inferred from Fig. 12-2).

The ECCS for the current BWR design (General Electric/BWR-6, 1978) employs separate high-pressure sprays rather than using injection on the feedwater line. The two low-pressure cooling functions are similar.

Heat and Radioactivity Removal

The initial post-accident heat removal function is facilitated by venting the LOCA-produced steam from the drywell to the pressure-suppression pool. Condensation of the steam reduces the temperature and pressure of the drywell atmosphere.

Continuing heat removal is accomplished by use of sprays in the drywell and pressure-suppression pool. Figure 17-7a shows the spray locations in the original BWR design. The RHR heat exhanger with its heat sink outside of the containment is employed to reduce the temperature of the water feeding the low-pressure injection and containment spray systems (Fig. 17-6).

The current BWR containment design shown in Fig. 17-7b is similar in concept to the original design. Steam from the drywell condenses as it vents to the suppression pool. In this design there are no containment sprays in the drywell, although they are employed in the annular region above the suppression pool.

The BWR radioactivity removal systems are similar to those of the PWR. Spray additives, reactive coatings, and adsorber–filter systems help reduce the concentrations of chemically reactive fission products.

Containment

The original BWR containment (Fig. 17-7a) employs a light-bulb-shaped, steel-lined drywell connected by vent pipes to the surrounding toroidal ["doughnut-shaped"] pressure-suppression pool. As was true for the PWR design, isolation valves and the pressure-reduction systems are the principle means of maintaining system integrity against airborne release of fission products. Prevention of core meltdown precludes containment melt-through failure.

The current BWR design employs a multiple containment (Fig. 17-7b) consisting of a drywell, steel containment shell, and concrete shield building. The drywell and the annular pressure-suppression pool are connected by horizontal vents separated from the drywell by a weir wall.

CANDU Safety Features

The engineered safety systems for the CANDU pressurized-heavy-water reactor bear some resemblance to those of the LWR designs. However, the unique features of the pressure-tube design also provide some major differences.

Reactor Trip

The primary reactor trip function is facilitated by gravity insertion of shutoff rods mounted above the calandria tank (Figs. 13-2 and 13-4). These rods are completely independent of those used for control purposes. Since the shutoff rods enter the tank rather than the fuel assemblies, accident-induced fuel damage is unlikely to restrict their operation.

The core protective system is designed to induce the trip when operating

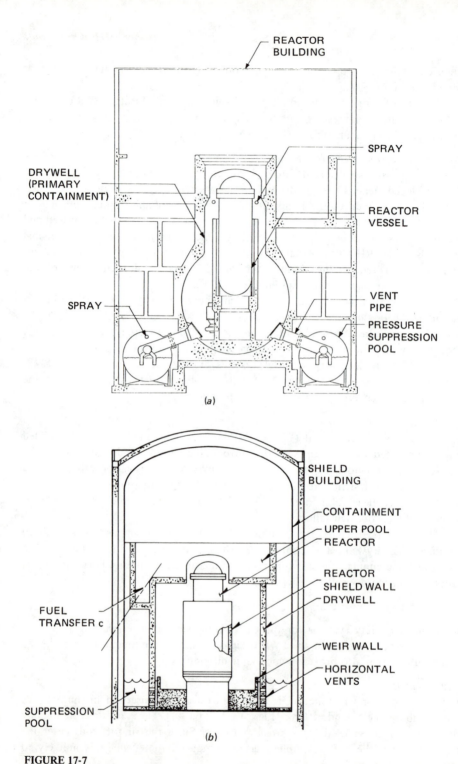

FIGURE 17-7
Containment structures for BWRs: (*a*) original design (WASH-1400, 1975); and (*b*) current design (courtesy of General Electric Company).

limits are exceeded. System reliability depends on redundant sensors, fail-safe design, in-core flux monitoring, and general computer control of reactor operations.

The backup shutdown system is based on injection of gadolinium poison into the heavy water in the calandria tank. Since this moderator volume is likely to be stable under accident conditions, one-time injection is expected to be adequate.

Emergency Core Cooling

The core cooling function is facilitated by an emergency coolant injection system [ECIS] which delivers water to the pressure tubes through the feeder pipes (Fig. 13-2). Multiple access points are located to assure cooling independent of break location.

The use of separated coolant systems for adjacent fuel channels may allow retention of flow in the half of the core not affected by the LOCA. The "cold" moderator volume also serves as an important passive cooling system.

Heat and Radioactivity Removal

Post-accident heat removal in the CANDU system is accomplished by several means. Heat exchangers and a recovery system allow viable water recycle in the emergency coolant injection system. From 10 to 15 air-to-water heat exchangers help to cool the containment atmosphere. The steam generators can provide a heat sink for half of the pressure tubes if the system is intact after the LOCA.

A containment dousing [spray] system is used for steam condensation as in the LWR designs. The water supply and distribution networks are located in the top of the containment building as shown by Fig. 17-8.

The heat exchangers for the moderator and light-water shield provide important backup heat removal capability. It is likely that they would be sufficient to prevent general fuel melting even if the emergency coolant injection and related systems failed to operate.

Post-accident radioactivity removal is facilitated by spray additives and filtration systems. These have performance characteristics which are similar to those of the LWR designs.

Containment

The containment building for a single CANDU reactor is shown in Fig. 17-8. It consists of either a steel or an epoxy-based liner in a prestressed, post-tensioned concrete shell. As in LWRs, isolation valves and the pressure-reduction systems are the principle means of maintaining system integrity against release of fission products. Prevention of core meltdown precludes containment melt-through failure.

The multiunit CANDU installations may employ one separate vacuum building (similar to that of Fig. 17-8, but without the reactor system) for primary containment purposes. Upon initiation of a LOCA in one reactor, its atmosphere is vented to the vacuum building through a piping system. The building (as its name implies) is maintained under vacuum conditions and actually allows the reactor containment to achieve a subatmospheric pressure following the LOCA. This design has the advantage of allowing spray and other energy removal functions to be conducted away from the reactor itself. Disadvantages are related to potential failure (e.g., seismic) of the connecting pipe runs.

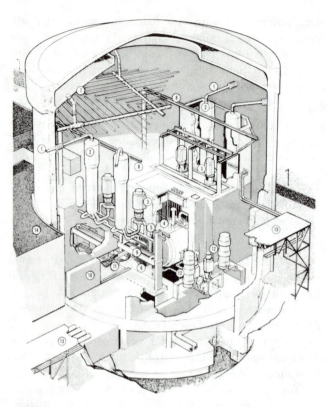

FIGURE 17-8

Containment structure for a single CANDU reactor: (1) main steam supply piping; (2) boilers; (3) main primary system pumps; (4) calandria assembly; (5) feeders; (6) fuel channel assembly; (7) dousing water supply; (8) crane rails; (9) fuelling machine; (10) fuelling machine door; (11) catenary; (12) moderator circulation system; (13) pipe bridge; (14) service building. (Courtesy of Atomic Energy of Canada Limited.)

HTGR Safety Features

Loss of coolant flow is part of the design-basis accident for the high-temperature gas-cooled reactor. Potential consequences of such an accident are mitigated by engineered safety systems in categories similar to those identified in Fig. 17-1.

Reactor trip is facilitated by gravity insertion of the control rods located above the core (Fig. 13-6). The trip signal is initiated by the core protective system as for the LWR designs. Full backup capability is available through the B_4C spheres of the reserve shutdown system (Fig. 13-10).

Since helium injection is not an effective counterpart of LWR water injection, the emergency core-cooling and post-accident heat-removal functions are essentially coincident in HTGR systems. Three auxilliary circulators and water-cooled heat exchangers (shown in Fig. 13-6) are employed to restore forced cooling. While the primary circulators are steam-driven (Fig. 13-5), the auxilliary systems rely on a backup electrical supply. Design specifications call for two of the three loops to provide adequate heat removal in either a pressurized or depressurized PCRV.

Radioactivity removal is dependent on adsorber and filtration systems like those employed in the LWR designs. Chemically reactive water sprays are *not* included among the HTGR safety features.

The primary containment for the HTGR coincides with the PCRV. Steel liners around the core, circulators, and heat exchangers are contained within and supported by the concrete vessel (Fig. 13-6). Under normal operating conditions, the helium leak rate is expected to be less than 0.01 percent per year. Since the coolant is entirely single-phase and since graphite heat-up rates are low, rapid overpressurization of the magnitude found in the LWRs is not credible. Timely restoration of forced cooling assures continued containment integrity.

LMFBR Safety Features

The engineered safety systems for liquid-metal fast-breeder reactors are designed to mitigate the hypothetical core disruptive accidents outlined by Fig. 16-2. Such systems can be classified in the same general areas identified on Fig. 17-1. Since the safety systems proposed for the Clinch River Breeder Reactor Plant [CRBRP] are similar to those included in, or under consideration for, most of the other LMFBR systems, they serve as a reasonable basis for further discussion of this chapter.

Since the HCDA scenarios of Fig. 16-2 are entirely dependent on loss of protective action, the reactor trip function is especially important to LMFBR safety. Two independent control-rod groups, each capable of producing a full shutdown, are employed as noted in Chap. 15 to enhance reliability. The core protective system is relied upon to initiate a trip in a manner similar to that for the other reference reactors.

Heat Removal

The emergency core-cooling and post-accident heat-removal functions coincide in the LMFBR. Both passive and active systems are employed to assure adequate heat removal.

Large sodium volumes enhance the potential for natural-convection cooling in both loop- (Fig. 15-3) and pool-type (Fig. 15-4) LMFBR's. This feature is enhanced further by arranging the reactor vessel and appropriate system components in a manner that limits the likelihood of the liquid sodium level dropping below that of the core. Figure 17-9 shows primary-loop elevation for the CRBRP design. The limited-volume guard vessels are intended to restrict the amount of coolant that can leave the reactor vessel, pumps, or heat exchangers. The arrangement in Fig 17-9 is also intended to facilitate continued operation of the cooling system if only two of the three primary loops are intact.

Recent experiments at the Prototype Fast Reactor [PFR]—a 250-MW(e) pool-type reactor—suggest that natural circulation can protect the core even if there is no pumping power available in the primary system. The initial results reported by Gregory (1979) indicate that during the transition from forced to natural circulation, excessive temperatures were not reached. Further experiments are expected to confirm that even during this transition stage, temperatures comparable to those for normal operation are not exceeded.

Separate pony motors coupled to the main CRBRP coolant pumps can provide $7\frac{1}{2}$ percent of full flow for decay-heat removal. If the heat sink of the main

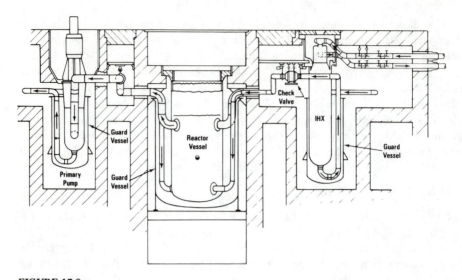

FIGURE 17-9
Primary-system elevation for a loop-type LMFBR. (Courtesy of Clinch River Breeder Reactor Plant.)

heat-transport system is unavailable, an auxilliary system may be used to provide cooling either through the steam generator (with the steam "dumped" rather than used for electrical generation) or through a set of separate, protected, air-cooled condensors.

The direct heat-removal system [DHRS] provides a third emergency cooling capability for CRBRP. It consists of an in-vessel heat exchanger, liquid sodium-potassium [NaK] coolant, an electromagnetic pump, and external air-blast heat exchangers.

Containment
Post-accident heat and radioactivity removal in CRBRP is facilitated by the containment design shown in Fig. 17-10. It consists of a 4.5-cm-thick free-standing steel shell, a 1.5-m annulus, and a concrete confinement building. The annulus cooling system removes accident-generated heat from the steel shell by air circulation. The containment venting and filter-scrubber systems collect sodium aerosols and fission products. Long-term containment integrity is enhanced by partially venting the shell atmosphere to the annulus, removing heat and fission products, and then purging from the annulus back to the shell.

The presence of sodium in the LMFBR provides up to half a million kilograms of an aerosol-forming medium for fission products which is not available in other reactor systems. Thus, the filtration system must be capable of handling very large mass loadings. Sand and gravel beds, cyclone separators, wet scrubbers, electrostatic precipitators, and acoustic agglomerators all have the potential for LMFBR use. After aerosol removal, more conventional adsorption and filtration systems can be used for removal of gaseous fission products.

Sodium fires are a major concern in LMFBRs since they can increase the aerosol loading, cause heat damage to other safety systems, and/or overpressurize the containment. Each primary coolant loop is contained in a separate steel-lined concrete cell (e.g., Fig. 17-9) which may be isolated when a fire occurs. Nitrogen purging and various extinguishers are available for fire-fighting.

ENVIRONMENTAL TRANSPORT

Once the containment fails, all fission products that have not already been removed by natural processes or engineered safety systems are available for release to the general environment. Their transport is a very complex process which depends on such factors as the containment failure mode and weather conditions.

Generalized exposure pathways for human and other populations are shown by Fig. 17-11. Gaseous effluents include the true gases as well as the particulates and aerosols carried with them. They provide a direct, external radiation dose by submersion in the "passing cloud" of radioactive material. Internal exposures occur as a result of actual inhalation.

Deposition of particulates and aerosols provides a long-term source of external radiation. Resuspension (not indicated on Fig. 17-11)—e.g., due to winds or combustion—can lift the material back into the atmosphere to restore the submersion and inhalation pathways. Contamination of the food chain is also an important consequence of fission-product deposition.

Drinking water and seafoods can also be important pathways to human populations (Fig. 17-11). Water supplies may be contaminated by direct liquid releases and by airborne deposition over water. Rain and the resulting runoff can

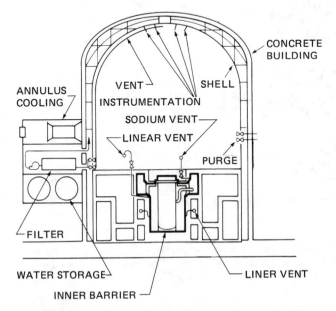

FIGURE 17-10
Annular containment structure for an LMFBR. (Courtesy of Clinch River Breeder Reactor Plant.)

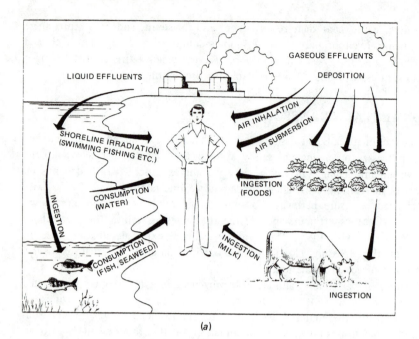

(a)

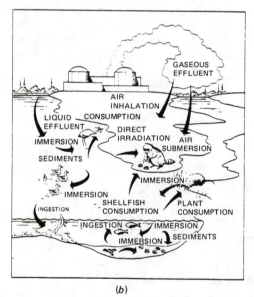

(b)

FIGURE 17-11

Generalized radionuclide exposure pathways for: (*a*) human populations and (*b*) other organisms. (Courtesy of "Nuclear Power in Canada: Questions and Answers," published by the Canadian Nuclear Association, 65 Queen St. W, Toronto, Ontario M5H 2M5.)

carry land-based radionuclides into bodies of water. Groundwater contamination is likely to follow a containment melt-through.

A number of the more significant fission products from reactor accidents are identified and described in Table 17-1. These materials are all subject to inhalation and are deposited in the indicated tissues. The ^{90}Sr [strontium-90], ^{137}Cs [cesium-137], and the isotopes of iodine generate special concern.

The ^{90}Sr nuclide has long radioactive and effective half lives and a relatively large fission yield. It and its short-lived ($T_{1/2} = 64$ h) ^{90}Y [yttrium-90] daughter combine to produce a very high dose per unit activity. Strontium behaves chemically like calcium, and thus tends to deposit in bone tissue after inhalation. While the marrow itself is not highly susceptible to radiation damage, the blood cells produced there are quite sensitive.

Another important pathway for ^{90}Sr is through the food chain. Because of its chemical behavior, it is concentrated in milk. The inherent potential to affect the younger members of the population makes this a special concern. As a matter of practice, ^{90}Sr levels are among the first to be measured following a suspected reactor accident (or, for that matter, a nuclear weapons test).

The ^{137}Cs nuclide, like ^{90}Sr, has a high fission yield and a long half-life.

TABLE 17-1
Fission Products of Significance in Internal Exposure from Reactor Accidents[†]

Isotope	Radio-active half-life $T_{1/2}$	Fission yield (%)	Deposition fraction[‡]	Effective half-life	Internal dose (mrem/μCi)	Reactor inventory[§] [Ci/kW(th)]	
						400 Days	Equilibrium
Bone							
^{89}Sr	50 d	4.8	0.28	50 d	413	43.4	43.6
^{90}Sr–^{90}Y	28 y	5.9	0.12	18 y	44,200	1.45	53.6
^{91}Y	58 d	5.9	0.19	58 d	337	53.2	53.6
^{144}Ce–^{144}Pr	280 d	6.1	0.075	240 d	1,210	34.7	55.4
Thyroid							
^{131}I	8.1 d	2.9	0.23	7.6 d	1,484	26.3	26.3
^{132}I	2.4 h	4.4	0.23	2.4 h	54	40.0	40.0
^{133}I	20 h	6.5	0.23	20 h	399	59.0	59.0
^{134}I	52 m	7.6	0.23	52 m	25	69.0	69.0
^{135}I	6.7 h	5.9	0.23	6.7 h	124	53.6	53.6
Kidney							
^{103}Ru–^{103m}Rh	40 d	2.9	0.01	13 d	6.9	26.3	26.3
^{106}Ru–^{106}Rh	1.0 y	0.38	0.01	19 d	65	1.8	3.5
^{129m}Te–^{129}Te	34 d	1.0	0.02	10 d	46	9.1	9.1
Muscle							
^{137}Cs–^{137m}Ba	33 y	5.9	0.36	17 d	8.6	1.2	53.6

[†]Adapted from T. J. Burnett, "Reactors, Hazard vs. Power Level," *Nucl. Sci. Eng.*, vol. 2, 1957, pp. 382–393.

[‡]Fraction of inhaled material that deposits in the indicated tissue.

[§]A somewhat typical average residence time for fuel in an LWR is 400 full-power days; equilibrium inventories are achieved at times that are long compared to the radionuclide half life.

Despite the relatively short effective half-life, ^{137}Cs could play an important role in imparting organ and whole-body doses following a reactor accident.

The five isotopes of iodine in Table 17-1 have sizeable fission yields and half lives which range from hours to days. They also have a very strong tendency to concentrate and remain in the thyroid gland. Small tumors called *thyroid nodules* may be expected to result from overexposure to radioiodine. Since the isotopes are relatively short-lived and are chemically reactive, containment time and removal efficiency are each important determinants of their potential hazard in reactor-accident scenarios.

RISK ASSESSMENT

The potential consequences of nuclear reactor accidents have been described in this and the previous chapter. Analysis of such consequences has been found to be valuable for designing engineered safety systems. However, it is also necessary to provide a perspective on the relative risk of reactor operation as compared to other human activities and natural events.

The most comprehensive study of relative reactor-accident risks is "Reactor Safety Study: An Assessment of Accident Risk in U.S. Commercial Nuclear Power Plants," WASH-1400 (NUREG-751014), October 1975. Although WASH-1400 is targeted to a population of 100 operating and soon-to-be-operating LWRs, its general methodology has been extended to more advanced designs and other reactor concepts.

Characteristics of Risk

Risk is generally considered to be the possibility of loss or injury to persons or property. Quantitatively, it is a function of *both* frequency and consequence magnitude according to the equation:

$$\text{Risk}\left[\frac{\text{consequences}}{\text{unit time}}\right] = \text{frequency}\left[\frac{\text{events}}{\text{unit time}}\right] \times \text{magnitude}\left[\frac{\text{consequences}}{\text{event}}\right]$$

As an example of typical societal risks, WASH-1400 (1975) notes that the United States experienced roughly 15 million automobile accidents in 1971. On the average, approximately 1 in 10 caused an injury and 1 in 300 resulted in a death. A population of 200 million persons, thus, was subjected to the following automobile accident risks:

$$\text{Risk (accident)} = \frac{15 \times 10^6 \text{ accidents/year}}{200 \times 10^6 \text{ persons}} = 0.075 \frac{\text{accident}}{\text{person} \cdot \text{year}}$$

$$\text{Risk (injury)} = 0.075 \frac{\text{accident}}{\text{person} \cdot \text{year}} \times \frac{1 \text{ injury}}{10 \text{ accidents}} = 0.0075 \frac{\text{injury}}{\text{person} \cdot \text{year}}$$

$$\text{Risk (death)} = 0.075 \frac{\text{accident}}{\text{person} \cdot \text{year}} \times \frac{1 \text{ death}}{300 \text{ accidents}} = 0.00025 \frac{\text{death}}{\text{person} \cdot \text{year}}$$

Noting that during the same year property damage from automobile accidents amounted to $15.8 billion and that there were 114 million drivers, it is found that

$$\text{Risk (property damage)} = \frac{\$15.8 \times 10^9/\text{year}}{114 \times 10^6 \text{ drivers}} = \$140/\text{driver} \cdot \text{year}$$

Table 17-2 displays risks from various human activities and natural events for the years 1967 and 1968. The trends contained therein are also representative of more recent years.

General Perception of Risks

While risks themselves may be somewhat readily calculated, human attitudes toward risks are much more complex. The work of Starr (1969) and others have identified several important features of risk perception.

A very rough correlation of risk magnitude and relative attitude is provided in Table 17-3 for death by *involuntary* societal activities. The reference risk is that from death by "natural causes" of about 10^{-2} per person·year based on an average lifetime of roughly 70 years. Other risks of 10^{-3} per person·year or higher are considered to be unacceptable and are, therefore, not found in most societies. Risks on the order of 10^{-4} per person·year have often been reduced from higher levels by expending effort and money (e.g., automobile accidents and falls as shown by Table 17-2). With decreasing risk, less action is taken until at 10^{-6} per person·year, the events are considered to be "acts of God" for which little or no special precaution appears to be warranted.

In contrast with the involuntary societal risks, individuals are found to accept voluntary risks which are greater by up to two orders of magnitude. Some sports, for example, have death risks as high as 10^{-2} per person·year. Individuals put themselves at risk willingly at levels they would find completely unacceptable if imposed upon them by society.

The perception of risk appears to be colored more by the consequence magnitude than by the frequency. Although the overall risk from air travel is found to be much less than that from the private automobile, for example, a plane crash that kills 200 persons tends to be viewed with greater alarm than a much higher

TABLE 17-2
Some U.S. Accident Death Statistics for 1967, 1968[†]

Accident	Total deaths		Probability of death per person per year	
	1967	1968	1967	1968
Motor vehicle	53,100	55,200	2.7×10^{-4}	2.8×10^{-4}
Falls	19,800	19,900	1.0×10^{-4}	1.0×10^{-4}
Fires, burns	7,700	7,500	3.9×10^{-5}	3.8×10^{-5}
Drowning	6,800	7,400	3.4×10^{-5}	3.7×10^{-5}
Firearms	2,800	2,600	1.4×10^{-5}	1.3×10^{-5}
Poisoning	2,400	2,400	1.2×10^{-5}	1.2×10^{-5}
Cataclysm	155	129	8×10^{-7}	6×10^{-7}
Lightning	110	162	6×10^{-7}	8×10^{-7}

[†]WASH-1400 (1975).

TABLE 17-3
General Correlation between Involuntary Societal Risk
of Death and the Perceived Attitude toward It[†]

Risk (deaths/person·year)	General attitude
10^{-2}	Natural-death reference
10^{-3}	Unacceptable, examples difficult to find
10^{-4}	Effort and money spent to reduce risk
10^{-5}	Mild inconvenience to avoid risk
10^{-6}	Considered an "act of God"

[†]Based on the work by Starr (1969).

death toll from automobile accidents on a typical holiday weekend. The former, in fact, may cause some to cancel future air travel plans. The latter is likely to have no identifiable impact on automobile travel.

Technological Risk Perception

The perception of risk from technological development is generally colored by the extent to which materials and processes are understood.[†] To the general public, familiar and long-established operations are often viewed as less threatening than those which are new or mysterious, even when the latter may have substantially lower risks. It is natural for individuals to feel most comfortable with those technologies of which they have the greatest understanding. As discussed briefly in Chap. 3, radiation tends to be viewed as "new, deadly, and silent" by many. This, of course, makes it, and nuclear energy in general, prime candidates for the existence of disparities between actual and perceived risks.

Viewed another way, there seems to be a "risk aversion syndrome" (or a psychological recognition of what is commonly called "Murphy's law"—"that if something can go wrong, it will"). When applied to very large, complex systems, this would lead inevitably to increased perception of risk.

The quantification of risk is often not meaningful as an isolated concept. Its real significance is in providing a basis for comparison of a proposed undertaking to its alternatives and to the general background of human and natural events. On this basis, risk assessment has become very important for development of nuclear reactor systems. The conclusions of such evaluations, although they may be technically meaningful, have more general significance only to the extent that they can provide for heightened public understanding. The remainder of this chapter is devoted to a description of reactor risk studies and their general conclusions.

The Reactor Safety Study: WASH-1400

The purpose of the "Reactor Safety Study" (WASH-1400, 1975) has been to provide a realistic assessment of the risks associated with the utilization of commercial nuclear power reactors. Its specific goals were to:

[†]These concepts are addressed further at the end of this chapter.

- Perform a quantitative assessment of the risk to the public from reactor accidents.
- Perform a realistic assessment rather than one that is overconservative.
- Develop and understand the limitations of the methodological approaches needed to perform the assessments.
- Identify areas for future safety research.
- Provide an independent check of the effectiveness of the reactor safety practice of industry and the government.

The second objective was intended to exclude evaluations like that in an earlier report entitled "Theoretical Possibilities and Consequences of Major Accidents in Large Nuclear Power Plants" (WASH-740, 1957). By including no assessment of accident probability and taking no credit for *any* engineered safety features, this early document tended to leave the general impression that very severe accidents (like those described in the previous chapter) might be expected to occur routinely. The WASH-1400 study, thus, was intended to address reactor-accident risk rather than merely potential consequences.

The WASH-1400 study was directed by Professor Norman Rasmussen (thus it is sometimes referred to as the "Rasmussen Report") under a general charter provided by the U.S. Atomic Energy Commission [AEC] and its successor, the U.S. Nuclear Regulatory Commission [NRC]. Although funding and staff assistance were provided by the commission, the study was essentially independent of the operating and regulatory organizations. Major portions of the study were undertaken by 17 contractors and national laboratories.

The WASH-1400 study limited its scope to the then-latest generation of light-water reactors with the Surry 1 [788-MW(e) PWR] and Peach Bottom 2 [1065-MW(e) BWR] units used for modeling purposes. With this limitation, the results of the study were extrapolated only to the 100 plants expected to be operable in the early 1980s. The study explicitly excluded consideration of the safety of other reactor concepts. It also did not address the potential effects of sabotage.

The following steps outline the basic flow of the reactor safety study:

1. definition of reactor accident sequences which have the potential for putting the public at risk
2. estimation of occurrence probabilities and radioactivity releases for the sequences
3. consequence modeling for health effects and property damage from the releases
4. overall risk assessment and comparison to non-nuclear risks

The primary contributors in the first category have been noted in the previous chapter to be the meltdown scenarios. The LWR sequences described there and in the earlier portion of this chapter are typical of those defined by WASH-1400.

Accident Sequences

Definition of the reactor accident sequences which can lead to core meltdown is the important first step of risk assessment. An *event tree* logic system provided the method for identifying various outcomes of given initiating events.

A simple example of event tree methodology is shown by Fig. 17-12 for

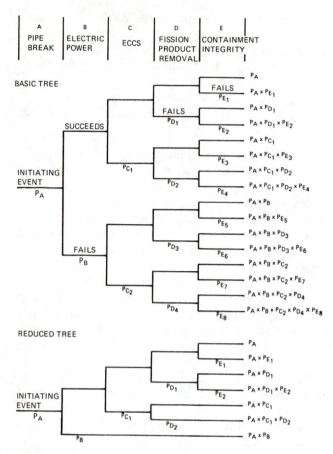

FIGURE 17-12
Simplified event tree logic diagrams for a design-basis LOCA in an LWR. (From WASH-1400, 1975.)

failure of engineered safety systems in a loss-of-coolant accident. The initiating event for the sequence is a double-ended, guillotine pipe break. Engineered safety system—electric power, emergency core cooling [ECCS], fission product removal, and containment integrity—successes or failures are then considered sequentially. (This formulation does *not* allow partial successes or failures in individual systems.) If all events were completely independent, there would be a total of 16 distinct possibilities (generally, 2^{i-1} outcomes for i events). This situation is shown by the "basic tree" on the upper portion of Fig. 17-12.

The failure probabilities P_i for system i are generally much less than unity. Thus, the success probabilities, $1 - P_i$, may be assumed to be equal to 1 as has been incorporated on Fig. 17-12. The overall probability associated with complete paths, then, is simply the product of the individual failure probabilities included in it.

Since the events considered in Fig. 17-12 are not all independent, system constraints reduce the number of distinct sequences as shown by the "reduced tree." Electric power failure, for example, automatically negates the operation of the ECCS (including heat removal) and fission-product-removal systems to assure violation of containment integrity. In a like manner, ECCS failure also leads to ultimate containment failure (by either of two pathways).

Event tree methods were employed in WASH-1400 for system-failure and containment-release analyses. Low risk paths were eliminated to leave the higher-risk paths for more detailed analysis.

Quantification of the system-failure probabilities was based on a *fault tree* logic system, which is essentially the reverse of the event tree system. Starting from the system as a whole, subsystems and then individual components are analyzed to identify the underlying failure mechanisms and to develop a basis for determining failure probabilities.

A simple fault tree for loss of electric power to the engineered safety features is shown in Fig. 17-13. The primary event is initiated by loss of either ac *or*[†] dc power, e.g., for the pumps and instrumentation respectively. Since failure data is not readily available for these functions, the next "lower" level needs to be considered. Failure of ac power implies loss of *both* off-site *and* on-site power.

[†] The symbols in the figure are common computer-logic representations. In the case at hand, the "or-gate" implies that *either* of the inputs is sufficient to produce the result in the box above it. The "and-gate" requires *both* inputs to produce the result. Or-gate and and-gate probabilities are the sum and product, respectively, of the constituent probabilities.

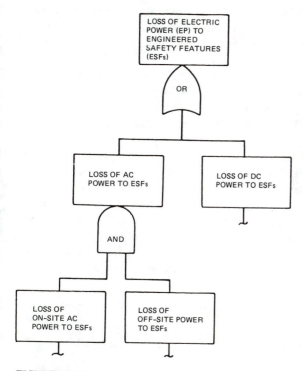

FIGURE 17-13

Illustration of fault tree logic development for loss of electric power to engineered safety systems in an LWR. (From WASH-1400, 1975.)

Further subdivision of the ac power function would account for multiple tie-ins to the off-site power grid, redundant diesel-generator systems for on-site power, and then individual components of each system. Similar procedures are appropriate for analysis of the dc network.

Data used with the fault trees included that for component failures, human error, and testing and maintenance time. The human error was found to have a probability of up to 100 times greater than that for component failure. The testing and maintenance was included in recognition that the related down-time is equivalent to system failure. One hour off-line per week, for example, is equivalent to a 6×10^{-3}/year "failure" rate.

Although the most reliable data comes from comparison of similar systems, the lack of reactor experience required use of data from fossil power plants and chemical operations. The unique LWR problems of radiation damage, activation, and high-temperature, wet steam required that the uncertainties be adjusted.

The effects of uncertainty in failure data (especially that related to common modes among components, human error, and testing) were analyzed using Monte Carlo techniques (e.g., as described in Chap. 4) and log-normal uncertainty distributions. Such analyses seemed to confirm that meaningful results could be obtained from available data.

Release Magnitudes

The magnitude of the fission-product release from the reactor containment building varies among the accident scenarios. The quantity of each species entering the general environment depends on both its prerelease inventory and the mode by which the containment is breached.

The fission-product inventory at the time of the accident is determined by the composition of the fuel, its burnup, and the power history of the system. The effectiveness of removal by decay and natural deposition processes as well as by the engineered safety systems controls the actual inventory available for release at any given time.

The timing and mode of containment failure is the final factor affecting fission-product-release magnitudes. The earlier the containment is breached, the less time there is for removal processes to reduce the inventory. Overpressure failure may be expected to result in an immediate, uncontrolled release to the atmosphere. If melt-through is the failure mode, however, the ground may be expected to produce both delay and filtration (as noted in the previous chapter).

The WASH-1400 study evaluated potential release magnitudes for each major accident scenario. Event tree methodology was used extensively for such purposes.

Consequence Modeling

The consequence model for the WASH-1400 study included detailed evaluations of:

- atmospheric dispersion of radionuclides
- population distribution
- evacuation
- health effects
- property damage

Since the models are quite extensive, only selected features are summarized in the following paragraphs.

The travel of fission products in the atmosphere was represented by the relatively simple, but well established, Gaussean plume model. The release "plume" is assumed to spread exponentially according to parametric representation of wind speed and general weather conditions (Slade, 1968). For the purposes of the WASH-1400 study, modifications were added to accommodate hour-by-hour changes in weather and rain-out effects. The weather data were selected systematically for representative sites with adjustments for time-of-day and seasonal variations. For each set of conditions, air and ground concentrations were calculated for each of 54 radionuclides.

The population model incorporated the characteristics of the 66 LWR sites in the United States which existed at the time of the study. Each site was divided into 16 sectors of $22\frac{1}{2}$ degrees each, with census data employed to correlate population and distances. Weather data was next correlated to each population distribution. Six typical data sets were selected to represent weather conditions of Eastern-Valley, East-Coast, Southern, Midwestern, Lakeside, and West-Coast sites. Two population exposure probabilities were calculated for each site—one a composite of the three most highly populated sectors, the other including the remaining 13 sectors.

Evacuation was assumed for the entire population within an 8-km [5-mi] radius of the reactor. For the two sectors along the direction of the fission-product plume, evacuation extended to a radius of 40 km [25 mi]. Population movement was assumed to follow a somewhat conservative pattern based on recent U.S. experience for other types of accidents. It calls for 30 percent of the population to average 11 km/h [7 mi/h], 40 percent to average 2 km/h [1.2 mi/h], and 30 percent to average 0 km/h, as measured radially outward from the plant. Overall, the evaluation suggested that staying indoors and breathing through a handkerchief during the time of passage of the plume would likely reduce consequences as much as attempted evacuation.

The WASH-1400 study evaluated health effects in three general categories—early [acute] fatalities, early illnesses, and latent effects. Acute fatalities are assumed to occur within 1 year of the accident. A dose-effect relationship for whole-body radiation called for 0.01 percent fatality for persons exposed at 320 rad, increasing linearly to 99.99 percent at 750 rad.

Early illnesses requiring medical treatment for varying lengths of time would be dominated by those in the respiratory tract. Lung impairment was calculated on the basis of 3000 rad causing 5 percent impairment, increasing linearly to 6000 rad for 100 percent impairment.

Latent effects considered in WASH-1400 included latent cancer fatalities, thyroid nodules, and genetic damage. Cancer incidence was assessed on an organ-by-organ basis from the internal doses predicted by the consequence model. Data from the BEIR (1972) report was modified to reflect dose-rate and dose-magnitude dependencies (Chap. 3). The net result was a predicted cancer fatality incidence of about 100 per 10^6 person·rem exposure spread over the 10–40-year period following the accicident.

As noted earlier in the chapter, thyroid nodules may result from exposure to radioiodine. They generally occur 10–40 years after exposure. Available data indicates that only about $\frac{1}{3}$ of thyroid nodules are malignant and that all can be treated medically with good success. (The roughly 10 percent fatality rate from malignancy was also included with the cancer fatality data).

Genetic mutations from accident radiation were initially assessed for a first generation during a 30-year period. The total effect was calculated approximately by assuming that the first-generation rate would persist for about 150 years.

The property-damage consequences extend from the time of the accident for an indeterminant period. Evacuation and relocation costs are the first incurred. Crop-loss and decontamination charges occur somewhat later. Long-term effects would be based on the necessity to "quarantine" land, buildings, and capital equipment. The property-damage model in WASH-1400 accounted for the costs in all of the above categories.

Reactor Accident Risk

In addition to the meltdown-accident sequences, the effects of earthquakes, tornadoes, floods, aircraft impact, and tidal waves were evaluated. All were found to have relatively low risks since they had been accommodated in the plant design basis.

Since a meaningful measure of overall risk was not found, the results were reported separately for each of the health-related categories and for property damage. Overall reactor accident risk in each area was calculated by integrating the spectrum of accident consequences and their frequency. Figure 17-14 shows the cumulative probability distribution (e.g., as defined by Eq. 4-38) for early fatalities per reactor per year for the light-water reactors considered in WASH-1400. The independent variable x is the number of fatalities, while the dependent variable is the probability per reactor per year that an accident will produce x fatalities *or more*. The uncertainties assigned to the consequence magnitudes and probabilities are noted on Fig. 17-14. Within these uncertainties, the difference in the PWR and BWR curves is not considered to be significant.

The 10^{-9}/year probability for the worst accident in Fig. 17-14 is an extremely low value. It did not appear to be unrealistic, however, since an accumulation of the partial probabilities—initiating event 10^{-3}, system failure 10^{-2}, containment failure 10^{-1}, worst weather 10^{-1}, highest population 10^{-2}—gave the same result. It was also noted that at the time of the study almost 2000 reactor-years of experience had been accumulated without a fatal accident. Since general industrial experience indicates that the low-consequence accidents should occur with higher frequency, the accident-free reactor experience suggests that the probability should be $\leqslant 10^{-3}$/year for the accident events considered by WASH-1400.

The consequences and frequencies for reactor accidents reported by the WASH-1400 study are shown in Table 17-4. The most likely core-meltdown accident was found to have very modest consequences. The more serious accidents have substantially lower frequencies. The natural incidence—the typical frequency in the general U.S. population—is included in Table 17-4 to allow comparison for appropriate consequence categories. Even in the most severe reactor accident, only the thyroid-nodule incidence matches the existing "background." (It should be

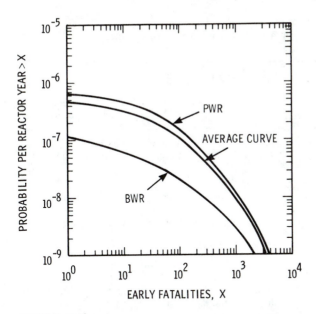

FIGURE 17-14

Cumulative probability distribution for early fatalities per reactor per year predicted by the "Reactor Safety Study" (WASH-1400, 1975). Note: Approximate uncertainties are estimated to be represented by factors of $\frac{1}{4}$ and 4 on consequence magnitudes and by factors of $\frac{1}{5}$ and 5 on probabilities.

recalled, however, that the small fraction of nodules which lead to fatalities are also included with the lethal cancers.)

Comparisons of reactor-accident fatality frequency to other accidents and natural events are shown in Figs. 17-15 and 17-16, respectively. A reactor population of 100—the basis for the WASH-1400 study—has been used. The notes accompanying the figures identify the uncertainties attributed to each of the curves.

The WASH-1400 study drew the following general conclusions for the reference 100-LWR population expected to be operating during the 1980s:

1. The most likely core meltdown accident has modest consequences to the public.
2. Reactor accidents have consequences which are no larger, and often much smaller, than those to which the population is already exposed.
3. The frequency of reactor accidents is smaller than that of most other accidents which have similar consequences.

The study also established methodology which has great potential for use in nuclear and other industries.

The WASH-1400 study specifically avoided assessing what level of risk should be accepted by society. Since such a decision is heavily weighted by individual perception, concensus on the matter would not be readily achieved.

Critical Review

The WASH-1400 Reactor Safety Study has been the subject of extensive critical review since the time it was issued in draft form in 1974. Major efforts have been

TABLE 17-4

Estimated Consequences of Reactor Accidents for Various Probabilities for One Reactor[†]

	Chance per reactor per year					
Consequences	1:2 × 10⁴ [‡]	1:10⁶	1:10⁷	1:10⁸	1:10⁹	Normal incidence
Early fatalities	<1.0	<1.0	110	900	3,300	–
Early illness	<1.0	300	3,000	14,000	45,000	4×10^5
Latent cancer fatalities (per year)[§]	<1.0	170	460	860	1,500	17,000
Thyroid nodules (per year)[§]	<1.0	1,400	3,500	6,000	8,000	8,000
Genetic effects (per year)[¶]	<1.0	25	60	110	170	8,000
Total property damage, $10⁹	<0.1	0.9	3	8	14	–
Decontamination area, km² [mi²]	<0.3 [<0.1]	5,000 [2,000]	8,000 [3,200]	8,000 [3,200]	8,000 [3,200]	–
Relocation area, km² [mi²]	<0.3 [<0.1]	340 [130]	650 [250]	750 [290]	750 [290]	–

[†]Data from WASH-1400 (1975).
[‡]This is the predicted chance of core melt per reactor year.
[§]These rates would occur in approximately the 10–40 year period following a potential accident.
[¶]This rate would apply to the first generation born after a potential accident. Subsequent generations would experience effects at a lower rate.

undertaken by the American Physical Society [APS] (APS, 1975), the U.S. Environmental Protection Agency [EPA] (EPA, 1975), the Electric Power Research Institute [EPRI] (Garcia & Erdmann, 1975; Leverenz & Erdmann, 1975), and the Union of Concerned Scientists [UCS] (Kendall, 1977). More recently, the Risk Assessment Review Group [RARG] (Lewis, 1977) conducted an extensive independent review for the U.S. Nuclear Regulatory Commission [NRC] for the purpose of clarifying the achievements and limitations of WASH-1400. It was also charged with recommending to the commission the proper uses for the risk-assessment methodology in the regulatory process.

The RARG (also called the "Lewis Commission" for its chairman, H. W. Lewis) included several members from the APS study group, took testimony from the (antinuclear) UCS and others, and generally benefited from the earlier evaluations. Thus, it represented something of a composite review. The Lewis Commission (Lewis, 1977) found WASH-1400 to:

• be a "conscientious and honest effort" and a substantial advance for quantitative analysis of the safety of a reactor
• be difficult to read and review
• contain technical faults leading to both underestimates and overestimates of risks such that uncertainties were surely substantially greater than stated

- leave "much to be desired" in the statistical analyses
- have sound methodology which can and should be used more widely by the NRC for assignment of priorities in research, inspection-time allocation, etc.

The Executive Summary, the most widely-read part of WASH-1400, was singled out for special criticism by the Lewis Report as not adequately indicating the full extent of the consequences of, and the uncertainties in the probabilities for, reactor accidents (e.g., Figs. 17-15 and 17-16). On this basis, the NRC (NRC, 1979) withdrew all explicit and implicit past endorsements of the Summary. Overall, the NRC has not repudiated the WASH-1400 study, but has instead recognized limitations to its general applicability.

An effective response of the electric-power industry to the Lewis Report was prepared by EPRI. Its own earlier reviews (Garcia & Erdmann, 1975; Leverenz &

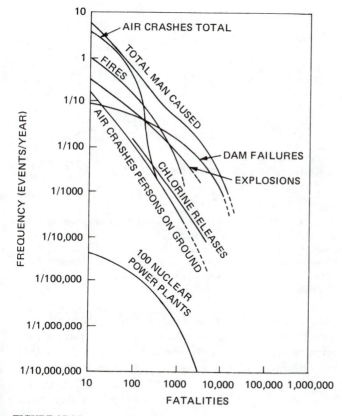

FIGURE 17-15

Frequency of fatalities due to man-caused accidents compared to the nuclear-reactor accidents predicted by WASH-1400 (1975). *Note:* Fatalities due to auto accidents are not shown because data are not available. Auto accidents cause about 50,000 fatalities per year. Approximate uncertainties for nuclear events are estimated to be represented by factors of $\frac{1}{4}$ and 4 on consequence magnitudes and by factors of $\frac{1}{5}$ and 5 on probabilities. For natural and man-caused occurrences the uncertainty in probability of largest recorded consequence magnitude is estimated to be represented by factors of $\frac{1}{20}$ and 5. Smaller magnitudes have less uncertainty.

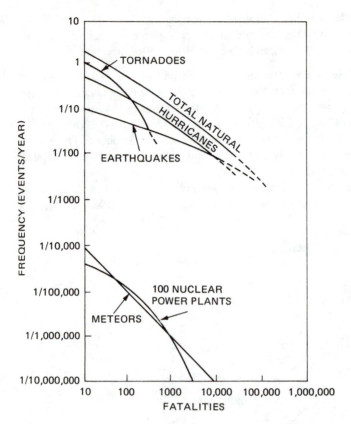

FIGURE 17-16

Frequency of fatalities due to natural events compared to the nuclear-reactor accidents predicted by WASH-1400 (1975). *Note:* For natural and man-caused occurrences the uncertainty in probability of largest recorded consequence magnitude is estimated to be represented by factors of $\frac{1}{20}$ and 5. Smaller magnitudes have less uncertainty. Approximate uncertainties for nuclear events are estimated to be represented by factors of $\frac{1}{4}$ and 4 on consequence magnitudes and by factors of $\frac{1}{5}$ and 5 on probabilities.

Erdmann, 1975) were found to address in greater detail essentially the same issues as did the Lewis study. General agreement was found on most points. However, the response document (Leverenz & Erdmann, 1979) demonstrates "that the WASH-1400 numerical results cannot be significantly optimistic and are most likely quite pessimistic." It concludes that quoted uncertainties should be substantially greater, but that this is effectively offset by mean values that are too high (i.e., that should be reduced).

One finding of WASH-1400 was that transients, small LOCA, and human errors can be important contributors to overall risk. The EPRI reviews and the Lewis Report each saw this as being inadequately reflected in NRC research and regulatory policies. Emphasis on hypothetical, design-basis accidents rather than on more likely events has been identified as an important contributor to the accident at the Three Mile Island [TMI] facility (the subject of the next chapter).

Experimental Verification

The scarcity of data related to reactor accidents has led to the formulation of many experimental programs aimed at enhanced understanding of both physical phenomena and entire accident sequences. The literature of the nuclear engineering field is filled with experimental and analytical results from such programs.

Of special current interest is the *Loss-of-Fluid Test* [*LOFT*] program sponsored by the NRC and DOE as part of an integrated research effort for confirming the safety of light-water reactors. LOFT was originally envisioned as a single test to melt a reactor core and to study, among other features, fission-product behavior in the vessel, the containment building, and, finally, the atmosphere. However, as the development of larger LWRs created a need for new safety information, LOFT's direction changed toward studying the performance of engineered safety systems, especially the emergency core cooling systems [ECCS].

Currently, LOFT is intended to support:

- assessment and improvement of analytical methods employed for predicting the LOCA behavior of PWRs
- evaluation of ECCS and other safety-feature performance
- assessment of safety margins inherent in the performance of the emergency safety systems

The components include a 55-MW(th) reactor modeled after a 3000-MW(th) [i.e., 1000-MW(e)] PWR. The core contains electrically heated rods or as many as 1300 zirconium-clad, UO_2 fuel pins, 1.7 m in length. The LOFT primary coolant system is shown in Fig. 17-17. One of its coolant loops is complete with pumps, steam generator, and pressurizer. The second loop, which has provision for simulation of the break, represents the pump and steam generator with "equivalent hydraulic resistances" (i.e., by passive components which offer the same resistance to the flow of the coolant). The expelled coolant is collected in a quench tank to minimize any contamination of the facility.

The proposed experimental series in the LOFT program are identified in Table 17-5. The L1 loss-of-coolant experiment [LOCE] series with the electrically heated (nonnuclear) core was completed between 1976 and 1978. The large, double-ended-break experiments in series L2 began in December 1978. In the aftermath of the Three Mile Island [TMI] reactor accident (Chap. 18), conduct of the L3 series on small pipe breaks was accelerated to a November 1979 starting date.

A primary goal of the LOFT program has been to use the experimental data as a benchmark for evaluation-model [EM] and best-estimate [BE] calculational procedures. The EM computer models used for licensing purposes are intentionally conservative. The BE models, as implied by the name, are designed to predict phenomena as closely as possible. A standard practice is to obtain pretest predictions for each series of experiments. One comparison of predicted and measured results for peak cladding temperature in the L2-3 experiment (36-MW power level, 39-KW/m maximum linear heat rate) is shown in Fig. 17-18. Of special note is the overconservatism of the NRC's water-reactor evaluation model [WREM-NRC] compared to the several best-estimate models. The BE computer codes (except for the TRAC-PIA + Iloeje) also give conservative results by failing to

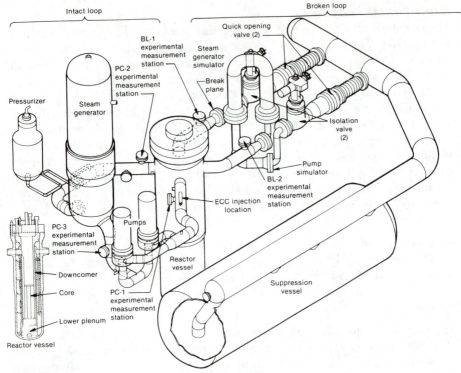

FIGURE 17-17
Loss-of-Fluid Test [LOFT] program major components. (Courtesy EG&G Idaho, Inc.)

account for the rewetting of the core (and resulting cooldown of the clad) that occurs about five seconds into the transient. This latter phenomenon, associated with the coolant water initially entrained in the upper plenum region with the control rods (e.g., as in the upper region of Fig. 17-2), was best handled by the Los Alamos's TRAC code when modified by a correlation attributed to Iloeje.

Although not intended as an experiment, the accident at the Three Mile Island [TMI] Reactor is also expected to provide a substantial data base for future LWR safety evaluations. Already-available information (described in the next chapter) should be supplemented greatly as the clean-up and renovation activities proceed.

TABLE 17-5
Proposed Experimental Series for the LOFT Program

Experimental series	Description
L1	Nonnuclear LOCE
L2	Nuclear power ascension LOCE, double-ended cold leg break
L3	Nuclear LOCE with small- and intermediate-size breaks
L4	Nuclear LOCE with alternate ECCS injection concepts
L5	Nuclear LOCE hot leg breaks
L6	Nuclear non-LOCE transients
L7	Nuclear LOCE with steam-generator tube rupture

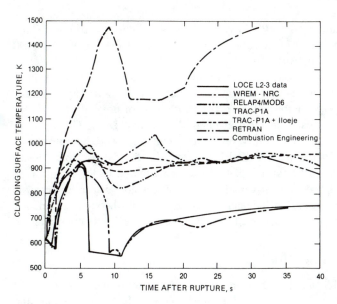

FIGURE 17-18
Comparison of measured peak cladding temperature for the LOFT L2–3 experiment with all pretest prediction methods. (Courtesy EG&G Idaho, Inc.)

Safety Studies of Other Reactors

Although WASH-1400 explicitly excluded consideration of systems other than LWRs, its methodology may be readily extended. Related studies of HTGR and LMFBR systems have been prepared. Although CANDU systems have received substantial attention by conventional safety-evaluation methods, the probabilistic methodology of WASH-1400 does not appear to have been applied [e.g , see van Erp (1977)] .

HTGR Study
The General Atomic Company has conducted a probabilistic safety study for a reference 3000-MW(th) HTGR. The "Accident Initiation and Progression Analysis" [AIPA] (General Atomic, 1976) had the major objectives of:

- providing guidance for safety research and development [R & D] programs
- considering alternative design options for safety
- developing risk assessment methodology

   Although it drew heavily on the WASH-1400 experience, the AIPA study was not a complete risk assessment since it examined only a limited number of sequences for a conceptual design (rather than operating systems). The study did identify R & D needs and in addition found the HTGR to be "exceptionally safe" under the conditions considered.

Clinch River LMFBR Study
The Clinch River Breeder Reactor Plant [CRBRP] was the subject of a probabilistic risk study intended to parallel the WASH-1400 effort. Since, however, one reactor rather than a group of 100 were under consideration, the consequence analysis in

the "CRBRP Safety Study" (CRBRP-1, 1977) was appropriate only to the specific site. Many of the contributing organizations and individuals from the WASH-1400 study were also part of the CRBRP effort.

The LWR accidents considered previously were found to proceed in a relatively orderly manner after a given set of initiating conditions. By contrast, LMFBR sequences are very complex due to the wide range of possibilities for core dispersal, recriticality, and post-accident heat removal (e.g., Fig. 16-2). The curves in Fig. 17-19 compare the CRBRP fatality results to those for a single reactor from WASH-1400. The lower values for the LMFBR system are due largely to its lower power level. The shape differences in Fig. 17-19 were attributed to the inclusion of highly energetic (and highly nonmechanistic) hypothetical core-disruptive-accident [HCDA] sequences with large-magnitude consequences. Overall, the study concluded that CRBRP should not present risks greater than those for the LWRs considered in WASH-1400.

Additional Perspectives on Risk

Two recent works provide additional perspective on risks from nuclear energy and other activities. Each has the prospect for providing valuable insight about the question of "what constitutes an acceptable level of risk?"

"A Catalog of Risks" prepared by Cohen & Lee (1979) is one of the most comprehensive collections of risk information. It includes evaluations of such diverse factors as radiation, accidents, diseases, overweight, tobacco use, alcohol and drugs, coffee, saccharin, oral contraceptives, occupational risks, socioeconomic factors, marital status, geography, serving in the U.S. armed forces in Vietnam, catastrophic events, energy production, and technology in general. Representative data from the study is provided in Table 17-6 in the form of the average loss in life expectancy from the specified cause or activity.

The catalog concludes that certain risks like those from radiation, oral

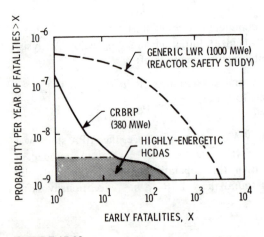

FIGURE 17-19
Frequency of fatalities predicted for an LWR by WASH-1400 (1975) and for the Clinch River LMFBR by CRBRP-1 (1977). (Adapted courtesy of Clinch River Breeder Reactor Plant.)

TABLE 17-6
Average Loss in Life Expectancy due to Various Causes[†]

Cause	Time (days)
Being unmarried—male	3500
Cigarette smoking—male	2250
Heart disease	2100
Being unmarried—female	1600
Being 30% overweight	1300
Being a coal miner	1100
Cancer	980
Cigarette smoking—female	800
Less than eighth-grade education	850
Living in unfavorable state	500
Serving in the U.S. Army in Vietnam	400
Motor vehicle accidents	207
Using alcohol (U.S. average)	130
Being murdered (homicide)	90
Accidents for average job	74
Job with radiation exposure	40
Accidents for "safest" job	30
Natural background radiation (BEIR, 1972)	8
Drinking coffee	6
Oral contraceptives	5
Drinking diet soft drinks	2
Reactor accidents (Kendall, 1975)	2[‡]
Reactor accidents (WASH-1400, 1975)	0.02[‡]
Radiation from nuclear industry	0.02[‡]
PAP test	−4
Smoke alarm in home	−10
Air bags in car	−50

[†]Reprinted with permission from B. L. Cohen and I. S. Lee, "A Catalog of Risks," *Health Phys.*, Vol. 36, June 1979, pp. 707–722, copyright © 1979, Pergamon Press, Ltd.
[‡]Assumes that all U.S. power is nuclear.

contraceptives and artificial sweeteners in soft drinks are overemphasized by society. In contrast, such factors as marital status, overweight, and socioeconomic factors seem to warrant increased concern.

The subject of differential risk perception has been addressed in a new study reported by Litai (1980).[†] Using a broad data base from the insurance industry and other sources, the eight paired concepts shown in Table 17-7 were evaluated to determine their effect on risk perception. The table also shows the multiplicative "risk conversion factor" by which perceived risk appears to exceed actual risk for the second of the paired concepts.

According to the study, the public *perception* of nuclear-power risk should exceed the actually expected risk by about a factor of 300 (within an order of

[†]These first results are from a doctoral dissertation supervised by Professor Norman Rasmussen, the director of the WASH-1400 study. The methods are likely to be pursued extensively in the near future.

TABLE 17-7
Factors Affecting the Perception of Risk[†]

Paired risk concepts	Risk conversion factor[‡]
Delayed–immediate	30
Necessary–luxury	1
Ordinary–catastrophic	30
Controllable–uncontrollable	5
Voluntary–involuntary	100
Natural–man-made	20
Occasional–continuous	1
Old–new	10

[†]From Latai (1980).
[‡]Uncertainty is estimated to be a factor of 10 (i.e., actual value should be between 0.1 and 10 times the stated value).

magnitude). Application of such a factor to the WASH-1400 fatality curve in Figure 17-15, for example, would raise the estimate of perceived risk to a level comparable to that of the other human activities.

EXERCISES

Questions

17-1. Identify the five categories of engineered safety systems. Considering the LWR, CANDU, HTGR, and LMFBR designs, list at least one unqiue feature for each system in four of the categories.

17-2. Describe at least three pathways by which an airborne release of radio-activity from a reactor accident can interact with the human environment.

17-3. Summarize Starr's theories of risk perception.

17-4. Explain the basic difference between fault tree and event tree methodologies.

17-5. Justify the 10^{-9} per-reactor-per-year probability assigned to the most severe LWR accident consequences by WASH-1400. Explain why the Clinch River Breeder Reactor may have lower overall risk than the LWR population.

17-6. Identify the major criticisms of WASH-1400 by the Lewis Committee.

17-7. Explain the role of the LOFT program in LWR safety.

17-8. Describe the status of probabilistic risk assessment applications to HTGR and LMFBR systems.

Numerical Problems

17-9. Identify the sensors on Fig. 12-16 that would be expected to give trip signals following a LOCA in a PWR.

17-10. Calculate the 1971 U.S. death and fatality risks per driver·year from the data in the text.

17-11. Explain explicitly why the complete event tree at the top of Fig. 17-11 may be reduced to that at the bottom.

17-12. Sketch two levels of a conceptual fault tree for loss of post-accident heat removal in an LMFBR.

17-13. Revise the data in Table 17-4 to state the chance per reactor·year as a fraction of normal incidence for the four applicable categories.

17-14. A PWR containment building has a dome with an 18-m radius. To first approximation, its volume may be assumed to be that of a sphere with the same radius. Based on data in Tables 16-3 and 17-1:

    a. Estimate the average concentrations of ^{131}I and ^{90}Sr in the containment following meltdown of a 3000-MW(th) core which has operated for 400 days.

    b. Calculate the attenuation factor (combined effect of decay, dilution, and removal) which would be required before release at MPC levels of 1×10^{-11} μCi/ml for ^{131}I and 3×10^{-11} μCi/ml for ^{90}Sr.

17-15. Calculate the time required for the *total* iodine activity in Table 17-1 to decay to 50 percent and 1.0 percent of the equilibrium reactor concentration.

SELECTED BIBLIOGRAPHY†

Reactor Safety (General)
    Farmer, 1977
    Lewis, 1977
    Nucl. Eng. Int., Aug. 1980, Nov. 1980
    Sesonske, 1973 (Chap. 6)
    Thompson & Beckerley, 1964, 1973

LWR Safety
    Brockett, 1972
    Cottrell, 1974
    Lewis, 1980
    Rippon, 1974
    WASH-740, 1957
    WASH-1250, 1973 (Chaps. 2, 5)
    WASH-1400, 1975
    (see also Selected Bibliography for Chap. 12)

LOFT Program
    Leach & McPherson, 1980
    Leach & Ybarrondo, 1978
    Linebarger, 1979
    McCormick-Barger, 1979
    Prassinos, 1979
    Reeder, 1978
    Reeder & Berta, 1979

CANDU Safety
    Rogers, 1979
    van Erp, 1977
    (see also Selected Bibliography for Chap. 13)

HTGR Safety
    Barsell, 1977
    General Atomic, 1976
    (see also Selected Bibliography for Chap. 13)

†Full citations are contained in the General Bibliography at the back of the book.

LMFBR Safety
    CONF-761001, 1976
    CRBRP-1, 1977
    Cybulskis, 1978
    Fauske, 1976
    Gregory, 1979
    (see also Selected Bibliography for Chap. 15)

Environmental Transport
    Eisenbud, 1973
    Sagan, 1974

Systems and Risks[†]
    Briggs, 1971
    Cohen & Lee, 1979
    Eicholz, 1976
    Erdmann, 1979
    IEEE, 1979*b*
    Lewins, 1978 (Chap. 7)
    Lewis, 1980
    Lish, 1972 (Chaps. 5, 9, 14)
    Litai, 1980
    Moeller, 1977
    Okrent, 1979
    Okrent, 1980
    Slade, 1968
    Starr, 1969
    WASH-1400, 1975

WASH-1400 Review
    APS, 1975
    EPA, 1975
    Garcia & Erdmann, 1975
    Kendall, 1977
    Leverenz & Erdmann, 1975
    Leverenz & Erdmann, 1979
    Lewis, 1978
    NRC, 1979
    Nucl. Eng. Int., May 1979

Individual Reactors
    Safety Analysis Reports [SAR] from the U.S. Nuclear Regulatory Commission docket
        (for every U.S. reactor)
    Nucl. Eng. Int. (see Selected Bibliographies for Chaps. 12, 13, or 15 for specific reactors)

Current Sources
    Nucl. Eng. & Des.
    Nucl. Eng. Int.
    Nucl. Safety
    Nucl. News
    Nucl. Sci. & Eng.
    Nucl. Tech.
    Science
    Tech. Rev.
    Trans. Am. Nucl. Soc.

[†]A potentially significant new assessment of reactor risks is contained in a paper by Levenson and Rahn, 1980. See the General Bibliography, page 587.

18

THE THREE MILE ISLAND REACTOR ACCIDENT

Objectives

After studying this chapter, the reader should be able to:

1. Identify several devices, operator actions, and inherent energy sources that contributed to the Three Mile Island reactor accident.
2. Identify the three earliest industry programs that evolved as a result of the TMI-2 accident.
3. List several of the seven major areas of concern identified in the Kemeny Commission report.

Prerequisite Concepts

Radiation Effects	Chapter 3
Pressurized-Water Reactors	Chapter 12
Accident Energy Sources	Chapter 16
Loss-of-Cooling Accidents	Chapter 16
PWR Meltdown Scenarios	Chapter 16
Fission-product Release	Chapter 16
PWR Engineered Safety Features	Chapter 17

The Three Mile Island [TMI] Nuclear Station is owned and operated by Metropolitan Edison Company, a member company of General Public Utilities [GPU]. It is located near Middletown, Pennsylvania, about 10 miles southeast of Harrisburg, the state's capital. The station consists of two Babcock & Wilcox PWRs rated at 792 MW(e) and 880 MW(e), respectively.

On March 28, 1979, Unit 2 at TMI experienced a series of events that resulted in the most serious accident in the history of commercial nuclear power in the United States. The reactor core was damaged extensively, leading to large releases of radioactivity to the containment building. Some atmospheric releases also occurred. Despite this, radiological consequences to operating personnel and the general public were found to be minor.

The accident has had and will continue to have a profound effect on the Metropolitan Edison Company and General Public Utilities, the nuclear industry, and the U.S. Nuclear Regulatory Commission. It is also likely to have an important impact on emerging federal energy policies.

THE ACCIDENT

The TMI-2 reactor is a standard Babcock & Wilcox PWR with a primary system arranged as shown in Fig. 18-1. It consists of a reactor vessel (similar to Fig. 12-7), a pressurizer (similar to Fig. 12-8), four coolant pumps, and two once-through steam generators (Fig. 12-10b). Hot-leg piping carries heated coolant from the two reactor outlet nozzles to the inlets of the steam generators. Two cold-leg pipe runs from each steam generator provide paths for coolant to be pumped to the reactor inlet nozzles.

A not-to-scale schematic layout of the TMI-2 facility is shown in Fig. 18-2. For simplicity it includes only one coolant loop of the standard PWR steam cycle (e.g., as in Fig. 12-6). Engineered safety system features (Fig. 17-2) include the control rods, high-pressure injection [HPI] emergency core-cooling system [ECCS], borated water storage tank, and the building sump (for ECCS recirculation water supply). The other components on Fig. 18-2 are described below as they relate to the TMI-2 accident sequence.

Sequence of Events

Before 4:00 a.m. on March 28, 1979, the TMI-2 reactor was operating at 97 percent of rated power under seemingly normal conditions. Three assumed-to-be-minor problems, however, did exist.

Pre-Accident Conditions
The first problem was based on the system losing small amounts of coolant through one or more of the valves on the pressurizer (Figs. 12-8 and 18-2). One pressure-relief valve and two safety valves are provided to prevent the pressure from building to a level that could damage the primary system and initiate a major LOCA. The electromatic pressure-relief valve[†] is in a normally closed configuration. It may be opened by energizing a solenoid, either automatically from a monitor set to detect an overpressure condition or manually from the reactor console. The mechanical safety valves are designed for pressure relief at levels above the set-point of their electromatic counterpart. Since valves do leak, good industrial practice calls for the escaping coolant to be piped to a drain tank (Fig. 8-2). A remotely actuated block valve is positioned so that it can be closed to stop the flow as desired.

Although the NRC has since stated that the pressurizer leakage exceeded specified regulatory limits, the reactor operators apparently believed it to be within

---

[†]Also referred to as a power-operated relief valve.

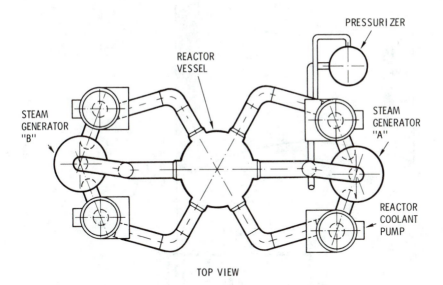

REACTOR
VESSEL

PRESSURIZER

STEAM
GENERATOR
"B"

STEAM
GENERATOR
"A"

REACTOR
COOLANT
PUMP

TOP VIEW

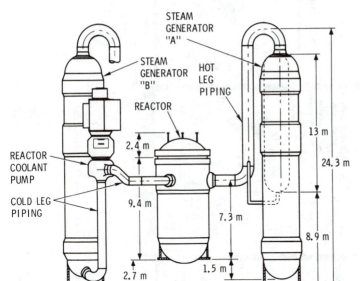

STEAM
GENERATOR
"A"

STEAM
GENERATOR
"B"

HOT
LEG
PIPING

REACTOR

REACTOR
COOLANT
PUMP

COLD LEG
PIPING

2.4 m

9.4 m

7.3 m

13 m

24.3 m

8.9 m

1.5 m

2.7 m

SIDE VIEW

FIGURE 18-1
Arrangement of primary-coolant system for the Three Mile Island [TMI] nuclear reactor. (From
Three Mile Island Final Safety Analysis Report, U.S.N.R.C. docket number 50-320, license
number DPR-73.)

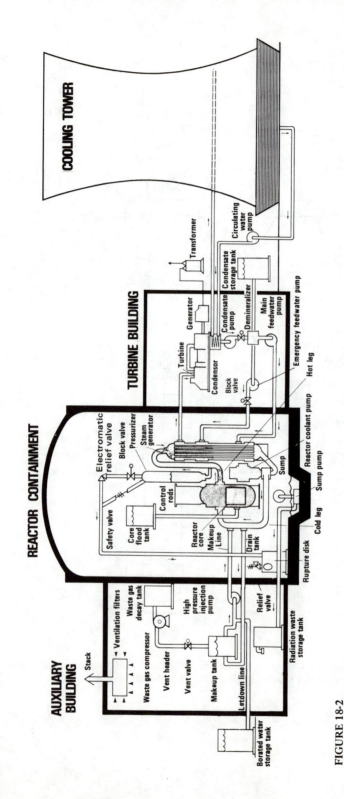

FIGURE 18-2

Schematic layout of the TMI-2 reactor. (Reprinted, and adapted with permission of IEEE, from *IEEE Spectrum*, November 1979 issue, special report on Three Mile Island.)

an acceptable range and, thus, controlled the primary system to maintain normal temperature and pressure conditions. The relatively minor loss of coolant had the effect of partially filling the drain tank. It also increased the temperature in the piping between the pressurizer and the tank. Although these latter factors were not significant by themselves, they helped lead to faulty conclusions when larger coolant losses occurred during the accident sequence.

A second condition (which retrospectively appears to have been of only minor significance to the accident) was that the operators were unaware that two valves on the emergency feedwater lines (Fig. 18-2) were closed. Although there are indications that the valves had been reopened following maintenance completed 2 days earlier, they were not open at the time of the accident. The net effect of the valve closure was to prevent emergency feedwater in the condensate storage tank from being pumped to the steam generators during the first 8 min of the transient.

The third and final pre-accident problem concerned the feedwater demineral- izers that are located in the turbine building (Fig. 18-2) and are designed to maintain very high purity in the secondary coolant loop. Each demineralizer consists of several condensate polishers and contains ion-exchange resins which perform the impurity-removal task. For roughly 11 h prior to the accident, shift foremen and auxiliary operators had been attempting to transfer spent resins to a resin regeneration tank. Under normal circumstances, compressed air can be used to fluff spent resins and allow their reuse. The resins are then transferred in demineralized water through a transfer line between tanks. Unsuccessful attempts by the operators to clear a blockage in this line are thought to have contributed to a condensate- pump trip. This, in turn, began the chain of events that progressed to become the TMI-2 accident.

Initiation

Loss of condensate flow at 4:00:36 a.m. on March 28, 1979, led to a trip of the main feedwater pumps within the next second. Nearly simultaneously, the turbine tripped off-line. Within another second of the feedwater-pump and turbine trips, the emergency feedwater pumps started automatically. Table 18-1 provides a chronology of the key events in the remainder of the TMI-2 accident.

The loss of feedwater to the steam generators reduced the rate of heat removal from the primary-coolant loop and the reactor core (a loss-of-heatsink accident [LOHA] as introduced in Chap. 16). As the coolant became hotter, its pressure increased enough to cause the electromatic relief valve on the pressurizer to open in response to an indicated level in excess of its 15.55 MPa [2255 psi] set-point. Continued heating 8 s into the accident led to a control-rod trip on high coolant-pressure signals in the core protective system (similar to Fig. 12-16). To this point in the sequence, all systems appeared to be operating as designed.

By 14 s into the accident the emergency feedwater pumps reached full design pressure, but, because the lines were blocked, did not deliver water to the steam generators. It was not until about 8 min later that an operator noted low water levels and pressures in the steam generators. When this led to the realization that the emergency feedwater block valves were closed, said valves were opened to restore coolant water flow to the steam generators.

TABLE 18-1
Chronology of Key Events in the TMI-2 Accident (Referenced to Feedwater Pump Trip at 4:00:37 a.m. on March 28, 1979)

Approximate time	Event
0	Main feedwater pump trips with simultaneous turbine trip
3 s	Primary pressure reaches relief-valve set point; valve opens
8 s	Reactor trips
13 s	Pressure drops below relief-valve set-point; valve remains open
14 s	Emergency feedwater pumps reach normal discharge pressure
38 s	Emergency feedwater directed to steam generators, but flow prevented by closed block valves
2 min	High-pressure injection ECCS starts automatically
2 min	Drain tank relief valve lifts
3–5 min	Operators throttle high-pressure injection ECCS
7 min	Coolant transfer from containment to auxiliary building begins
8 min	Block valves on emergency feedwater lines are opened
15 min	Drain tank rupture disk lifts
73 min	Main coolant pumps in Loop A are tripped off-line
100 min	Main coolant pumps in Loop B are tripped off-line
1.5–3.5 h	Core uncovery results in zirconium–water reactions (with hydrogen evolution and fuel-assembly damage)
142 min	Block valve on pressurizer drain line is closed
150 min	In-core thermocouple readings go off scale
3 h	Site emergency is declared
3.5 h	General emergency is declared
9.5 h	Hydrogen detonation produces pressure spike
16 h	Main coolant pumps in Loop A restart to restore forced cooling

The extended absence of feedwater would seem to have a significant impact on the reactor system. This is, indeed, the contention of the Babcock & Wilcox Company, the manufacturer. However, analyses performed by the utility and the Nuclear Safety Analysis Center [NSAC] —an organization operated by the Electric Power Research Institute [EPRI] —indicated that the 8-min absence of feedwater had no significant effect on the progression of the accident.

Loss of Primary Coolant
The primary system began to cool somewhat following the insertion of the control rods. This led to a pressure reduction. The pressure fell below 15.21 MPa [2205 psi] —the set-point for closure of the electromatic relief valve—about 13 s into the accident. Although the solenoid was deenergized on cue, the valve actually stayed open. A loss of coolant accident [LOCA] existed at this time even though there were no pipe breaks involved. Continuing coolant discharge reduced the primary pressure and rapidly filled the drain tank in the bottom of the containment building (Fig. 18-2). The tank pressure built to a sufficient level to lift the safety valve at about 3 min and burst the rupture disk at about 15 min. Thus, primary coolant flowed from the electromatic relief valve to the drain tank, and finally to the containment building sump. The process continued unabated for approximately

2.4 h in the accident before the pressurizer block valve was closed to stop the loss of coolant.

The control-panel indicator for the electromatic relief valve told the reactor operators only that the actuating solenoid was deenergized. No direct reading of actual valve closure was available. Temperature sensors on the drain pipe gave ambiguous readings since leakage was occurring before the accident and the relief valve had, indeed, opened for a few seconds early in the accident.

The relief-valve discharge could also be inferred from a pressure sensor on the drain tank. The readout was on a meter located on a panel behind the reactor console. The location away from the main instrument clusters resulted in infrequent monitoring by the operators. Since a meter was used rather than a chart recorder, only instantaneous readings were available. This was particularly significant as the meter indicated atmospheric pressure for an unfilled tank as well as at all times after the drain tank rupture disk was broken (about 15 min into the accident). The system's data-acquisition computer did contain a time history of the tank pressure in its memory. However, data printout was lagging significantly during the intense activity associated with the accident.

Engineered Safety Systems

At approximately 2 min into the accident the high-pressure injection [HPI] system began pumping borated water from the refueling water storage tank (Figs. 18-2 and 17-2). This occurred as the primary system pressure dropped below the 11.31 MPa [1640 psi] set-point. At approximately $4\frac{1}{2}$ min the operator turned off one of the HPI pumps and trottled back the other. This injection rate was later found to be less than the rate of coolant loss through the stuck electromatic relief valve.

The high-pressure injection system was cut back in response to indications of a high coolant level in the pressurizer. Previous training had stressed to the operators that a "solid' (i.e., all liquid water) pressurizer was to be avoided. Under normal operating conditions a solid pressurizer does not allow the device to function for pressure control. The condition also violated the technical specifications in the reactor's operating license from the U.S. Nuclear Regulatory Commission. On these bases, the emergency cooling was cut back to what turned out to be an excessively low level.

The configuration of the primary system was such that no direct relationship existed between the coolant levels in the pressurizer and in the primary system as a whole. With continuing loss of coolant, the system reached saturation conditions where steam voids coexisted with the liquid (e.g., as shown in Fig. 18-3a). The void content increased with the pressurizer still indicating a filled system. Since the operators were still unaware that a LOCA was in progress, use of the ECCS appeared to be inappropriate.

Reactor Coolant Pumps

The reactor pumps, designed for pressurized water, experienced substantial vibrations in handling the increasing steam content of the remaining coolant. At about 73 min into the accident, both pumps in loop B (Fig. 18-1) were turned off in response to indications of low system pressure, high vibration, and low coolant flow. This action was predicated on preventing potentially serious damage to the

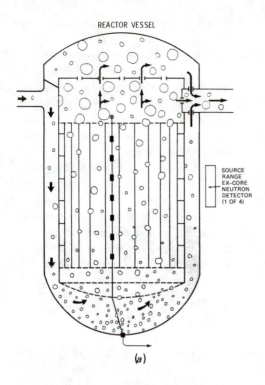

(a)

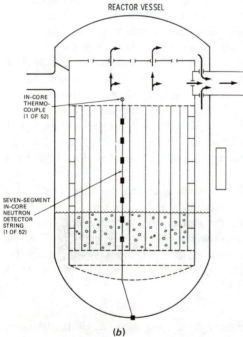

(b)

FIGURE 18-3
Reactor vessel: (a) with forced circulation of coolant with a relatively uniform distribution of voids; and (b) following coolant pump shutdown. [Courtesy of Nuclear Safety Analysis Center, operated by the Electric Power Research Institute (NSAC-1, 1979).]

pumps and associated piping. More importantly, it was considered necessary to protect the pump seals since their failure would provide a substantial pathway for coolant loss. The net effect of the pump shutdown was to allow the steam and water to separate in loop B and apparently preclude further circulation through that steam generator. At about 100 min into the accident, the loop A pumps were shutdown for similar reasons to those noted above.

Since the condition of the primary system was not recognized at the time of pump shutdown, the operators expected natural circulation to keep the core from overheating. The steam–water mixture in Fig. 18-3*a* apparently provided adequate cooling before the pumps were stopped. However, after all of the pumps stopped, the steam and water separated to reduce the effective coolant volume and uncover the core. Continued core heating driven by decay heat boiled off more of the liquid. Figure 18-3*b* shows the situation conceptually at a time much later in the accident.

Within about 10 min of the pump shutdown (approximately 111 min into the accident), reactor coolant outlet temperatures started to rise rapidly. Strip-chart records show that temperatures exceeded the 325°C [620°F] instrument-scale range about 40 min later and remained there for about the next 7.5 h (until nearly 10 h into the accident). This suggests an environment of superheated steam and, presumably, noncondensible hydrogen above the core and in the coolant outlet piping.

As the core became uncovered, the clad temperatures became high enough for exothermal Zr–steam reactions to occur. They added energy to the system as well as producing hydrogen. A survey of the postaccident hydrogen inventory suggested that about one-third of all of the zircaloy in the core participated in such chemical reactions. It is not believed, however, that any melting occurred in the UO_2 fuel.

Cooling Restoration

The operators closed the block valve on the pressurizer drain line (Fig. 18-2) at approximately 142 min into the accident. They then spent the next 13 h trying to reestablish stable cooling in the core based on either:

1. natural or forced circulation of coolant water with the steam generators as a heat sink; or
2. use of the low-pressure-injection ECCS with its associated heat removal function (not shown in Fig. 18-2, but as described in general terms on Fig. 17-2) requiring primary system pressures below 2.2 MPa [320 psi]

For roughly 5 h after block-valve closure, numerous attempts were made to establish heat removal through the steam generators. All attempts at establishing forced circulation or promoting natural convection of reactor coolant were unsuccessful due to hydrogen blockage on the vessel outlet piping (which forms the "candycanes" at the steam-generator inlets shown in Fig. 8-1). The primary system pressure varied substantially during this period of time depending on the position of the block valve on the pressurizer relief line and actuations of the high-pressure injection system. At one point, a sustained attempt was made to reestablish heat removal through the steam generators by pressurizing the primary coolant system to approximately 14.5 MPa [2100 psi] with continuous operation of the high pressure injection system. The procedure proved to be unsuccessful.

The next attempt at cooling restoration involved what was essentially an induced LOCA. During a 4-h period, the operators reduced the system pressure by opening the pressurizer block valve. The goal was to reduce the pressure far enough for the low-pressure coolant injection system to flood the core. Heat removal would have been facilitated through the residual-heat-removal [RHR] heat exchangers (e.g., Fig. 17-2) whose ultimate heat sink is the Susquehannah River flowing by Three Mile Island. Although the core flood tanks [accumulators] (e.g., Fig. 17-2) did inject some water directly into the vessel when the pressure dropped below 4.1 MPa [600 psi], the primary coolant pressure remained too high to activate the low-pressure injection system. One major impact of the operation was to vent a large fraction of the hydrogen from the coolant system.

The pressurizer block valve was closed when the operators were not able to depressurize the primary coolant system any further. During the next 2 h no effective heat removal mechanism was functioning, as both steam generators were still blocked and coolant injection was at a very low level. The primary coolant system was repressurized approximately $13\frac{1}{2}$ h into the event by sustained high-pressure injection. Although the steam generators were still blocked by hydrogen, the venting that had occurred during the depressurization was sufficient to allow restart of a coolant pump. This reestablished forced circulation of coolant and allowed heat removal through the loop A steam generator (Fig. 18-1). By about 16 h the second pump on the loop was started. Core stabilization proceeded in an orderly manner from this point on.

Radioactivity

With the bursting of the drain tank rupture disk, primary coolant water began to enter the building sump. Initial radiation levels were low since the coolant contained only the fission and activation products which were characteristic of normal operations. As the core heated and clad integrity decreased, however, more fission products entered the coolant and radiation levels increased.

The makeup and letdown systems (Fig. 18-2) are used to balance the inventory of the primary system. Small amounts of leakage during normal operation are of little concern, since the coolant usually has low levels of radioactivity. During the accident, however, some of the high-activity coolant water posed a problem when it escaped into the auxiliary building. A later venting of fission gases from the makeup tank to the waste gas decay tank resulted in leakage through a vent header.

The water collecting in the building sump was automatically pumped to radioactive-waste storage tanks in the auxiliary building (Fig. 8-2). A blown rupture disk on these tanks allowed the water to collect on the floor of the auxiliary building.

Automatic isolation of the containment building would have prevented the transfer of the waste water and subsequent release of radioactivity to the atmosphere. However, in the TMI-2 design such isolation is predicated on a 0.03 MPa [4 psi] overpressure characteristic of a large-break (design-basis) LOCA. Sump-line pressure limits were similarly large. The LOCA associated with the stuck electromatic pressurizer relief valve was too small to activate these important engineered safety features.

Fission gases from the several sources in the auxiliary building were picked up by the ventilation system. After filtration they were discharged from the stack (Fig. 18-2). The filters removed much of the chemically active gas like iodine, but had no real effect on the inert noble gases (as described in Chaps. 16 and 17).

Rising radiation levels led to successive declarations of site and general emergencies, respectively, at roughly 3 h and $3\frac{1}{2}$ h into the accident. At about 6 h the control room was evacuated of all nonessential personnel, and by 7 h the remaining operators were wearing respirators to limit their uptake of airborne radioactivity.

At the time of the accident, the core inventory of fission products included several hundred megacuries each of noble gases and iodine. The noble gas released as a result of the accident has been estimated at about 10 MCi (NUREG-0600, 1979), mainly of ^{133}Xe. The large release fraction relates to the chemically inert nature of these radionuclides. Fortunately, this same characteristic also allows it to pass through human tissue without being retained.

By contrast to the very large noble gas releases, estimates for ^{131}I are only in the range of 14 to 27 Ci (NUREG-0600, 1979). The reduction of five to six orders of magnitude between core inventory and release[†] is fortunate because of the tendency for iodine to concentrate in the thyroid. The natural solubility of iodine in water plus the action of additives and filtration systems (Chap. 17) are attributed with preventing much larger escape.

Hydrogen Behavior

During normal reactor operations, the high radiation levels cause decomposition of water into hydrogen and oxygen. Natural recombination tends to occur too slowly to prevent large buildups of these gases. The addition of excess hydrogen, however, speeds the recombination and stabilizes the overall oxygen and hydrogen inventories. Some of this normal hydrogen inventory was vented to the TMI-2 containment through the open electromatic relief valve.

When the core became uncovered between about $1\frac{1}{2}$ and $3\frac{1}{2}$ h after the start of the accident, Zr–steam reactions produced large amounts of hydrogen (equivalent to the reaction of as much as one-third of all the zircaloy in the core, as noted previously). The hydrogen distributed itself between a gas–steam bubble in the reactor vessel (above the inlet and outlet nozzles in Fig. 18-1), gas dissolved in the remaining coolant, and gas which escaped to the containment building. At about $9\frac{1}{2}$ h into the accident, the concentration in the containment became high enough to support combustion. Ignition occurred with a resulting 0.19 MPa [28 psi] pressure pulse recorded in the reactor control room. This pressure is well within the design capability of the containment building. The fact of the hydrogen burn was later

[†]Until now, many design-basis accident analyses as well as the WASH-1400 study have assumed that up to 50 percent of the iodine would escape containment. The actual TMI-2 ("experimental") results suggest that iodine may present a substantially lower hazard than had been thought.

confirmed by sampling the containment atmosphere and establishing that its oxygen content was depleted to about 16 percent from its normal 20 percent value.

The accumulation of hydrogen in the top portion of the reactor vessel led to postulation of a "hydrogen bubble." In reality, the hydrogen coexisted in an equilibrium with steam. The limited amount of oxygen in the vessel precluded the chance for explosion. Thus, the great concern in the media for a vessel-rupturing explosion was unfounded. The hydrogen was eventually removed during the first week. Advantage was taken of its variable solubility with water temperature and pressure. The cyclic process involved dissolution by pressurization in the core, followed by depressurization and release through the pressurizer.

Reactor Core Status

The in-core instrumentation which monitors core status indicated high temperatures and some intermittent coolant voiding early in the accident in at least the upper region of the core. The in-core thermocouples monitoring core outlet temperatures and the self-powered, rhodium neutron detectors indicated high temperatures beginning about 135 min into the accident. (One of 52 instrument strings, with a single thermocouple and seven detector segments each, is shown in Fig. 18-3. The operation of the rhodium detectors is described in Chap. 12.) One set of temperature measurements taken between 4 and $5\frac{1}{2}$ h into the accident indicate some temperatures near 1370°C [2500°F] in the center of the core but below 370°C [700°F] around the outside. All temperatures above the core center exceeded 370°C [700°F] (the scale limit of the recording instrumentation) for about 8 h into the accident, with some remaining there for up to 30 h.

The neutron detectors outside the reactor vessel at the core mid-plane indicated possible intermittent voiding of part of the core coolant from about $1\frac{1}{2}$ to $3\frac{1}{2}$ h into the accident. (The location of one such detector is shown in Fig. 18-3. Four are positioned radially around the vessel. Each has two axial segments to provide general flux-shape information.) Variations in neutron moderation and gamma and neutron shielding caused by the voiding led to substantial changes in the detector signals.

The sequence of events, instrumentation responses, core exit thermocouple readings, hydrogen production, and fission-product release to the containment building all suggest extensive clad damage. This may include failure of (i.e., fission-product release from) essentially all cladding tubes with some being heavily deformed. The quenching effect of emergency core-cooling water is likely to have caused substantial fragmentation of the hot cladding and of some of the UO_2 fuel pellets. There is no evidence at present to indicate that any general fuel melting occurred. Localized melting of some of the nonfuel materials, especially the relatively low-melting Ag–In–Cd control rods (Fig. 12-12), can reasonably be inferred. Despite the extensive damage, the core configuration has been coolable by natural circulation since April 27, as indicated by the low coolant outlet temperatures measured since then.

Radiological Consequences

Following the TMI-2 accident, the radiological consequences were evaluated by the Ad Hoc Interagency Dose Assessment Group (NUREG-0558, 1979) comprised of personnel from the Nuclear Regulatory Commission [NRC], Department of Health, Education, and Welfare [HEW], and the Environmental Protection Agency [EPA]. Final results incorporated data from Metropolitan Edison's normal environmental surveillance devices (required for routine operation by federal regulations) and those emplaced during the accident, NRC surveys, Department of Energy [DOE] aerial monitoring, and other sources.

Taking no credit for either shielding or relocation effects, it was estimated that the 2 million persons within an 80-km [50-mi] radius of the plant were exposed to 3300 person·rem or an average individual dose of about 1.5 mrem. The maximum potential off-site dose was estimated to be 83 mrem. (An actual individual known to be on a nearby island during the accident was believed to have received at most 37 mrem.)

Long-term health-effect estimates were made using BEIR-report (BEIR, 1972) correlations. One excess cancer fatality was predicted in the population as a result of the TMI-2 accident. This compares to 325,000 such fatalities otherwise expected in the population of two million. A 100 mrem dose, an upper bound on individual dose, is estimated to result in a 1-in-50,000 chance of cancer, compared to the 1-in-7 value associated with normal incidence.

Extensive food sampling activities were also undertaken. Concentrations of ^{131}I were found to be below Food and Drug Administration [FDA] maximum levels in milk by a factor of 300. No clearly detectable levels (i.e., above those present from fallout from early atmospheric tests of nuclear weapons) of other nuclides, such as ^{137}Cs were found in a wide variety of food samples.

POSTACCIDENT RESPONSE

Even with the establishment of stable cooling in the TMI-2 core, the effects of the accident were far from over. Metropolitan Edison and General Public Utilities faced the task of recovering use of the TMI-2 facility and modifying the similar TMI-1 unit to allow resumption of service. The nuclear industry began taking steps to reduce the likelihood of further accidents in reactors of the same and different designs. In addition, a wide range of governmental investigations, including that of the President's Commission on the Accident at Three Mile Island, were begun.

Recovery Operations

The TMI-1 reactor was completing a refueling operation at the time of the accident in its neighboring unit. Prior to that time it had been in operation for $4\frac{1}{2}$ years. It remained shut down during the time of the accident and for a period of time afterward pending modifications to preclude its having a similar accident. Metropolitan Edison and General Public Utilities had hoped to be ready to restart the unit in summer 1980. Regulatory actions, including new ASLB hearings (described in the next chapter), have delayed this indefinitely.

The eventual restart of TMI-2 is also under consideration (GPU, 1979). Table 18-2 summarizes the *preliminary* work plan and timetable. The decontamination, reentry, and core removal activities are considered necessary whether or not the reactor is to return to commercial operation. If, as indicated by preliminary

TABLE 18-2
Preliminary Schedule for Decontamination and Reactivation of TMI-2

Activity	Date
Begin decay heat removal through steam generator "A"	April 1979
Complete construction of Epicore-II system for treatment of radioactive water	November 1979
Complete general area and sump decontamination in auxiliary building, begin individual area decontamination	January 1980
Begin construction of submerged-demineralizer system for decontamination of reactor building water	April 1980
Vent ^{85}Kr from containment building	July 1980
First personnel entry in containment building	July 1980
Completion of mini-decay-heat system for long-term heat removal	October 1980
Open reactor vessel, remove head, and assess fuel damage	To be determined
Decontaminate reactor coolant system	To be determined
Inspect component systems and prepare for requalification	To be determined
Evaluate feasibility/advisibility of return to commercial operation	To be determined
Restart plant	To be determined

analyses, the primary system (i.e., reactor vessel, steam generators and piping) is undamaged, replacement of the core internals and fuel assemblies could allow for restart. Total costs for the recovery efforts could run in the neighborhood of $1 billion, exclusive of replacement power costs during the extended shutdown period.

The schedules and costs are subject to revision as the recovery operations proceed and the status of the reactor is established with greater certainty. Extraordinary legal, political, or regulatory problems are also recognized for their potential to cause delays and increase costs.

Concurrently with the recovery operations, cooperative efforts among the operating utilities, NRC, and DOE were designed to collect and evaluate the multitude of available data. Since the TMI-2 accident was a unique event at the time of its occurrence, the information obtained in its aftermath is likely to have great long-term value to a wide range of nuclear safety programs.

Nuclear Industry Reactions

From the first day of the accident, Metropolitan Edison and General Public Utilities received technical support from other segments of the nuclear industry. Over 1400 personnel from utilities, reactor vendors, architect–engineer firms, and other segments of the industry were involved at the peak of the accident-control effort.

Lessons Learned

The TMI-2 accident has pointed out a number of general problems related to reactor designs and operations. Among the most important are:

- ineffective reactor-safety information exchange
- difficult person–machine interfaces in the control room
- inadequate reactor operator training.

It is quite possible that the accident at TMI-2 could have been avoided had information been available on an earlier incident at the similarly designed Davis-Besse station. Here the pressurizer relief valve also stuck in an open position with resulting coolant loss. However, since the system was operating at only 9 percent of full power, the resulting progression led to no substantial consequences.

Smooth person–machine interfaces in the reactor control room appeared to be lacking at TMI-2. Necessary information was not always readily available in a convenient form. The large number of visual and audible alarms that occur during the course of even a relatively minor incident were noted to have the potential for unnerving the operators and obscuring important events that occur at later times.

The TMI-2 operators had been trained in compliance with NRC regulations. This seemed to have included unquestioning compliance with technical specifications and with reacting to large design-basis LOCA. The training apparently did not include sufficient flexibility to handle unfamiliar situations.

Response

The electric utilities as a group have already taken several positive steps as a result of the lessons learned from the TMI-2 accident. First, they have established the Nuclear Safety Analysis Center [NSAC] operated under the general charter of the Electric Power Research Institute [EPRI]. The Center is intended to develop strategies for minimizing the possibility of future reactor accidents and to answer generic questions of reactor safety. It will also recommend changes in safety systems and operator training as well as acting as a clearinghouse for technical information.

The utilities have also formed the Institute of Nuclear Power Operations [INPO]. The Institute is primarily intended to establish industry-wide qualifications, training requirements, and testing standards for nuclear plant managers, operators, and engineers. It is also expected to serve audit and testing functions for utility staffs. In general, INPO will train those with responsibility for training programs, rather than individual operating personnel.

An additional major step taken by the utilities is a proposal for a self-sponsored insurance program to provide coverage for replacement power costs in the event of a prolonged postaccident reactor shutdown. This would limit the financial consequences of accidents and provide some stability on an industry-wide basis.

Kemeny Commission

The President's Commission on the Accident at Three Mile Island—called the "Kemeny Commission" for its chairman, John Kemeny—has the potential for being the most influential of the governmental review organizations[†] operating in the

---

[†]Two other studies of note were both initiated by the NRC. The Special Inquiry Group—or Rogovin Commission (Rogovin & Frampton, 1980) after its director, lawyer Mitchell Rogovin—operated on a charter similar to that of the President's Commission that preceded it. The Lessons Learned Task Force (NUREG-0585, 1979) consisted of an in-house group from the Office of Nuclear Reactor Regulation [NRR] that addressed those concerns that were most germane to the NRC itself. Since both studies supported the general findings and recommendations of the Kemeny Commission, they are not addressed further here. (Both are recommended to the interested reader, however; the smooth, journalistic style used by the Rogovin report in its account of the TMI-2 accident is especially interesting.)

aftermath of the TMI-2 accident. The 12-person group included a wide range of professional backgrounds and interests (with only three having any explicit experience in nuclear energy fields).

Findings

The commission's final report (Kemeny, 1979) was released to the public and discussed with the President on October 31, 1979. Its findings and recommendations took to task the reactor operators, the utility, the nuclear industry, and especially the U.S. Nuclear Regulatory Commission. The TMI-2 reactor design itself was, of course, found to have contributed to the accident, but much less than the human factors and attitudes involved. The report also emphasized many of the deficiencies that had previously been identified by the NRC and the industry and for which corrective actions were already in progress.

The radiological consequences of the accident were considered insignificant with perhaps 1 to 10 cancer fatalities expected over the lifetime of the 2 million affected population [as consistent with an upper-bounding correlation applied to the assessment in NUREG-0558 (1979) described previously]. The major health consequence was perceived "to have been on the mental health of the people living in the region," including "immediate short-lived mental distress produced by the accident."

It was noted during testimony before the Congress that a majority of the Kemeny Commission supported a moratorium on the licensing of new nuclear power plants while its recommendations were being implemented. However, the provision was not part of their final report due to a lack of concensus on guidelines for lifting the moratorium once it was put into force. General acceptance of the intent of the commission's recommendations imposed a de facto moratorium during the inherent transition period. This occurred, in fact, as a result of the NRC's decision to delay granting of reactor licenses pending resolution of relevant issues and lessons learned from the TMI experience.

Recommendations

The recommendations section of the Kemeny Commission report is extensive. It calls for sweeping changes in the operation and regulation of nuclear reactors. Its seven parts include the following important observations and recommendations:

1. The Nuclear Regulatory Commission—has "a number of inadequacies" such that the Kemeny Commission "proposes a restructuring" to replace the five NRC commissioners by a single administrator and concentrate the agency's responsibilities more on reactor safety.
2. The Utility and Its Suppliers—"must dramatically change its attitudes toward safety and regulations" and "must also set and police its own standards of excellence to ensure the effective management and safe operation of nuclear power plants."
3. Training of Operating Personnel—need the "establishment of agency-accredited training institutions for operators and immediate supervisors of operators" with utility and NRC responsibility for adequate reactor-specific training.
4. Technical Assessment—improve the person–machine interface, probably using computers, perform risk assessment with emphasis on small-break LOCAs

including multiple failures and special attention to human failure,[†] and research LOCA-scenario areas where data is insufficient.

5. Worker and Public Health and Safety—research low-level and other radiation effects and assure that the utilities conduct advance planning for emergency radiation response.

6. Emergency Planning and Response—"must detail clearly and consistently the actions public officials and utilities should take in the event of off-site radiation doses resulting from release of radioactivity" and should educate the public about nuclear power, radiation, and protective actions.

7. The Public Right to Information—federal and state agencies as well as the utilities "should make adequate preparations for a systematic public information program so that in time of a radiation-related emergency, they can provide timely and accurate information to the news media and the public in a form that is understandable."

A substantial portion of the Kemeny Commission recommendations are related to the need for restructuring the regulatory process in general, and the NRC in particular. These are addressed further in the next chapter.

The recommendations for the nuclear industry are largely related to a feeling by the commission that the existing attitude[‡] that plants are "sufficiently safe" "must be changed to one that says nuclear power is by its very nature potentially dangerous, and . . . one must continually question whether the safeguards already in place are sufficient to prevent major accidents." That the previously described industry responses to the accident indicated constructive changes in attitude was noted in the report by one commissioner.

EXERCISES

Questions

18-1. Explain the role of the following devices in the TMI-2 accident:
   a. feedwater pumps
   b. electromatic pressurizer relief valve
   c. pressurizer block valve
   d. sump pumps

18-2. Describe the effect of the following operator actions on the TMI-2 reactor accident:
   a. early shutoff of the high-pressure injection system
   b. failure to recognize that the electromatic pressurizer relief valve was stuck open
   c. full shutdown of coolant pumps

18-3. Explain the role of the following energy sources in the TMI-2 reactor accident:

[†]That is, conduct *belatedly* some important safety activities identified by both WASH-1400 and the Lewis Report (Chap. 17).

[‡]The word *attitude* is a modified version of the term *mind-set* (from George Orwell's *1984*), which was apparently used repeatedly by a member of the NRC staff in testimony before the Kemeny Commission.

      a. stored energy

      b. decay heat

      c. Zr–steam reaction

      d. hydrogen evolution

18-4. Identify the three earliest industry programs that evolved as a result of the TMI-2 accident.

18-5. List the seven major areas of concern in the recommendations of the Kemeny Commission.

SELECTED BIBLIOGRAPHY†

TMI-2 Accident Description
      Babcock & Wilcox, 1979
      Collier & Davies, 1980
      GPU, 1979
      IEEE, 1979
      Kemeny, 1979
      Lapp, 1979 (App. B)
      Lewis, 1980
      NSAC-1, 1979
      Nucl. News, 1979
      NUREG-0600, 1979
      Rogovin & Frampton, 1980

TMI-2 Accident Analysis and Recovery
      Babcock & Wilcox, 1979
      Collier & Davies, 1980
      GPU, 1979
      IEEE, 1979
      Kemeny, 1979
      Lewis, 1980
      Met. Ed., 1979
      NSAC-1, 1979
      Nucl. Eng. Int., March 1980, Aug. 1980, Nov. 1980
      NUREG-0558, 1979
      NUREG-0585, 1979
      NUREG-0600, 1979
      Olds, 1980*a*
      Rogovin & Frampton, 1980
      Technology Review, 1979

Current Sources
      Nucl. Eng. Int.
      Nucl. Ind.
      Nucl. News
      Nucl. Safety
      Science

†Full citations are contained in the General Bibliography at the back of the book.

19

THE REGULATORY PROCESS

Objectives

After studying this chapter, the reader should be able to:

1. Describe the effects of the Atomic Energy Act of 1954, Energy Reorganization Act of 1974, and the Department of Energy Organization Act of 1977 on federal regulation of nuclear energy.
2. Explain the purpose of 10CFR and identify several of its parts.
3. Identify six federal agencies which have an interest in nuclear power plant siting according to the National Environmental Policy Act of 1969 [NEPA].
4. Explain the roles of each of the following NRC components in the licensing process: commission, regulatory staff, ACRS, ASLB, and ASLAB.
5. Identify the basic features and potential impacts of proposed regulatory reform legislation.

Prerequisite Concepts

Radiation Dose Limits	Chapter 3
Nuclear Power Economics	Chapter 8
Design-Basis Accidents	Chapter 16
Kemeny Commission	Chapter 18

Nuclear energy is subject to governmental control in all nations of the world. Although the processes vary significantly from country to country, most elements of regulatory policy are encompassed by the extensive practices that are currently employed in the United States.

Major constituents of the regulatory process include both the legal bases and the body of the established practices for reactor siting and licensing. Recent events, especially the TMI-2 accident, may lead to substantial modifications of existing regulatory policies.

LEGISLATION AND ITS IMPLEMENTATION

The great destructive power of the nuclear weapons used at the end of World War II provided a strong basis for continued United States' control of nuclear energy. The *Atomic Energy Act of 1946*—the "McMahon Act"—legislated complete federal control of all materials, operations, and facilities related to the nuclear fuel cycle.

The later *Atomic Energy Act of 1954* served as the basis for the commercialization of nuclear power in this country and, through the "Atoms for Peace" program, in other parts of the world. The Act established a civilian *Atomic Energy Commission* [AEC] to oversee the development of regulation of the production and utilization of nuclear energy.

Controversies over nuclear safety and safeguards that developed in the 1960s and early 1970s called into question the dual role of the AEC as both promoter and regulator of nuclear energy. The *Energy Reorganization Act of 1974* served to separate the functions between the *Nuclear Regulatory Commission* [NRC] and the *Energy Research and Development Administration* [ERDA].

The Nuclear Regulatory Commission has responsibility for all of the activities which were formerly conducted by the AEC's Divisions of Licensing and Compliance. The NRC also develops standards and contracts confirmatory research in subject areas of direct importance to its regulatory duties.

The Energy Research and Development Administration acquired the AEC's basic research and development activities, as well as those from several other energy-related federal agencies. The ERDA mandate also included operation of the national laboratories and other government-owned research facilities.

The *Department of Energy Organization Act of 1977* ended the brief existence of ERDA. The newly-formed *Department of Energy* [DOE] encompassed all of the activities which were formerly the province of ERDA, the Federal Energy Administration [FEA], and the Federal Power Commission [FPC]. Also incorporated were energy-related activities of the Interstate Commerce Commission [ICC] and the Departments of Interior, Defense, Commerce, and Housing and Urban Development.

In general, DOE facilities are exempt from the regulations that the NRC applies to commercial power reactors and fuel-cycle facilities. Certain joint-jurisdiction ventures, however, such as those with DOE funding but proposed commercial applications (e.g., certain waste management operations and the Clinch River Breeder Reactor Plant) may be licensed according to standard NRC procedures.

Although the remainder of this chapter focuses explicitly on the NRC, it should be noted that DOE uses many comparable (or, in some cases, identical) regulations and practices. There is, however, no explicit licensing process for the federally controlled facilities. The regulatory processes in other countries also have some similarities with national preferences and constraints providing the differences.

Code of Federal Regulations

Since most legislation is written in general terms, the affected agency promulgates its own rules and regulations that become part of the *Code of Federal Regulations* [CFR]. The code—the master manual for the operations of the executive departments and agencies of the federal government—has the force and effect of law.

The Code of Federal Regulations is divided into 50 "titles" each of which represents a broad subject area. Title 10 was reserved for the Atomic Energy Commission and is currently shared by the NRC (Chapter I) and DOE (Chapters II, III, and X). The NRC chapter is, in turn, divided into parts 0–199, of which about 32 are presently in use. Parts of special interest for commercial reactors and fuel-cycle facilities are identified and described in Table 19-1. Standard nomenclature is, for example, to refer to "title 10, CFR, part 20" as 10CFR20.

TABLE 19-1
Parts of Title 10 of the Code of Federal Regulations [10CFR] Affecting Commercial Reactor and Fuel Cycle Operations (10CFR, 1979)

Part	Name	Summary of scope
0	Conduct of employees	
1	Statement of organization and general information	
2	Rules of practice for domestic licensing proceedings	Identifies NRC's own rules of procedure for: issuing, amending, and revoking licenses; imposing fines; and making rules
20	Standards for protection against radiation	Establishes standards for protection against radiation hazards
50	Licensing of production and utilization facilities	Establishes requirements for obtaining construction permits and operating licenses for fuel-cycle facilities and reactors
51	Licensing and regulatory policy and procedures for environmental protection	Identifies procedures through which the NRC complies with the National Environmental Policy Act of 1969 [NEPA]
55	Operators' licenses	Established procedures and criteria for issuance of licenses to operators of reactors and reprocessing facilities
70	Domestic licensing of special nuclear material	Establishes procedures and criteria for issuance of licenses to receive, own, and use special nuclear material
73	Physical protection of plants and materials	Prescribes requirements for physical protection of special nuclear materials and facilities against acts of industrial sabotage or theft
100	Reactor site criteria	Describes criteria for evaluation of the suitability of proposed reactor sites
170	Fees for facilities and materials licenses	Sets fees charged for licensing services

10CFR20—Standards for protection against radiation—contains the regulations relating to occupational dose limits and maximum permissible nuclide concentrations [MPC] for air and water in and around nuclear facilities. Such limits are established by the NRC in conjunction with the Environmental Protection Agency [EPA]. Since 1971, part 20 has included the provision that exposures and releases should be "as far below specified limits as reasonably achievable" [the ALARA criterion].

10CFR50—Licensing of production and utilization facilities—establishes the requirements for licensing of fuel-cycle facilities and reactors. The two-step process requiring a construction permit and an operating license is considered in some detail for reactors later in this chapter. Appendixes to 10CFR50 spell out specific procedural requirements and provide guidance in areas that include:

- general design criteria for nuclear power plants with statements of the general principles of multiple barrier containment, core protective and control systems, fluid systems, containment structures, and fuel radioactivity control
- quality assurance [Q/A] of materials, components, and systems
- emergency planning
- siting of reprocessing plants and related waste management facilities
- meeting "as low as reasonably achievable" [ALARA] criterion
- acceptance testing for containment leakage
- modeling for emergency core-cooling system [ECCS] evaluation
- standardization of reactor designs
- pre-approval review of reactor sites

10CFR51—Licensing and regulatory policy and procedures for environmental protection—considers implementation of the National Environmental Policy Act [NEPA] of 1969. The impact of NEPA on plant siting and licensing is considered later in this chapter.

10CFR55—Operator licenses—requires that individual licenses be obtained by each person who manipulates the controls of nuclear power reactors, fuel reprocessing plants, and certain other production facilities. Examinations for these licenses include written, oral, and operational components.

10CFR70—Special nuclear material—establishes procedures and criteria for the issuance of licenses for the commercial use of special nuclear materials [SNM]— generally defined as all plutonium and uranium enriched in ^{235}U and/or ^{233}U—outside of reactors. Authorized use, transfer, and record-keeping requirements are spelled out in some detail. (The concepts of SNM and material safeguards are addressed in the next chapter.)

10CFR73—Physical protection of plants and materials—sets requirements for protecting special nuclear materials in fuel-cycle facilities and reactors against acts of industrial sabotage or theft. Similar protection is also required in transportation of such materials. (The next chapter describes representative procedures and systems for implementation of 10CFR73 conditions.)

10CFR100—Reactor site criteria—provides guidelines for the evaluation of the suitability of sites proposed for construction of nuclear power plants. Seismic, geologic, and other considerations are covered in detail in its appendixes. Typical requirements are examined later in this chapter.

Regulatory Guides and Standards

Regulatory guides (formerly designated "safety guides") are issued from time to time by the NRC to describe methods which the staff finds generally acceptable in implementing 10CFR regulations, to discuss techniques used by the staff to evaluate specific problems or nuclear accidents, and/or to provide general guidance to applicants for various NRC licenses. Since the guides are not substitutes for the regulations, they do not require compliance. They are becoming increasingly important from a practical standpoint, however, to the extent that they can shorten (or at least prevent extension of) the overall license-review times.

The NRC has worked with various technical societies and organizations in developing standards that represent a codification of sound industrial practice. They are typically developed by committees whose membership assures a wide range of scientific and industrial experience. Wherever possible, the proven national standards and codes that are suitable for nuclear applications have been incorporated into NRC regulatory guides. Examples include:

- Institute of Electrical and Electronics Engineers [IEEE] Criteria for Nuclear Power Plant Protective Systems
- American Society of Mechanical Engineers [ASME] Boiler and Pressure Vessel Code, Section III
- American Welding Society [AWS] methods and specifications
- standards approved by the American National Standards Institute [ANSI]

National Environmental Policy Act

Although not explicitly directed to nuclear facilities, the *National Environmental Policy Act of 1969* [NEPA] has had a profound effect on the licensing of power reactors. The Act requires all agencies of the federal government to prepare a detailed *environmental impact statement* [EIS] for every "major federal action significantly affecting the quality of the human environment." Since the issuance of a reactor license falls within the guideline, the NRC must prepare an EIS with each application.

Specifically, NEPA requires statements addressed to:

- the environmental impact of the proposed action
- adverse environmental effects which cannot be avoided should the proposal be implemented
- alternatives to the proposed action
- the relationship between local short-term uses of the human environment and the maintenance and enhancement of long-term productivity
- any irreversible and irretrievable commitments of resources which would be involved in the proposed action should it be implemented

The statements must be submitted to all federal agencies "having an interest in" environmental protection. Each agency, in turn, is required by NEPA to prepare a detailed analysis of the project. The following agencies are among the multitude that represent specific environmental viewpoints:

1. Environmental Protection Agency
2. Forest Service and National Park Service

3. Bureau of Mines
4. Bureau of Sport Fisheries and Wildlife
5. Federal Aviation Administration
6. Coast Guard
7. National Oceanic and Atmospheric Administration
8. Air Force
9. Water Quality Office
10. Army Corps of Engineers
11. Geological Survey
12. Department of Housing and Urban Development

The important factors of environmental impact for an electric power plant may be divided into the following categories (from ERDA-69, 1975):

> In evaluating long-range plans, conducting preliminary site reviews, and evaluating the application for certification of bulk power supply facilities, the certifying agency shall give consideration to the following factors where applicable:
>
> a. *Electric Energy Needs* (major emphasis of long-range plan reviews)
>   1. Growth in demand and projection of need.
>   2. Availability and desirability of non-electric alternative sources of energy.
>   3. Availability and desirability of alternative sources of electric power to this facility or to this type of facility.
>   4. Promotional activities of the electric entity which may have given rise to the need for this facility.
>   5. Socially beneficial uses of the output of this facility, including its use to protect or enhance environmental quality.
>   6. Conservation activities which could minimize the need for more power.
>   7. Research activities of the electric entity or new technology available to it which might minimize environmental impact.
> b. *Land Use Impacts* (major emphasis of preliminary site reviews)
>   1. Area of land required and ultimate use.
>   2. Consistency with any State and regional land use plans.
>   3. Consistency with existing and projected area land use.
>   4. Alternative uses of the site.
>   5. Impact on population already in the area; population attracted by construction or operation of the facility itself; impact of availability of power from this facility on growth patterns and population dispersal.
>   6. Geologic suitability of the site or route.
>   7. Seismologic characteristics.
>   8. Construction practices.
>   9. Extent of erosion, scouring, wasting of land—both at site and as a result of fossil fuel demands of the facility.

10. Corridor design and construction precautions for transmission lines.
11. Scenic impacts.
12. Effects on natural systems, wildlife, plant life.
13. Impacts on important historic, architectural, archeological, and cultural areas and features.
14. Extent of recreation opportunities and related compatible uses.
15. Public recreation plan for the project.
16. Public facilities and accommodation.

c. *Water Resources Impacts* (major emphasis during preliminary site reviews and facility certification)

1. Hydrologic studies of adequacy of water supply and impact of facility on stream flow, estuarine and coastal waters, and lakes and reservoirs.
2. Hydrologic studies of impact of facilities on ground water.
3. Cooling system evaluation including consideration of alternatives.
4. Inventory of effluents including physical, chemical, biological, and radiological characteristics.
5. Hydrologic studies of effects of effluents on receiving waters, including mixing characteristics of receiving waters, changed evaporation due to temperature differentials, and effect of discharge on bottom sediments.
6. Relationship to water quality standards.
7. Effects of changes in quantity and quality on water use by others, including both withdrawal and in situ uses; relationship to projected uses; relationship to water rights.
8. Effects on plant and animal life, including algae, macroinvertebrates, and fish population.
9. Effects on unique or otherwise significant ecosystems; e.g., wetlands.
10. Monitoring programs.

d. *Air Quality Impacts* (major emphasis during preliminary site reviews and facility certification)

1. Meterology—wind direction and velocity, ambient temperature ranges, precipitation values, inversion occurrence, other effects on dispersion.
2. Topography—factors affecting dispersion.
3. Standards in effect and projected for emissions, design capability to meet standards.
4. Emissions and controls
   a. Stack design
   b. Particulates
   c. SO_x
   d. NO_x
5. Relationship to present and projected air quality of the area.
6. Monitoring program.

    e. *Solid Wastes Impact* (major emphasis during facility certification)
1. Solid waste inventory.
2. Disposal program.
3. Relationship of disposal practices to environmental quality standards.
4. Capability of disposal sites to accept projected waste loadings.

    f. *Radiation Impacts* (major emphasis during preliminary site review and facility certification)
1. Land use controls over development and population.
2. Wastes and associated disposal program for solid liquid and gaseous wastes—criteria set by AEC and EPA.
3. Analyses and studies of the adequacy of engineering safeguards and operating procedures—determined by AEC.
4. Monitoring—adequacy of devices and sampling techniques.

    g. *Noise Impacts* (major emphasis during facility certification)
1. Construction period levels.
2. Operational levels.
3. Relationship of present and projected noise levels to existing and potential stricter noise standards.
4. Monitoring-adequacy of devices and methods. (pp. 181-183)

For a nuclear plant, water resource and radiation impacts are most likely to receive special attention.

REACTOR SITING

The selection of a site for a nuclear reactor must include analyses of the impact of the reactor on the site as well as the potential impact of the site on the reactor. This latter category includes the site-specific energy sources introduced in Chap. 16 as having the potential to cause or enhance reactor accidents.

    The complex problems of siting any electric power plant are identified in the following statement attributed to G. O. Wessenaur of the Tennessee Valley Authority:

> An ideal site for a nuclear reactor plant is one for which there is no evidence of any seismic activity over the past millenia, is not subject to hurricanes, tornadoes, or floods; is an endless expanse of unpopulated desert with an abundant supply of cold water flowing nowhere and containing no aquatic life. Most important it should be located adjacent to a major population center.

It points out that the desire to minimize transmission losses by close-in siting must be tempered by the need to minimize environmental impacts. For nuclear plants this may include reducing perceived risks by distant-site selection.

    Current regulations as delineated in part 100 of 10CFR (1979) call for consideration of the following factors in determining the acceptability of a proposed nuclear reactor site:

- characteristics of the design and proposed modes of operation for the reactor as they bear on accident risk
- population density and land use characteristics in the vicinity of the site with provisions for an exclusion area, low-population zone, and population-center distance
- physical characteristics of the site, including geology, hydrology, meteorology, and seismology

Where unfavorable physical characteristics exist, the site may still be found to be appropriate providing that adequate compensation is made in the form of modifications of, or additions to, the engineered safety systems (Chap. 17).

Population

The results of analyses of design-basis accidents (Chap. 16) are used extensively in qualifying reactor sites. Specifically, the radiological consequences[†] are referenced to a whole-body dose of 25 rem and a thyroid dose (from radioiodine) of 300 rem.[‡] The former corresponds to the recommended [Table 3-4 and NCRP/39 (1971)], once-in-a-lifetime emergency dose that is judged to have no detectable medical consequences. It is emphasized that the two dose values are intended as standards of comparison and *not* as design targets.

The reactor site is required to have an *exclusion area* which is under the full control of the licensee. It is assumed that prompt and orderly evacuation of personnel within this area can be assured, if necessary. The exclusion area must be sized such that a "fencepost person" anywhere on its outer boundary would not receive more than the 25 rem whole-body *or* 300 rem thyroid doses during the first 2 h of fission product release following a design-basis accident.

Although the exclusion area normally coincides with the designated boundaries of the site itself, a highway or a rail line crosses some sites. The San Onofre site in Southern California, for example, is adjacent to a major freeway. Since it is also surrounded by a military base, it has an exclusion area judged to afford adequate control.

The outer boundary of the exclusion area coincides with the inner boundary of the *low-population zone*. This zone should have a low enough total population and population density that timely evacuation is feasible. While its inner boundary is determined by calculated radiation doses during the *first two hours* of fission-product release, the outer boundary is to be selected such that the "fencepost person" here would not exceed either of the dose limits during the *entire time* of the accident.

The *population center distance* is measured from the reactor to the nearest boundary of a densely populated center containing more than 25,000 residents. This is based on actual population distributions rather than merely on boundaries of

---

[†]Based on NRC prescribed assumptions on reactor operating history, fission-product behavior, safety-system effectiveness, and weather conditions [e.g., as in Regulatory Guides 1.3 and 1.4 (NRC, 1974) for BWR and PWR systems respectively].

[‡]The radionuclide release from the TMI-2 accident (Chap. 18) suggests that the whole-body dose limit would be exceeded long before the thyroid limit is reached. It is possible that the latter may eventually be deleted from the regulations.

incorporated cities or towns. The current guideline is that the population center distance should be at least $1\frac{1}{3}$ times as great as the distance from the reactor to the outer boundary of the low-population zone.

The population-based siting guidelines must be somewhat flexible to accommodate regional variations in site availability. The exclusion-area criteria are adhered to quite strictly. The remaining two are more subject to negotiation.

The least inhabited low-population zone will generally favor one site over otherwise comparable sites. However, other factors like evacuation effectiveness must also be considered.

Geology and Hydrology

The geological structure of a proposed site must be shown to be adequate to provide firm support for the entire reactor system including the containment and all auxiliary buildings. Flood protection requirements are determined through hydrological studies of rainfall, runoff patterns, and tidal wave and/or dam failure potentials.

The siting of the Clinch River Breeder Reactor Plant provides an example of the application of hydrological considerations. The final location was selected to be about 2 m above the "maximum probable water level." This level was determined by first assuming that rain and its runoff produce flooding to a level half as great as the probable maximum for the site. Then, simultaneously, failure of Norris Dam is assumed to occur due to an earthquake and to cause sequential failure of the Melton Hill Dam. In general terms, the Clinch River design calls for continued operation during all site-related occurrences that are expected more frequently than once in 100 years. Safe shutdown is the goal for the more severe events postulated as design-basis accidents.

Meteorology

One aspect of the meteorological evaluation of a site is the general pattern of weather. Since local conditions are an important determinant of reactor accident consequences, frequently unfavorable weather may exclude a site from further consideration. Atmospheric "inversions," for example, are undesirable because they would tend to hold a fission-product plume near to the ground and cause human radiation exposures to be relatively large. By contrast, prevailing off-shore winds would be expected to be an advantage to a coastal reactor site.

Weather conditions which could provide the energy to initiate or enhance the progression of a reactor accident are also of concern. Hurricanes and tornadoes are the most important of these.

Hurricanes are large storms of up to 1000 km in diameter. The circulating winds generally have speeds which range from 100 to 300 km/h. Direct damage results from direct impact of the storm. Tidal waves and flooding are secondary effects which should also be considered. Reactors must be designed to withstand the "probable maximum hurricane" postulated for the specific site.

Tornadoes are small but very intense storms. These storms are generally shaped like groundward-pointing funnels and have rapidly rotating winds. Upper diameters in excess of a kilometer and lower diameters of a few meters have been observed (although not necessarily for the same storm). Tangential speeds for the horizontally rotating winds generally range from about 150 to 500 km/h. Vertical updraft winds have been found to exceed 300 km/h on occasion. Typical translational speeds range from as low as 15 km/h to over 110 km/h.

Severe damage can result from direct wind force, rapid pressure drop, and/or missile impacts resulting from a tornado. If tornadoes are predicted to occur more frequently than once in 4000 years at a proposed site, the reactor must be designed to withstand a reference storm with a 480 km/h [300 mi/h] tangential velocity, a 97 km/h [60 mi/h] translational velocity, and a 0.02 MPa [3 psi] pressure drop occurring in 3 s. In addition to the direct effects of the wind assault and the pressure drop, secondary effects from missiles (e.g., telephone poles or automobiles) are also important considerations.

Seismic Concerns

Of all the site-specific concerns, that of seismic activity is the most problematical. This is because magnitudes, frequencies, and general locations for earthquakes cannot be predicted with any great accuracy. Although the most serious events generally occur along geological fault lines, there are also cases of earthquakes in areas with no known active surface faults. Current seismic monitoring procedures limit the likelihood of major quakes in totally new areas. However, human activities including deep-well waste injection and the use of dams must now be recognized as potential earthquake initiators.

The severity of earthquakes is usually referenced on one of two scales. The *Modified Mercalli Intensity Scale* [M.M.°] records damage done by earthquakes on a scale from 1 (not felt) to 12 (nearly total damage) as indicated by Table 19-2. The Mercalli categories are also referenced to the maximum acceleration in units of standard gravitational force [g]. The *Richter Magnitude Scale* measures energy release on a logarithmic scale. The two scales are compared roughly in Table 19-2. Since the amount of damage for a given seismic energy is dependent on soil characteristics, the nature of the underlying bedrock, and types of building construction, the Mercalli and Richter Scales are not equivalent.

Earthquake risk estimates are derived from knowledge of proximity to known active faults and historic earthquake activity. Figure 19-1 shows a map of seismic risk for the contiguous United States. The four zone designations are related to the Mercalli intensities as shown.

Seismic siting criteria for nuclear reactors are set forth in some detail in an appendix to 10CFR100. The seismic vibratory motion or g-force that could be reasonably expected during the lifetime of the plant becomes the "operating basis earthquake" for which the facility design should assure continued safe operations. The *maximum* potential vibratory motion defines the "safe shutdown earthquake" (formerly referred to as the design-basis earthquake). For the latter, certain systems, structures, and components are to remain functional such that they assure:

1. the capability for initial and continued neutronic shutdown
2. the integrity of the reactor coolant systems
3. the capability to prevent or mitigate accident consequences to within predetermined guidelines

TABLE 19-2

Approximate Relationships between Seismic Intensity, Acceleration, Magnitude, and Energy Release[†]

Modified Mercalli intensity scale	Description of effects[‡]	Maximum acceleration (g)	Richter magnitude	Energy release (ergs)
			M2	
I	Not felt; marginal and long-period effects of large earthquakes evident			10^{14}
II	Felt by persons at rest, on upper floors, or favorably placed		M3	10^{15}
III	Felt indoors; hanging objects swing; vibration like passing of light trucks occurs; duration estimated; might not be recognized as an earthquake	0.003 to 0.007		10^{16}
IV	Hanging objects swing; vibration occurs that is like passing of heavy trucks, or there is a sensation of a jolt like a heavy ball striking the walls; standing motor cars rock; windows, dishes, and doors rattle; glasses clink; crockery clashes; in the upper range of IV, wooden walls and frame creak	0.007 to 0.015	M4	10^{17}
V	Felt outdoors; duration estimated; sleepers waken; liquids become disturbed, some spill; small unstable objects are displaced or upset; doors swing, close, and open, shutters and pictures move; pendulum clocks stop, start, and change rate	0.015 to 0.03		10^{18}
VI	Felt by all; many are frightened and run outdoors; persons walk unsteadily, windows, dishes, glassware break; knickknacks, books, etc., fall off shelves; pictures fall off walls; furniture moves or overturns; weak plaster and masonry D crack; small bells ring (church, school); trees, bushes shake	0.03 to 0.09	M5	10^{19}
VII	Difficult to stand; noticed by drivers of motor cars; hanging objects quiver; furniture breaks; damage occurs to masonry D, including cracks; weak chimneys break at roof line; plaster, loose bricks, stones, tiles, cornices fall; some cracks appear in masonry C; waves appear on ponds, water turbid with mud; small slides and caveins occur along sand or gravel banks; large bells ring	0.07 to 0.22	M6	10^{20}
VIII	Steering of motor cars affected; damage occurs to masonry C, with partial collapse; some damage occurs to masonry B, but none to masonry A; stucco and some masonry walls fall; twisting, fall of chimneys, factory stacks, monuments, towers, and elevated tanks occur; frame houses move on foundations if not bolted down; loose panel walls are thrown out; changes occur in flow or temperature of springs and wells; cracks appear in wet ground and on steep slopes	0.15 to 0.3		10^{21}
IX	General panic; masonry D is destroyed; masonry C is heavily damaged, sometimes with complete collapse; masonry B is seriously damaged; general damage occurs to foundations; frame structures shift off foundations, if not bolted; frames crack; serious damage occurs to reservoirs; underground pipes break; conspicuous cracks appear in ground; sand and mud ejected in alluviated areas; earthquake fountains and sand craters occur	0.3 to 0.7	M7	10^{22}
X	Most masonry and frame structures are destroyed, with their foundations; some well-built wooden structures and bridges are destroyed; serious damage occurs to dams, dikes, and embankments; large landslides occur; water is thrown on banks of canals, rivers, lakes, etc.; sand and mud shift horizontally on beaches and flat land; rails are bent slightly	0.45 to 1.5		10^{23}
XI	Rails are bent greatly; underground pipelines are completely out of service	0.5 to 3	M8	
XII	Damage nearly total; large rock masses are displaced; lines of sight and level are distorted; objects are thrown into air	0.5 to 7		10^{24}
			M9	

[†]Courtesy of Oak Ridge National Laboratory, operated by the Union Carbide Corporation for the U.S. Department of Energy (Lomenick, 1970).

[‡]Masonry A: Good workmanship, mortar, and design; reinforced, especially laterally, and bound together by using steel, concrete, etc.; designed to resist lateral forces; Masonry B: Good workmanship and mortar; reinforced, but not designed in detail to resist lateral forces; Masonry C: Ordinary workmanship and mortar; no extreme weaknesses like failing to tie in at corners, but neither reinforced nor designed against horizontal forces; Masonry D: Weak materials, such as adobe; poor mortar; low standards of workmanship; weak horizontally.

FIGURE 19-1
Seismic risk map for the contiguous United States: Zone 0–no damage; zone 1–minor damage, corresponds to intensities V and VI of the M.M.° Scale; zone 2–moderate damage, corresponds to intensity VII of the M.M.° Scale; zone 3–major damage, corresponds to intensity VII and higher of the M.M.° Scale. (Adapted from "United States Earthquake, 1968," U.S. Department of Commerce, U.S. Coast and Geodetic Survey, 1970.)

Secondary effects such as flooding and water-wave effects produced by tsunamis or seiches—seismic-induced tidal waves and lake motion, respectively—must also be included in the operating and safe-shutdown evaluations.

Seismic design of power reactors depends on detailed site evaluation followed by appropriate selection of components, equipment, and structures. Among the most visible site-specific portions of the seismic design are the amount of reinforcement used in the concrete containments (Figs. 13-6, 17-4, 17-7, 17-8, and 17-10) and the application of seismic-constraint and vibration-limiting fixtures to vessels, heat exchangers, and especially exposed piping (e.g., as for the PWR primary system of Fig. 12-9). Extensive testing or qualification programs for these and safety-system components (e.g., the isolation valve in Fig. 17-3) are used to assure adequate margin for both operating-basis and safe-shutdown purposes.

Alternative Siting Concepts

The orderly expansion of nuclear (and nonnuclear) electric power generation depends on the continuing availability of appropriate plant sites. One goal favors remote siting in low population areas for safety reasons. On the other hand, near-load-center siting is desirable to minimize transmission losses. Reasonable compromises on these competing goals are becoming more difficult as population expansion limits the general availability of land.

It has been suggested that traditional reactor siting practices could be supplemented by use of underground or off-shore locations. Table 19-3 provides a general comparison of these alternatives to the standard above-ground siting employed for current LWR systems.

The off-shore siting concept is based on a floating plant anchored on an artificial island which serves as a breakwater. The large increase in potential reactor sites for coastal communities and the ability to construct such systems at a single location are major advantages. It has the disadvantage of being an unproven concept with new potential operational and safety problems.

The underground siting option is particularly attractive for close-in siting near densely populated areas. A number of countries in Western Europe, for example, are considering this option in some detail. The major advantage of underground siting is related to the additional safety assumed to be afforded by the inherent shielding of the surrounding rock and earth. The potential for new accident scenarios and different fission-product release pathways are disadvantages.

REACTOR LICENSING

Commercial nuclear reactors and fuel-cycle facilities must be licensed by one or more government agencies before they are allowed to operate in the United States and most other countries of the world. U.S. activities in this regard are within the jurisdiction of the Nuclear Regulatory Commission under the terms of 10CFR50.

TABLE 19-3
Comparison of Off-Shore and Underground Siting Alternatives with Current Above-Ground Siting of LWR Systems

Consideration	Off-Shore		Underground	
	Advantages	Disadvantages	Advantages	Disadvantages
Site availability	Additional sites	Coastal geography limitations	May allow more close-in urban siting	May be water-table limited
Construction	Reactor built at one site, towed to final site	Build artificial island	Simplified containment structure	Evacuation technology; water-table seal; entryways
Environmental	Reduced seismic concern	Increased hurricane, tsunami concern	Reduced tornado, falling object, sabotage concern	Increased flood, hydrological concern
Safety systems	Reduce existing systems; heat sink for LOCA	New safety problems	Reduced containment, other	New accidents possible
Population dose	Distance	Prevailing winds	Shielding, filter effect	New access routes possible
Economics	Standardization, mass production	New technology; island construction	Near load center	Construction cost increase of 5–10%

Current features of the licensing process for power reactors are examined in the following pages. Potential changes in the process are considered from the standpoint of long-standing legislative proposals as well as more recent recommendations from the Kemeny Commission.

Nuclear Regulatory Commission

The current structure of the Nuclear Regulatory Commission [NRC] is shown by the organization chart in Fig. 19-2. The keystone is *the commission*[†] consisting of five members, each serving a fixed five-year term. They are appointed by the president and confirmed by the Senate. The president also appoints a chairperson from among the Commissioners. Otherwise, the commission and the NRC staff are independent of the other executive departments and agencies and, thus, responsible directly to the president.

The Executive Director for Operations has responsibility for the day-to-day functions of the NRC. Various offices reporting directly to the commission and others reporting to the executive director further aid the conduct of NRC business.

Regulatory Staff

The *Regulatory Staff* is organized to carry out the legislative and policy mandates of the NRC. It is divided into five offices for:

- standards development
- nuclear regulatory research
- nuclear material safety and safeguards
- nuclear reactor regulation
- inspection and enforcement

The first two activities are quite general in scope while the latter three are specifically related to the licensing process. The offices are subdivided as shown by Fig. 19-2. The Office of Standards Development has responsibility for directing, or otherwise encouraging, the formulation of the standards and codes employed in design and operation of nuclear facilities (as described earlier in this chapter). The Office of Nuclear Regulatory Research [RES] conducts confirmatory research activities deemed to be necessary to the regulation of commercial facilities. (This is in contrast to the role of the DOE in sponsoring more basic or conceptual research and development activities.) Important programs have included the WASH-1400 study as well as continuing support of the LOFT program (Chap. 17) and certain analyses of the TMI-2 accident (Chap. 18).

The Office of Nuclear Material Safety and Safeguards [NMSS] has complete responsibility for the safety and safeguards of all facilities and activities involving the processing, handling, and transportation of nuclear materials licensed under the Atomic Energy Act of 1954, as amended. As of late 1979, NMSS also assumed responsibility for safeguards at licensed reactors. This includes material accountability and/or physical security functions (as described more fully in the next chapter).

[†]The usual convention refers to the entire organization as "the NRC" and its five directors as "the NRC Commissioners," or simply "the commission."

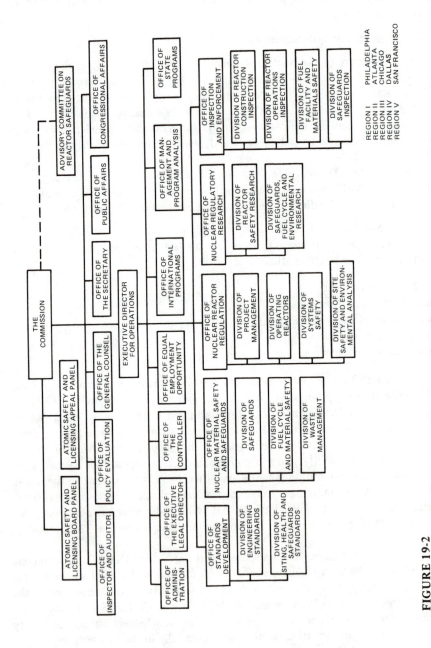

FIGURE 19-2

Organization chart for the U.S. Nuclear Regulatory Commission [NRC].

The Office of Nuclear Reactor Regulation [NRR] is responsible for the safety of all licensed nuclear reactors. It must evaluate the safety and the potential environmental impact of each reactor before a license document is actually issued.

The Office of Inspection and Enforcement [OIE or I&E] is responsible for enforcing licensee compliance with NRC requirements. Separate inspection activities are directed toward reactor construction, reactor operations, fuel facilities and materials safety, and safeguards. Where noncompliance with a regulatory requirement is found, various enforcement actions are taken which may result in civil penalties such as fines or license revokation and criminal penalties, as appropriate.

The I&E personnel who perform the inspections are distributed among five regional offices as noted on Fig. 19-2. Each of these field offices is organized with the same divisions as the I&E headquarters operations.

Review and Hearing Organizations

The NRC organization chart in Fig. 19-2 shows three organizations which report directly to the commission. Each is constituted independently from the usual staff functions.

The *Advisory Committee on Reactor Safeguards* [ACRS] is a review organization consisting of no more than 15 members. It was established by the Congress in 1957 to conduct independent safety reviews for all power reactor license applications. Although the ACRS reports to the commission, it is not actually a part of the NRC.

The *Atomic Safety and Licensing Boards* [ASLB] are empaneled to conduct public hearings and make decisions with respect to granting, suspending, revoking, or amending any NRC-authorized license. Each three-person board generally consists of an attorney and two technically oriented members (with knowledge of reactor safety and environmental impacts, respectively). Separate *Atomic Safety and Licensing Appeal Boards* [ASLAB] are similarly constituted for the purpose of handling appeals of ASLB decisions.

Current Power Reactor Licensing Program

Federal regulatory authority over commercial nuclear energy activities is implemented through the issuance of various NRC licenses. Since power reactors produce electricity and have various environmental impacts, they also require an additional 25–50 licenses granted by federal, state, and local agencies with overlapping jurisdictions. The latter requirements vary substantially on a regional basis.

The licensing process as currently constituted begins with an application for a construction permit. Upon issuance of this permit and then completion of construction, an operating license must be obtained before power generation may begin.

Construction Permit

The major chronological steps for obtaining a construction permit are identified in the top portion of Fig. 19-3. The first step for a utility desiring to build a nuclear reactor is to hold a preliminary site review with regulatory staff personnel.

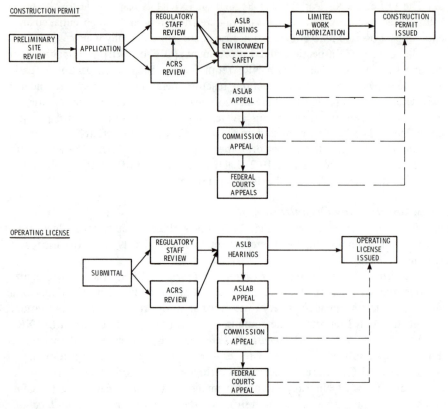

FIGURE 19-3
Flow diagram for reactor licensing by the U.S. Nuclear Regulatory Commission [NRC].

The formal construction permit application is quite extensive.[†] It must include organizational and financial data, antitrust information, an Applicant Environmental Report/Construction Stage [AER], and a Preliminary Safety Analysis Report [PSAR]. When the application materials are judged to be complete, they are entered on the NRC docket, i.e., assigned an identifying number and scheduled for review by the regulatory staff and the ACRS.

The PSAR is a comprehensive document including detailed site analyses, facility description with emphasis on safety features, design-basis accident analyses, personnel and operational procedures, and emergency plans. The site and reactor system must be

[†]The *initial* submittal for the Palo Verde Nuclear Generating Station under construction near Phoenix, Arizona, for example, consisted of 34 volumes (in 7-10-cm-thick binders). The breakdown of the volumes is: application–1; antitrust information–3; environmental report–7; PSAR exclusive of the nuclear steam supply system [NSSS] design–17; and PSAR NSSS sections [Combustion Engineering generic safety analysis report (CESSAR)] –6.

well established by this time, although some changes may be made during construction to the extent that they maintain or enhance safety margins.

The PSAR must also contain technical specifications which are intended to govern the operations. These "tech specs" are divided among:

1. safety limits, limiting safety settings, and limiting control settings
2. limiting conditions for operation
3. surveillance requirements including testing, calibration, and maintenance
4. design features
5. administrative provisions relating to organization and management, procedures, record keeping, review and audit, and reporting

(These ultimately serve as a major basis for judging whether or not a facility is operated in compliance with regulatory requirements.)

The predominate concern of the regulatory staff is assuring that the proposed plant can be operated "without undue risk" to the public. As the initial application is evaluated, the applicant must respond to various questions and occasionally implement design changes. Ultimately, a Safety Evaluation Report [SER] is drafted to present the staff's assessment of the proposed facility.

The regulatory staff also reviews the AER and prepares its own Draft Environmental Statement [DES] for distribution to all concerned federal agencies as required by the National Environmental Policy Act [NEPA]. Responses from the various agencies are then incorporated in a Final Environmental Statement [FES].

The Advisory Committee on Reactor Safeguards [ACRS] conducts a review of safety issues only. A subcommittee prepares a detailed report which then with the SER serves as the basis for evaluation by the full ACRS. After the findings are reported directly to the Commission chairperson, the regulatory staff prepares a Supplement to the Safety Evaluation Report [SSER].

Upon completion of the staff and ACRS reviews, an Atomic Safety and Licensing Board [ASLB] conducts hearings on the construction permit application in the vicinity of the proposed reactor site. Safety and environmental issues are considered in separate sessions as indicated by Fig. 19-3. The application, FES, SER, and SSER are the primary documents in support of the permit.

Opposition to the issuance of the construction permit may be offered by *intervenors* who become party to the ASLB proceedings. They must file a petition which identifies *specific* aspects of the project for intervention and which sets forth *facts* of interest and the bases for contention. Acceptance of the intervention by the ASLB results in a *contested* hearing.

If the hearings were uncontested (a situation which has not occurred in recent history), testimony would consist of all applicant and NRC documentation. The (usual) contested hearings are based on the same documents plus additional testimony from the applicant, regulatory staff, and intervenors. The ASLB issues an initial decision on the basis of "findings of fact" and "conclusions of law." If the environmental hearings return a decision favorable to the applicant, a limited work authorization [LWA] may be issued to allow certain site preparation activities to begin. A successive favorable verdict on safety allows issuance of the construction permit in ten days, pending appeal by the intervenors. Decisions unfavorable to the applicant, of course, preclude issuance of the permit and require reworking all or part of the application (or, perhaps, a decision not to build the plant).

The ASLB decisions may be appealed to the Commission which then orders Atomic Safety and Licensing Appeal Board [ASLAB] hearings. These hearings consider only issues contested to be *errors* of fact or law. The ASLAB may, by law, formulate its decisions on the basis of the proceedings from the ASLB hearings plus written briefs submitted by the applicant, NRC staff, and intervenors. However, in practice, open hearings are normally conducted.

An unfavorable decision of the ASLAB may be appealed directly to the commission for resolution by this five-member group. Additional appeals beyond this point must be made to the federal courts.

Operating License

After receipt of the construction permit, the utility is allowed to begin the actual construction of the facility. It also initiates the progression for obtaining an operating license as summarized on Fig. 19-3.

The applicant's submittal for the operating license consists of three basic parts. The first of these is a version of the application amended to reflect the construction progress. Then the PSAR is updated to produce a Final Safety Analysis Report [FSAR] which describes the final design and operation of the reactor system. An Applicant Environmental Report/Operating License Stage is the third part of the submittal.

According to 10CFR50, the license will be issued only if the NRC has reasonable assurance that the applicant:

- will comply with license regulations including those that assure that the health and safety of the public will not be endangered
- is "technically and financially qualified to engage in the proposed activities"
- will not use the license in a manner that is "inimical to the common defense and security or to the health and safety of the public"
- has satisfied applicable requirements of NEPA

The review by the regulatory staff at this stage is concentrated mainly on new safety and/or environmental information. The reports from the NRC inspectors on construction progress and permit compliance are also crucial to the evaluation of the license application. A new Safety Evaluation Report is issued upon completion of the staff review. The ACRS also conducts a final safety evaluation, the findings of which are again reported directly to the commission chairperson.

The application as amended through all of the review processes becomes the bulk of the final operating license. A cover document issued by the NRC is used to set forth any license conditions that are not already spelled out explicitly in the application. The technical specifications ["tech specs"] —first set forth and justified in the PSAR (as described earlier) and revised in the FSAR—are also incorporated in the license. The conditions and specifications are important parts of the "inspectables" against which license compliance will ultimately be judged by the I&E inspectors.

Public notice is required prior to issuance of an operating license. Although hearings are not required by law, the Commission may order them if sufficient public interest is indicated. The hearings and appeal processes are conducted in much the same manner as they are during the construction permit phase (see Fig. 19-3).

Licensing and Regulation Costs

Based on decisions by the U.S. Supreme Court, fees for regulatory activity should:

- "not be based solely on full cost recovery but should allocate part of the cost to public benefit"
- be assessed to "specific person or company for specific measurable services to each" and be set according to the "value to the recipient" of the services

The former recognizes that the NRC's research and development [R&D] costs are for the general public good and, thus, are not appropriately charged to licensees.

The original fee structures were geared to reactor size as shown for a recent period in Table 19-4. The current schedule is based on fixed fees that reflect the tendency for utilities to order new units at approximately 3880 MW(th) (a de facto power ceiling imposed by the NRC). Distinction is also made between custom reviews and those with varying degrees of standardization. (In the latter case, of course, site-specific safety issues and environmental considerations must still be addressed in full detail.)

Potential Changes

The licensing process as currently implemented in the United States extends 10–12 years. Since each of these years costs up to $120 million (primarily from carrying charges and equipment cost escalations), the NRC and the Congress have been considering a number of long-standing proposals to reduce the licensing time while con-

TABLE 19-4
Maximum Fees Set Forth in 10CFR170 for NRC Regulatory Activities for Nuclear Power Reactors

| | | Maximum fees[†] | |
| | | Effective 3/34/78 | |
Fee for	Prior to 3/24/78	Custom	Standardized[‡]
Application	$125,000	$125,000	$125,000
Construction permit	$250,000 + $170/MW(th)	$944,000	$721,800
Operating license	$250,000 + $185/MW(th)	$1,024,500	$829,100
Routine annual inspection	$65/MW(th) ($20,000 min)	Health, safety & environment Safeguards	$75,700/y $11,800/y

[†]Fees are assessed for actual NRC expenditures for professional staff and support services but will not exceed the maximum listed.

[‡]For a utility referencing a standardized nuclear steam supply system and standardized balance of plant for both the construction permit and the operating license.

tinuing to assure that the public and the environment are adequately protected. Recent decisions by the U.S. Supreme Court may also serve to cut licensing times.

In the aftermath of the TMI-2 accident, the Kemeny Commission has recommended sweeping changes in the regulatory process in general and in the NRC in particular. The extent to which the recommendations are adopted will have a major influence on future licensing requirements and schedules.

NRC Proposals

Over the past several years the Nuclear Regulatory Commission has developed a series of proposals for streamlining the reactor licensing process. These range from procedural modifications that could be implemented administratively to major revisions that would require legislative reform.

Among the more readily implemented proposals would be the use of informal meetings between the regulatory staff and the applicant to assure that the latter has an adequate understanding of such features as:

- quality assurance [Q/A] requirements and inspection procedures
- regulations, standards, and guides
- documentation requirements
- technical issues and staff positions related to applications for construction permits

Additional measures include standard review plans for safety evaluation which would serve to improve quality, uniformity, and predictability of regulatory staff reviews. Standards development could lead to identification of acceptable methods to achieve the performance requirements for nuclear plants. Research programs may be aimed at achieving "better quantification of the degree of conservatism" built into current safety requirements.

More extensive proposals are based on pre-approved sites and standardized plant designs. Sites would be qualified independently from a specific reactor design. Staff evaluation, public hearings, and a partial ASLB decision would be employed to approve one or more sites for use during the following 5-year period.

Standardized designs for nuclear steam supply systems [NSSS] have been developed by the reactor vendors. Standardized balance-of-plant designs have also been developed by architect–engineer firms. Both have been described in generic safety analysis reports (including, for example, the reactor volumes identified in the Selected Bibliography of Chap. 1). Taking credit for such standardization would shorten the licensing process to the extent that redundant safety reviews among the various applications could be eliminated.

The impact of the combined NRC proposals is indicated by the chart in Fig. 19-4. With a pre-approved site, limited work could begin within about 6 months of the application and construction of the standardized plant within $1\frac{1}{2}$-2 years. The total elapsed time of 7 years or less would be more equivalent to that now required in many other nations.

Legislative Proposals

A series of specific legislative proposals aimed at streamlining the licensing process have been under consideration by the U.S. Congress since early 1978. The program's basic features are based on:

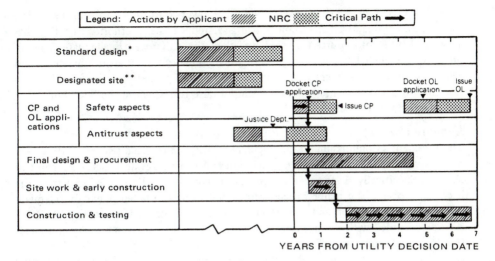

FIGURE 19-4
Proposal for reducing NRC reactor licensing time: *submitted by vendor, architect-engineer, etc.;
**submitted by utility, state, etc. (Reproduced from *Nuclear Engineering International* with
permission of the editor.)

- pre-approval of sites and standardized plant design
- combined construction permit and operating license
- an increased role of the states and assurance of ample opportunity for the public
  to participate and intervene in the licensing process at the earliest possible times
- a 5-year pilot program to fund or reimburse intervenors for expenses incurred
- preclusion of constant relitigation of old issues unless new information becomes
  available

Use of a single permit issued at the time construction begins would allow the project
to proceed to operation pending the results of the on-site inspection program and
assuming no new health and safety issues appeared in the interim. Overall, the proposals were designed to facilitate energy planning, assure the viability of nuclear
power as an energy option, and reduce costs to consumers. It is also intended to
assure that intervenors have adequate input to the licensing process.

The nuclear industry has expressed concern that the provisions for increased
state involvement and intervenor funding could actually extend the licensing time.
On the other hand, critics of nuclear power have expressed concern that the streamlining would severely limit the opportunities for the "public" to be an effective part
of the licensing process.

Supreme Court Actions
During 1976, separate federal court rulings remanded NRC permits issued for the
Midland (Michigan) and Vermont-Yankee nuclear reactors. The action resulted in a
temporary de facto moratorium on reactor licensing.

The U.S. Supreme Court overturned these lower-court rulings in April 1978. The 7-0 decision noted that these courts had overstepped their authority in ruling that duly-constituted procedures (of what was then the AEC) were inadequate. The Supreme Court chided the appeals justices for "Monday morning quarterbacking" and warned them against "engrafting their own notions of proper procedures." The decision also directed that intervenors not simply raise issues before the NRC but rather present substantive aspects of the issues.

Kemeny Commission Recommendations

The President's Commission on the Accident at Three Mile Island (Kemeny, 1979) was particularly critical of the NRC organization and procedures described earlier in this chapter. Specifically, it found the NRC Commissioners to be too isolated from the licensing process. The major offices were found to be too independent for adequate exchange of information and experience. An important conclusion was that "the NRC is so preoccupied with the licensing of plants that it has not given primary consideration to overall safety issues."

The recommendations of the Kemeny Commission with respect to the NRC were divided into general categories of organization and management, mandate, and procedures. Implementation by administrative and legislative means would result in the changes summarized in the following paragraphs.

From an organization standpoint, the five-member commission would be replaced by a single administrator appointed by the president and confirmed by the Senate. This individual would be subject to removal at the pleasure of the president, as is the practice with other executive departments and agencies. Management would be enhanced by locating the administrator and all major staff in a single location (as has *not* been the situation in the past). Steps would be taken to assure adequate communication of research results, operating experience, and inspection and enforcement results.

An oversight committee on nuclear reactor safety of diverse makeup (i.e., like that of the Kemeny Commission itself) would be empaneled to monitor the performance of the agency and the nuclear industry. The ACRS would receive increased staffing to allow its reviews to be more independent of the regulatory staff.

The agency's mandate would emphasize the upgrading of reactor operator and supervisor training, greater attention to a wide range of safety-related matters, and enforcement of higher organizational and management standards for licensees. Increased emphasis would be placed on remote siting for power reactors.

Procedural changes would be devised to require development of a public agenda, resolution of generic safety issues, and reevaluation of the agency's existing rules. Licensing procedures would "foster early and meaningful resolution of safety issues before major commitments in construction can occur." Steps would also be taken to ensure that "safety receives primary emphasis in licensing" and repetitive consideration of some issues would be eliminated.

The agency's inspection and enforcement functions would "receive increased emphasis and improved management," including an improved safety evaluation program and systematic evaluation of operating reactors. Substantial penalties would be assessed against licensees for "failure to report new 'safety-related' information" and for rule violations. Improved inspection and review procedure would also be implemented.

The ultimate impact of the Kemeny Commission and other recommendations[†] will depend on the extent to which the president and members of Congress are effective in promoting their implementation. Evaluations of the Kemeny Commission report by an intergovernmental panel appointed by the president (i.e., to "review his review"), by the NRC itself, and probably by a multitude of other organizations may also provide important inputs. Regardless of specific conclusions, substantial changes in regulatory policy became a virtual certainty in the aftermath of the TMI-2 accident.

EXERCISES

Questions

19-1. Explain the effect of each of the following pieces of legislation on federal regulation of nuclear energy:
  a. Atomic Energy Act of 1954
  b. Energy Reorganization Act of 1974
  c. Department of Energy Organization Act of 1977

19-2. Explain the purpose of 10CFR. Identify the scopes of 10CFR20 and 10CFR50.

19-3. Consider the list of federal agencies with interest in reactor siting under the terms of the National Environmental Policy Act [NEPA]:
  a. Describe the overall impact of NEPA on reactors.
  b. From the list, postulate a reason for interest in a reactor site for the six whose numbers are closest to that of your birth month.

19-4. Explain the roles of each of the following NRC components in the reactor licensing process:
  a. Commission
  b. Regulatory Staff
  c. ASLB and ASLAB
  d. ACRS

19-5. List the basic features of the regulatory reform package and identify those features which may shorten and lengthen, respectively, the licensing process. Explain the reasons for your answers.

19-6. Describe the recommendations from the Kemeny Commission related to the NRC's:
  a. Commission
  b. ACRS
  c. Reactor siting policies
  d. Licensing process
  e. Inspection and enforcement function
  What role would a proposed oversight committee on reactor safety play?

[†]Including those from the Lessons Learned Task Force (NUREG-0585, 1979), the Special Inquiry Group (Rogovin & Frampton, 1980), and a recent study by the Advisory Committee on Reactor Safeguards (ACRS, 1980).

Numerical Problems

19-7. A typical containment building consists of a 20-m-diameter cylinder extending 30 m above grade and having a hemispherical cap. Assuming atmospheric pressure inside, calculate the force associated with the maximum differential pressure of a design-basis tornado.

19-8. Using the old and new criteria, calculate the total-license and annual-inspection fees for a 1200-MW(e) reactor which has a thermal efficiency of 0.32. What fraction of reactor plant cost (Chap. 8) would the license fees be for each case?

SELECTED BIBLIOGRAPHY[†]

Legislation and Regulations
    10CFR
    WASH-1250, 1973 (Chap. 3)

Environmental Impact
    Algermissen, 1969
    Eicholz, 1976
    Eisenbud, 1973
    ERDA-69, 1975
    Lish, 1972 (Chap. 6, App.)
    Lomenick, 1970
    Sagan, 1974

Alternative Reactor Sites
    Anderson, 1971
    Cramer, 1973
    Crowley, 1974
    Perla, 1973
    Yadigoraglo & Anderson, 1974

Regulatory Process
    ACRS, 1980
    Hendrie, 1977
    IEEE, 1979c
    Kemeny, 1979
    LeDoux & Rehfuss, 1978
    Levine, 1978
    Negin, 1979
    Nucl. Eng. Int., Nov. 1977, Oct. 1979
    Nucl. News, May 1978
    NUREG-0585, 1979
    O'Donnell, 1979
    10CFR
    Ward, 1978
    WASH-1174-74, 1974

Current Sources
    Nucl. Eng. Int.
    Nucl. Ind.
    Nucl. News
    Nucl. Safety
    Science

[†]Full citations are contained in the General Bibliography at the back of the book.

Other Sources with Appropriate Sections or Chapters
  Glasstone & Jordan, 1980
  Lamarsh, 1975
  Lewis, 1977
  Nero, 1979
  Sesonske, 1973
  WASH-1250, 1973

20

NUCLEAR SAFEGUARDS

Objectives

After studying this chapter, the reader should be able to:

1. Identify the special nuclear materials and the bases for assigning strategic significance to each.
2. Describe the roles of physical protection, material control, and material accounting in an integrated facility safeguards system.
3. Explain the protection-in-depth concept and each of the five subsystems generally employed for physical protection.
4. Describe several desirable characteristics of a conceptual transportation system for SNM of high strategic significance.
5. Define proliferation.
6. Differentiate between the goals of domestic and international safeguards, respectively.
7. Describe the roles of inspectors, inventory verification, containment, and surveillance in IAEA safeguards.
8. Explain the potential safeguards advantages of several alternative fuel cycle features.

Prerequisite Concepts

In recent times the term *nuclear safeguards*[†] has become synonymous with those measures designed to deter, prevent, delay, detect, and report actions aimed at diversion or theft of nuclear materials and sabotage of nuclear facilities. Such measures are applied in fundamentally different manners for threats classified as domestic and international, respectively.

Domestic safeguards are devised to counter potential actions by subnational or terrorist groups. The major focus is prevention of theft of materials and sabotage of facilities by groups using external force and/or insider assistance.

International safeguards are directed toward prevention of proliferation, i.e., the spread of nuclear weapons. In this case the concern is that national governments may divert nuclear materials from their own commercial facilities for other than peaceful uses.

A common denominator for both domestic and international safeguards is the concern that materials from commercial nuclear facilities could be employed to build and detonate an explosive device. The device could range from crude to sophisticated depending on the capabilities of the terrorist or national group, respectively. Even a credible threat of a nuclear explosion could be expected to have serious consequences.

Safeguards measures may be designed and implemented to match the separate domestic and international concerns. It has also been proposed that certain fuel-cycle alternatives could provide positive deterrence to diversion of nuclear materials.

SPECIAL NUCLEAR MATERIALS

Any of the fissile nuclides in the nuclear fuel cycle can sustain a neutron chain reaction under appropriate conditions. In recognition that such a chain reaction could be the basis for a nuclear explosion, any material artificially enriched in ^{235}U, ^{233}U, or plutonium is classified as *special nuclear material* [SNM].[‡] Reference SNM quantities for criticality are shown in Table 20-1. Since the design of a nuclear-fission explosive is based on achieving supercriticality for a "long enough" period of time that a "significant" amount of energy is released before feedback effects (Chap. 5) terminate the chain reaction, solid (metal and oxide) forms are of most interest. (Solutions can, of course, sustain chain reactions, but would not be expected to produce a sizeable explosive energy yield.) Comparison of a given SNM quantity to the reflected critical masses in Table 20-1 provides a rough basis for assessing its *strategic significance*, i.e.,

[†]As noted in the introduction to Chap. 17, an earlier use of the term *safeguards* is applied to the engineered safety systems of reactors.

[‡]Uranium and thorium, which fall outside of the definition of SNM, are classified as *source materials* (since enrichment of uranium and irradiation of uranium and thorium are *sources* of the special nuclear materials).

TABLE 20-1
Reflected Critical Masses, Commercial Fuel Cycle Locations, and Other Features
of the Special Nuclear Materials [SNM]

| Material | Approximate reflected critical mass (kg) | | Potential commercial fuel-cycle locations | Other features related to nuclear safeguards |
	Metal	Oxide		
Enriched ^{235}U				
93 wt %	17	20	HTGR Fuel	
20 wt %	250	375		
10 wt %	1000	1500		
< 5 wt %	–	–	LWR Fuel cycle	Requires further enrichment for use in nuclear explosive
^{233}U	~ 8		HTGR, Thorium-breeder cycles	Contains ^{232}U
Plutonium				
⩾ 95 wt % ^{239}Pu	4	8	LMFBR Blanket	Plutonium toxicity
"Commercial"	8	10	LWR, LMFBR Cycles	Plutonium toxicity; high ^{240}Pu, ^{242}Pu content

its direct potential for use in a nuclear explosive. Both isotopic composition and chemical form may be factored into the evaluation. As a practical matter, the NRC has defined a *formula quantity* of SNM as that equal to or exceeding 5 kg as computed by the formula:

$$\{\text{Effective mass}\} = \{\text{mass contained } ^{235}\text{U}\}$$
$$+ 2.5 \times \{\text{mass } ^{233}\text{U} + \text{mass plutonium}\}$$

As would be expected, the critical mass for ^{235}U increases sharply as it is diluted with nonfissile ^{238}U (Table 20-1). Criticality is not even possible for enrichments below 5 wt % ^{235}U for the (unmoderated) metal or oxide forms. The commercial nuclear fuel cycle has highly enriched uranium [93 wt % ^{235}U] present only if the HTGR concept (Chap. 13) is employed. Otherwise, the low-enriched uranium would require isotope-separation technology before it would be of use in a nuclear explosive.

Uranium-233 occurs only in thorium fuel cycles like those for the HTGR and the thermal breeder reactors (Chap. 14). The ^{233}U content of an operating reactor or spent fuel in storage is accessible only after reprocessing. Even in its separated form, the presence of ^{232}U with the ^{233}U poses a potential radiation hazard (e.g., Fig. 6-5) that might necessitate remote handling.

All plutonium that exists in commercial nuclear fuel cycles is in the form of isotopic mixtures (e.g., as inferred from Fig. 6-2). The so-called "weapons grade" material [⩾ 95 wt % ^{239}Pu] is present in ^{238}U-bearing fuel irradiated to low burnup. This is available outside of a reactor only following very early discharge of fuel assemblies or, perhaps, in certain LMFBR blanket elements.

Plutonium which is part of usual LWR and LMFBR fuel discharge batches is considered "commercial grade" because of relatively large fractions of the nonfissile ^{240}Pu and ^{242}Pu. These isotopes dilute the effectiveness of the fissile constituents, as may be inferred from the increase in critical mass. Spontaneous fission in the ^{240}Pu also provides a source of neutrons that is undesirable in nuclear explosives and is a potential radiation hazard. Access to any plutonium in spent fuel requires use of reprocessing technology. Even with separated material, the toxicity of plutonium leads to requirements for glovebox operations.

DOMESTIC SAFEGUARDS

Domestic safeguards are designed to nullify the potential threat from subnational or terrorist groups that might seek political, financial, or other rewards from successful adversary actions at nuclear facilities. Physical protection, material control, and accounting measures are integrated to meet safeguards requirements.

Threats

The possible terrorist acts that are of primary concern to the nuclear industry and the NRC may be classified as:

- theft or diversion of nuclear materials
- sabotage of facilities
- hoaxes related to alleged theft or diversion of materials and to threatened sabotage of facilities

Each may take a number of forms and lead to requirements for different safeguards measures.

Theft or diversion of separated SNM would be of most concern if quantities were sufficient for use in a nuclear explosive. The actual design and construction of such a device would require substantial technological expertise as well as relatively sophisticated equipment and facilities. Because of the nature of the SNM itself, highly enriched uranium might be favored as a theft or diversion target as compared to the more hazardous ^{233}U and plutonium. The task of building and detonating any nuclear explosive would exceed or severely strain the capabilities of most terrorist groups.[†] A verified theft or diversion of a formula quantity of SNM would certainly place a terrorist group in a strong bargaining position whether or not construction of a crude nuclear weapon were ever attempted.

Successful theft or diversion of plutonium or spent fuel in any quantities could also lead to the possibility of its dispersal. A dispersal-type weapon might employ conventional explosives to divide the nuclear material into small pieces for distribution over a presumably wide area. Related scenarios might include contamination of water supplies or building ventilation systems. As for the case of a nuclear explosive, the bona fide threat of dispersal could be as important as its actual deployment.

Potential sabotage of facilities is a major concern to the extent that radioactivity could be released to the general environment and pose a hazard to populations. At

---

[†]The article entitled "The Homemade Bomb Syndrome" (Meyer, 1977) delineates the major concerns in some detail.

reactors, for example, the fission-product content of the reactor core, spent-fuel pool, and waste handling systems would be the most likely targets. Core sabotage would have the greatest consequences if a LOCA were induced and important safety systems were disabled (Chaps. 16-17). The storage pool and the radwaste systems have lower radionuclide contents but are also somewhat more accessible than the reactor core.

Hoaxes involving alleged theft or diversion of nuclear materials and the threatened sabotage of facilities form the final category of potential safeguards concerns. The extent to which related claims could or could not be substantiated would determine their ultimate effectiveness from the standpoint of the terrorist.

System Elements

An effective safeguards system must be able to:

- deter potential adversary actions through public awareness of the general capability of safeguards
- detect unauthorized activities and material balance discrepancies
- delay unauthorized activities until appropriate response can be made
- respond to unauthorized activities and discrepancies in an adequate and timely manner

Ideally the system should include physical protection, material control, and material accounting functions that are fully integrated into a single system.

The physical protection function consists of detection, delay, and response capabilities that, in combination with entry-control and zone-operations-control elements, are used to:

- exclude all unauthorized personnel and contraband
- allow only essential personnel to enter sensitive areas
- monitor all significant activities and prevent those that are unauthorized
- prevent unauthorized removal of SNM

The material control function is predicated on real-time material surveillance and accounting at operational interfaces between people, vital equipment, and the SNM. The major challenge is to be able to monitor and enable authorized plant activities while delaying those unauthorized activities that could result in theft or sabotage.

The material measurement and accounting function provides verification of quantities and locations of SNM within the facility. Measurements include bulk and sample analytic chemistry, nondestructive assay techniques, and determination of weights, volumes, dimensions, and locations. Effective material accountability requires both:

- "single-theft" material detection based on material balance calculations (on as near a real-time basis as possible)
- "long-term" detection of continuing diversions of amounts too small to be seen in a single material balance

The latter is based on analysis of trends in successive balances.

The protection, control, and accounting functions augment each other in providing facility safeguards. For example, weakness in ability to detect long-term diversion may suggest the need for an increase in intensity of physical protection measures (e.g., personnel searches). Following an emergency evacuation (produced by a real, spurious, or otherwise-induced signal) which circumvents one or more physical protection barriers, real-time material control and accounting can assure that diversion of a large quantity of SNM has not occurred.

Safeguards System Design

The design of an effective system of domestic safeguards begins with specification of the threat against which the nuclear material and the facility is to be protected. For commercial operations in the United States, NRC regulations provide performance objectives that address this threat. Elsewhere in the world local conditions and situations are evaluated to make a similar determination.

The elements of safeguards are applied to fixed sites as well as to transportation activities. The concept of protection in depth is employed to minimize the impact of individual component failures. System concepts may be evaluated in detail by employing fault tree and event tree logic systems and adversary-simulation models.

The basic safeguards regulations of the U.S. Nuclear Regulatory Commission are contained in 10CFR70 and 10CFR73 (e.g., as noted in Table 19-1). Part 70 addresses the licensing of special nuclear materials. It sets forth basic requirements for material control and accounting, including practices for material balance, inventory, record-keeping, and reporting.

The regulations in 10CFR73—Physical Protection of Plants and Materials—were first directed toward protection against theft, diversion, and sabotage at certain fuel-cycle facilities and later extended to include protection of power reactors. Effective in March 1980, these rules were revised to include *design-basis threat* specifications for radiological sabotage and for theft or diversion of formula quantities of SNM, independent of the type of facilities or activities to be protected. The design-basis threat for radiological sabotage is identified in 10CFR73.1 (in a manner equivalent to the design-basis accidents described in Chaps. 16 and 19). Specifically, the external sabotage threat includes a determined, violent assault by several well-trained persons with inside assistance, hand-held automatic weapons, incapacitating agents, and explosives. The internal sabotage threat involves an insider in *any position* (including station manager, security chief, etc.).

The design-basis threat for theft of formula quantities of SNM is similar to that for sabotage except the external assault group is assumed to have the ability to operate as two or more teams. The insider threat includes a conspiracy among insiders. The objective of physical protection against both design-basis threats is to assure that licensed activities involving SNM are not inimical to the nation's common defense and security and do not constitute an unreasonable risk to public health and safety.

The original 10CFR73 regulations were highly prescriptive in nature; i.e., licensees were told what their safeguards systems should consist of rather than what they should accomplish. The newer requirements emphasize performance against specified threats applicable to fixed sites, reactors, and material in transit. These regulations are intended to allow flexibility in the licensee's approach to physical security.

The specific characteristics of the SNM in terms of type and quantity determine the nature of NRC-imposed physical protection requirements. Nuclear materials are assigned strategic significance as shown in Table 20-2. The classifications are based

TABLE 20-2
Categorization of Special Nuclear Material

Material	Form	Strategic significance		
		High	Moderate	Low
Plutonium	Uniradiated	2 kg or more	500 g to 2 kg	15 g to 500 g
Uranium-233	Uniradiated	2 kg or more	500 g to 2 kg	15 g to 500 g
Uranium-235 (contained isotopic mass)	Uniradiated			
	Enriched to 20% or more	5 kg or more	1 kg to 5 kg	15 g to 1 kg
	Enriched to 10% or more but less than 20%	–	10 kg or more	1 kg to 10 kg
	Enriched above natural but less than 10%	–	–	10 kg or more
Combinations of ^{235}U, ^{233}U, and Pu	^{235}U Portion enriched to 20% or more	5 kg or more computed from the formula: (mass contained ^{235}U) + 2.5 (mass ^{233}U + mass Pu)	1 kg to 5 kg computed from the formula: (mass contained ^{235}U) + 2.0 (mass ^{233}U + mass Pu)	15 g to 1 kg computed from the formula: (mass contained ^{235}U) + 2.0 (mass ^{233}U + mass Pu)

roughly on the concerns for each material type as addressed earlier in this chapter. Formula quantities of SNM (plutonium, ^{233}U, and ^{235}U enriched to 20 *percent or more*), for example, are assigned high strategic significance and must be protected against the design-basis threats described above. SNM of moderate and low strategic significance is subject to less stringent requirements (as spelled out in 10CFR73.67).

Appendixes to 10CFR73 cover qualifications of security personnel, requirements for safeguards contingency plans, and training programs related to transportation of spent fuel. Various regulatory guides provide assistance on selection and use of specific components and personnel in physical security systems.

Fixed Sites

Several fundamental features of safeguards at fixed facilities are shown conceptually in Fig. 20-1. Protection against malevolent insiders is provided by controlling access and detecting unauthorized activities. Persons determined to be involved in such activities would be detained. Overt attack must be detected and analyzed in the control center so that a response force can act to neutralize the theft or sabotage attempt. Both passive and active delays may be employed to provide adequate time for assessment and response.

Protection in Depth

Facility safeguards are based on a *protection-in-depth* concept (with similar underlying principles to multiple-barrier containment of fission products in reactors). Reliance is placed on multiple components with overlapping functions such that one or several failures due to environmental conditions or adversary actions will not disable the entire system.

Physical protection systems generally provide the following functions:

- entry control
- deterrence
- detection

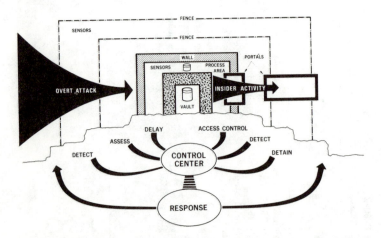

FIGURE 20-1

Conceptual features of a safeguards system at a fixed facility. (Courtesy of Sandia National Laboratories.)

TABLE 20-3
Physical Security Subsystem Functions

Subsystems	Functions	
	Primary	Secondary
Entry control systems	Access control	Delay
	Detection	Deterrence
	Communication	
Intrusion detection systems	Detection	Deterrence
	Communication	
Communication systems	Communication	
Barriers	Delay	Deterrence
Protective force	Response	
	Neutralization	
	Access control	
	Detection	
	Communication	
	Deterrence	

- delay
- communications
- response

In turn, these functions may be accomplished by the subsystems identified in Table 20-3.

Entry control systems are designed to permit entry of only authorized personnel, prevent entry of contraband (mainly guns and explosives), and assist the security force in its assessment function. Authorization for access is usually accomplished with picture badges that may be augmented by use of computer-based information on physical characteristics (ranging from simple parameters like weight or a memorized identification number to sophisticated input like hand geometry, fingerprints, or voice) and specific-area access authorizations. The latter is employed at some reactors, for example, to exclude an individual "insider" from having access to more than one of the redundant safety-system components (all of which would need to be disabled for a LOCA-based sabotage attempt to be successful).

Entry control systems should also include metal and explosive detectors to identify contraband. Detectors for SNM must be capable of surveying personnel, packages, and vehicles in order to preclude insider theft and diversion. Secondarily, the entry control systems introduce delay into adversary actions. They may also serve to deter such actions by convincing would-be adversaries that their probability of detection is high.

Intrusion detection systems consist of a variety and multiplicity of devices to identify violation of boundaries and motion or proximity of individuals. Microwaves, infrared beams, electric fields, seismic vibrations, and televised pictures can all serve to identify intrusion threats. Since each system has different vulnerabilities to environmental nuisance alarms (e.g., weather and wildlife) and adversary defeat, use of several types in overlapping configurations enhances detection probabilities. Alarm displays and closed-circuit television [CCTV] pictures communicate the sensor signals to the protective force and to computer-based assessment systems. The mere appearance and perceived capabilities of the intrusion-detection systems may also serve to deter would-be adversaries.

Communications between the entry control and intrusion detection functions and the security force may be accomplished by a combination of land lines and radio. Provisions for display, storage, and retrieval of transmitted information in one or more central control rooms are important to threat assessment and response. Communications among the guards by radio and telephone are also necessary features.

Barriers in and of themselves cannot prevent the success of a determined adversary. However, they do delay such actions and provide more time for response from the rest of the system. Conventional barriers like fences, vaults, locks, and walls may be designed to delay personnel and/or vehicles. Surveillance equipment including television cameras and lighting may also serve as a barrier to the extent that attempted avoidance slows adversary progress. It is also possible to design a facility with barriers that can be activated following detection of intrusion. Examples of the latter include closing doors or bars, obscuring smokes, and sticky foams. Perception of real or imagined barriers may serve as another deterrent to adversary action.

The protective force consists of the on-site protective force and off-site law enforcement personnel who can be called in the event of an adversary action. It plays a primary role in access control, intrusion detection, and communications, as noted above. The force has principle responsibilities for response to the attack by tracking, intercepting, interrupting, delaying, and eventually neutralizing the adversaries. The skills and training of personnel must be commensurate with the perceived threat and the other system components. The abilities to assess adversary actions and to devise and implement contingency plans are crucial attributes for the protective force as a whole.

Security may be further enhanced by proper design and use of process equipment, storage containers, and facilities to limit the quantities of SNM that are available at any one time. A conceptual design for a plutonium storage facility shown in Fig. 20-2 illustrates one way of implementing the principle. Each storage module consists of a carousel housed in a vault-type cubicle. The containers are located in such a manner that only one may be removed for a given position of the cylinder. Continuous monitoring of material in storage is coordinated with remote access verification through the material control and accountability center. The system is designed to provide a safeguards mechanism for both insider and outsider threats.

Evaluations

Important interplays among the safeguards functions and subsystems are summarized on the "decision diagram" in Fig. 20-3, which serves as a starting point for effectiveness modeling. Beginning with a contemplated attack, progressions are identified based on success or failure of various constituents. Even if the attack on the facility is successful, recovery-force or use-denial actions may limit achievement of the adversary's goal. The recovery force, trained to search for and locate stolen SNM, may accomplish retrieval before an explosive or dispersal device is constructed and/or employed. Use denial consists of measures like isotopic dilution (e.g., of ^{235}U with

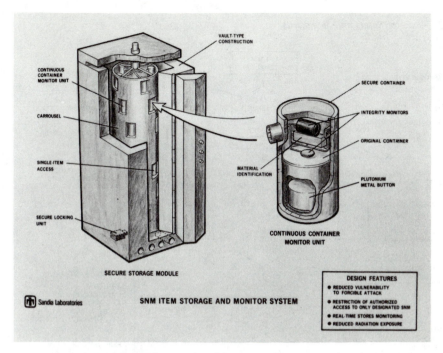

(a)

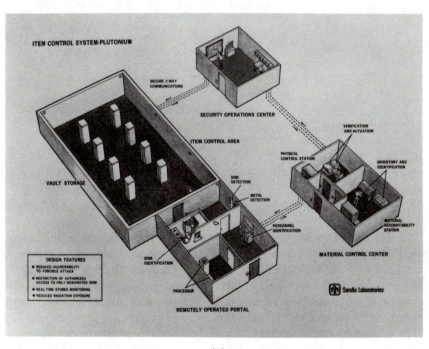

(b)

FIGURE 20-2

Conceptual design of a plutonium storage facility: (*a*) SNM item storage and monitor modules; and (*b*) system components. (Courtesy of Sandia National Laboratories.)

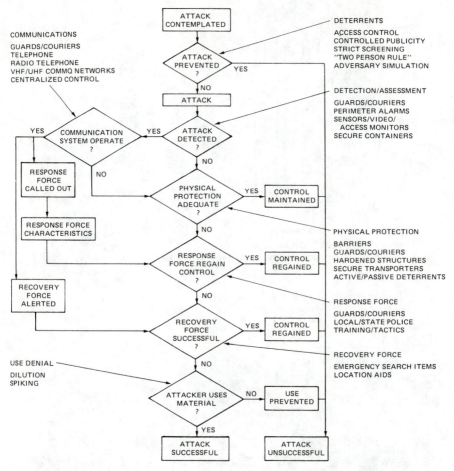

FIGURE 20-3
Decision diagram for modeling a physical security system. (Courtesy of Sandia National Laboratories.)

^{238}U) or spiking with highly radioactive constituents to limit usefulness to terrorist groups.

Risk assessment for safeguards systems proceeds along the same lines described in Chap. 17 for the Reactor Safety Study (WASH-1400, 1975). The major components are shown in Fig. 20-4. Although the probability of an attempt cannot be determined, the other variables may be evaluated. Fault tree and event tree methodologies have proven valuable for identifying vulnerabilities and potential attack sequences, while tests have provided a data base for risk evaluation.

The principles of one computer model for adversary simulation are shown by Fig. 20-5. Site characteristics are examined so that sequencing and number of barriers, sensors, on-site guards, and off-site guards can be diagrammed as shown. A computer can then be programmed to identify critical attack paths, i.e., those with

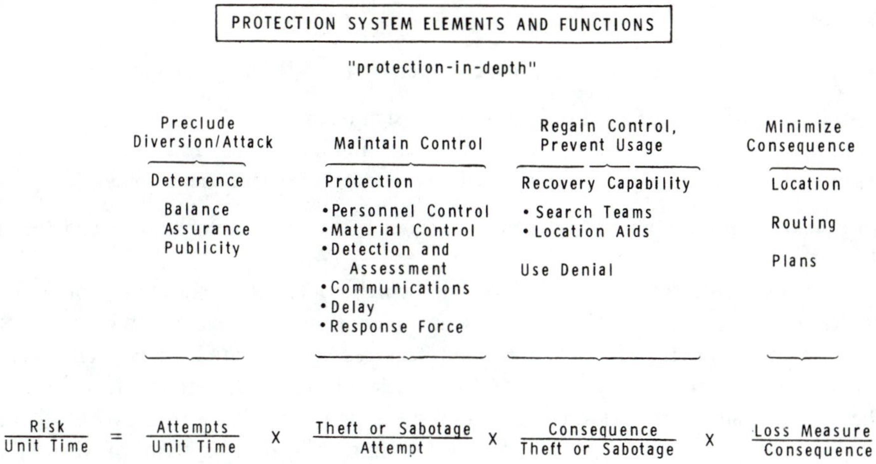

```
┌─────────────────────────────────────────────────┐
│  PROTECTION SYSTEM ELEMENTS AND FUNCTIONS         │
└─────────────────────────────────────────────────┘
```

"protection-in-depth"

Preclude Diversion/Attack	Maintain Control	Regain Control, Prevent Usage	Minimize Consequence
Deterrence	Protection	Recovery Capability	Location
Balance Assurance Publicity	•Personnel Control •Material Control •Detection and Assessment •Communications •Delay •Response Force	• Search Teams • Location Aids Use Denial	Routing Plans

$$\frac{\text{Risk}}{\text{Unit Time}} = \frac{\text{Attempts}}{\text{Unit Time}} \times \frac{\text{Theft or Sabotage}}{\text{Attempt}} \times \frac{\text{Consequence}}{\text{Theft or Sabotage}} \times \frac{\text{Loss Measure}}{\text{Consequence}}$$

FIGURE 20-4
Elements of risk assessment for physical security. (Courtesy of Sandia National Laboratories.)

the shortest potential delay times after detection or with the fewest number of detection elements. Further examination of such paths can then be used to identify additional protective measures which may be appropriate.

Extensive safeguards testing programs have been carried out by Sandia National Laboratories to obtain data on reliability and effectiveness of entry control and intrusion detection equipment. Mock attacks by well-trained military personnel and demolition teams have generated valuable data on barrier penetration times and have provided an overall better understanding of physical security requirements. Another useful procedure, called "black hatting," uses technically trained staff members to think like adversaries and devise ways to defeat various safeguards system concepts.

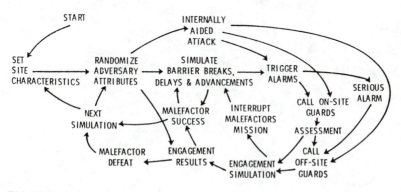

FIGURE 20-5
Schematic representations of computer modeling procedures for physical security systems: (*a*) barrier and guard force conceptual arrangement; and (*b*) logic flow pattern. (Courtesy of Sandia National Laboratories.)

Transportation

Transportation safeguards regulations developed by the NRC are aimed at:

- prevention of theft of SNM of high strategic significance
- prompt detection of theft of SNM of moderate and low strategic significance
- protection of public health and safety from sabotage of spent fuel shipments

Requirements for shipments are then predicated on the specific quantities and forms of SNM.

A safe-secure transport [SST] vehicle, which could serve as a model for future application in the commercial sector, is under continuing development by the U.S. Department of Energy. Several important features are indicated on Fig. 20-6. The cab is armored and employs bullet-resistant glass to protect the drivers from attack. The special wall construction of the trailer is designed to resist attack by both hand tools and explosives. A command-activated immobilization system provides the capability for locking all wheels to prevent easy removal of the vehicle following an incident. Other features include dead-bolt latches on the trailer door (to prevent removal by hinge destruction) and a deterrent system which could be activated during an attack. Since 1972, DOE has transported government-owned SNM by means of a nationwide SST system employing trucks similar to that of Fig. 20-6, couriers in armored escort vehicles (which look like standard recreational vehicles), and a dedicated high-frequency communication system for monitoring the status of all shipments.

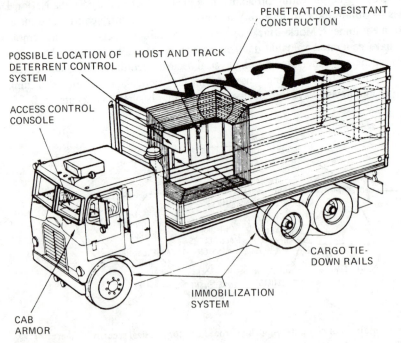

FIGURE 20-6
Basic features of a conceptual safe-secure transport [SST] vehicle for special nuclear materials. (Courtesy of Sandia National Laboratories.)

INTERNATIONAL SAFEGUARDS

The domestic safeguards considered previously are conducted by individual nations to protect their own materials and facilities from indigenous terrorist threats. International safeguards are fundamentally different since they address the concern in the world as a whole that more nations will proliferate, or add nuclear weapons capability to their military programs.

The potential pathways and motivations for proliferation are rather diverse. However, commercial nuclear power facilities present a possible concern. The desire by many nations to decouple energy production from proliferation has led to the formation of the International Atomic Energy Agency [IAEA] as the primary organization for international safeguards.

Proliferation

Proliferation is generally considered to have occurred when a nation detonates a nuclear explosive for the first time. From a practical standpoint, however, acquisition of the ability to assemble and detonate such a device on a short time scale would also be considered as proliferation.

The overall likelihood and rate of proliferation are controlled by an individual nation's perceptions of the incentives and opportunities as balanced against the disincentives and barriers. Status, increased autonomy, or political gain may be incentives. They must, however, be balanced against the cost of national resource diversion and unfavorable international reactions, ranging from diplomatic pressure to embargoes.

Alternative routes to proliferation are seen to be:

- diversion of SNM from commercial nuclear reactors and/or fuel cycle facilities
- production of SNM at indigenous, dedicated facilities
- purchase or theft of SNM or actual weapons

In the absence of international or other agreements, a nation could divert material at will (from *its own*) commercial facilities at some cost in terms of ultimate energy production. The use of dedicated facilities, on the other hand, would be less disruptive of the commercial fuel cycle and, perhaps, less visible to the outside world. Purchase or theft could, of course, be connected with subnational, terrorist, or military action in or against another nation. All known weapons development at this time in history is believed to have been based on the full or partial use of dedicated facilities.

Motivation for proliferation is a complex problem. While commercial nuclear power may provide an important source of SNM and a partial technology base, it may also discourage proliferation to the extent that an assured energy supply is provided. Related concerns are the subject of detailed discussions in the international political arena.[†]

[†]An article entitled "Nuclear Power, Nuclear Weapons and International Stability" (Rose & Lester, 1978) provides some valuable insight, as do others listed in the Selected Bibliography at the end of this chapter.

International Atomic Energy Agency

Concurrently with passage of its own Atomic Energy Act of 1946, the United States proposed that nuclear materials be under international control. The idea, known as the Baruch Plan, was presented to the United Nations but was rejected when agreement could not be reached on the form of an international safeguards system. The Soviet Union in particular rejected the principle of on-site inspection.

When the Atoms for Peace Program was first proposed by the United States in 1953, it recommended establishment of an International Atomic Energy Agency [IAEA] under the auspices of the United Nations. Passage a year later of the Atomic Energy Act of 1954 essentially cleared the way for commercial nuclear development in the United States and elsewhere.

The IAEA came into existence in 1957 following approval of its statute a year earlier by the State Members of the United Nations. It is now constituted as an intergovernmental organization with ties to the United Nations similar, for example, to those of the World Bank, World Health Organization [WHO], and the Food and Agricultural Organization [FAO]. The IAEA is headquartered in Vienna, Austria.

Mandate

According to its statute, the general objective of the IAEA is to "accelerate and enlarge the contribution of atomic energy to peace, health and prosperity throughout the world. It shall ensure, so far as it is able, that assistance provided by it or at its request or under its supervision or control is not used in such a way as to further any military purpose."

The IAEA international safeguards have been developed on the basis of both its statutory provisions and a number of treaties with or among nations. The most important of the treaties is the Non-Proliferation Treaty of 1970 [NPT], under the terms of which over 100 nations have agreed to allow IAEA inspections of all commercial nuclear facilities and materials. Nonsignatories of the NPT, however, may also accept IAEA safeguards on specific facilities by unilateral treaty or under the terms of a bilateral agreement (e.g., related to purchase of SNM or facilities) with another nation.

The resulting agreements between the IAEA and a nation are designed to implement safeguards on all source and special nuclear material "for the exclusive purpose of verifying that such material is not diverted to nuclear weapons or other nuclear explosive devices" (INFCIRC/153, 1971).

The international safeguards conducted by the IAEA are designed to verify the *absence* of diversion, rather than to prevent it by assuming any physical control of materials or facilities. Since little can be done to stop the overt seizure of a nation's own SNM, however, timely detection of diversion is desirable to allow ample time for diplomatic response. An important underlying principle of IAEA safeguards is that the commercial nuclear fuel cycle should be an undesirable pathway to proliferation.

Safeguards Goals

The specific safeguards objective of the IAEA has been set forth as being "the timely detection of diversion of significant quantities of nuclear material from nuclear activities . . . and the deterrence of such diversion by risk of early detection" (INFCIRC/153, 1971). Further, the safeguards are to be employed "in a manner designed to avoid hampering a state's economic and technological development" and to be "consistent with prudent management practices required for the economic and safe conduct of nuclear activities" (INFCIRC/66, 1968). Under these guidelines, *verification* of the nation's system of nuclear material accountancy has become a safeguards measure of primary importance. This is augmented by *containment* and *surveillance* measures. Effective containment reduces the necessity for continuous reverification of inventories. Surveillance can identify significant movements of material for which prompt inventory verification may be appropriate.

The overall effectiveness of IAEA safeguards must be correlated to the risks of proliferation associated with the various types, quantities, and forms of nuclear materials present in particular facilities. Safeguards objectives may be formulated in terms of SNM quantities and timeliness of detection based on the required additional processing. Low-enriched uranium, for example, would require isotopic enrichment before it could be used in a nuclear explosive. Similarly, spent fuel would require reprocessing, while mixed-oxide fuel would only require chemical separation.

Significant quantities of nuclear materials are estimated by the IAEA to be the values shown in Table 20-4. These are the quantities which should be detected by IAEA safeguards if diverted or otherwise missing within a nation. The timeliness goals for detecting diversion are predicated on the form of the material as indicated by Table 20-4. For plutonium, ^{233}U, and enriched ^{235}U ($\geqslant 20$ wt %), the 1–3-week time span is to be adjusted according to the chemical form and purity of the materials.

The extent to which the objectives in Table 20-4 can be attained at any time is a function of such factors as the resources available to the IAEA safeguards system and the state of development of safeguards technology. Comparison with current capabilities provides the impetus for defining increased resource requirements and research and development needs.

Safeguards Measures

The IAEA has safeguards responsibilities at power reactors and fuel-cycle facilities. The primary measures employ some combination of on-site inspectors, inventory verification procedures, and containment/surveillance devices.

TABLE 20-4
Approximate National Detection Objectives of the International Atomic Energy Agency [IAEA]

Material	Contained isotopic mass (kg)	Detection time
Plutonium	8	1–3 weeks
Uranium-233	8	1–3 weeks
Uranium-235		
$\geqslant 20$ wt % enriched	25	1–3 weeks
< 20 wt % enriched (including natural and depleted U)	75	1 year
Thorium	20,000	1 year
Spent fuel	–	1–3 months

The multinational inspection force is the backbone of the IAEA safeguards program. In addition to having flexibility in personal observation, individual inspectors are responsible for coordinating applications of the various instruments and devices that have been designed to assist them. They must also perform their duties in a manner that minimizes interference with normal operations (as directed by the IAEA mandate).

Verification of facility inventories is generally accomplished by a combination of sampling procedures and nondestructive assay [NDA] methods. Chemical samples collected by the inspectors are generally sent to designated IAEA laboratories for detailed analysis. These conventional destructive assay methods, however, have always been accompanied by inherent time delays and problems of drawing representative samples.

Nondestructive assay [NDA] techniques and instrumentation have been developed to circumvent some of the problems of chemical analysis. Modern NDA methods provide rapid, accurate assay capability for essentially all types of SNM in a wide variety of forms. The NDA techniques fall into two major categories, passive and active. The passive methods identify different SNM species through their unique natural "signatures" (i.e., the types, energies, and intensities of the radiations they emit). The active methods rely on irradiation followed by detection of the induced-radiation signatures. A few NDA instruments of particular current interest are identified in appropriate later sections. These and other devices are described fully in various references (e.g., as listed in the Selected Bibliography at the end of this chapter).

Whenever nuclear material is to be stored or otherwise immobilized for a period of time, containment/surveillance devices may be employed to minimize the need for routine reverification of inventory. Containment is most readily assured by use of seals with unique markings and methods of application and removal designed to indicate attempts at tampering. These range from relatively simple paper or metal seals to sophisticated electronic or fiber-optic devices.

One means of accomplishing the surveillance function is to use self-contained photographic or video cameras which take and store pictures of nearby activities on a periodic basis. Other surveillance systems may be designed to monitor such parameters as equipment use patterns or electric power output.

The interfaces between inspectors and equipment is quite important from the standpoint of costs, resource utilization, and effectiveness. Inspector training and deployment requirements, for example, must be compared to the costs, availability, and proven reliability of various devices. Technical measures for inventory verification and containment/surveillance must be tested thoroughly to assure their usefulness to the inspectors as well as their reliability under potentially adverse environmental conditions during transport, handling, and use.

The SAM [Stabilized Assay Meter] and SAM-2—portable electronic signal processing packages designed flexibly enough to be used in several different NDA methods—are examples of units that have proven to be both reliable and easy to use in routine inspection operations. Favorable attributes of containment/surveillance devices must also include very high resistance to possible mistreatment and tampering (e.g., as tested by "black-hatting" in the laboratory and at on-site locations).

System Application Examples

Safeguards applications to facilities generally employ some combination of inspection devices. Since the first commercial nuclear operation for most countries has been a reactor, IAEA measures are relatively well developed in this area. The expansion to fuel fabrication, reprocessing, and enrichment facilities poses somewhat newer and more complex safeguards problems.

Safeguards for conventional power reactors reduce to verifying the locations of individual fuel assemblies. Receipt, storage, and placement into the reactor are readily confirmed by any of several means. Reactor power operation provides secure containment in LWR systems, since extended shutdown is required before fuel can be removed. A simple monitor of power output can be quite effective for tracking the operating history. Inspection during refueling operations can serve to verify assembly counts and/or serial numbers, as appropriate. One NDA method employing detection of Cherenkov radiation[†] from spent fuel, for example, is now proving to be very useful for in situ inspection and qualitative verification of burnup. Since in any event low-enriched uranium and spent fuel are relatively undesirable targets, a detection time for diversion of 1–3 months to a year (Table 20-4) allows substantial flexibility for power reactors.

The on-line refueling capability of the CANDU system (Fig. 9-12) poses somewhat more complex problems than found in LWRs. However, a system is now under development which appears to be capable of logging the number and direction of all spent-fuel-bundle transfers.

Safeguards for fuel fabrication facilities are substantially more complex than those for reactors, since SNM is in the form of powder, fuel pellets, and finished fuel assemblies. Important components of the inspection program include a combination of NDA capability and availability of an IAEA Safeguards Analytical Laboratory.

The passive gamma-ray signatures from slightly enriched uranium in LWR fuel, for example, provide a basis for NDA with a fuel-rod scanner for finished pins and a segmented gamma scanner for powder or scrap. An active NDA system for mixed oxides employs a ^{252}Cf (spontaneous fission) neutron irradiation source and distinguishes among the constituents by their delayed-neutron signatures.

Reprocessing plants are coming under active IAEA safeguards for the first time. It is assumed that they can be adequately safeguarded provided that the following conditions are met:

- All accountancy vessels (located between the dissolution and extraction steps in Fig. 10-1) are carefully calibrated.
- Reliable samples can be taken of local inputs and outputs as well as of all streams that leave the plant.
- Frequent inventories with controlled cleanouts are taken.

[†]Blue *Cherenkov radiation* is present whenever a beta-active source like spent fuel is surrounded by water. The relativistic electrons, emitted with speeds close to the speed of light *in a vacuum* [c], set up shock waves and slow down when they enter the water, where the speed of light is lower. The shock waves result in production of electromagnetic radiation in the visible-light range (i.e., the "blue glow" characteristic of irradiated fuel in water.)

- Appropriate NDA methods (passive and active) can provide rapid verification of the contents of the product-output and waste streams.
- The output can be put under an easily checked seal or continuous surveillance.

Enrichment plants present a special problem because of the sensitivity of the operators to the commercial value of the plant design. Thus, a "black-box" approach may be called for where safeguards activities are carried out on the perimeter of sensitive areas without access to the inner workings. Procedures would be directed toward careful verification of all material fed to the cascades and of all material withdrawn, both with regard to quantity and enrichment. One currently available NDA method, for example, can provide a direct measurement of enrichment via the passive gamma-ray signature of ^{235}U. Examination of the material balance under these circumstances not only indicates whether material is missing, but also indicates the mode of operation (since tails and feed ratios change with increasing enrichment, as may be inferred, for example, from Table 8-4). Since power requirements vary with the separative work provided, monitoring of this parameter may also be of value.

ALTERNATIVE FUEL CYCLES

In recent times a great deal of attention has been focused on the nature of the nuclear fuel cycle as it affects both domestic and international safeguards. Studies, including the major efforts of the U.S. Nonproliferation Alternative Systems Assessment Program [NASAP] and the U.S.-inspired International Nuclear Fuel Cycle Evaluation [INFCE], have identified options in the following categories:

- structural modifications of existing fuel cycles
- fuel-cycle process modifications
- co-location of fuel cycle facilities
- advanced reactor concepts

Since the possible permutations of the alternative concepts are large in number, only one or a few representative examples are considered in each category.

As a point of perspective before proceeding, it should be noted that little short of an act of war can really prevent a determined country with the necessary resources from developing a nuclear weapons capability (using reactors, enrichment facilities, or particle accelerators). Weapons detonated to date have been based on such dedication. Thus, the alternative fuel cycles could be expected, at most, to make commercial nuclear power a less desirable proliferation pathway than dedicated facilities, purchase, or theft. The ultimate motivation for proliferation, of course, lies well outside of strictly technical considerations.

A second important recognition is that near-term safeguards concerns must be centered squarely on facilities and materials that are already or soon to be in existence. The alternatives could only have an impact on time scales sufficiently long for their wide-spread implementation.

Existing Cycles

The once-through ["throw-away"] LWR fuel cycle has been touted for its proliferation resistance because neither highly enriched uranium nor separated plutonium appear. Since plutonium occurs only in spent fuel, the fission-product inventory produces a self-protecting mechanism which makes reprocessing difficult. However, the decay of the fission-product activity makes the plutonium more accessible with passing time. According to Marshall (1976), the buildup of spent-fuel inventory can actually be thought of as providing a growing "plutonium mine."

Recycle is one means of reducing the overall inventory, but, of course, at the expense of having separated plutonium more readily available. An alternative involves use of plutonium by certain nations, while a supply of low-enriched uranium coupled with spent-fuel collection is provided to those nonweapons nations which are assumed to have significant proliferation potential.

Since the once-through cycles have relatively low resource utilization, modifications are possible to extend burnup in the reactor itself or in tandem cycles with other reactors. The CANDU–PHW, for example, could use spent-fuel from an LWR to extend burnup by up to 30 percent.

Process Modifications

Modifications in fuel-cycle processes have been identified which leave nuclear materials in relatively difficult-to-handle forms. The *co-processing* option as applied to LWR fuel, for example, calls for incomplete separation of plutonium from uranium. *Spiking* results in product material that is radioactive either from incomplete fission-product removal or by specific introduction of fission-product or other radionuclides. Although either method would act as a deterrent to domestic adversaries, their effect in combating proliferation would be much less.

The *Civex* reprocessing concept (Civex, 1978) would combine spiking and co-processing in a single system. Based on a flowsheet like that in Fig. 20-7, the process has attractive features in that there would be:

- no pure plutonium in storage or at any intermediate point
- no way to produce pure plutonium without extensive equipment modifications requiring new components and extensive plant decontamination
- a lengthy time for successful diversion that should allow adequate detection time for national and/or international response
- technical credibility

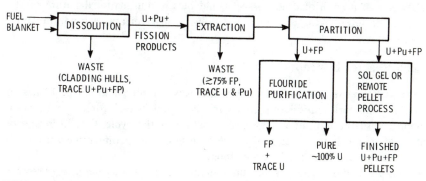

FIGURE 20-7
Simplified flowsheet for the Civex reprocessing concept.

One major difference between the Civex process and the conventional Purex (e.g., Fig. 10-1) process is that the former is sized such that criticality would occur if the plutonium concentration approached 50 percent. Further purification could be accomplished only in a separate facility or by making very substantial equipment modifications. Further, the fluoride volatility process is amenable to uranium purification, but is inoperable for plutonium. At a minimum, the Civex concept has constituted a first attempt at defining a proliferation-resistant reprocessing scheme.

Inherent problems with fuel that is spiked by any method include the need for remote fabrication and handling activities. Spiking may also have the effect of increasing potential radiological hazards from sabotage threats. An additional concern relates to the impact of the high radiation levels on the nondestructive assay methods that are now readily applied to fresh fuel.

Denatured fuel cycles would rely on isotopic dilution of fissile material as a safeguards measure. All ^{235}U and ^{233}U would be mixed with enough ^{238}U to reduce their effective enrichment below 20 wt %. This concept would limit the diversion value of the fuel substantially, provided that the bred plutonium content were handled in an appropriate manner.

Co-Location

It has also been proposed that *co-location* on the same site could have safeguards advantages for certain fuel-cycle facilities. The concept as shown by Fig. 20-8 confines the SNM of highest strategic significance to a single site while shipping out only low-enriched uranium and receiving natural uranium, thorium, and spent fuel. Enrichment, reprocessing, fuel fabrication, and in-reactor plutonium utilization would be concentrated at single, heavily-safeguarded sites. Figures 20-8b and 20-8c show specific applications to uranium and thorium fuel cycles, respectively. In practice, the "safeguarded site" could consist of more than one site located close enough together that transportation requirements would be limited.

The co-location of fuel-cycle facilities at international service centers is a natural extension of the concept. If such centers were located in weapons states or other stable, technically advanced nations, IAEA safeguards could treat the enrichment, reprocessing, and fabrication at a single site while dealing only with intact fuel assemblies elsewhere. An additional advantage would be the multinational nature of the venture and the interlocking economic and political investments each nation would wish to protect.

Advanced Reactor Concepts

The molten-salt breeder reactor [MSBR] which was considered in Chap. 14 has an advantage for nonproliferation due to its relatively low net plutonium production rate coupled with the continuous, in-process nature of the cycle. Careful inventory control could assure that operation of the reactor would be incompatible with diversion of sufficient material to produce a weapon.

A somewhat similar (but, as yet, undeveloped) concept is a *gaseous-core reactor* which would employ UF_6 fuel, a beryllium moderator/reflector, a molten thorium-salt blanket, and an energy transport gas. Since the system could have a fissile inventory as

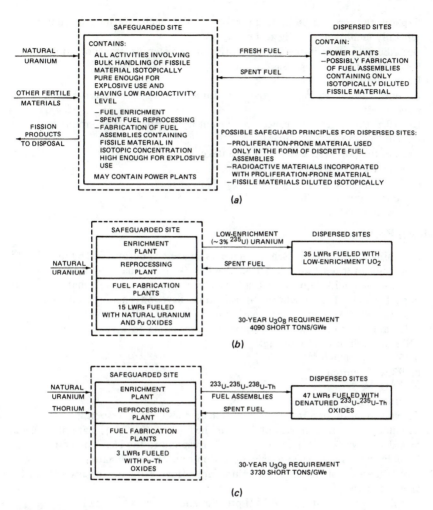

FIGURE 20-8

Principle features for co-location of nuclear facilities: (*a*) in concept; (*b*) for a uranium fuel cycle; and (*c*) for a thorium fuel cycle. (Courtesy of *Nuclear Engineering International* with permission of the editor.)

low as tens of kilograms, large-scale diversion and electrical generation would be incompatible.

EXERCISES

Questions

20-1. Identify the special nuclear materials and the bases for assigning strategic significance to each.

20-2. Describe the roles of physical protection, material control, and material accounting in an integrated facility safeguards system.

20-3. Explain the protection-in-depth concept and each of the five subsystems generally employed for physical protection.

20-4. Describe several desirable characteristics of a conceptual transportation system for SNM of high strategic significance.

20-5. Define proliferation.

20-6. Differentiate between the goals of domestic and international safeguards, respectively.

20-7. Describe the roles of inspectors, inventory verification, containment, and surveillance in IAEA safeguards.

20-8. Explain the potential safeguards advantages of:
   a. once-through LWR fuel cycle
   b. Civex reprocessing
   c. co-location of facilities
   d. gas-core reactor

Numerical Problems

20-9. Assuming an LWR fuel assembly achieves an average burnup of 16,000 MWD/T uniformly over a 12-month period, use Fig. 6-2 to estimate the maximum time the reactor could operate and still produce "weapons-grade" material. What impact would fuel discharge at this time have on energy production and economics? How would the scenario differ for a CANDU–PHW reactor?

20-10. Calculate the total mass of material corresponding to a formula quantity of SNM for each of the following:
   a. 5 wt % enriched ^{235}U
   b. 93 wt % enriched ^{235}U
   c. ^{233}U
   d. plutonium
   To what fraction of a reflected critical mass does each correspond?

20-11. Calculate the total masses of low-enriched [3 wt % ^{235}U], natural, and depleted [0.3 wt % ^{235}U] uranium that correspond to the IAEA detection goals. Explain the basis for the assigned detection time for these materials.

20-12. Assuming a constant natural uranium input flow, estimate the mass flows for an enrichment plant with:
   a. full production at 3 wt % ^{235}U product and 0.3 wt % ^{235}U tails
   b. 75 percent production as in (a) with the remainder at 93 wt % ^{235}U at the same tails assay

   Explain how the difference between the facilities would be most readily detected by IAEA safeguards. Which of the five enrichment technologies introduced in Chap. 8 would be most easily converted from case (a) to case (b)?

SELECTED BIBLIOGRAPHY[†]

General
   Deitrick, 1977
   Kovan, 1978
   Meyer, 1977
   Smith & Waddoups, 1976
   Willrich & Taylor, 1974

[†]Full citations are contained in the General Bibliography at the back of the book.

Accountability and Nondestructive Assay
    Augustson, 1979
    Augustson & Reilly, 1974
    Dayem, 1979
    IAEA, 1978
    Keepin, 1978
    Keepin (in press)
    Lovett, 1974
    Reilly & Evans, 1977

Physical Production
    Blake & Leutters, 1976
    Carpency & Rittenhouse, 1980
    Herrington, 1979
    INFCIRC/225, 1976
    Jones, 1975
    McCloskey, 1977
    Ney, 1976
    Nucl. Eng. Int., May 1978

Nonproliferation and International Safeguards
    ASME, 1978
    Atlantic Council, 1978
    Bantel, 1980
    Greenwood, 1977
    IAEA, 1976
    IAEA, 1977
    INFCE, 1980
    INFCIRC/66, 1968
    INFCIRC/153, 1971
    Kratzer, 1980
    Marshall, 1978
    Nye, 1979
    Olds, 1980c
    OTA, 1977
    Rose & Lester, 1978
    U.S. Govt., 1977
    Wilson, 1977b

Alternate Fuel Cycles
    APS, 1978
    Chang, 1977
    Civex, 1978
    Hafemeister, 1979
    Heising & Connolly, 1978
    Hutchins, 1975
    INFCE, 1980
    INFCE/SEC/11, 1979
    NASAP, 1979
    Nucl. Eng. Int., Nov. 1979
    Shapiro, 1977
    Spiewak & Barkenbus, 1980
    Williams & Rosenstrock, 1978

Current Information Sources

Nuclear Materials Management [*Nucl. Mat. Man.*] —published four times a year (three regular issues plus the proceedings of the annual INMM meeting) by the Institute of Nuclear Materials Management [INMM]; reports on current research, development, and application activities in all areas of nuclear safeguards.

Nucl. Tech.

Trans. Am. Nucl. Soc.

VI

NUCLEAR FUSION

Goals

1. To introduce the fundamental concepts of nuclear fusion.
2. To describe the basic features of conceptual engineered systems for controlled thermonuclear fusion.
3. To compare fusion and fission technologies in terms of designs, safety, and other impacts.

Chapter in Part VI

Controlled Thermonuclear Fusion

21

CONTROLLED THERMONUCLEAR FUSION

Objectives

After studying this chapter, the reader should be able to:

1. Explain the term *controlled thermonuclear fusion*.
2. Describe the DT, DD, and ^{11}B-p fusion reactions and identify relative advantages and disadvantages of each for potential commercial applications.
3. Identify the similarities and differences between fuel cycles for DT fusion and for fission.
4. Describe the bases for the Lawson criterion.
5. Sketch the components for a generic fusion reactor which uses DT fuel.
6. Differentiate among the tokamak, magnetic-mirror, pinch, laser, and particle-beam fusion concepts.
7. Identify several potential limitations and safety issues for commercial applications of fusion power.

Prerequisite Concepts

Applications of the nuclear fission process to commercial power generation have been described in some detail in the previous chapters. Fusion, the other source of nuclear energy, is the subject of this chapter. Even though it still requires more energy to produce such reactions than is produced, an immense energy resource would become available if the requisite scientific and engineering advances could be achieved.

Use of the phrase *controlled thermonuclear fusion* emphasizes several important concepts. First, the term thermonuclear emphasizes that net energy production must be based on chain reactions driven by *thermal energy* rather than by a neutron population. Neutrons do play a secondary role in fusion.

Net energy production from nuclear fusion has been achieved, but only transiently in thermonuclear weapon detonations. A much more *controlled* energy release is, of course, the important goal for commercial applications.

The temperatures required for controlled thermonuclear fusion are far too large to allow containment by material structures. One major confinement method would employ magnetic fields to hold the nuclei in an evacuated space as they undergo the fusion reactions. The main alternative method would use lasers or particle beams to produce short bursts of fusion in small fuel pellets. These approaches are known as magnetic-confinement fusion [MCF] and inertial-confinement fusion [ICF], respectively.

Assuming that break-even energy production is achieved, a number of severe engineering problems must be solved before commercial power can be generated from fusion. Some concerns are similar to those that must be addressed in fission reactors. Others are unique.

FUSION OVERVIEW

Fusion is essentially the reverse of the fission process (as described in Chap. 2). Light nuclei [$A \lesssim 20$] have relatively low binding energies per nucleon (Fig. 2-1) such that their combination can produce a release of energy.

It may be recalled that fission occurs most readily when an uncharged neutron strikes a heavy nucleus like that of ^{235}U. Fusion reactions are achieved with somewhat more difficulty because two positively charged nuclei must essentially come into contact with each other. The strongly repulsive Coulomb forces between the nuclei necessitate a large energy expenditure to achieve such contact.

Reactions

Four potentially useful fusion reactions are identified in Table 21-1. Reaction equations and energies (i.e., the Q values from Eq. 2-8) are included for each entry. The reactant nuclides are hydrogen [$^1_1$H or ^1_1p], deuterium [$^2_1$H or $^2_1$D], tritium [$^3_1$H or $^3_1$T], helium-3 [$^3_2$He], and boron-11 [$^{11}_5$B], with neutrons and alpha particles [$^4_2$He] also included among the products. Common shorthand notations for the various reactions are included in Table 21-1.

Reaction cross sections for two nuclei undergoing fusion are defined in the same manner as for a neutron and a nucleus (e.g., Eq. 2-12). In the case of fusion, however, the reference energy may belong to either nucleus or be shared between them. Energy-dependent cross sections for several important reactions are shown in Fig. 21-1.

TABLE 21-1
Four Potentially Useful Fusion Reactions

Reaction	Shorthand notation	Reaction energy (MeV)	Threshold plasma temperature (keV)	Maximum energy gain per fusion
$^2_1D + ^3_1T \rightarrow ^4_2He + ^1_0n$	D-T	17.6	10	1800
$^2_1D + ^2_1D \rightarrow ^3_2He + ^1_0n$	D-D	3.2	50	70
$^2_1D + ^2_1D \rightarrow ^3_1T + ^1_1p$	D-D	4.0	50	80
$^2_1D + ^3_2He \rightarrow ^4_2He + ^1_1p$	D-^{3}He	18.3	100	180
$^{11}_5B + ^1_1p \rightarrow 3^4_2He$	^{11}B-p	8.7	300	30

The most viable method for obtaining net energy production from fusion appears to be to elevate the temperature of the fuel materials to the point where their
average thermal energies are sufficient for fusion to occur. Figure 21-1 suggests that
energies on the order of tens of keV are required. Under such conditions, the fuel would
be in a *plasma* state where the atoms are completely ionized with the nuclei and electrons coexisting as separate entities in a highly energetic "sea" of charged particles. An
equilibrium plasma with average particle energy E has a mean temperature T given by

$$E = kT$$

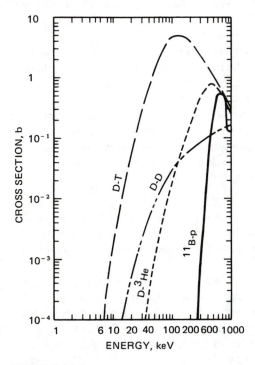

FIGURE 21-1
Reaction cross sections for selected fusion reactions.

where k is the Boltzmann constant (equal to 8.6×10^{-8} keV/K in one set of useful units). Accordingly, each 10 keV energy increment corresponds to a temperature somewhat greater than 100 million K! The fusion chain reaction under these conditions is a *thermal* phenomenon (as opposed to the neutron chain reaction in fission systems).

Table 21-1 includes approximate threshold energies for each of the reactions. Since these energies are average values, it should be noted that some of the nuclei in the plasma will have higher energies while others have lower energies. The maximum energy gain per fusion is simply the ratio of the exothermal reaction energy to the average plasma energy.

The first generation of fusion reactors are being developed around the deuterium-tritium [D-T or, simply DT] reaction, since it has the lowest threshold energy and a large energy gain. The presence of the deuterium suggests that some D-D reactions also occur, but with much lower probability (Fig. 21-1). They, in turn, produce $^3_2$He, which reacts with other deuterium. Thus, the first four reactions in Table 21-1 contribute to DT fusion.

The neutrons produced in DT fusion are both an advantage and a disadvantage. They make a positive contribution in providing a method for producing the tritium fuel which does not occur naturally. On the other hand, the 14.7-MeV energy of the neutrons complicates energy conversion and leads to potential neutron activation and damage problems (the latter as described in Chap. 3).

The DD reactions are based on more readily available fuel and give off a smaller fraction of their energy to neutrons. These advantages, however, are offset by their higher threshold energies.

The ^{11}B-p reaction is of special interest because it does not produce any neutrons at all. Nor does it yield nuclei that undergo secondary reactions (i.e., $^4_2$He is already very tightly bound as shown by Fig. 2-1). Unfortunately, the energy threshold is a factor of 30 higher than that for the DT reaction.

Despite its drawbacks, the deuterium–tritium reaction is the clear choice for the pioneering effort in controlled thermonuclear fusion. Its success could well open the way for future DD and ultimate ^{11}B-p systems. On these bases, the remainder of the chapter addresses DT fusion to the near exclusion of the other concepts.

DT Fuel Cycle

There is a fusion fuel cycle that has several similarities to that considered previously for fission systems. Isotope-enrichment, breeding, and recycle steps are important components (even if these terms are not always employed by the practitioners in the fusion community).

Deuterium is a constituent of ordinary water in a $1:6500$ ratio with $^1_1$H atoms (or, equivalently, 1 kg $^2_1$D in 30,000 kg water). Isotopic separation is accomplished by the well-established process employed to produce heavy water for the CANDU-PHW systems. In fact, the current capacity of 800 metric tons per year for the Bruce heavy-water plant in Ontario would be sufficient to fuel a million-MW(e) fusion economy (compared to a total U.S. electrical capacity of slightly over 600,000 MW(e) in 1980). Deuterium for fusion is obtained from heavy water by a standard electrolysis process.

Tritium is an isotope of hydrogen that does not occur in nature. However, it is readily produced from lithium by employing the reaction pair

$$_3^7 \text{Li} + {}_0^1 \text{n} \text{ (fast)} \rightarrow {}_1^3 \text{T} + {}_2^4 \text{He} + {}_0^1 \text{n}$$

$$_3^6 \text{Li} + {}_0^1 \text{n} \text{ (thermal)} \rightarrow {}_1^3 \text{T} + {}_2^4 \text{He} + 4.8 \text{ MeV}$$

The neutron from the DT fusion would react with $^7 \text{Li}$ to give one $^3 \text{T}$. The neutron from the latter reaction would be thermalized to allow production of a second $^3 \text{T}$ by interaction with $^6 \text{Li}$. A process equivalent to breeding occurs when the average number of $^3 \text{T}$ nuclei produced per fusion exceeds *unity*. (Since the fusion reaction is driven by thermal energy rather than neutrons, the fission breeding requirement of more than *two neutrons* per reaction is not applicable.)

Tritium breeding would most likely be facilitated by using a lithium blanket around the outside of the fusion reaction region (as a fertile blanket is employed with a fissile core in a breeder reactor). Lithium in a molten-salt form could serve the additional functions of heat removal and, because of its low mass, neutron moderation to enhance the $^6 \text{Li}$ reaction. The tritium and helium gases produced by the reactions are readily separated from the salt and from each other. Potential leakage of tritium and neutron activation of system components appear to constitute the major radiological hazards of the fuel cycle for DT fusion.

Deuterium is so abundant that use for fusion of only 1 percent of the amount contained in the world's oceans would provide enough energy to exceed the total postulated energy demand for the year 2000 by a factor of 100 million. Lithium resources available with current technology and at roughly current prices appear to be more than adequate for the conceptual million-megawatt DT-fusion economy. Future lithium supplies could be extracted from seawater. An eventual transition to DD fusion would, of course, remove lithium requirements completely.

Controlled Fusion

Sustained or controlled thermonuclear fusion reactions require that

1. a plasma be formed and held at a high temperature T
2. the plasma have a high density n
3. the plasma be confined for a sufficient time τ

The temperature determines the average plasma energy and, thus, the microscopic reaction cross section (e.g., as in Fig. 21-1). The energy and plasma density, in turn, set the macroscopic reaction rate. The product of the effective confinement time and the reaction rate determine the total energy output.

The first goal of controlled fusion is to exceed *break-even*, i.e., produce more energy than that required to cause the reactions. The ultimate utilization of controlled fusion for economic and reliable power production is included in more long-term planning. For both, appropriate combinations of T, n, and τ are sought by employing interrelated processes for heating and confining the plasma.

Heating

Plasma heating up to about 1 million degrees may be accomplished by ohmic or resistance heating when an electric current is passed through the plasma. Above this temperature the resistance of the plasma falls too low for the method to have further use. Supplemental or complete heating of the plasma to the higher temperatures required for fusion may be accomplished by one or more of the following methods:

- magnetic compression
- neutral-atom injection
- magnetic pumping
- laser-beam impact
- electron-beam impact
- ion-beam impact
- microwave or radio-frequency [RF] radiation
- shock waves

As the plasma heats to a level where fusion occurs, the reaction-product kinetic energy provides an important heat source to make the reaction self sustaining. The fusion energy must be sufficient to balance the radiation losses (bremsstrahlung from electron deceleration and cyclotron radiation if magnetic fields are employed).

Confinement

The high temperature requirement demands that the plasma be prevented from contacting material surfaces. This, plus the necessities to achieve the specified density and confinement time, may be achieved by either the magnetic or inertial confinement methods.

Magnetic confinement is based on the fact that a magnetic field can be employed to shape and orient a plasma (since it consists of moving charged particles). Particles which move randomly in the absence of a magnetic field become constrained to helical paths around the lines of force when such a field is employed as shown in Fig. 21-2.

The pressure of the plasma particles is compared to the pressure from the magnetic field through the parameter beta β defined as

$$\beta = \frac{\text{Plasma particle pressure}}{\text{magnetic field pressure}}$$

Beta may vary from zero for overpowering magnetic fields to unity for balanced pressures. (Above unity, of course, no confinement would result.) Present magnetic-confinement studies tend to be centered around high-beta ($\beta > 0.2$) and low-beta ($\beta < 0.05$) devices.

Inertial confinement employs very rapid heating of a DT pellet to thermonuclear temperatures. When a pulse of high-intensity energy is absorbed by the pellet, the outer material forms a plasma which blows off to implode the remaining pellet mass. The implosion in turn causes compression and heating to the point where fusion occurs and the pellet explodes. The energy should be added quickly enough so that inertia holds the pellet together for a "long enough" time to allow break-even energy production to be exceeded. The most promising energy sources have been lasers, electron beams, and ion beams.

Lawson Criterion

A frequently used measure of break-even requirements for fusion is known as the *Lawson criterion*. It is expressed in terms of the plasma temperature T and the product

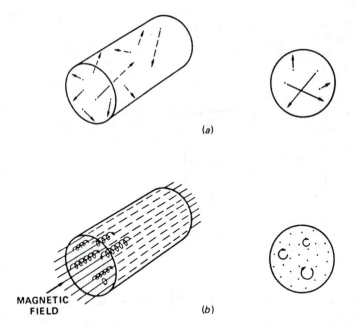

MAGNETIC
FIELD   (b)

FIGURE 21-2
Principles of magnetic confinement fusion: charged particle motion (a) without a magnetic field and (b) in a homogeneous magnetic field. (Courtesy of U.S. Department of Energy.)

of the plasma particle density n and the confinement time τ. A range of such conditions for DT fusion is shown by the curve in Fig. 21-3. The most usual statement of the Lawson criterion for DT fusion is that break-even requires an 8.6 keV [100 × 10⁶ K] temperature at $n\tau \approx 10^{14}$ cm⁻³·s.

The magnetic- and inertial-confinement methods would achieve the Lawson conditions by fundamentally different paths. In the former, relatively low density would be maintained, but for long periods of time. Initial or continuous plasma heating would be required to initiate the thermonuclear "burn."

Inertial confinement would employ laser or particle-beam energy for heating and compression. Very high densities would be achieved for extremely short confinement times.

Conceptual Fusion Reactor

The ultimate commercial application of controlled thermonuclear fusion is dependent on exceeding break-even in a system capable of long-term energy production and conversion. A conceptual DT fusion reactor is shown in Fig. 21-4. The fusion mechanism itself may be based on either magnetic or inertial confinement. The remainder of the system would have somewhat similar features for either mechanism.

The fusion plasma in the center of Fig. 21-4 emits its energy primarily in the form of $^4_2$He ions and neutrons which would strike the first wall. (In practice, a buffer material of gas, or perhaps lithium, might be employed to limit damage to a material wall.) The blanket region would contain molten lithium salt for heat removal fro᎓ the first wall and for breeding tritium. Deuterium and tritium fuel would need to be introduced into the chamber while spent fuel materials would be discharged.

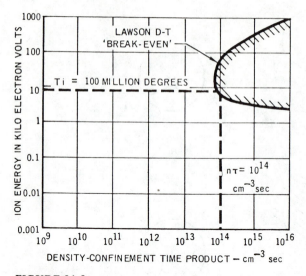

FIGURE 21-3
Break-even plasma conditions for fusion power. [Courtesy of Electric Power Research Institute (DeBellis & Sabri, 1977).]

The still-early state of development of controlled thermonuclear fusion reactors includes an extensive group of design concepts. The remainder of this chapter considers the basic features of several types of systems (without attempting to provide a complete overview of state-of-the-art technology). Selected factors related to the ultimate commercial application of fusion reactors are also addressed toward the end of the chapter.

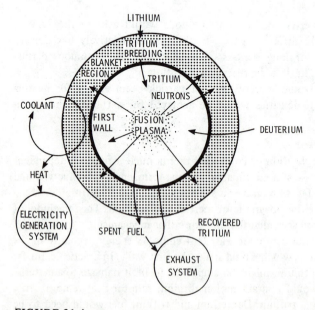

FIGURE 21-4
Principal features of a conceptual DT fusion reactor. (Adapted courtesy of *Technology Review.*)

MAGNETIC CONFINEMENT

A large number of fusion concepts employ magnetic fields for confinement and/or heating of plasmas. Differences are based on magnetic field strength and configuration, heating method, fueling procedures, and operating mode. Three substantially different concepts are

1. the tokamaks
2. the magnetic mirrors
3. the pinch-type systems

Many of the other concepts have some similarities to one or more of the above.

Tokamaks

The nature of the *tokamak* is well described by this word of Russian origin, roughly translated as: TO—toroidal; KA—chamber; MAK—magnetic. The principal features of tokamak designs include:

- plasma confinement using toroidal and transformer coils to produce the necessary magnetic fields
- neutral-beam injection to heat the plasma to the required temperatures
- fueling by injection of DT pellets or neutral beams
- burn times limited by the available magnetic flux through the center of the torus
- a diverter system to remove spent fuel

The bases for plasma confinement in a tokamak are shown by the schematic diagram in Fig. 21-5. The metal-walled torus is wrapped with field coils to produce a toroidal magnetic field in the direction shown by the figure. The primary coils induce a current in the plasma to generate a poloidal field. The combined effect of the toroidal and poloidal fields is a helical or spiral magnetic field which enhances stability for plasma confinement. In typical systems, the toroidal field has a strength about 10 times greater than the poloidal field.

The tokamak devices in operation around the world represent more than half of the effort toward controlled thermonuclear fusion. Although none have yet satis-

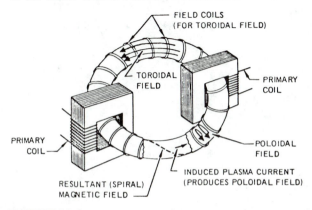

FIGURE 21-5
Schematic representation of a common type of tokamak. [Courtesy of Electric Power Research Institute (DeBellis & Sabri, 1977).]

fied the Lawson criterion, or have even used a DT fuel mixture, valuable experience has been gained in system design and operation. One such device is the Princeton Large Torus [PLT] shown in Fig. 21-6. Since it began operation in late 1975, the PLT has been used to test confinement and heating in plasmas with characteristics near those required for operating reactors. The relatively complex system in Fig. 21-6 does not include many of the features required of ultimate commercial systems.

The conceptual commercial reactor for DT fusion would require features noted in the schematic diagram in Fig. 21-7. The first wall encloses the plasma-containing vacuum chamber and absorbs up to 20 percent of the plasma energy. The moderating-blanket region provides necessary cooling for the first wall as well as moderating and reflecting neutrons to enhance breeding of tritium. The magnet and personnel shield of iron, lead, and probably boron is designed to protect the superconducting magnets and operating personnel from the effects of electromagnetic radiation and neutrons. The magnets must be superconducting to preclude unacceptable power requirements. Other necessary systems not shown on Fig. 21-7 include those for fueling, neutral-beam or other heating, tritium removal and recycle, and conversion of fusion energy to electrical energy.

There would be tremendous temperature differences in reactors like that in Fig. 21-7. The extremes of 10^8 K in the DT plasma and 4 K required in the superconducting magnets would present many difficult problems of thermal insulation.

Among the other problem areas to be investigated in research fusion devices are

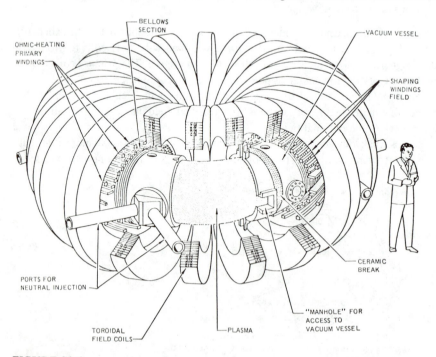

FIGURE 21-6
Princeton Large Torus [PLT] schematic. [Courtesy of Electric Power Research Institute (Debellis & Sabri, 1977).]

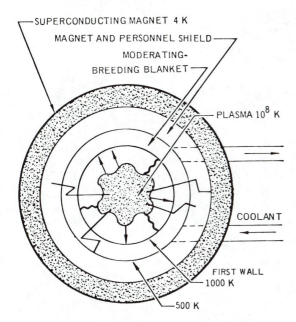

FIGURE 21-7
Schematic of a controlled thermonuclear reactor based on a DT fuel cycle. [Courtesy of Electric Power Research Institute (DeBellis & Sabri, 1977).]

those related to the first wall. If plasma interactions lead to vaporization of the first wall, impurities are introduced which have the effect of cooling the plasma and reducing reaction efficiency. Damage and activation of the wall by DT neutrons are also of concern. Selection of first-wall material is, thus, quite important as are design provisions for effective remote repair and maintenance operations.

Since the tokamak is a low-beta device, the fusion energy per unit volume is low, and tokamak sizes must be relatively large. Beta can be increased somewhat by reducing the ratio of the major radius to the minor radius (i.e., "fatten the doughnut") and by using a noncircular plasma cross section. Overall it is necessary to balance decreased system size (increased beta) against confinement stability (favored by low beta).

A conceptual design for the General Atomic Doublet Reactor is shown in Fig. 21-8. The noncircular plasma cross section is designed to increase power density and decrease system size. This reactor is about 21 m in diameter and 23 m high. The Lawson criterion would be met with a 13 keV temperature, $\sim 10^{14}$ cm^{-3} density, and ~ 1 s effective energy confinement times. Proposed operation would allow "burns" of 10–15 min followed by a short shutdown for chamber evacuation and refueling. The power rating for this unit would be on the order of 1500–2000 MW(th).

Magnetic Mirrors

The magnetic mirror concept is the primary alternative to the tokamak from the standpoint of experimental activity. These high-beta devices employ either an open or closed configuration for magnetic confinement.

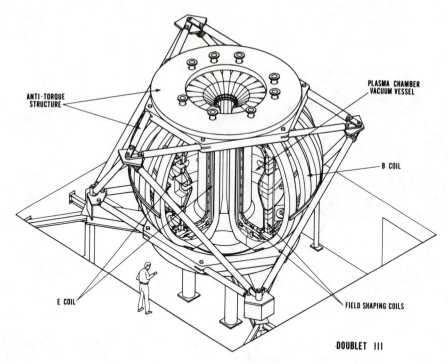

FIGURE 21-8
Conceptual design for the Doublet fusion reactor. (Courtesy of General Atomic Company.)

The principle of the *simple magnetic mirror* is illustrated by Fig. 21-9. The two current-carrying coils set up a magnetic field which is strongest at the ends (where the field lines are most dense). Charged particles moving toward the ends tend to reverse their directions as a result of the stronger field. Thus, reflection occurs as a consequence of this "mirror" effect. The simple-mirror reactor would operate in a driven

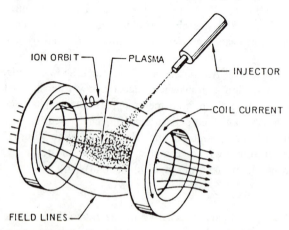

FIGURE 21-9
Schematic representation of a simple magnetic mirror. (Courtesy of U.S. Department of Energy.)

steady-state mode as compared to the cyclic operation of a tokamak. The plasma burn would be maintained by continuous injection of fuel into the plasma, while spent fuel (unreacted fuel and reaction products) is removed. Natural plasma end-losses provide an inherent spent-fuel-removal mechanism. Neutral beam injection provides both a startup heat source and a refueling mechanism.

The main advantages of a simple mirror reactor would be steady-state operation, high power density, and automatic impurity control through end losses. The disadvantages are high leakage and low energy gain (near unity).

Excessive plasma losses make the simple mirror an undesirable container. These can be reduced by employing a *minimum-B mirror* configuration where the magnetic field increases from the center outward in all directions. Such conditions may be produced as represented in Fig. 21-10a by a pair of coils and Ioffe bars where current is oppositely directed in adjacent bars. They may also be produced in a baseball coil as shown by Fig. 21-10b. The baseball coil, so named for its similarity in shape to the

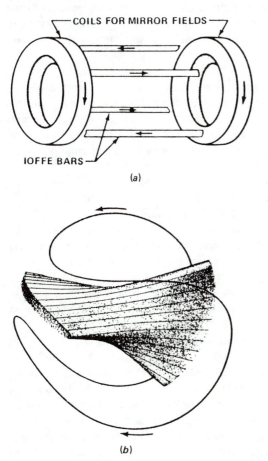

(a)

(b)

FIGURE 21-10
Schematic representation of magnetic-mirror systems using (a) Ioffe bars and (b) baseball coil. (Courtesy of U.S. Department of Energy.)

seams on a baseball (or on a tennis ball), forms two fan-shaped mirror fields at right angles to each other in the regions where parts of the coil come closest together.

Two other methods for enhancing stability have also been proposed. The *field-reversed mirror* would use neutral beam injection to set up magnetic field lines oppositely directed to those of the coils. This would have the effect of closing the lines on themselves to form a toroidal field region which would greatly reduce end losses. The *tandem mirror* would consist of a long central solenoid that is "plugged" by baseball coils (Figure 21-10b) at each end. The plasma in the solenoid region is confined electrostatically at its ends due to the difference in plasma potential between the plugs and the central cell. The net result is greatly improved confinement.

The magnetic fields and fuel injection would need to be maintained on a continuous basis in all of the mirror concepts. As is also true for the tokamak, superconducting magnets are a requirement if total power requirements are not to be excessive.

Conceptual mirror designs appear to be consistent with development of relatively small units with capacities on the order of 100 MW(e) or of large units in the multi-1000-MW(e) range. This occurs because capital costs would be relatively insensitive to unit size. This flexibility is a potential advantage as compared to the tokamak.

Pinch-Type Systems

The pinch-type magnetic confinement systems differ substantially from both the tokamaks and the mirrors by operating in a pulsed mode. They rely on shock heating the plasma by application of a very fast-rising magnetic field. In the *Z-pinch* concept, a time-varying current running through the plasma itself (i.e., in the axial or Z direction) produces the changing magnetic field. The *theta-pinch* concept employs an external current circumferential to the plasma (i.e., in the θ direction).

The Z-pinch is among the simplest concepts for magnetic-confinement fusion because the necessary magnetic field would be produced by a current carried in the plasma rather than by external coils. The situation is represented in Fig. 21-11. A primary coil would be used to induce a current in the plasma confined in the toroidal chamber (according to the same principle employed in electric power transformers). A poloidal magnetic field would be produced by the current. A very rapid rise in primary current would cause the magnetic field to squeeze the plasma, shock heat it, and provide confinement. The resulting plasma burn would produce an energy pulse. Thus, power applications would have to be based on frequent repetition of the sequence.

The basic principles of the theta-pinch concept are shown in Fig. 21-12. The tube would be filled with low-temperature plasma. Then a capacitor bank discharged to the single-turn coil would create a circumferential current and a sharply increasing magnetic field parallel to the axis. The field would form a cylindrical sheath about the surface of the plasma and then drive it inward. The plasma would be heated first by shock from the sheath motion and later by compression as the magnetic field continues to increase (at a decreasing rate). After the field reaches its maximum strength, the plasma would exist in a dense state until it escaped through the ends. End losses could be reduced by forming linear sections into a toroidal shape.

Both pinch concepts have been explored, although the Z-pinch is the recipient of the most current attention. Instabilities of the Z-pinch to kinking or excessive

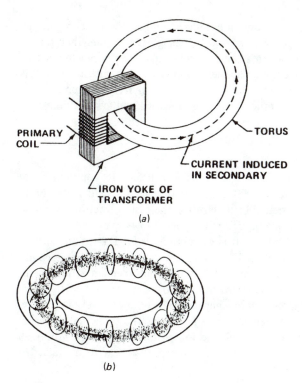

FIGURE 21-11

Schematic representation of a Z-pinch: (a) principle of current induction in toroidal plasma and (b) representation of pinch effect by magnetic field (circles) around the plasma. (Courtesy of U.S. Department of Energy.)

pinching in the current channel have been major problems. Inclusion of a second magnetic field in the reverse direction (i.e., field reversal) is under consideration as a partial remedy.

Continued interest exists in the Z-pinch concept since extremely high plasma densities ($n \approx 10^{21}$ cm^{-3}) are possible. Under this condition, a confinement time as

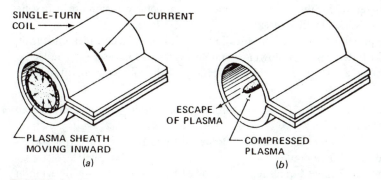

FIGURE 21-12

Schematic representation of a theta-pinch in (a) shock-heating phase and (b) compression phase. (Courtesy of U.S. Department of Energy.)

short as $\tau \approx 10^{-7}$ s would satisfy the Lawson criterion for break-even energy production (at, of course, a plasma temperature of 100 keV or greater). Other than the inherent instabilities, the greatest problem associated with either of the pinch concepts is the highly energetic pulsing that the components would need to withstand. Viable power systems would depend on development of energy delivery equipment and chamber walls which could hold up to the cyclic stresses caused by the short time-scale repetitions.

INERTIAL CONFINEMENT

Inertial confinement fusion would use lasers or particle beams to compress small thermonuclear DT pellets to between 1000 and 10,000 times liquid density for an extremely short period of time. At such high densities, the fusion reactions occur so rapidly that substantial "burn" is achieved before the pellet blows apart and cools.

It has been noted that the Lawson break-even criterion for DT fusion calls for $n\tau \approx 10^{14}$ at $T \approx 10^8$ K. Although all viable systems must achieve the same conditions, there is substantial latitude in meeting the density-confinement requirement. Inertial confinement systems are characterized by $n \approx 10^{25}$ cm^{-3} and $\tau \approx 10^{-11}$ s = 10 ps. In contrast, tokamak magnetic confinement systems have relatively low densities ($n \approx 10^{14}$ cm^{-3}) and long confinement times ($\tau \approx 1$ s). The three major inertial confinement concepts employ laser, electron, and ion beams, respectively. Although each method has its differences, the following requirements appear to be valid for any "commercial" system:

- 1–10-MJ pulse energy at 100–1000 TW
- 1–10 pulses/s
- net efficiency $\gtrsim 2$ percent
- focus at 5–20 meters

Pellets

The small-scale explosions of the pellets have similarities to those from thermonuclear weapons. The fusion neutrons would also provide a capability for studying the physics of nuclear weapons and for simulating their effects on a laboratory scale. Thus, many of the details of the technology development are classified for reasons of national security. It is, therefore, highly possible that inertial-confinement fusion might first be achieved with a "classified" pellet design.

The basic principles involved in the design of an inertial-confinement-fusion pellet may be illustrated by the following example addressed to a laser system. Under optimum conditions, a 1-MJ laser could produce any of the following conditions in 1 mg of DT:

- heating at liquid density to 10^8 K, a resulting fusion burn of 0.1 percent, and an energy output of 0.3 MJ
- compression to 1000 times liquid density followed by heating to 10^8 K, a 10 percent burn, and 30 MJ of energy
- compression to 1000 times liquid density followed by heating of a small region at the center of the high density DT to 10^8 K with generation of a thermonuclear burn front and ultimate production of 300 MJ of energy

These latter two cases of *isentropic compression* would expend only about 1 percent of the total energy on compression as long as the fuel is kept "cool" until the final density is achieved. The advantages of such compressions from the standpoint of energy output are obvious.

Multilayer fusion pellets conceptually similar to that of Fig. 21-13 are designed to allow isentropic compression conditions to be approached. In this example, symmetric laser heating of the pellet causes the surface to be ablated [boiled off]. Pressures up to 10 TPa [10^8 atm] cause the pusher and DT regions to implode into the pellet void. Appropriate variation of laser power with time allows near-final compression to be completed before a substantial temperature increase occurs. Further compression then causes the DT region to reach a temperature of 10^8 K for plasma formation and thermonuclear ignition.

Primary concerns for pellet implosion efficiency are:

- symmetry—laser- or particle-beam pressures must be closely balanced (although *spherical* symmetry is not necessarily a requirement)
- stability—surface smoothness, heating uniformity, and shell thickness must be tightly controlled
- entropy—pulse shapes must be matched carefully to the pellet design and preheat-shield layers must prevent early heating of the DT region by x-rays and electrons (i.e., allow near-isentropic compression)

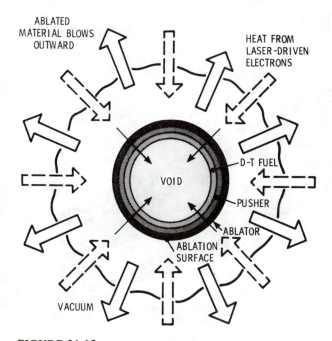

FIGURE 21-13
Principles of laser-induced pellet implosion. (Adapted courtesy of University of California Lawrence Livermore Laboratory.)

Sophisticated computer codes (some of which are classified for security reasons) have been developed to provide accurate modeling of the important phenomena involved in the implosion of fusion pellets.

Laser-fusion pellets employed in current research efforts are quite small, as commensurate with the relatively low power that can be delivered on target. One such pellet shown in Fig. 21-14 consists of a 50 μm hollow glass microballoon which could contain a DT mixture at pressures up to 10 MPa [100 atm].

As beam energies increase, layered pellets of up to about a 5-mm diameter are anticipated. Production of such pellets would be expected to be complex but feasible. For economical production of commercial power, the targets must be produced at rates of 10 to 600 per minute at costs from a few cents to a few dollars each.

Laser Fusion

In the early 1970s, two lasers appeared to have the potential for scaling to the high-peak-power, short-pulse regime required for laser fusion. The neodymium-doped glass laser, operating at 1.06 microns, had the advantage of demonstrated short-pulse operation but was relatively inefficient and was not capable of high repetition rates. The carbon dioxide gas laser, at 10.6 microns, had an intrinsically high efficiency and was well-suited to repetitive operation, but its long wavelength may inhibit its ability to deliver power on target. A variety of other laser systems are also under study for inertial-confinement-fusion applications. These include HF and iodine lasers which have wavelengths in the 1–3 micron range. More recently some attention has been directed toward a KrF laser with a wavelength of about 0.25 μm.

Both the Nd-glass and CO_2 lasers are well-developed technologies. Major U.S.

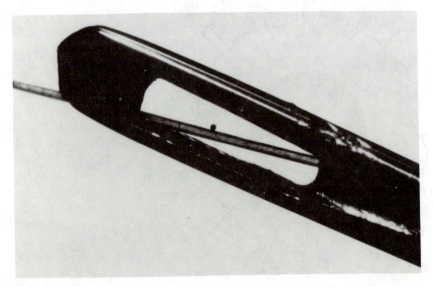

FIGURE 21-14
Glass microballoon used as a target for laser-fusion research, compared to the eye of a needle and a strand of human hair. (Photograph courtesy of University of California Los Alamos Scientific Laboratory.)

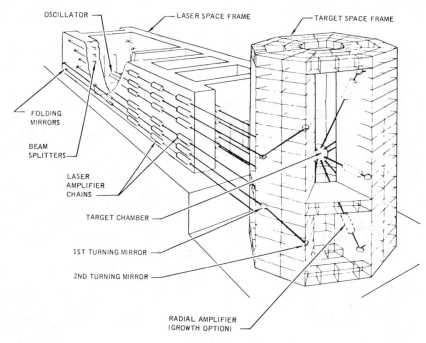

FIGURE 21-15

Artist's conception of Shiva facility for laser-fusion research. [Courtesy of Electric Power Research Institute (DeBellis & Sabri, 1977).]

commitments are to build and operate by 1984 the Antares 100-kJ CO_2 laser at the Los Alamos Scientific Laboratory and the Shiva Nova 300-kJ Nd-glass laser at the Lawrence Livermore Laboratory.

The experimental Shiva system, shown in Fig. 21-15, employs a single initial laser beam, beam splitters, and amplifier chains to deliver 20 beams on target. Although the figure indicates illumination with relatively spherical symmetry, the current arrangement calls for a balanced, two-sided illumination. The ultimate configuration of the Shiva Nova system is to be based on adding 20 more amplifier chains (in what is essentially a mirror image of the existing system as viewed from the center of the target chamber).

Although inertia serves to confine the DT plasma during the burn, the eventual pellet explosion releases large amounts of energy in a very short period of time. No material could sustain more than a few of these bursts without experiencing very severe damage. The lithium-wetted-wall concept shown by Fig. 21-16 has been proposed as one method of protecting the chamber. The DT fusion pellets would be ignited by laser beams in a central cavity. Liquid lithium would be contained between two concentric spherical shells. The inner shell made of porous niobium would allow lithium to flow through the wall and form a thin film designed to protect the wall from pellet debris. Energy deposition in the film would produce lithium vapor to be exhausted from the chamber for heat removal. Energy deposited in the lithium blanket by neutrons and electromagnetic radiation would be removed in a heat transport

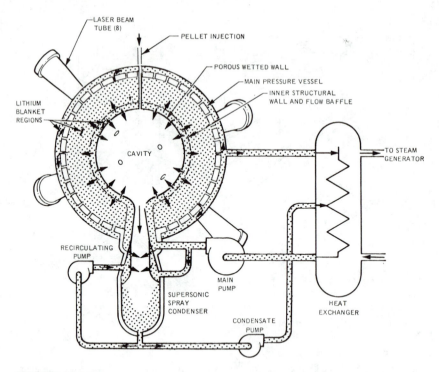

FIGURE 21-16
Lithium-wetted-wall concept for laser fusion. [Courtesy of Electric Power Research Institute (DeBellis & Sabri, 1977).]

loop. Tritium breeding and recycle would be accomplished as described previously (e.g., Fig. 21-4).

Particle-Beam Fusion

Fusion microexplosions can be induced by charged-particle beams in a manner similar to that employed in laser fusion. Possible sources include electrons (1–5 MeV), light ions like protons or deuterium (≈ 10 MeV), and ions as heavy as uranium (≈ 100 MeV).

Particle beams are readily available, and their technology is comparatively well understood. They have major advantages over lasers in terms of system costs and efficiencies (up to 50 percent as compared to the 7 percent projected for advanced CO_2 lasers). Problem areas include formation and focusing of short-pulse beams, beam transport, and optimized beam–pellet interaction.

The first successful pellet implosions with particle beams were achieved in the U.S.S.R. in 1975 using electrons. Similar results were obtained a year later in the United States. Both countries have since conducted extensive research into electron-beam fusion. However, the United States has shifted its emphasis to light-ion sources based on very encouraging research progress since 1978. Heavy-ion fusion has been pursued at a much lower level of effort.

Accelerators

Although configurations may differ dramatically, the major components of electron-beam accelerators are Marx generators (capacitive energy storage systems which function as voltage multipliers), a pulse-forming line, and a diode. The self-focusing effect of the electric and magnetic fields around a moving electron-beam may be employed to facilitate beam transmission between the poles of the diode. One early mechanism for focusing with a dual diode is shown in Fig. 21-17. Since a concentrated beam destroys metal anodes, repetitive-pulse operation would require use of an injected plasma anode that could be replaced after each firing. An alternative approach begins with pinching the beam between two solid electrodes. The pinched beam is then injected through an aperture in the electrode into a preformed plasma channel. This plasma channel permits the beam to be transported over long distances and can be used to guide the beam from the diode to the target. Preliminary research efforts string thin wires and explode them to form the plasma channel just before firing the electron beam. Other methods to produce the plasma anode, such as by use of a laser, must be developed before the needed repetition rates for power applications would be possible.

Fortunately, ion beams may be generated from essentially the same apparatus used for electron beams. By reversing the polarity of the diode, the electrons can generally be self-focused into a hollow, spherical anode while the ion flux is focused back into the center of the evacuated cavity. A comparison of electron-beam and ion-beam accelerator concepts is shown in Fig. 21-18. This complementary nature allowed the programs in the United States to shift emphasis from electron beams to light-ion beams quite readily.

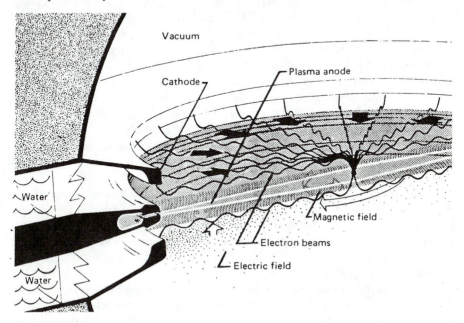

FIGURE 21-17
Mechanism for electron-beam focusing in a conceptual dual diode employing a plasma anode. (Courtesy of Sandia National Laboratories.)

Electron beam accelerator

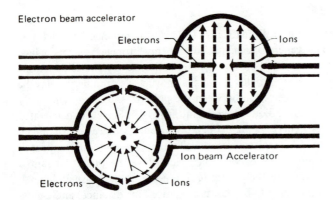

FIGURE 21-18
Comparison of methods for generating electron beams and ion beams using the same accelerator but reversed diode polarity. (Courtesy of Sandia National Laboratories.)

The program in the U.S.S.R. has approached the generation of intense electron beams in a somewhat different manner from that described above. Multiple beams are transported and focused at a large distance from the beam generator. The procedure has the advantage of allowing the use of higher yield pellets (since pellet debris would spread over a larger area). Disadvantages are related to problems with concentrating the beam on target. Current problems of diode and transmission-line destruction on each shot must also be solved.

The DT pellets for particle beam fusion would employ multiple-layer designs similar to those for laser fusion. Specific differences would be predicated on energy deposition characteristics. Relativistic electrons, for example, are highly penetrating and can produce significant amounts of bremsstrahlung. These characteristics lead to requirements for a thick outer shell with a high-density inner shell to limit bremsstrahlung production and the ensuing radiative preheat effects (which would inhibit the desired isentropic compression). Proton beams, on the other hand, have shorter path lengths per unit energy and do not generate any appreciable amounts of bremsstrahlung. As shown by Fig. 21-19, protons also deposit relatively more energy toward the end of their travel path, a fact which would tend to enhance isentropic pellet compression and reduce operating power requirements. Pellet designs for proton-fusion would reflect these unique energy-deposition features.

The major U.S. effort in light-ion-beam fusion is the Particle Beam Fusion Accelerator [PBFA] project at Sandia Laboratories. The first-generation system concept shown in Fig. 21-20 was designed for electron beams but has now been converted for light ions. The 24 Marx-generator lines are targeted for operation at an energy of 1 MJ and a peak power of 30 TW. The second generation PBFA II would double the number of lines by adding a second level and directing all of the lines toward the target. This system would be envisioned as approaching fusion break-even on a mid-1980s time scale commensurate with the Antares and Shiva Nova laser-fusion efforts.

Commercial particle-beam fusion facilities would need to be designed under constraints similar to those for the laser-based systems. Protection of the first wall from DT debris and neutrons might be accomplished using a buffer gas. (Gas-protection concepts may, in fact, be equally applicable to laser beams.)

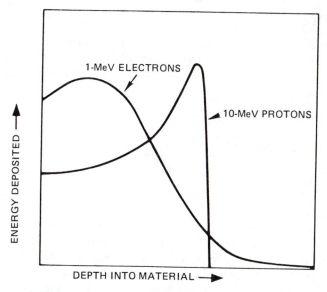

FIGURE 21-19
Comparison of energy deposition patterns for 1-MeV electrons and 10-MeV protons. (Courtesy of Sandia National Laboratories.)

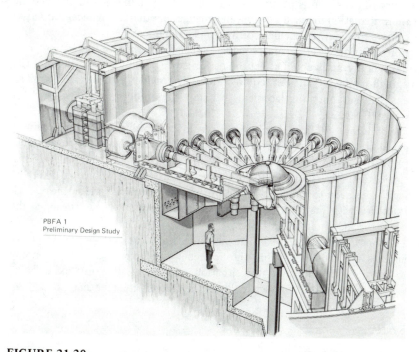

FIGURE 21-20
Conceptual design for the Particle Beam Fusion Accelerator [PBFA] facility. (Courtesy of Sandia National Laboratories.)

COMMERCIAL ASPECTS

Assuming break-even is achieved by one or more of the controlled thermonuclear fusion concepts, a number of monumental engineering problems would require solution before commercially viable power systems would be available. Several concerns selected for brief review relate to:

- materials
- control
- energy conversion, storage, and costs
- construction and maintenance
- safety and environmental impact

Materials

The selection of materials for fusion reactors must be based on considerations similar to those described in Chap. 11 for fission systems. Although thermal and radiation stability tend to be dominant concerns, strength, other mechanical properties, and general compatibility with the remaining components are also very important.

The first wall and blanket required in all conceptual DT systems (Fig. 21-4) would be subjected to intense radiation and resultingly large heat loads. Ions, neutral particles, neutrons, and electromagnetic radiation are all available to strike the first wall unless protective systems (e.g., employing liquid lithium or gas) are used to limit impacts. Neutron interaction is planned for the blanket region in order to produce the needed tritium. Radiation-induced swelling and changes in other mechanical properties would necessitate frequent replacement of first-wall components.

Neutron radiation entering superconducting magnets would cause heating problems, but more importantly could affect the performance of alloys and insulators. The high-purity, high-symmetry lattices required for superconducting devices suggests an extreme sensitivity to both displacement and impurity-production mechanisms for neutron damage (Chap. 3).

The performance of laser mirrors and windows also depends on maintaining precise dimensions and compositions. High susceptibility to radiation damage would challenge protective shield designs, especially in the vicinity of chamber penetrations.

The major material compatibility problems for fusion reactors would likely be associated with the need to employ lithium for breeding tritium. Lithium has a high chemical reactivity with many other elements. It also burns in oxygen (in much the same manner as the sodium in an LMFBR).

In this conceptual design phase of fusion-reactor development, no excessive resource impacts are anticipated for the reference "million megawatt" economy (DeBellis & Sabri, 1977). However, the situation may change if more "exotic" materials are incorporated into first-wall alloys, superconducting-magnet coils, laser optics, etc.

Control Systems

An operating fusion reactor would be a very complex device with a large energy content and high thermal-stress and radiation environments. A sophisticated, computer-based control system would most certainly be required for data acquisition and routine on-line control. The computer would also need to initiate prompt protective action (up to and including shutdown) response to abnormal conditions.

Control requirements would, of course, vary greatly among the fusion-reactor concepts. In a tokamak, for example, operating steps include initial ohmic heating followed by neutral-beam heating of the DT plasma, densification, on-line refueling for periods of an hour of operation, impurity control, shutdown, and finally reloading of DT fuel to begin a new cycle. Inertial-confinement systems, by contrast, would be expected to operate in a pseudosteady-state mode of particle injection, implosion, and chamber purging.

Energy Conversion

The first generation of fusion reactors would need to be based on a DT fuel cycle for reasons identified at the beginning of the chapter. Neutrons carry 80 percent of the DT reaction energy, while charged particles account for the remainder. Employing present technology, the large neutron component can be converted to electrical energy only through a steam cycle with its inherently low (Carnot-limited) efficiency. It may be possible for the charged-particle energy to be converted directly with an efficiency up to 90 percent. However, since the neutrons do diminate the DT reaction, overall conversion efficiencies would be strictly limited in the early reactors.

A later generation of fusion reactors based on DD reactions, characterized by a 2/3 energy contribution from charged particles, would offer much better direct conversion efficiencies. The ultimate ^{11}B-p system, of course, would have the major advantage of producing entirely charged particles. Since the development of DD and ^{11}B-p systems would depend heavily on operating experience gained from DT systems, substantial direct energy conversion would be relegated to the somewhat distant future.

Fission Interfaces

Fusion reactors must breed tritium in order to operate in a steady-state mode with DT. Since more neutrons are produced than the number necessary to meet tritium requirements, it is possible for the extra neutrons to be used for alternative purposes.

The *fusion–fission hybrid reactor* concept would employ excess DT neutrons to cause fissions in an otherwise subcritical array of fissile and other fissionable materials. This would provide an energy multiplication effect augmenting the fusion output. In fact, it is suggested that the hybrid could lead to net energy production even if the fusion reaction itself is slightly below break-even. The fission–fusion hybrid has now become the official target for a first-generation fusion device in the U.S.S.R.

Alternative uses for the excess neutrons in the fusion fuel cycle include breeding and transmutation. A blanket of depleted uranium or thorium would employ neutrons to breed plutonium and ^{233}U, respectively (while not primarily producing energy). The use of excess DT neutrons to transmute actinide wastes (as described in Chap. 10) is another possibility.

Costs

Economic considerations will ultimately determine the viability of controlled fusion as an electrical energy source. Plants are expected to be highly capital-intensive with extremely low fuel costs. Operating and maintenance costs would tend to be particularly sensitive to the specific system design.

One early projection of fusion costs (DeBellis & Sabri, 1977) suggests the following breakdown for DT fusion systems:

1. capital costs 80-95 percent
2. operating and maintenance costs 5-20 percent
3. fuel costs 0.1 percent

Assuming that the *total* costs would make fusion viable in comparison to its major alternatives, the fuel component would dictate base-load operation (for reasons identified in Chap. 8).

Construction and Maintenance

Construction and siting requirements for fusion reactors are expected to be somewhat similar to those for fission systems. It is likely that a hot cell will be a necessary part of the fusion-reactor containment for handling activated first-wall and blanket segments. In the inertial-confinement systems, somewhat more stringent seismic and geologic conditions may be necessary to assure operation with close-tolerance alignment.

Since the very high neutron flux levels appear to preclude design of first-wall and blanket elements which can last the lifetime of the fusion reactor, the most critical maintenance activities would be related to replacement of such elements. Requirements for remote maintenance of the highly activated components are expected to be extremely important considerations in overall plant design and operation strategies.

Maintenance of other subsystems would also be challenging. Components like superconducting magnets, beam heaters, vacuum pumps, and tritium-handling systems may develop problems in the high radiation environments. Viable maintenance programs should include detailed inspection procedures as well as provisions for automatic detection and location of failures.

Safety and Environmental Impact

The potential safety and environmental concerns for controlled thermonuclear fusion appear to be lower than those for fission systems. Specific advantages of conceptual DT fusion reactors over present fission reactors appear to include:

- lower radionuclide inventory and relative biological hazard
- reduced hazard from long-lived wastes
- low radioactive afterheat (i.e., decay heat)
- limited energy release from reactivity accidents
- minimal material safeguards requirements

These assume, of course, that the fusion reactors are not operated in a hybrid mode with fissionable materials. In the latter case, the systems would share the potential hazards of both systems (although the subcritical configuration of fissile material would preclude reactivity transients).

In the early stages of development of fusion systems, safety and environmental impact evaluations are somewhat difficult to make. They are very useful, however, in directing the design process toward minimizing potential deficiencies. For fusion to be an attractive power source, a number of issues must be addressed adequately.

Substantial inventories of tritium will be associated with the first-generation fusion reactors. Escape of tritium in water (as HTO) or air (as HT) is problematical since isotope separation is not practical. Procedures which prevent tritium from entering the liquid or gaseous streams would be the most effective control mechanisms. Since tritium has a 12.3-year half-life and is readily transported by air and water, it is a contributor to global radiation exposures as well as to local exposures.

The other major source of radioactivity in fusion systems is activation. Under normal operating conditions, the products are expected to be fairly well immobilized in the structural materials. Disposal of activated first-wall and blanket elements would constitute one of the major environmental impacts.

The strong magnetic fields employed in tokamaks and other systems may produce subtle effects on biological tissue. Likewise, stray laser light might also have an impact. These and other concerns must be addressed more fully to assure minimal effects on operating personnel.

Major accident scenarios for fusion reactors would be based on:

- stored energy in the plasma
- stored energy in superconducting magnets
- lithium fires

The first two of these appear to provide self-limiting consequences that would tend to be localized. Lithium fires are generally considered to be the basis for fusion's "maximum hypothetical accident."

Lithium, like sodium, reacts vigorously and exothermally with air, water, and concrete. Although the energy production itself might cause significant damage, potentially more serious would be the immediate release of contained tritium and the probable later release of activation products from structures.

Overall worst-accident consequences from fusion-reactor accidents appear to be lower than those for fission reactors by about a factor of 100 (see DeBellis & Sabri, 1977). Waste management problems appear, likewise, to be of lower order than for current fission-based systems.

EXERCISES

Questions

21-1. Explain the term *controlled thermonuclear fusion.*

21-2. Describe the DT, DD, and ^{11}B-p fusion reactions and identify the relative advantages and disadvantages of each for potential commercial applications.

21-3. Identify the similarities and differences between fuel cycles for DT fusion and for fission.

21-4. Describe the bases for the Lawson criterion. Explain the fundamental difference between attempts to satisfy it by magnetic confinement and inertial confinement systems.

21-5. Sketch the components for a generic fusion reactor which uses DT fuel.

21-6. Differentiate among the tokamak, magnetic mirror, and pinch concepts for magnetic-confinement fusion.

21-7. Describe the isentropic compression mechanism that is desired for inertial-confinement fusion. Compare lasers, electron beams, and light-ion beams as inertial-confinement fusion drivers.

21-8. Explain the following for potential commercial applications of controlled fusion:

    a. maintenance requirements due to radiation damage of the first wall

    b. fusion-fission hybrid

    c. maximum hypothetical accident

Numerical Problems

21-9. Estimate the energy release from DT, DD, and ^{11}B–p fusion, using Fig. 2-1.

21-10. Compute the energy split between the products of DT fusion based on conservation of energy and conservation of mass. (Assume their net momentum is zero prior to the reaction.)

21-11. Repeat Prob. 2-8 for DD and ^{11}B–p reactions. Use 11.009305 amu as the mass for ^{11}B.

21-12. Estimate the fraction of thermal neutron adsorptions that would be expected in the ^{6}Li and ^{7}Li components of natural lithium. Would isotopic enrichment of lithium be of value in DT reactors? Why?

21-13. Compare projected fusion systems to present LWRs in terms of percentage cost components for capital, operations and maintenance, and fuel. Explain why a fusion system would be preferred for base-load operation over a fission system with the same total power costs.

SELECTED BIBLIOGRAPHY[†]

Fusion (General)

    Blake, 1980

    Chen, 1974

    Clarke, 1980

    CONAES, 1978*b*

    DeBellis & Sabri, 1977

    Dingee, 1979

    Glasstone, 1974

    Glasstone & Loveberg, 1960

    Gough & Eastlund, 1971

    Kammash, 1975

    Kenton, 1977

    Krall & Trivelpiece, 1973

    Olds, 1978

    Rose & Clark, 1961

    Rose & Feirtag, 1976

    Veziroglu, 1978

    Zaleski, 1976

Magnetic-Confinement Fusion

    Calrson, 1979

    Chen, 1977, 1979

    Coppi & Rem, 1972

    Hubbard, 1978

    Murakami & Eubank, 1979

    Nucl. Eng. Int., Feb. 1977, March 1977, March 1978

    Rose, 1979

    Steiner & Clarke, 1978

[†]Full citations are contained in the General Bibliography at the back of the book.

Inertial-Confinement Fusion
>   Booth & Franke, 1977
>   Emmett, 1974
>   Freiwald & Frank, 1975
>   LA-79-29, 1979
>   LLL Fusion, 1977
>   Sandia, 1980
>   Schwarz & Hora, 1974, 1977
>   Stickley, 1978
>   Varnado, 1977
>   Yonas, 1976, 1978

Applications
>   Bethe, 1979
>   Crocker, 1980
>   Easterly, 1977
>   Holdren, 1978
>   Parkins, 1978

Current Information Sources
>   *Nucl. Eng. Int.*
>   *Nucl. Sci. Eng.*
>   *Nucl. Tech.*
>   *Phys. Today*
>   *Science*
>   *Trans. Am. Nucl. Soc.*

Other Sources with Appropriate Sections or Chapters
>   Connolly, 1978
>   Foster & Wright, 1977
>   Lamarsh, 1975
>   Lapedes, 1976

APPENDIXES

APPENDIXES

I

NOMENCLATURE

The following list identifies variables with typical units, indexes, and uniquely nuclear units employed in this text. It is generally confined to those terms used in more than one chapter and those employed routinely in nuclear engineering practice. The following notations are employed:

(sub)—subscript
(sup)—superscript
(unit)—unit of measure

A	atomic mass number; area, m^2; arbitrary constant; arbitrary dimension
a	absorption (sub)
amu	atomic mass unit
at%	atom percent
at	atom; atom (sub)
B	arbitrary dimension; bulk (sub)
B^2	buckling, cm^{-2}
B_g^2	geometric buckling, cm^{-2}
B_m^2	material buckling, cm^{-2}
B.E.	binding energy, MeV
Bq	Becquerel (unit)
b	barn (unit)
C	concentration, at/cm^3; compound nucleus; arbitrary dimension
c	speed of light, m/s; heat capacity, $W \cdot s/g \cdot °C$; capture (sub)
c	centerline (sub)
D	diffusion coefficient, cm; distribution coefficient
DB^2	"leakage cross section," cm^{-1}
D-D	deuterium fusion reaction

D-T	deuterium–tritium fusion reaction
d	density, g/cm^3; differential; deuteron
dis	disintegration
E	energy, J
$\bar{E}$	average energy, J
e	electron; charge on electron (unit); elastic scattering (sub); enriched fraction
eV	electron volt (unit)
F	peaking factor
F_E	engineering factor
F_Q	heat flux factor
$F_{\Delta H}$	"hot channel" or enthalpy rise factor
f	fraction; void fraction; fuel (sub); thermal utilization; fission (sub); fast neutron (sub); fractional energy transfer
Gy	Gray (unit)
g	acceleration due to gravity, m/s^2; energy group number; energy group number (sub)
H	height, m; enthalpy [energy], J
h	convective heat transfer coefficient, $W/cm^2 \cdot °C$
i	index (sup, sub); delayed neutron group number (sub)
J	neutron current density, n/cm^2
j	index (sup, sub); nuclide ID (sup)
KE	kinetic energy, J or MeV
k	Boltzmann constant, J/K; heat transfer coefficient, $W/cm \cdot °C$; index (sup); effective multiplication factor
k_{eff}	effective multiplication factor
k_∞	infinite multiplication factor
L	thermal diffusion length, cm
l	neutron lifetimes; length, cm
l^*	prompt neutron lifetime, s
m	mass, kg; particle mass, amu; moderator (sub); metastable state (sup)
max	maximum (sub)
min	minimum (sub)
mix	mixture (sup)
M	nucleus mass, amu
M^2	migration area, cm^2
MO_2	mixed (U + Pu) oxide
MOX	mixed (U + Pu) oxide
N	neutron density, n/cm^3
n	neutron
n	neutron (sub); atom density, at/cm^3
n'	neutron from scattering reaction
nvt	neutron fluence, n/cm^2
P	power density, W/cm^3; power, W; cumulative probability distribution function; probability
P_{nl}	nonleakage probability
P_{tnl}	thermal neutron nonleakage probability

P_{fnl}	fast neutron nonleakage probability
p	pressure, MPa; proton; particle; resonance escape probability; probability density function
$p\text{-}^{11}B$	proton–boron-11 fusion reaction
Q	Q value, MeV; activity, dis/s
QF	quality factor
q	electric charge, C; heat rate, W
q'	linear heat rate, W/m
q''	heat flux, W/cm^2
q'''	volumetric heat rate, W/cm^3
R	radius, m; range, cm; dose rate, rad/h
R	Roentgen (unit)
r	radius, m; arbitrary nuclear reaction (sub)
rad	radiation absorbed dose (unit)
rem	Roentgen equivalent man (unit)
S	surface; surface area, cm^2
S_o	radiation source, #/cm^3
Sv	Sievert (unit)
S_n	discrete ordinates formulation
SF	spontaneous fission
s	neutron scattering (sub)
T	temperature, °C or K; period, s
T_S	surface temperature, °C
T_B	bulk temperature, °C
$T_{1/2}$	half-life, s
t	time, s; total (sub)
th	thermal; thermal neutron (sub)
u	neutron lethargy
V	volume, m^3
v	speed or velocity, m/s; void (sub)
vol%	volume percent
wt%	weight percent
x	distance
Z	atomic number
X, x; Y, y; Z, z	arbitrary position coordinates, dimensions, distances; arbitrary reaction products
α	alpha particle [$^4_2$He^{2+}]; capture-to-fission ratio; E_{max}/E ratio; reactivity coefficient; enrichment separation factor; reprocessing separation factor
β	beta particle; delayed neutron fraction; magnetic confinement fusion ratio
β^-	electron [$^0_{-1}e$]
β^+	positron [$^0_{+1}e$]
β_{eff}	effective delayed neutron fraction
γ	gamma radiation; fission yield
Δ	mass defect, amu
δ	extrapolation distance, cm

ϵ fast fission factor

η thermodynamic efficiency; average number of neutrons produced per neutron absorbed in fuel

λ radioactive decay constant, s^{-1}; mean free path, cm^{-1}

Λ effective prompt neutron lifetime, s^{-1}

μ linear attenuation coefficient, cm^{-1}

μ/ρ mass attenuation coefficient, cm^2/gm

ν average number of neutrons per fission; neutrino

ν^* antineutrino

$\nu\Sigma_f$ "neutron production cross section," cm^{-1}

ρ density, g/cm^3; reactivity

σ microscopic cross section, b

Σ macroscopic cross section, cm^{-1}

Σ summation

τ mean lifetime, s; Fermi age, cm^2; plasma confinement time, s

θ polar angle

Φ neutron flux (units vary according to functional dependence)

ϕ azimuthal angle; group neutron flux, $n/cm^2 \cdot s$

χ fission neutron energy spectrum

Ω direction vector

ω inverse period, s^{-1}

$ dollar of reactivity (unit)

¢   cent of reactivity (unit = 0.01$)

0   initial condition or constant value (sub)

0( )   thorium isotope with A number ending in ( ) (e.g., 02 $\rightarrow$ ^{232}Th)

2( )   uranium isotope with A number ending in ( )

4( )   plutonium isotope with A number ending in ( )

$\langle\ \rangle$ mean [average] value

II

UNITS
AND CONVERSION FACTORS

The International System of Units [SI], a modernized metric system, is becoming the world's most common language for expressing scientific and technical data. The SI consists of seven base units, a series of consistent derived units, and approved prefixes for the formulation of multiples and submultiples of the various units. Base units, prefixes, and some derived units for the SI are shown in Tables II-1, II-2, and II-3, respectively.

Since much of the literature base for nuclear engineering still contains U.S. Customary Units (and other units), Table II-4 provides some useful conversion factors. The remaining Tables II-5 and II-6 contain physical constants and approximate equivalents for various energy sources.

A comprehensive reference on SI is "Metric Practice Guide," E 380-74, American National Standard Z210.1, American Society for Testing and Materials, 1974 (updated periodically). This and other works (e.g., GE/Chart, 1977) also contain useful conversion tables.

TABLE II-1
SI Base Units

Quantity	Unit	SI symbol
Length	meter[†]	m
Mass	kilogram	kg
Time	second	s
Electric current	ampere	A
Temperature	kelvin	K
Luminous intensity	candela	cd
Amount of substance	mole[‡]	mol

[†]Also spelled "met*re*."
[‡]Mole is the amount of substance that contains as many elementary entities (atoms, ions, electrons, other particles, or specified groups of such particles) as there are atoms in 0.012 kg of carbon-12. The basis for the ^{12}C reference is discussed in Chap. 2.

TABLE II-2
SI Prefixes

Prefix	SI symbol	Multiplication factor
tera	T	10^{12}
giga	G	10^{9}
mega	M	10^{6}
kilo	k	10^{3}
hecto[†]	h	10^{2}
deca[†]	da	10^{1}
deci[†]	d	10^{-1}
centi[†]	c	10^{-2}
milli	m	10^{-3}
micro	μ	10^{-6}
nano	n	10^{-9}
pico	p	10^{-12}
femto	f	10^{-15}
atto	a	10^{-18}

[†]These prefixes that are not integral multiples of three are to be avoided where possible. However, their use for areas and volumes is acceptable.

TABLE II-3
Some Derived SI Units

Quantity	Formula	SI symbol	Unit
Acceleration	m/s^2		
Activity (radioactive source)	(dis)/s	Bq	becquerel
Area	m^2		
	cm^2		
Density	kg/m^3		
	g/cm^3		
Electric potential difference	W/A	V	volt
Energy	N·m	J	joule
Force	kg·m/s^2	N	newton
Frequency	s^{-1}	Hz	hertz
Length	cm		centimeter
	km		kilometer
Mass[†]	g		gram
	Mg		megagram
Power	J/s	W	watt
Pressure	N/m^2	Pa	pascal
Quantity of electricity	A·s	C	coulomb
Quantity of heat	N·m	J	joule
Specific heat	J/kg·K		
Thermal conductivity	W/m·K		
Velocity [speed]	m/s		
Voltage	W/A	V	volt
Volume	m^3		
	dm^3	l	liter[‡]
Work	N·m	J	joule

[†]Note that the *kilo*gram is the base unit, while the gram [1 g $= 10^{-3}$ kg] and megagram [1 Mg $= 10^3$ kg] are derived units.

[‡]Also spelled "li*tre*."

TABLE II-4
Useful Conversion Factors

Quantity	Unit	Symbol	Equivalent of 1 unit
Base units			
Length	inch	in	25.4 mm
	foot	ft	0.305 m
	mile	mi	1.61 km
Mass	atomic mass unit	amu	1.6605×10^{-27} kg
	pound-mass	lbm	0.454 kg
	ton (short)	ton	907.1 kg = 2000 lbm
	metric ton [tonne]	MT[t]	1 Mg = 1.102 ton
			= 2204 lbm
Time	minute	min	60 s
	hour	h	3600 s
	day	d	86.4×10^3 s
	year	y	31.6×10^6 s
			($\approx \pi \times 10^7$ s)
Temperature[†]	degree Celsius	°C	1 K
	degree Fahrenheit	°F	(5/9) K
	degree Rankine	°R	(5/9) K
Derived units			
Activity	curie	Ci	3.7×10^{10} (dis)/s = 3.7×10^{10} Bq
Area	barn	b	10^{-24} cm^2
Density	lbm/ft^3		16.02 kg/m^3
Energy	atomic mass unit (equivalent)	amu	1.492×10^{-10} J = 931.116 MeV
	calorie	cal	4.186 J
	British thermal unit	btu	1.055×10^3 J
	electron-volt	eV	1.602×10^{-19} J
	erg	erg	1×10^{-7} J
Force	pound-force	lbf	4.448 N
	dyne	dyne	1×10^{-5} N
Power		btu/h	0.2931 W
	horsepower	hp	746.0
		erg/s	1×10^{-7} W
Pressure	atmosphere	atm	1.013×10^5 Pa = 14.70 lb/in^2
	bar	bar	1.000×10^5 Pa
		lb/in^2	6.894×10^3 Pa
Volume	cubic foot	ft^3	28.32 l = 0.0283 m^3
	gallon	gal	3.785 l
	quart	qt	0.946 l
Work (see energy)			

[†]Conversion among the temperature scales $T(x)$ use the following formulas:

$$T(°C) = T(K) - 273.16 \text{ K} = [T(°F) - 32]/1.8$$

$$T(°F) = 1.8\ T(°C) + 32 = T(°R) - 459.67$$

$$T(°R) = 1.8\ T(K) = T(°F) + 459.67$$

$$T(K) = T(°C) + 273.16 \text{ K} = [T(°F) + 459.67]/1.8 = T(°R)/1.8$$

TABLE II-5
Physical Constants

Name	Symbol	Value
Avagadro's number	A_o	6.025×10^{23} molecules/mol
Boltzmann constant	k	1.380×10^{-23} J/K
Electron rest mass	m_e	9.11×10^{-31} kg $= 5.5 \times 10^{-4}$ amu
		$= 0.511$ MeV
Elementary charge	e	1.602×10^{-19} C
Neutron rest mass	m_n	1.008665 amu $= 939.53$ MeV
Proton rest mass	m_p	1.00727 amu $= 938.23$ MeV
Speed of light (in vacuum)	c	3.00×10^8 m/s

TABLE II-6
Approximate Energy Content of Various Alternative Energy Sources

| Form | Energy equivalent | | |
	J	kW·h	btu
1 barrel [42 gal] oil	6.1×10^9	1.70×10^3	5.8×10^6
1 ton Eastern coal	28×10^9	7.62×10^3	26×10^6
1 ton low-sulphur Western coal	19×10^9	5.27×10^3	18×10^6
1 cord [2.5 ton] wood	$14-29 \times 10^9$	$3.8-7.9 \times 10^3$	$13-27 \times 10^6$
1 ft^3 natural gas	1.13×10^6	0.314	1070
1 lb natural uranium metal	210×10^9	58.6×10^3	200×10^6
1 langley solar energy	4.19×10^4 J/m^2	1.16×10^{-2} kW·h/m^2	3.69 btu/ft^2
1 joule	1	2.778×10^{-7}	9.48×10^{-4}
1 kW·h	3.60×10^6	1	3415
1 btu	1.055×10^3	2.930×10^{-4}	1
1 quad [q]	1.055×10^{18}	2.930×10^{11}	10^{15}
1 Quad [Q]	1.055×10^{21}	2.930×10^{14}	10^{18}

THE IMPENDING ENERGY CRISIS:
A PERSPECTIVE
ON THE NEED
FOR NUCLEAR POWER
IN THE UNITED STATES

Objectives

After studying this appendix, the reader should be able to:

1. Compare use rates and supplies for presently available fuels.
2. Identify advantages and problems associated with energy production from each of the fossil fuels.
3. Differentiate between the potential impacts of voluntary and enforced conservation measures.
4. Identify advantages and problems related to energy production from new technology energy sources.
5. List several conclusions of the report by the National Research Council's Committee on Nuclear and Alternative Energy Systems [CONAES].
6. Identify the four major energy alternatives that are considered necessary to avert a severe energy crisis.
7. Perform basic calculations related to energy demands, supplies, and costs.

The main body of the book is intended to provide a technical description of current applications of nuclear energy. It does not specifically address the proper role of this energy form among the available alternative sources.

When viewed in isolation, nuclear technology may appear to be too complex to justify its wide-scale implementation. Comparison with alternative energy sources, however, serves to modify this perception.

The intent of this appendix is to provide perspective on the potential role of nuclear energy in meeting present and near-future energy requirements. The necessary policy·decisions must be based on many factors other than technical evaluation.

ENERGY CRISIS

The term *energy crisis* generally implies a severe shortfall of energy supply with respect to demand. Actual consequences of such a situation would depend on the nature and use of the resources involved, the timing, the affected population, and other factors. Since some turmoil would be associated with any energy crisis, avoidance is desirable (except, perhaps, to those who might benefit politically or economically from the anarchy that would result from such disruption).

Problems

Problems related to energy supplies and utilization may be divided roughly into three interrelated categories—domestic, international, and technological. Although the concerns vary with location and time, several fundamental features may be identified.

Domestic energy concerns are related largely to the availability of specific resources at reasonable and competitive costs. Occupational health and safety considerations, as well as overall environmental impact, provide increasingly important constraints to the development and use of resources.

The international dimension of the energy picture relates largely to each country's desire for national security (military and economic). Costs (as they relate to unfavorable balance of payments), supply uncertainties, and foreign policy limitations have become increasingly important concerns in recent times.

The technologies responsible for converting resources into usable energy are constrained by considerations of total costs, reliability, and environmental impact. Even the use of "free" fuel (e.g., solar energy) may be impractical if conversion costs and impacts are unacceptably high. Energy conservation in new and/or existing systems can be a very important technological contribution to reducing energy demand.

Resources and Demands

The major currently used energy resources are the fossil fuels—natural gas, petroleum, and coal—and uranium. Recent estimates of potentially recoverable energy from these and several other U.S. resources are shown by Fig. III-1. The reference unit for the figure is the *quad* ($\equiv 10^{15}$ btu $= 1.06 \times 10^{18}$ J). The white areas represent resources considered recoverable with current technology. Shaded areas represent those which would require technological development. The inset allows comparison of 1975 and projected 1975-2000 requirements to the resources. Although changes in specific values in Fig. III-1 will occur with the passage of time, the relative rankings are unlikely to change substantially. One additional current resource not shown on the figure is hydroelectric, estimated at a maximum of about 5 quad/year.

According to Fig. III-1, geothermal, oil shale, solar, and fusion are large potential resources. None is currently employed on a large commercial scale. Geothermal and oil shale uses will require new technological developments as well as necessary policy decisions. Although solar applications are developing rapidly, widespread use of this diffuse energy source is still in the somewhat distant future.

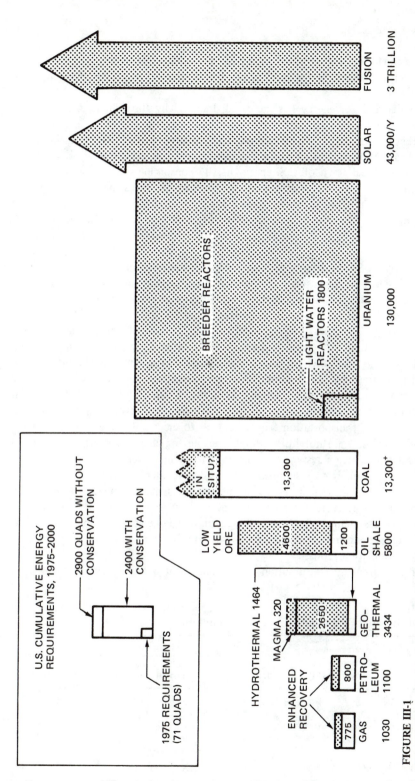

FIGURE III-1

Potentially recoverable domestic energy resources shown graphically by area. (Data from ERDA-76-1, 1976.)

Nuclear fusion offers the prospect of a nearly inexhaustible energy supply but has not yet been developed to the level of laboratory break-even (as described in Chap. 21).

The demand for energy has increased dramatically throughout history. Table III-1 shows approximate per capita consumption as a function of time. Energy for food gathering has been supplemented sequentially by that for household use (initially heating), organized agriculture, industry, and transportation.

A recent breakdown of U.S. energy use by consuming sector is shown by Fig. III-2. Comparison with the resource estimates shown on the same figure shows the striking fact that roughly three-quarters of the current energy supply is drawn from the least abundant of the nonrenewable resources, i.e., from oil and natural gas.

Balance

Any smoothly operating energy economy depends on a reasonable balance between supply and demand. This requires that both total-energy and individual-resource supplies and demands be in an acceptable equilibrium at nearly all times.

The fossil fuels are currently employed in two distinct ways. Direct chemical use occurs when the fuel is burned at the site of ultimate energy use, e.g., as gasoline in an automobile engine, natural gas in a furnace, or coal in a steel mill. In electrical use, the fossil fuel (or any fuel) is burned at a central location with the energy transmitted by conducting lines for final consumption. Chemical and electrical energy supplies and demands must both be balanced to avert problems (at least within the time span required for conversion of existing systems as, for example, changing transportation from chemical to electrical energy sources). Even the three fossil fuels are not fully interchangeable for chemical energy applications without relatively major system modifications (e.g., in vehicles and space heating).

Historically, energy consumption has tended to increase year by year due to growing population as well as escalating per capita demand. Population growth in the United States is currently about 1.2 percent. Although the birth rate is now equivalent to less than 2 children per woman, the large female population below child-bearing age assures continued increase. Immigration is an even larger contributor to population growth.

Per capita energy consumption was increasing at a rate of about 2.4 percent

TABLE III-1
Historical Growth of Per Capita Energy Consumption

Time	Average energy consumption per person (MJ/d)	Principal uses				
		Food	House-hold	Agriculture	Industry	Transportation
1,000,000 B.C.	8	X				
100,000 B.C.	20	X	X			
5,000 B.C.	40	X	X	X	X	
A.D. 1400	150	X	X	X	X	X
A.D. 1875	360	X	X	X	X	X
A.D. 1970	960	X	X	X	X	X

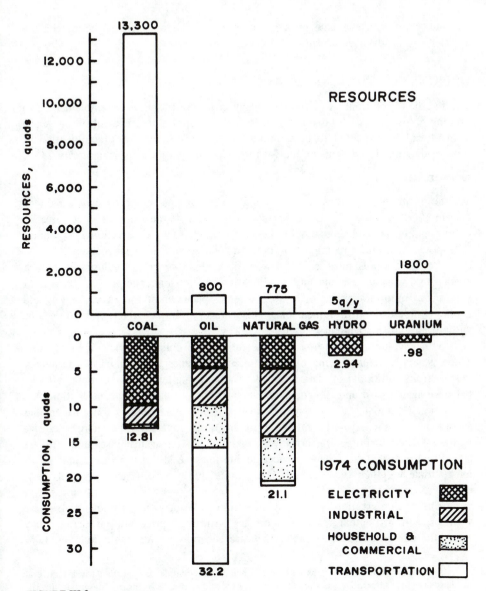

FIGURE III-2
Available domestic energy resources compared to 1974 consumption rates. (Data from
ERDA-76-1, 1976 and EEI, 1976.)

per year as recently as 1978. This was due partially to increased use of labor-saving
equipment. It may also relate to the smaller family size since, for example, existing
homes and appliances now serve fewer individuals. Per capita consumption has
begun to level off as energy costs have escalated into the 1980s.

No serious energy imbalances would occur if supplies and demands were to
remain stable at current levels. Demand growth, however, must be compensated by
increased supply. Conversely, supply reduction must be balanced by decreased

demand. Only with adequate prior planning can supply and demand be matched to avert energy crises of either short or long duration.

OPTIONS

The so-called "solutions" to potential energy crises are the measures implemented to prevent supply and demand mismatch. It does not appear that control of any single element can be expected to be adequate in this regard. Instead, a combination of supply adjustment and demand control through conservation appear necessary.[†]

Conservation

The term *energy conservation*[‡] is generally used by the public to signify reduced consumption of one or more specific resources. It also implies decreases in corresponding demand sectors. The degree to which such conservation is voluntary or imposed determines its impact on both the energy and financial economies.

Conservation is most effective when individuals or groups voluntarily reduce energy use without making any modifications to existing equipment, devices, or structures. Important examples include lowered space conditioning (i.e., heating and cooling) and less motor vehicle use. This category allows reductions in one area without any apparent increase elsewhere.

Other voluntary measures include those that enhance efficient energy use. These generally require some expenditure of both energy and money. The energy savings from additional insulation added to an existing home, for example, must be balanced against cost and the energy expenditure of production and installation.

The potential routine energy savings from a voluntary shift to smaller automobiles and energy-efficient buildings are quite apparent. The energy-intensive nature of their manufacture, however, generally dictates against discarding vehicles and structures that still have serviceable lifetimes. Additionally, most individuals must "trade in" their old cars and houses to be able to afford new, energy-efficient ones.

Energy conservation is most readily imposed on a society through high prices with or without actual resource shortages. Although price-induced savings may affect the energy balance favorably, they are generally less desirable than voluntary

---

[†]A small computer known as an Energy-Environment Simulator is particularly effective at emphasizing the interactive nature of the supply and demand measures described in the remainder of this appendix. It has been an integral part of the Citizens' Workshops on Energy and Environment that have been sponsored by the U.S. Department of Energy and its predecessors. The Simulators are currently available at about 100 sites across the United States under the general supervision of Oak Ridge Associated Universities in Oak Ridge, Tennessee. (Anticipated withdrawal of funding by DOE in the fall of 1980, however, makes the future of the units uncertain.)

[‡]It may be noted that a scientist applies the name *energy conservation* to a fundamental law of nature which states that energy can neither be created nor destroyed. Energy can, however, be converted from one form to another. Conservation as applied to public use is actually based on decisions *not* to convert *potential* energy (e.g., chemical potential energy in fossil fuels, or nuclear potential energy in uranium) to *heat* energy (e.g., for space heating or electrical generation).

measures. The major concern is the potential for inequitable distribution of resources within the society. Very severe shortages, especially those which could appear suddenly (e.g., as from an internationally based embargo), could be expected to cause economic recession and societal instability.

Conservation must be an extremely important measure in any attempts to avert future energy crises. Both voluntary and imposed conservation measures should contribute. However, the recognition that it is not necessarily "free" must temper any policy decisions.

Natural Gas

Natural gas is a clean-burning fuel that is very convenient for cooking, heating, and a variety of industrial applications. If its supply were inexhaustible, other energy sources would receive little attention. Existing supplies of this finite resource are predicted to last 40 years or less at current use rates. In addition, both discovery and production rates have been declining since the mid-1970s.

Rising costs have caused some reduction in the use of natural gas. Supply shortages have not been uncommon, especially during extremely cold weather. During the harsh winter of 1977, for example, severe natural gas shortages in portions of the United States led to widespread discomfort, layoffs of over half-a-million workers, and several deaths.

Domestic natural gas production may be augmented by imports, usually in the form of liquified natural gas [LNG]. In the longer time frame, gasification should make it feasible to merge the large coal resource (Fig. III-1) with that for natural gas.

Petroleum

Petroleum, or simply oil, is currently the world's most widely used energy resource. Like natural gas, it is convenient and relatively clean burning. Current oil supplies are predicted to last less than another century at current rates of consumption.

Oil supply in the United States is divided about evenly between domestic production and imports at the beginning of the 1980s. As land-based production increased in cost with declining productivity, off-shore drilling offered the most promise for new reserves. The potential for serious environmental impact, however, provides a substantial basis for opposition to off-shore oil operations.

Since the advent of the Organization of Petroleum Exporting Countries [OPEC] in the early 1970s, oil prices have risen continuously, and international power has shifted substantially toward its member nations. Unfavorable trade balances among the industrialized nations could often be attributed directly to large expenditures for oil. Expenditures at a rate of $90-100 billion per year in the United States in early 1980 promised continuing trade deficits as well as increasing economic power for OPEC nations.

Coal

Among the fossil fuels, coal resources are by far the largest (Fig. III-1). Current utilization is relatively low (Fig. III-2), although calls for doubling coal use have been common since the mid-1970s.

The general problems of coal use are summarized by the statement: "There are two things wrong with coal today. We can't mine it and we can't burn it."[†] It relates, of course, to the hazards of mining operations and environmental pollution.

Coal mining is conducted by both surface and underground operations (similar to those for uranium as described in Chap. 8). Surface [strip] mining causes unsightly changes in the land, but is relatively safe for mine workers. Underground mining has much less visual impact, but poses dangers ranging from black-lung disease to cave-in fatalities for miners. The risks to mine workers have led to relatively high wages and to formation of strong labor unions that attempt to protect worker interests.

The use of coal for electric power production is accompanied by evolution of gaseous (SO_2, nitrous oxides, and hydrocarbons) and particulate ash emissions. Typical current technology removes about 90 percent of the SO_2 and 99.7 percent of particulates at increments of about 20 percent on capital cost and 5 percent in energy consumption. The wastes from these antipollution devices result in a disposal problem. (As a matter of fact, all such wastes contain naturally occurring radioactivity in the form of uranium, radium, potassium-40, and other species. The concerns are not unlike those for uranium mill tailings described in Chap. 8).

Expansion of coal production would require opening new mines and recruiting new miners. It would also necessitate new transportation systems and equipment manufacture. As an example, a recent study by the National Academy of Engineering (NAE, 1975) concluded that representative requirements for doubling coal production would include:

- In the eastern United States—
  140 new underground mines (2×10^6 t/year)
  30 new surface mines (2×10^6 t/year)
  80,000 new coal miners
  60 new rail-barge systems (150–800 km each)
- In the western United States—
  100 new surface mines (5×10^6 t/year)
  45,000 new coal miners
  70 new rail-barge systems (1500–2000 km each)
- Plus—
  2400 continuous mining machines
  140 75-m^3 [100-yd^3] shovels and draglines
  8000 railroad locomotives
  150,000 100-t gondola and hopper cars
  4 1600-km slurry pipelines (25×10^6 t/year)
  2 1600-km gas pipelines

It has been proposed that coal seams which are too deep for surface mining and too narrow for conventional underground mining techniques may be converted by in situ coal gasification to enhance the supply of natural gas. Full development

---

[†]This statement has been attributed to several individuals, including S. David Freeman, the current director of the Tennessee Valley Authority.

of the technology will require a major commitment by private or governmental concerns. Water requirements could be a major problem for large-scale gasification of otherwise favorable coal deposits in the Rocky Mountains and southwestern United States.

Hydroelectric

The potential energy of falling water has long been used for generating electricity. This hydroelectric energy, derived from both natural falling water (e.g., Niagara Falls) and artificial dams, is essentially "free" from a fuel standpoint and does not produce any pollutants.

One major disadvantage of hydro power is its limited availability. For example, something over half of the United States resource is already in use, as shown by Fig. III-2. The scenic beauty of many of the potential sites (e.g., the Grand Canyon and Canyonlands areas of the Southwest) is also likely to dictate against fully expanded use of hydroelectric generation.

Since the energy potential of a specific dam is proportional to the volume of water it impounds, this potential decreases with time as silt accumulates in the reservoir. Annual precipitation levels that are far below normal also reduce the amount of energy available from hydro sources. (The latter was particularly evident in the Pacific Northwest during the late 1970s.)

Recognition of the potential environmental impacts on the land flooded by dam construction has restricted the expansion of this energy source. Even for existing facilities, it is now common to restrict water discharges so that fish populations are not endangered. The potentially severe consequences of dam failures have also been more clearly recognized (e.g., as a result of serious incidents in Idaho and Georgia in the late 1970s).

New Technologies

Energy sources that are not currently available for large-scale use are commonly classified as new technologies. Even though the principles may be well established, extensive engineering research and development activities are required before economy and reliability are commensurate with the use of other resources.

Two important new technologies—solar and geothermal—are already in use on a limited scale. Nuclear fusion has not yet reached energy break-even on a laboratory scale and, thus, has only long-term potential (as described in detail in Chap. 21). Other technologies may have valid local applications.

Direct Solar
In the most general terms, solar energy sources encompass all direct and indirect applications of the sun's energy. General classifications include:

1. solar-thermal—utilize sun's heat directly
2. solar-electric—convert sunlight directly to electricity
3. wind—employ air currents produced by differential atmospheric heating
4. thermal gradients—utilize temperature differences from uneven heating of large bodies of water
5. biomass conversion—employ sunlight to cultivate fast-growing plants that can be burned to produce energy

The first two constitute direct solar applications, while the remaining entries are more indirect. Although the sun's energy constitutes a truly free resource, its widespread utilization still awaits adequate technological development and reduction in the costs thereof. The diffuseness of the solar energy may also constrain its ultimate contribution to the total energy balance.

Solar-thermal applications to space heating and water heating are becoming increasingly economical as fossil fuel costs escalate. However, capital expenditures for such active systems were still high in 1980 (even in "sun-belt" areas of the Southwest). Enhanced manufacturing capacity and an increasingly well trained labor force should augment viability with each passing year.

Attention to design and construction can enhance the "passive solar" heating that already occurs in all buildings. Proper use of such features as insulation, glass windows and doors, curtains, and overhangs can increase heating effects during the winter months while reducing them during the summer.

Both the active and passive systems are more applicable to new construction than to retrofit on existing buildings. Since most of the buildings that will be used in the year 2000 have already been built, full solar heating in half of all new construction from 1980 to 2000 would be estimated to reduce U.S. heating demand by 8 percent or less.

Another major solar-thermal application is to the generation of electricity. Typical systems employ a large number of mirrors to focus the sun's energy on a collector from which a conventional steam cycle is operated. Current research and development activities are directed toward units which would produce tens of MW(e). Large systems with expectedly lower capital costs will be relegated to the more distant future. Construction material, land, and cooling-water requirements are all potential concerns. Another major problem to electric utilities (general requirements for which are discussed in Chap. 8) may result from the daily and seasonal variations in available solar energy. Resolution of these concerns will depend on improved technologies for storage of electrical energy.

The solar-electric concept employs photosensitive cells which are capable of converting sunlight directly to electricity. Although steam cycle problems would be avoided, concerns related to conversion efficiency, land area, and energy variation would still exist. Some promising materials for solar cells also pose environmental impact questions (e.g., currently popular gallium arsenide cells contain the toxic element arsenic).

It has been proposed that solar cells on an orbiting satellite could provide an essentially constant source of electricity. Although the concept could eliminate the land area and energy variation concerns, severe technological problems appear to preclude near-term implementation. Deployment of solar collectors in arrays with multikilometer dimensions would be a major engineering feat. Proposed energy transport back to earth via microwave radiation would have the potential for serious environmental impacts.

Wind

Simple wind generators were used successfully for pumping water and minor electrical applications in agricultural areas during the early to middle 1900s.

Widespread availability of natural gas and centrally-generated electric power, however, decreased wind use dramatically.

The energy available from wind at any given location depends on the time history of the wind speed. Local average wind speeds of 10-60 km/h are common. Variations about these averages range from a minimum of essentially zero to local maxima of about 80–240 km/h. Relatively large and constant wind speeds are most advantageous for electric power generation.

Wind provides a more diffuse energy form than sunlight, but has the advantage of being generally present during the night as well as the day. The Heronemus wind generator concept, for example, calls for a 220-m-high tower with 20 16-m-long propeller-like blades each fastened to a turbine generator. It has been estimated that application in the Great Plains would require 1800 such towers spaced about 1.6 km [1 mi] apart to generate 1000 MW(e).

The potential hazards of premature use of new technologies are well illustrated by a wind generator installed in Clayton, New Mexico, as a joint project of the Department of Energy [DOE] and the National Aeronautics and Space Administration [NASA]. The unit, configured roughly like a large airplane propeller, has been inoperable for several relatively long periods of time. On a number of occasions, it has been disassembled and shipped back to its point of manufacture in Ohio for needed repairs.

Thermal Gradients

Temperature differences produced by uneven solar heating in large bodies of water have the potential for energy applications. Many parts of the oceans, for example, exhibit a temperature difference of more than 20°C over 1000 m of depth.

A low-temperature "steam cycle" employing freon or ammonia could exploit the temperature difference with 2–4 percent thermal conversion efficiency. However, the systems would be very large and would strain the limits of current technology. A major concern is the relatively inhospitable ocean environment which would tend to produce both corrosion and fouling with marine life.

Biomass

Since early in human history, biomass in the form of wood has been employed as an energy source. Reforestation efforts ["tree farming"] are common, but have been generally directed toward nonenergy uses. Development of fast-growing wood fiber plants has been proposed as a method of providing an important energy resource. Direct burning or an intermediate conversion to alcohol are possible modes of use.

A typical biomass application could provide energy equivalents to about 3 percent of the local incident solar energy. A typical region could be expected to produce about 1 MW(th) per 2.5 km^2 [1 mi^2].

Major problems with biomass conversion include water resource impacts and hydrocarbon releases (during growth as well as combustion). Many fibrous plants will emit pollutants when burned which are similar to those from coal. During the winter of 1979-1980, for example, many communities requested their citizens to refrain from burning wood in fireplaces because of serious pollution problems. More efficient fireplaces and wood-burning stoves would reduce but not completely eliminate the problems.

Geothermal

The high temperatures in the interior of the earth's crust provide a potentially valuable energy source. Natural geothermal reservoirs deliver steam directly to the surface in a number of areas. Subsurface hydrothermal reservoirs may be tapped by drilling. Steam from these sources has been employed for electrical generation in Italy, New Zealand, and the Geysers Field in northern California.

In the absence of underground water supplies, the energy stored in hot, dry rock may be tapped by water injection. The steam produced by the process would then be returned to the surface for use in a steam cycle. Although the concept would, in principle, be workable anywhere, preliminary research and development activities are centered on areas where the hot, dry rock is relatively near to the earth's surface.

Some general disadvantages of geothermal energy have been identified as a result of experience with the Geyers Field, which currently provides about 400 MW(e) to northern California. One major drawback is the hydrogen sulfide [H_2S] gas which accompanies the natural steam. This gas has a bad smell (like rotten eggs) and causes atmospheric haze. Another concern relates to potential seismic activity. For example, one steam well in the Geysers Field remains uncapped for electrical production because attempts to do so begin to cause a landslide.

Nuclear

The potential energy available from nuclear sources increases from the current generation of light-water reactors [LWR], to breeder reactors, and to nuclear fusion as shown by Fig. III-1. The major disadvantages are all related directly or indirectly to the presence of radiation and radioactivity.

The energy available from nuclear fission sources hinges on the supply of uranium. With light-water reactors (Chaps. 1 and 12), the ^{235}U resource determines the energy potential. The introduction of a plutonium breeder reactor like the liquid-metal fast-breeder reactor [LMFBR] extends the fuel supply to include the more abundant ^{238}U (Chap. 15). Thorium would also be an energy resource with the introduction of ^{233}U–Th fuel cycles in the high-temperature gas-cooled reactor [HTGR] (Chap. 13) or thermal breeder reactors (Chap. 14).

Expansion of nuclear fission as a major U.S. energy source has been slowed substantially by perceived hazards related to:

- low-level radiation (Chap. 3);
- radioactive waste disposal (Chap. 10);
- reactor safety (Chaps. 16–18);
- nuclear safeguards (Chap. 20)

Other nations are being far more aggressive in pursuit of the nuclear energy option (as shown, for example, by Table 8-1) despite their recognition of the same potential hazards.

Controlled nuclear fusion would provide an essentially limitless energy supply. However, energy break-even on a laboratory scale is a required first step which must be followed by at least several decades of engineering technology development (Chap. 21).

PROPOSED SOLUTIONS

Each energy option has the potential for reducing severe mismatches in energy supply and demand. Each also has certain risks associated with large-scale dependence.

Conservation and solar form the basis for the "small is beautiful" (Schumacher, 1973) syndrome and the "soft energy path" (Lovins, 1977). These are particularly appealing scenarios because they would seem to provide "free" contributions to both sides of the energy balance. Although applications to new construction and community planning may be quite appropriate, the costs of transition for the world as it exists now could be extremely high in terms of both dollars and uncertainty-related risks (e.g., see Stiefel, 1979 and Rossin, 1980). Additionally, health and resource impacts of conservation and solar energy may be quite significant (e.g., see Inhaber, 1978).

A number of comprehensive studies including a recently concluded four-year, $4.1 million effort by the National Academy of Sciences' Committee on Nuclear and Alternative Energy Systems (CONAES, 1978, 1979) have concluded that a combination of *all* options is necessary. Conservation is extremely important, but should not be imposed to the extent of crippling the economy. Continued dependence on natural gas and oil is necessary, although reduction is desirable. The new technologies, especially solar and geothermal, can make important but relatively small contributions until their costs, reliability, and impacts can be established and their use can become widespread. Coal and uranium, despite their potential hazards, appear to be necessary well into the 21st century if energy crises are to be avoided:

> As fluid fuels are phased out of use for electricity generation, coal and nuclear power are the only economic alternatives for large-scale application in the remainder of this century. . . . A balanced mix of coal and nuclear-generated electricity is preferable to the predominance of either (CONAES, 1979).

Energy Policy

Serious problems with energy supply and demand can be avoided only if a coherent policy is developed around the various options. However, the inputs to the political decision-making process are often conflicting and may lead to inaction. A stated goal of increased coal production, for example, is seriously compromised by regulatory actions which prevent issuance of new coal leases and which restrict power-plant construction and operation. Energy actions perceived to have either increased cost or environmental consequences are bound to cause some public outcry and, thus, are politically unpopular.

Ideally, energy policy should be based on comparisons of costs, benefits, and risks. Decisions should provide a reasonable balance among available alternatives with adequate attention to the special risks of supply shortages. Direct effects of such shortages may be dramatic,[†] e.g., electrical blackouts affect hospitals, refrigera-

---

[†]The popular novel *Overload* (Hailey, 1979) provides a considerable amount of "food for thought" on this subject.

tion, or even availability of drinking water. The potential for political instability is of even more concern (recalling, for example, the short tempers and occasional shootings that accompanied relatively minor gasoline shortages during the Arab oil embargo in 1973 and the Iranian situation in 1979).

Regulatory activities must be responsive to the continuing needs for energy production. It must be recognized that unnecessary delays are costly (e.g., as discussed for nuclear plants in Chap. 8). Extended delays may also result in industry decisions to cut back on the manufacturing capabilities that would be necessary for later response to serious energy crises.

Many proposed energy policies seem to be characterized by wishful thinking that easy solutions exist or that the problems will go away by themselves if ignored. Unfortunately, the simple and neat answers also tend to be *wrong* answers. The bumper-sticker statetment "turn off the nukes, turn on the sun" is a case in point. Shutdown of nuclear plants would have unacceptable consequences in many parts of the country (e.g., in the New England states, where 70 percent of the December 1979–January 1980 electricity was drawn from such sources). Additionally, sound engineering practice suggests that the sun should be "turned on" (in an economic and reliable manner) *before* any existing technology is "turned off." Shutdown of nuclear plants would, thus, be most feasible if solar first proved to be capable of providing adequate replacement power.

Four-Part Solution

The ability of the United States to meet future energy requirements, and thereby avert a severe energy crisis, depends on a well-balanced energy policy. Such a policy should include all of the following energy alternatives:

1. continued dependence on natural gas and petroleum from both domestic production and imports
2. voluntary and involuntary conservation
3. near-term conversion to and increased reliance on coal and uranium
4. near- and long-term development of the new technologies

Current reliance on oil and gas is too great for large cut-backs to be made immediately. Instead, gradual decreases should follow successful conservation efforts and switchover of specific demands to other energy sources.

Well-planned, voluntary conservation efforts are preferred modes for reducing overall energy consumption. Although conservation measures imposed by arbitrary domestic or international actions are likely to be disruptive, they too may be expected to play a role in energy policy.

Despite potential problems, coal and uranium are needed energy resources, primarily because the technologies for their use are immediately available. Neither by itself appears to be capable of meeting the nation's differential energy requirements through the remainder of the twentieth century.

New technologies in the form of direct solar, wind, and geothermal are currently employed on a relatively small scale. Although the resource potential of these and related concepts warrants extensive research and development efforts, substantial energy payoff is at least several decades away. Nuclear fusion offers a tremendously large potential resource, but its utilization will require breakthroughs in basic research followed by a number of decades of engineering development.

Sound energy policy considers both short and long time scales. Continued use of oil and natural gas, conservation, and expansion of coal and nuclear are all necessary in the short term. The new technologies must be developed aggressively so that they may make small but significant near-term contributions while holding the prospect for very large impacts in the more distant future.

EXERCISES

Questions

III-1. Rank coal, oil, natural gas, and uranium in terms of
   a. present use rates
   b. resource availability

III-2. Identify advantages and problems associated with U.S. energy production from each of the fossil fuels.

III-3. Differentiate between the potential impacts of voluntary and enforced conservation measures.

III-4. Describe the concepts involved in the "new technologies"—geothermal, direct solar, wind, thermal gradients, and biomass. Identify one disadvantage of each energy form.

III-5. List several conclusions of the report by the National Research Council's Committee on Nuclear and Alternative Energy Systems [CONAES].

III-6. Identify the four energy alternatives which must be used in concert to meet future energy requirements, and thereby avert a severe energy crisis.

Numerical Problems[†]

III-7. During February 1980 the United States spent $7.7 billion on oil imports. Consider the potential impact of such expenditures by calculating the equivalent cost in terms of:
   a. number of $6,000 automobiles
   b. number of F-15 jet fighter planes ($17 million each)
   c. number of 1000-MW(e) nuclear power plants (see Chap. 8 for cost estimate)
   d. number of $60,000 residential homes
   e. number of days required to buy all of the common stock in each of the following corporations (at 3/28/80 prices):
      (1) American Telephone and Telegraph (652,884,000 shares/$48 each)
      (2) Exxon (447,400,000 shares/$57 each)
      (3) General Motors (287,705,000 shares/$46 each)
      (4) Playboy Enterprises (9,211,000 shares/$12.50 each)
      (5) one enterprise of local interest (at current share number and cost).

III-8. Repeat the previous problem for more recent oil costs and other updated data.

III-9. Exponential energy growth obeys equations similar to those used for

[†]Appendix II contains energy conversion factors that may be of use for some of these exercises.

radioactive decay, radiation attenuation, nuclear fuel depletion, and neutron kinetic growth or decay.

a. Explain why the following "rule of thumb" is a good approximation:

$$\text{Energy use doubling time in years} \approx \frac{70}{\text{percentage growth rate}}$$

b. Estimate the doubling time associated with a 1.2% annual population growth, a 2.4% annual per capita energy consumption, and the combination of the two.

c. Repeat (b) for current population and per capita consumption growth rates.

III-10. The following expression may be used to estimate the home energy consumption per heating season:

$$E = \frac{DH}{T}$$

where E = energy consumed by average home [J]

$D = \Sigma_{i=1}^{I} \Delta T_i$ = "Degree-days" for season of I days [°C·d]

where ΔT_i = mean temperature difference inside-to-outside on day i [°C]

$H = (\Delta H)_{\text{ave}}$ = Daily heat load for home at average temperature difference [J/d]

$T = (\Delta T)_{\text{ave}}$ = Average temperature difference [°C]

a. For a cold climate country with

$$(\Delta T)_{\text{ave}} = 40°C$$

$$\Sigma \Delta T_i = 2000°C·d$$

$$(\Delta H)_{\text{ave}} = 10^8 \text{ J/h}$$

calculate the total energy consumption for 1,000,000 homes.

b. If identical new homes were added at the rate of 20,000 per year, what would be the heat load in each of the next five years?

c. Assume that the results in (a) and (b) would represent a constant 11% of the country's total energy consumption for the given year. Calculate the potential energy savings (in J, percentage of heat load, and percentage of total national energy use) that would result from a 50% heat load reduction in all *new* construction.

d. Replace the condition in part (c) by solar heating which handles 75% of the load in 50% of the *new* homes.

III-11. Many newspapers now include information on solar energy in their weather forecast sections. An example is the following information from the *Albuquerque Journal* for December 30, 1978:

SOLAR ENERGY

The amount of solar energy received Saturday in Albuquerque 45 langleys; normal daily amount for the current month is 252 langleys.

a. If a solar electric power system were to be based on "normal daily" energy values, what land area would be required to meet Albuquerque's average electrical requirement of 400 MW? Assume a solar-thermal system with 60% collection efficiency and 30% thermal-electric conversion efficiency. Also assume that the energy can be stored and used as needed over a 24-h period.

b. If the energy could be stored for no more than 24 h, what land area would be required to meet the total electricity requirements for two or more consecutive days like Saturday, December 30, 1978?

c. What fraction of the area ($\sim$200 km^2) of the city of Albuquerque do results (a) and (b) represent?

III-12. Repeat Prob. III-11 for data on your own local area.

III-13. Assume a goal of 6000 MW(e) (1% of the current U.S. electrical capacity) is set for wind power by the year 2000.

a. Calculate the number of Heronemus generators required to meet the goal.

b. Determine the average daily rate from January 1, 1980 at which the generators would need to be completed to meet the goal.

c. Assuming one device per km^2, what area would the total population cover?

d. What fraction is (c) of the area of a Great Plains state like Kansas?

III-14. The hot, dry rock concept for geothermal energy production would employ the earth's $\sim$20°C/km temperature change with increasing depth.

a. Calculate the depth required for a geothermal temperature equivalent to that of the steam temperature for a light-water reactor [LWR]. (Assume a surface temperature of 25°C).

b. Repeat (a) for the steam temperature of a liquid-metal fast-breeder reactor [LMFBR].

SELECTED BIBLIOGRAPHY†

General References
Fowler, 1975
Hammond, 1973
Lapedes, 1976
Lapp, 1976
Sci. Am., 1971
U. Oklahoma, 1975

Energy Options
Benemann, 1978
Buckley, 1979
Carmichael, 1979
Catron, 1980
CONAES, 1978*b*
CONAES, 1979*a*
CONAES, 1979*b*
CONAES, 1979*c*
CONAES, 1979*d*
ERG, 1976
Glaser, 1977

†Full citations are contained in the General Bibliography at the back of the book.

Hartline, 1980
Hayes, 1979
Hopkinson, 1979
Johnson, 1977
Kreith & Kreider, 1978
Lichtblau & Frank, 1978
NAE, 1975
Nesbit, 1980
Olds, 1979*c, d*
Pruce, 1979*a*
Pruce, 1979*b*
Pruce, 1980
Rittenhouse, 1980
Ross, 1980
Smith, 1976
Whitaker, 1980
Williams, 1974
Wilson, 1977*a*

Cost/Risk/Benefit
AMA, 1978
Cazalet, 1978
Cohen & Lee, 1979
ERDA-69, 1975
Erdmann, 1979
Hailey, 1979
Inhaber, 1978
Keeny, 1977
Lave & Freeburg, 1973
Litai, 1980
McBride, 1978
O'Donnell & Mauro, 1979
Okrent, 1979
Okrent, 1980
Starr, 1969
Technology Review, 1979
Whipple, 1980

Energy Policies
Berman, 1980
Bethe, 1976
CONAES, 1979*a*
CONAES, 1979*b*
EEI, 1976
EEI, 1977
EEI, 1979
ERDA-76-1, 1976
Hopkinson, 1980
Lovins, 1977
Nesbit, 1979
Rossin, 1980
Rudman & Whipple, 1980
Schumacher, 1973
Stiefel, 1979
Stobaugh & Yergin, 1979

IV

REFERENCE REACTOR CHARACTERISTICS

Table IV-1 contains parameters that are typical of the five reference power reactor types identified in Chap. 1. Since the evolution of each design is essentially continuous, specific values may vary slightly in current offering. Data provided in various forms by the manufacturers is the basis for most entries in the table. Additional information was extracted from "Power Reactors 1978," *Nucl. Eng. Int.,* vol. 23, no. 274, July 1978 supplement.

The five reference reactor designs are:

- boiling-water reactor [BWR]
- pressurized-water reactor [PWR]
- pressurized heavy-water reactor [PHWR]
- high-temperature gas-cooled reactor [HTGR]
- liquid-metal fast-breeder reactor [LMFBR]

Specific systems were included as representative of each design concept (and on the basis of data availability to the author). The table may be most readily related to the text by noting that:

1. Reactor descriptions are contained in Chap. 1 plus the reference chapter listed.
2. Figures appropriate to specific data sections are identified [e.g., (1-10) for Fig. 1-10].

TABLE IV-1
Typical Characteristics for Five Reference Power Reactor Types

	BWR	PWR(B&W)	PWR(CE)	PWR(W)	PHWR	HTGR	LMFBR[†]
Reference design							
Manufacturer	General Electric	Babcock & Wilcox	Combusion Engineering	Westinghouse	Atomic Energy of Canada, Ltd.	General Atomic	Novatome[‡]
System (reactor station)	BWR/6	Babcock-241	System 80	(Sequoyah)	CANDU-600	(Fulton)	(Superphenix)
Reference chapter	12	12	12	12	13	13	15
General							
Steam cycle	(1-4)	(1-4)	(1-5)	(1-5)	(1-5)	(1-5)	(1-6)
No. loops	1	1	2	2	2	2	3
Primary coolant	H_2O	H_2O	H_2O	H_2O	D_2O	He	Liq. Na
Secondary coolant	–	–	H_2O	H_2O	H_2O	H_2O	Liq. Na/H_2O
Moderator	H_2O	H_2O	H_2O	H_2O	D_2O	Graphite	–
Neutron energy	Thermal	Thermal	Thermal	Thermal	Thermal	Thermal	Fast
Fuel production	Converter	Converter	Converter	Converter	Converter	Converter	Breeder
Energy conversion							
Gross thermal power, MW(th)	3,579	3,818	3,817	3,411	2,180	3,000	3,000
Net electrical power, MW(e)	1,178		1,287	1,150	638	1,160	1,200
Efficiency, %	32.9		33.7	33.7	29.3	38.7	40
Heat transport system	(12-1)	(12-6) (12-10b)	(12-6) (12-10b)	(12-6) (12-9) (12-10a)	(13-1)	(13-5) (13-7)	(15.4)[‡]
No. primary loops and pumps	2	4	4	4	2	6	4
No. of intermediate loops	–	–	–	–	–	–	8
No. steam generators	–	2	2	4	4	6	8
Steam generator type	–	Once-through	U-tube	U-tube	U-tube	Helical coil	Helical coil

	(1-7) (9-5)	(9-5) (12-11)	(1-8) (9-5) (12-11)	(12-11) (9-5)	(1-9)	(1-10) (9-7) (13-8)	(1-11)‡ (9-6)‡ (15-6)
Fuel §							
Particles							
Geometry	Cylindrical pellet	Cylindrical pellet	Cylindrical pellet	Cylindrical pellet	Cylindrical pellet	Coated microspheres	Cylindrical pellet
Dimensions, mm	$10.4\,D \times 10.4\,H$	$8.2\,D \times 9.5\,H$	$8.3\,D \times 9.9\,H$	$8.2\,D \times 13.5\,H$	$12.2\,D \times 16.4\,H$	$400\text{--}800\ \mu m\,D$	$0.70\,D$
Chemical form	UO_2	UO_2	UO_2	UO_2	UO_2	UC/ThO_2	PuO_2/UO_2
Fissile (1st core ave.), wt%	$1.7\ ^{235}U$	$2.79\ ^{235}U$	$2.38\ ^{235}U$	$2.6\ ^{235}U$	$0.711\ ^{235}U$	$93\ ^{235}U$	15–18 Pu
Fertile	^{238}U	^{238}U	^{238}U	^{238}U	^{238}U	Th	Depl. U
Pins							
Geometry	Pellet stack in clad tube	Pellet stack in clad tube	Pellet stack in clad tube	Pellet stack in clad tube	Pellet stack in clad tube	Cylindrical fuel stick	Pellet stack in clad tube
Dimensions, mm	$12.27\,D \times 4.1\ m\,H$	$9.6\,D \times 4\ m\,H$	$9.7\,D \times 4.1\ m\,H$	$9.5\,D \times 4\ m\,H$	$13.1\,D \times 490\,L$	$15.7\,D \times 62\,L$	$8.5\,D \times 2.7\ m\,H$ [C] $15.8\,D \times 1.95\ m\,H$ [BR]
Clad material	Zircaloy-2	Zircaloy-4	Zircaloy-4	Zircaloy-4	Zircaloy-4	Graphite	Stainless steel
Clad thickness, mm	0.813	0.6	0.64	0.57	0.42	–	0.7
Assembly							
Geometry	8 × 8-Square pin array	17 × 17-Square pin array	16 × 16-Square pin array	17 × 17-Square pin array	Concentric circles	Hexagonal graphite block	Hexagonal pin array
Pin pitch, mm	16.2	12.7	12.9	12.6	14.6	132 [SA]/ 76 [CA]*	9.7 [C]/17.0 [BR]
No. pin locations	64	289	256	289	37		271 [C]/
No. fuel pins	62	264	236	264	37		91 [BR]
Outer dimensions, mm	139	217	203	214	$102\,D \times 495\,L$	$360\,F \times 793\,H$	173 F
Channel	Yes	No	No	No	No	No	Yes
Total weight, kg	273	652					
Core	(12-5)	(9-11)	(9-11)	(9-11)		(13-9)	(15-7)‡
Axis	Vertical	Vertical	Vertical	Vertical	Horizontal	Vertical	Vertical

(*See footnotes on page 270.*)

TABLE IV-1
Typical Characteristics for Five Reference Power Reactor Types (*Continued*)

	BWR	PWR(B&W)	PWR(CE)	PWR(W)	PHWR	HTGR	LMFBR†
Core (*continued*)							
No. of assemblies							
Axial	1	1	1	1	12	8	1
Radial	748	241	241	193	380	493	364 [C] / 233 [BR]
Assembly pitch, mm	152	218	207		286	361	179
Active fuel height, m	3.81	3.63	3.81	3.66	5.94	6.30	1.0 [C] / 1.6 [C + BA]
Equivalent diameter, m	4.70	3.82	3.81	3.37	6.29	8.41	3.66
Total fuel weight, kg	156 UO_2	125 UO_2	117 UO_2	101 UO_2	98.4 AUO_2	1.72 U / 37.5 Th	32 MO_2
Performance							
Equilibrium burnup, MWD/T	27,500	33,000	34,400	27,500	7,500	95,000	100,000
Average assembly residence, full-power days					470	1,170	
Refueling							
Sequence	1/4 per y	1/3 per y	1/3 per y	1/3 per y	Continuous, on-line	1/4 per y	Variable
Outage time, d	30	30	21–35	30		14–20	32
Thermal-hydraulics							
Primary coolant							
Pressure, MPa	7.17	15.5	15.5	15.5	10.0	4.90	~0.1
Inlet temp., °C	278	301	296	292	267	318	395
Average outlet temp., °C	288	332	327	325	310	741	545
Core flow rate, Mg/s	13.1	20.1	20.7	18.0	7.6	1.42	16.4
Volume, l		4.02×10^5	3.30×10^5		1.20×10^5	(9550 kg)	(3200 Mg)

							Na/H_2O
Secondary coolant							
Pressure	—	7.83	7.38		4.7	17.2	~0.1/17.7
Inlet temp., °C	—	244	232		187	188	345/235
Outlet temp., °C	—	313	289		260	513	525/487
Power density							
Core average, kW/l	54.1	91.2	95.6	105	12	8.4	280
Fuel average, kW/l	54.1	91.2	95.6	105	60	44	280
Linear heat rate							
Core average, kW/m	19.0	16.0	17.5	17.8	25.7	7.87	29
Core maximum, kW/m	44.0	42.5	41.0	42.7	44.1	23.0	45
Design peaking factors							
Radial (Total)	1.4	1.55	1.55	(2.5)	1.21	(2.9)	(1.55)
Axial	1.6	1.67	1.47		1.41		
Moderator							
Volume, l	Same as primary coolant	Same as primary coolant	Same as primary coolant	Same as primary coolant	2.17×10^6	Graphite blocks	—
Inlet temp., °C					43		—
Outlet temp., °C					71		—
Reactivity control							
Control rods	(12-5)	(12-12)	(12-13) (12-14)	(12-12)	(13-4)		(15-8)‡
Geometry	Cruciform	Rod clusters	Rod clusters	Rod clusters	See Chap. 13	Rod pairs	Hexagonal bundles
No. drives	177	84	89	61		73	21 Primary
Absorber materials	B_4C	Ag–In–Cd	B_4C	Ag–In–Cd		B_4C/graphite	B_4C
Absorber length, m	3.66	3.6	3.8	3.6		6.35	1.3
Trip mechanism	Hydraulic	Electro-magnetic, gravity	Electro-magnetic, gravity	Electro-magnetic, gravity		Electro-static, gravity	Mechanical, gravity

(See footnotes on page 570.)

TABLE IV-1
Typical Characteristics for Five Reference Power Reactor Types (*Continued*)

	BWR	PWR(B&W)	PWR(CE)	PWR(W)	PHWR	HTGR	LMFBR[†]
Reactivity control (*continued*)							
Burnable poisons							
Composition	Gadolinia in fuel pellets	Al_2O_3/B_4C	Al_2O_3/B_4C	Borosilicate glass	—	B_4C/graphite	—
Number		3,552	1,792	1,400	—	22,540	—
Length, m		3.20	3.45		—		—
Other systems	Voids { 44% core average 79% exit average }	Soluble boron	Soluble boron	Soluble boron	See Chap. 13	Reserve shutdown	3-Bundle secondary
Reactor vessel	(12-2)			(12-7)	(13-2) (13-3)	(13-6)	(15-4)[‡]
Inside dimensions, m	6.05 D × 21.6 H	4.95 D × 13.1 H	4.68 D × 12.3 H	4.83 D × 13.4 H	7.6 D × 4 L	11.3 D × 14.4 H 4.72 m min	21 D × 19.5 H
Wall thickness, mm	152	255	216	224	28.6		25
Material	SS-clad carbon steel	SS-clad carbon steel	SS-clad carbon steel	SS-clad carbon steel	Stainless steel	Prestressed concrete	Stainless steel
Other features					Pressure tubes	Steel liners	Pool-type

[†] LMFBR–core [C], radial blanket [BR], axial blanket [BA].
[‡] LMFBR drawings not specific to the Novatome/Superphenix design.
[§] Fuel dimensions–diameter [D], height [H], length [L], (across the) flats [F], (width of) square [S].
[*] HTGR–standard assembly [SA], control assembly [CA].

GENERAL BIBLIOGRAPHY

*ACRS, 1980: "A Review of NRC Processes and Functions," Advisory Committee on Reactor Safeguards, NUREG-0642, January.

AECL, 1976: "CANDU 600," Atomic Energy of Canada Limited, Ottawa, Ontario, May.

Algermissen, S. T., 1969: "Seismic Risk Studies in the United States," Proceedings of the Fourth World Conference on Earthquake Engineering, Santiago, Chile.

AMA Council on Scientific Affairs, 1978: "Health Evaluation of Energy-Generating Sources," *Journal of the American Medical Association,* vol. 240, no. 20, November 10, pp. 2193–2195.

Anderson, T. D., 1971: "Offshore Siting of Nuclear Energy Stations," *Nuclear Safety,* vol. 12, no. 1, January–February, pp. 9–14.

APS, 1975: "Report to the American Physical Society by the Study Group on Light-Water Reactor Safety," *Reviews of Modern Physics,* vol. 47, supplement No. 1, Summer.

APS, 1978: "Report to the APS by the Study Group on Nuclear Fuel Cycles and Waste Management," *Reviews of Modern Physics,* vol. 50, no. 1, part II, January. (Summarized in "The Nuclear Fuel Cycle: An Appraisal," *Physics Today*, vol. 30, no. 10, October 1977, pp. 32–39.)

Archer, V. E., 1980: "Effects of Low-Level Radiation: A Critical Review," *Nuclear Safety,* vol. 21, no. 1, January–February, pp. 68–82.

Ash, M. 1979: *Nuclear Reactor Kinetics,* McGraw-Hill, New York.

ASME, 1978: "Non Proliferation: Reality and Illusion of a Plutonium-Free Economy," 18th Annual American Society of Mechanical Engineers Symposium, Albuquerque, NM, March 16–17.

Astrom, K. A., and K. J. Eger, 1978: "Spent Fuel Receipt and Storage at the Morris Operation General Electric Company," General Electric Company, NEDG-21889, June.

Atlantic Council of the United States, 1978: "Nuclear Power and Nuclear Weapons Proliferation," Westview Press, Boulder, CO.

Note: Reports of most work performed by U.S. government agencies, national laboratories, and other contractors (as well as some Canadian and other international sources) are available from: *National Technical Information Service [NTIS], U.S. Department of Commerce, 5285 Port Royal Road, Springfield, VA 22161*. Most reports prepared by the Electric Power Research Institute are available from NTIS as well as from: *Electric Power Research Institute, 3412 Hillview Avenue, Palo Alto, CA 94304.*

Such reports are indicated in this bibliography by an asterisk (*).

*Augustson, R. H., 1979: "DYMAC: A Dynamic Materials Accountability System for the LASL Plutonium Facility," Los Alamos Scientific Laboratory, LASL-79-6.

*Augustson, R. H., and T. D. Reilly, 1974: "Fundamentals of Passive Nondestructive Assay of Fissionable Material," Los Alamos Scientific Laboratory, LA-5651-M, September.

Azimov, I., and T. Dobzhansky, 1966: "The Genetic Effects of Radiation," U.S. Atomic Energy Commission Office of Information Services.

Babcock & Wilcox, 1975: "Babcock-205 Basic Training Course NSR-111," Babcock and Wilcox Company, Lynchburg, VA, October.

Babcock & Wilcox, 1978: "Babcock-205 NSS Design Summary," Babcock and Wilcox Company, Lynchburg, VA.

Babcock & Wilcox, 1979: "The Impact of TMI: Our Energy Future—Will Nuclear Play a Part?," Babcock and Wilcox Company, Lynchburg, VA, October.

Babcock & Wilcox, 1980: *Steam—Its Generation and Use*, Lynchburg, Virginia.

Bandtel, K. C. et al., 1980: "Proliferation Resistant Technology Assessment," Electric Power Research Institute, EPRI NP-1306, June.

*Barnes, C. H., L. Geller, and D. R. Hill, 1977: "Uranium Data," Electric Power Research Institute, EPRI-EA-400, June.

Barsell, A. W., V. Joksimovic, and F. A. Silady, 1977: "An Assessment of HTGR Accident Consequences," *Nuclear Safety*, vol. 18, no. 6, November–December, pp. 761–773.

*Barthold, W. P., et al., 1979: "Optimization of Radially Heterogeneous 1000-MW(e) LMFBR Core Configurations," Electric Power Research Institute, EPRI NP-1000, November.

Bebbington, W. P., 1976: "The Reprocessing of Nuclear Fuels," *Scientific American*, vol. 235, no. 6, December, pp. 30–41.

BEIR, 1972: "The Effect on Populations of Exposures to Low Level of Ionizing Radiation" (BEIR Report), National Academy of Sciences and National Research Council, November. (An updated study is complete and was published July 29, 1980.)

Bell, G. I., and S. Glasstone, 1970: *Nuclear Reactor Theory,* Van Nostrand and Reinhold Company, New York.

Benedict, M., and T. H. Pigford, 1957: *Nuclear Chemical Engineering,* McGraw-Hill, New York.

Benedict, M., T. H. Pigford, and H. W. Levi, 1981: *Nuclear Chemical Engineering,* 2nd ed., McGraw-Hill, New York.

*Benemann, J. R., 1978: "Biofuels: A Survey," Electric Power Research Institute, EPRI ER-746-SR, June.

Berman, I. M., 1980: "New Generating Capacity: When, Where and by Whom," *Power Engineering,* vol. 84, no. 4, April, pp. 70–78.

Bethe, H. A., 1976: "The Necessity of Fission Power," *Scientific American*, vol. 234, no. 1, January, pp. 21–31.

Bethe, H. A., 1979: "The Fusion Hybrid," *Physics Today,* vol. 32, no. 5, May, pp. 44–51.

Billington, D. S., and J. H. Crawford, 1961: *Radiation Damage in Solids,* Princeton University Press, Princeton, NJ, 1961.

Blake, E. B., 1980: "Fusion in the United States," *Nuclear News,* vol. 23, no. 8, June, pp. 59–67; vol. 23, no. 9, July, pp. 45–48; vol. 23, no. 10, August, pp. 64–68; vol. 23, no. 11, September, pp. 38–41.

*Blake, V. E., Jr., and F. A. Luetters, 1976: "Special Nuclear Materials Transportation System Goals," Sandia Laboratories, SAND76-9183, Albuquerque, NM.

*Booth, L. A., and T. G. Frank, 1977: "Commercial Applications of Inertial Confinement Fusion," Los Alamos Scientific Laboratory, LA-6838-MS, Los Alamos, NM, May.

Briggs, G. A., 1971: "Plume Rise: A Recent Critical Review," *Nuclear Safety,* vol. 12, no. 1, January–February, pp. 15–24.

Brockett, G. F., R. W. Shumway, J. O. Zane, and R. W. Griebe, 1972: "Loss of Coolant: Control of Consequences by Emergency Core Cooling," Proceedings of the 1972 International Conference on Nuclear Solutions to World Energy Problems, Washington, DC, November 13-17.

B-SAR-241 (Generic Safety Analysis Report for Pressurized Water Reactor), Babcock and Wilcox Company, Lynchburg, VA.

Buckley, S., 1979: *Sun Up to Sun Down*, McGraw-Hill, New York.

Bull. At. Sci.: *The Bulletin of the Atomic Scientists* (see description in Chap. 1 Selected Bibliography).

Burcham, W. E., 1963: *Nuclear Physics: An Introduction*, McGraw-Hill, New York.

*Carlson, G. A., K. R. Shultz, and A. C. Smith, Jr., 1979: "Definition and Conceptual Design of a Small Fusion Reactor," Electric Power Research Institute, EPRI ER-1045, April.

*Carmichael, A. D., 1979: "Ocean Thermal Energy Conversion: A State-of-the-Art Study," Electric Power Research Institute, EPRI ER-1113-SR, July.

Carpency, F. M., and R. C. Rittenhouse, 1980: "Power Plant Security: Meeting an Expanded Need," *Power Engineering*, vol. 84, no. 9, September, pp. 46–52.

*Carter, L. L., and E. D. Cashwell, 1975: "Particle Transport Simulation with the Monte Carlo Method," U.S. Energy Research and Development Administration, TID-26607.

Catron, J., 1980: "Putting Baseload to Work on the Night Shift," *EPRI Journal*, vol. 5, no. 3, April, pp. 6–13.

*Cazalet, E. G., C. E. Clark, and T. W. Keeling, 1978: "Costs and Benefits of Over/Under Capacity in Electric Power System Planning," Electric Power Research Institute, EPRI EA-927, October.

Cember, H., 1969: *Introduction to Health Physics*, Pergamon Press, New York.

CESSAR (Generic Safety Analysis Report for Pressurized Water Reactor), Combustion Engineering, Inc., Windsor, CT.

*Chang, Y. I., C. E. Till, R. R. Rudolph, J. R. Deen, and M. J. King, 1977: "Alternative Fuel Cycle Options: Performance Characteristics and Impact on Nuclear Power Growth Potential," Argonne National Laboratory, ANL-77-70, September.

Chen, F. F., 1974: *Introduction to Plasma Physics*, Plenum, New York.

*Chen, F. F., 1977: "Alternate Concepts in Controlled Fusion," Electric Power Research Institute, EPRI-ER-429-SR, May.

Chen, F. F., 1979: "Alternate Concepts in Magnetic Fusion," *Physics Today*, vol. 32, no. 5, May, pp. 36–42.

Civex, 1978: Starr, C.: "The Separation of Nuclear Power from Nuclear Proliferation"; Culler, F. L., Jr.: "Precedents for Diversion Resistant Fuel Cycles"; Levenson, M. and E. Zebroski: "A Fast Breeder Reactor Concept: A Diversion Resistant Fuel Cycle"; Flowers, R. H., K. D. B. Johnson, J. H. Miles, and R. K. Webster: "Possible Long-Term Options for the Fast Reactor Plutonium Fuel Cycle," Fifth Energy Technology Conference, Washington, D.C., February 27. (The first three papers are also contained in *Nucl. Eng. & Des.*, vol. 51, no. 2, January 1979.)

Clarke, J. F., 1980: "The Next Step in Fusion: What It Is and How It Is Being Taken," *Science*, vol. 210, November 28, pp. 967–972.

Clark, M. and K. F. Hansen, 1964: *Numerical Methods of Reactor Analysis*, Academic Press, New York.

*Clayton, E. D., 1979: "Anomalies of Nuclear Criticality," Battelle Pacific Northwest Laboratory, PNL-SA-4868 Rev. 5, June.

Cohen, B. L., 1974: *Nuclear Science and Society*, Anchor Books, Garden City, NY.

Cohen, B. L., 1977: "The Disposal of Radioactive Wastes from Fission Reactors," *Scientific American*, vol. 236, no. 6, June, pp. 21–31.

Cohen, B. L., and I. Lee, 1979: "A Catalog of Risks," *Health Physics*, vol. 36, June, pp. 707–722.

Collier, J. G., and L. M. Davies, 1980: "The Accident at Three Mile Island," *Heat Transfer Engineering*, vol. 1, no. 3, January–March, pp. 56–67.

Combustion Eng., 1978: "System 80: Nuclear Steam Supply System," Combustion Engineering Power Systems, Windsor, CT.

CONAES (Committee on Nuclear and Alternative Energy Systems), 1978*a*: "Problems of U.S. Uranium Resources and Supply to the Year 2010," National Academy of Sciences, Washington, DC.

CONAES (Committee on Nuclear and Alternative Energy Systems), 1978*b*: "Controlled Nuclear Fusion: Current Research and Potential Progress," National Academy of Sciences, Washington, DC.

CONAES (Committee on Nuclear and Alternative Energy Systems), 1979*a*: "Alternative Energy Demand Futures to 2010," National Academy of Sciences, Washington, DC.

CONAES (Committee on Nuclear and Alternative Energy Systems), 1979*b*: "U.S. Energy Supply Prospects to 2010," National Academy of Sciences, Washington, DC.

CONAES (Committee on Nuclear and Alternative Energy Systems), 1979*c*: "Geothermal Resources and Technology in the United States," National Academy of Sciences, Washington, DC.

CONAES (Committee on Nuclear and Alternative Energy Systems), 1979*d*: "Domestic Potential of Solar and Other Renewable Energy Sources," National Academy of Sciences, Washington, DC.

*CONF-740501, 1974: "Gas-Cooled Reactors: HTGR and GCFBR," Proceedings of the American Nuclear Society Topical Meeting, CONF-740501, Gatlinburg, TN, May 7–10.

*CONF-761001, 1976: "Proceedings of the International Meeting on Fast Reactor Safety and Related Physics," CONF-761001, Chicago, October 5–8.

Connolly, T. J., 1978: *Foundations of Nuclear Engineering,* Wiley, New York.

*COO-4057-6, 1978: "German Pebble Bed Reactor Design and Technology Review," U.S. Department of Energy, COO-4057-6, Washington, DC, September.

Coppi, B. and J. Rem, 1972: "The Tokamak Approach in Fusion Research," *Scientific American,* vol. 229, no. 2, July, pp. 61–75.

Cottrell, W. B., 1974: "The ECCS Rule-Making Hearing," *Nuclear Safety,* vol. 15, no. 1, January–February, pp. 30–54.

Cowan, G., 1976: "A Natural Fission Reactor," *Scientific American,* vol. 235, no. 1, July , pp. 36–47.

Cramer, E. N., 1973: "Evaluation of Floating Nuclear Plants Off the Coast of California," *Nuclear News,* vol. 16, no. 13, October, pp. 52–56.

*CRBRP-1, 1977: "CRBRP Risk Assessment Report," Clinch River Breeder Reactor Plant, (Summarized in Piper, H. B., et al.: "Overview of CRBRP Safety Study," Clinch River Breeder Reactor Plant, CRBRP-PMC-78-01, Winter 1978; and Piper et al.: "Clinch River Breeder Reactor Plant Safety Study," *Nuclear Safety,* vol. 19, no. 3, May–June 1978, pp. 316–329.)

*CRBRP-ARD-0230, 1978: "1978 CRBRP Technical Progress Report," Clinch River Breeder Reactor Plant.

CRBRP/PSAR, Clinch River Breeder Reactor Project, Preliminary Safety Analysis Report.

Crocker, J. G., 1980: "Fusion Reactor Safety," *Nuclear Safety* (to be published).

Crowley, J. H., P. L. Doan, and D. R. McCreath, 1974: "Underground Nuclear Plant Siting: A Technical and Safety Assessment," *Nuclear Safety,* vol. 15, no. 5, September–October, pp. 519–534.

Cybulskis, P., 1978: "Effect of Engineered Safety Features on the Risk of Hypothetical LMFBR Accidents," *Nuclear Safety,* vol. 19, no. 2, March–April, pp. 190–204.

*Dahlberg, R. C., R. F. Turner, and W. V. Goeddel, 1974: "HTGR Fuel and Fuel Cycle Summary Description," General Atomic Company, GA-A12801 (rev.).

*Davidovitz, P., et al., 1979: "Uranium Isotopic Separation by Aerodynamic Methods," Electric Power Research Institute, EPRI NP-1069, June.

*Davies, S. M., 1978: "Pool-Type LMFBR Plant, 1000 MWe Phase A Design," Electric Power Research Institute, EPRI NP-646, April.

Dayem, H. A., et al., 1979: "Dynamic Materials Accounting in the Back End of the Light Water Reactor Fuel Cycle," *Nuclear Technology,* vol. 43, Mid-April, pp. 222–243.

*DeBellis, R. J., and Z. A. Sabri, 1977: "Fusion Power: Status and Options," Electric Power Research Institute, EPRI-ER-510-SR, June.

Deffeyes, K. S., and I. D. MacGregor, 1980: "World Uranium Resources," *Scientific American,* vol. 242, no. 1, January, pp. 66–76.

Deitrich, J. R., 1977: "Safeguarding the Nuclear Fuel Cycle," *Nuclear Engineering International,* vol. 22, no. 264, November, pp. 57–61.

de la Garza, A., 1977: "Uranium-236 in Light Water Reactor Spent Fuel Recycled to an Enriching Plant," *Nuclear Technology,* vol. 32, February, pp. 176–185.

Dingee, D. A., 1979: "Fusion Power," *Chemical and Engineering News,* April 2, pp. 32–47.

*DOE/EDP-0061, 1979: "Environmental Development Plan Advanced Isotope Separation," U.S. Department of Energy, DOE/EDP-0061, May.

*DOE/EIS-0046-D, 1979: "Draft Environmental Impact Statement, Management of Commercially Generated Radioactive Waste," U.S. Department of Energy, DOE/EIS-0046-D, April. (Final report DOE/EIS-0046-F was issued October 1980.)

*DOE/ET-0089, 1979: "Fission Energy Program of the U.S. Department of Energy: FY 1980," U.S. Department of Energy, DOE/ET-0089, April.

Duderstadt, J. J., and L. J. Hamilton, 1976: *Nuclear Reactor Analysis,* Wiley, New York.

Duderstadt, J. J., and W. Martin, 1979: *Transport Theory,* Wiley, New York.

Dukert, J. M., 1975: "Atoms on the Move—Transportating Nuclear Material," U.S. Energy Research and Development Administration.

Easterly, C. E., K. E. Shank, and R. L. Shoup, 1977: "Radiological and Environmental Aspects of Fusion Power," *Nuclear Safety,* vol. 18, no. 2, March–April, pp. 203–215.

EEI (Edison Electric Institute), 1976: *Economic Growth in the Future,* McGraw-Hill, New

EEI (Edison Electric Institute), 1977: *The Transitional Storm . . . Riding It Out from One Energy Epoch to Another.* Edison Electric Institute, New York.

EEI (Edison Electric Institute), 1979: *Ethics and Energy,* Edison Electric Institute, Washington.

Eicholz, G. G., 1976: "Cost-Benefit and Risk Benefit Assessment for Nuclear Power Plants," *Nuclear Safety,* vol. 17, no. 5, September–October, pp. 525–539.

Eisenbud, M., 1973: *Environmental Radioactivity,* Academic Press, New York.

Elliott, D. M., and L. E. Weaver, 1972: *Education and Research in the Nuclear Fuel Cycle,* University of Oklahoma Press, Norman, OK.

El-Wakil, M. M., 1979: *Nuclear Heat Transport,* American Nuclear Society, Hinsdale, IL.

Emel'yanov, J. Ta., 1977: "On the Development of Nuclear Power," *Soviet Power Engineering,* vol. 15, no. 5, pp. 10–26. (English translation from *Izvestiya AN SSSR Energetika i transport.*)

Emmett, J. L., J. Nuckolls, and L. Wood, 1974: "Fusion Power by Laser Implosion," *Scientific American,* vol. 230, no. 6, June, pp. 24–37.

Engel, J. R., W. A. Rhoades, W. R. Grimes, and J. F. Dearing, 1979: "Molten-Salt Reactors for Efficient Utilization Without Plutonium Separation," *Nuclear Technology,* vol. 46, November, pp. 30–43.

*EPA, 1974: "Environmental Radiation Dose Commitment: An Application to the Nuclear Power Industry," U.S. Environmental Protection Agency, EPA-520/4-73-002, June (revised).

*EPA, 1975: "Reactor Safety Study (WASH-1400): A Review of the Draft Report," U.S. Environmental Protection Agency, EPA-520/3-75-012.

EPRI Journal (see description in Chap. 1 Selected Bibliography).

*ERDA-1, 1975: "Report of the Liquid Metal Fast Breeder Reactor Program Review Group," U.S. Energy Research and Development Administration, ERDA-1, January.

*ERDA-69, 1975: "The Environmental Impact of Electrical Power Generation: Nuclear and Fossil," U.S. Energy Research and Development Administration, ERDA-69.

*ERDA-76-1, 1976: "A National Plan for Energy Research, Development and Demonstration: Creating Energy Choices for the Future," U.S. Energy Research and Development Administration, ERDA-76-1.

*ERDA-76-43, 1976: "Alternatives for Managing Wastes from Reactors and Post-Fission Operations in the LWR Fuel Cycle," U.S. Energy Research and Development Administration, ERDA-76-43, May.

*ERDA-76-162, 1976: "The Management and Storage of Commercial Power Reactor Wastes," U.S. Energy Research and Development Administration, ERDA-76-162.

*ERDA-1535, 1975: "Final Environmental Statement, Liquid Metal Fast Breeder Reactor Program," U.S. Energy Research and Development Administration, ERDA-1535, December. (Succeeds "Proposed Final Environmental Statement, Liquid Metal Fast Breeder Reactor Program," U.S. Atomic Energy Commission, WASH-1535, December 1974.)

*ERDA-1541, 1975: "Light Water Breeder Reactor Program, Draft Environmental Statement," U.S. Energy Research and Development Administration, ERDA-1541, July.

*Erdmann, R. C., et al., 1979: "Status Report on the EPRI Fuel Cycle Accident Risk Assessment," Electric Power Research Institute, EPRI NP-1128, July.

ERG, 1976: "New Energy Sources: Dreams and Promises," Energy Research Group, Framingham, MA.

Etherington, H., 1958: *Nuclear Reactor Handbook,* McGraw-Hill, New York.

Evans, R. D., 1955: *The Atomic Nucleus,* McGraw-Hill, New York.

Farmer, F. R. (ed.), 1977: *Nuclear Reactor Safety,* Academic Press, New York.

Fauske, H. K., 1976: "The Role of Core Disruptive Accidents in Design and Licensing of LMFBRs," *Nuclear Safety,* vol. 17, no. 5, September–October, pp. 550–567.

Fisher, A., 1978: "What Are We Going to Do About Nuclear Waste?," *Popular Science,* December.

Foster, A. E., and R. L. Wright, Jr., 1977: *Basic Nuclear Engineering,* Allyn and Bacon, Boston.

Fowler, J. M., 1975: *Energy and the Environment,* McGraw-Hill, New York.

Freiwald, D. A., and T. G. Frank, 1975: "Introduction to Laser Fusion," Los Alamos Scientific Laboratory.

French Atomic Energy Commission, 1980: "The Creys–Malville Power Station" (translation of "Bulletin d'Informations Scientifiques et Techniques," No. 227, January 1978), April.

Frigerio, N. A., 1967: "Your Body and Radiation," U.S. Atomic Energy Commission Office of Information Services (revised).

*Garcia, A. A., and R. C. Erdmann, 1975: "Summary of the AEC Reactor Safety Study (WASH-1400)," Electric Power Research Institute, EPRI 217-2-1, April.

GASSAR (Generic Safety Analysis Report for High Temperature Gas-Cooled Reactor), General Atomic Company, San Diego, CA.

GCFR/PSAR, "Gas-Cooled Fast Breeder Reactor, Preliminary Safety Analysis Report," General Atomic Company, San Diego, CA.

GE/BWR-6, 1978: "BWR-6: General Description of a Boiling Water Reactor," General Electric Company, San Jose, CA.

GE/Chart, 1977: "Chart of the Nuclides," General Electric Company, San Jose, CA. (Available as booklet or 130 cm × 75 cm wall chart.)

*General Atomic, 1976: "HTGR Accident Initiation and Progression Analysis Status Report," General Atomic Company, GA-A13617, San Diego, CA, January.

General Electric, 1979: "General Description, IF 300, Irradiated Fuel Shipping Cask," General Electric Company, NEDO-10864-2, San Jose, CA, July.

*GFHT, Gesellschaft fur Hochtemperaturreaktor–Technik mbH, 1976: "Development Status and Operational Features of the High Temperature Gas-Cooled Reactor," Electric Power Research Institute, EPRI NP-142, April.

GESSAR (Generic Safety Analysis Report for Boiling Water Reactor), General Electric Company, San Jose, CA.

Gibbs & Hill, 1980: "Economic Comparison of Coal and Nuclear Electric Power Generation," Gibbs & Hill, Inc., January (distributed by Atomic Industrial Forum, Washington, DC).

Ginoux, J. J. (ed.), 1978: *Two-Phase Flows and Heat Transfer with Application to Nuclear Reactor Design Problems,* Hemisphere, Washington, DC.

Glaser, P. E., 1977: "Solar Power from Satellites," *Physics Today,* vol. 32, no. 2, February, pp. 30–38.

Glasstone, S., 1974: "Controlled Nuclear Fusion," U.S. Atomic Energy Commission Office of Information Services.

Glasstone, S., and R. H. Loveberg, 1960: *Controlled Thermonuclear Fusion,* D. Van Nostrand, New York.

Glasstone, S., and A. Sesonske, 1967: *Nuclear Reactor Engineering,* D. Van Nostrand Reinhold, New York. (New edition available, 1981.)

Glasstone, S., and W. H. Jordan, 1980: *Nuclear Power and Its Environmental Effects,* American Nuclear Society, LaGrange Park, IL.

Goldstein, H., 1959: *Fundamental Aspects of Reactor Shielding,* Addison-Wesley, Reading, MA.

Gough, W. C., and B. J. Eastlund, 1971: "The Prospects for Fusion Power," *Scientific American,* vol. 224, no. 2, February.

GPU, 1979: Keaton, R. W., T. Van Witbeck, T. G. Broughton, and P. S. Walsh: "Analysis of the

TMI-2 Sequence of Events and Operator Response"; Long, R. L., T. M. Crimmins, and W. W. Lowe: "Three Mile Island Accident Technical Support"; and Wilson, R. F., J. W. Thiesing, and R. W. Heward, Jr.: "Plan for Recovery of TMI-2," General Public Utilities, unpublished papers presented at the American Nuclear Society Winter Meeting, San Francisco, CA, November 12.

Graves, H. W., Jr., 1979: *Nuclear Fuel Management,* Wiley, New York.

Gray, J. E., et al., 1978: "International Cooperation on Breeder Reactors," The Rockefeller Foundation, New York, May.

Greenspan, H., C. N. Kelber, and D. Okrent, 1968: *Computing Methods in Reactor Physics,* Gordon and Breach Science Publishers, New York.

Greenwood, T., G. W. Rathjens, and J. Ruina, 1977: "Nuclear Power and Weapons Proliferation," Adelphi Papers, no. 130, International Institute for Strategic Studies, London.

Gregory, C. V., et al., 1979: "Natural Circulation Boosts Safety of Pool-Type Fast Reactor," *Nucl. Eng. Int.,* vol. 24, no. 293, December, pp. 29–33.

Grella, A. W., 1977: "A Review of the Department of Transportation (DOT) Regulations for Transportation of Radioactive Materials," U.S. Department of Transportation, Washington, DC, October.

Hafemeister, D. W., 1979: "Nonproliferation and Alternate Nuclear Technologies," *Technology Review,* vol. 81, no. 3, December/January, pp. 58–62.

Hailey, A., 1979: *Overload,* Doubleday, New York.

Haley, J., 1980: "Nuclear Waste Management–An Overview," *Nuclear Digest,* Westinghouse Electric Corporation, pp. 17–24.

Hammond, A. L., et al., 1973: *Energy and the Future,* American Association for the Advancement of Science, New York.

Hammond, R. P., 1979: "Nuclear Wastes and Public Acceptance," *American Scientist,* vol. 67, March–April, pp. 146–150.

*Hanson, J. E., 1978: "Safety Related Fuel Pin Design," Hanford Engineering Development Laboratory, HEDL-SA-1443, February.

*Harmon, K. M., 1979: "Summary of National and International Radioactive Waste Management Programs 1979," Pacific Northwest Laboratory, PNL-2941, March.

Hartline, B. K., 1980: "Tapping Sun-Warmed Ocean Water for Power," *Science,* vol. 209, August 15, pp. 794–796.

Hayes, E. T., 1979: "Energy Resources Available to the United States, 1985 to 2000," *Science,* vol. 203, January 19.

*Haywood, L. R., 1976: "The CANDU Power Plant," Atomic Energy of Canada Limited, Ottawa, Ontario, AECL-5321, January.

*Heising, C. D., and T. J. Connolly, 1978: "Analyzing the Reprocessing Decision: Plutonium Recycle and Nuclear Proliferation," Electric Power Research Institute, EPRI NP-931, November.

Hendrie, J. M., 1977: "How the NRC Is Working for a More Up-to-Date Licensing Process," *Nuclear Engineering International,* vol. 22, no. 264, November, pp. 51–52.

Henry, H. R., 1969: *Fundamentals of Radiation Protection,* Wiley-Interscience, New York.

Henry, A. F., 1975: *Nuclear Reactor Analysis,* MIT Press, Cambridge, MA.

*Henry, K. H., and D. A. Turner, 1978: "Storage of Unreprocessed Fuel (SURF)," Rockwell Hanford Operations, RHO-SA-40, March.

*Herrington, P. B. (ed.), 1979: "Class Notes from the First International Training Course on the Physical Protection of Nuclear Facilities and Materials," Sandia Laboratories, SAND79-1090, May.

Hetrick, D. L., 1971: *Dynamics of Nuclear Reactors,* University of Chicago Press, Chicago.

HEW, 1970: "Radiological Handbook," U.S. Department of Health, Education and Welfare, U.S. Government Printing Office.

Hiebert, R., and R. Hiebert, 1970: "Atomic Pioneers, Book 1, From Ancient Greece to the 19th Century," U.S. Atomic Energy Commission Office of Information Services.

Hiebert, R., and R. Hiebert, 1974: "Atomic Pioneers, Book 2, From the Mid-19th to the Early 20th Century," U.S. Atomic Energy Commission Office of Information Services (revised).

Hiebert, R., and R. Hiebert, 1973: "Atomic Pioneers, Book 3, From the Late 19th to the Early 20th Century," U.S. Atomic Energy Commission Office of Information Services.

*Hinchley, E., 1975: "On-Line Control of the CANDU-PHW Power Distribution," Atomic Energy of Canada Limited, AECL-5045, Ottawa, Ontario, March.

Hogerton, J. F., 1970: "Nuclear Reactors," U.S. Atomic Energy Commission Office of Information Services (revised).

Holden, A. N., 1958: *Physical Metallurgy of Uranium,* Addison-Wesley, Reading, MA.

Holdren, J. P., 1978: "Fusion Energy in Context: Its Fitness for the Long Term," *Science*, vol. 200, April 14, pp. 168–200.

*Honeck, H. C./ENDF, 1966: "ENDF/B–Specifications for an Evaluated Nuclear Data File for Reactor Applications," Brookhaven National Laboratory, BNL-50066.

Hopkinson, J., 1979: "Does the United States Waste Energy?," *EPRI Journal,* vol. 4, no. 9, November, pp. 26–32.

Hopkinson, J., 1980: "Quality of Life: An International Comparison," *EPRI Journal,* vol. 5, no. 3, April, pp. 26–31.

*Hubbard, E. L., 1978: "Doublet III," General Atomic Company, GA-A15026, San Diego, CA, May.

*Hughes, D. J., et al./BNL-325, 1955: "Neutron Cross Sections," Brookhaven National Laboratory, BNL-325. (Revisions and additions in 1960, 1964, and 1965.)

*Hutchins, B. A., 1975: "Denatured Plutonium–A Study of Deterrent Action," Electric Power Research Institute, EPRI-310, July.

Hyde, E. K., 1964: *The Nuclear Properties of the Heavy Elements III: Fission Phenomena,* Prentice-Hall, Englewood Cliffs, NJ.

IAEA, 1976: "IAEA Safeguards Technical Manual: Introduction and Part A–Safeguards Objectives Criteria and Requirements," IAEA-174, International Atomic Energy Agency, Vienna, Austria.

IAEA, 1977: *International Atomic Energy Agency Bulletin* (International Safeguards Issue), vol. 19, no. 5, Vienna, Austria, October.

IAEA, 1978: "Nuclear Safeguards Technology," Proceedings of an IAEA Symposium, Vienna, October 2–6.

ICRP/26, 1977: International Commission on Radiological Protection: "Recommendations," ICRP no. 26, Pergamon Press, Elmsford, NY. (Summarized in "Recommendations of the International Commission on Radiological Protection," *Nuclear Safety,* vol. 20, no. 3, May–June 1979, pp. 330–341.)

IEEE, 1979: "Special Issue: Three Mile Island and the Future of Nuclear Power," IEEE *Spectrum*, vol. 16, no. 11, November, pp. 29–111: (a) "An Analysis of Three Mile Island," pp. 32–57; (b) "Nuclear Power and the Public Risk," pp. 59–79; (c) "Institutional Constraints," pp. 81–95; and (d) "International Outlook," pp. 96–109.

INFCE/SEC/11, 1979: "IAEA Contribution to INFCE: The Present Status of IAEA Safeguards on Nuclear Fuel Cycle Facilities," International Atomic Energy Agency, Vienna, Austria, INFCE/SEC/11, February 1.

INFCE, 1980: "Report of the First Plenary Conference of the International Nuclear Fuel Cycle Evaluation (INFCE)," International Atomic Energy Agency, Vienna, Austria.

INFCIRC/66, 1968: "The Agency's Safeguards System (1965, As Provisionally Extended in 1966 and 1968)," International Atomic Energy Agency, INFCIRC/66/Rev. 2, September 16.

INFCIRC/153, 1971: "The Structure and Content of Agreements Between the Agency and States Required in Connection with the Treaty on the Non-Proliferation of Nuclear Weapons," International Atomic Energy Agency, INFCIRC/153, May.

INFCIRC/225, 1976: "The Physical Protection of Nuclear Material," International Atomic Energy Agency, INFCIRC/225/Rev. 1, June.

*Inhaber, H., 1978: "Risk of Energy Production," Atomic Energy Control Board, Ottawa, Ontario, Canada, AECB-1119 (Rev. 1), May. (Summarized in Inhaber, H.: "Risk with Energy from Conventional and Nonconventional Sources," *Science*, vol. 203, February 23, 1979, pp. 718–723.)

*IRG, 1979: "Report to the President by the Interagency Review Group on Nuclear Waste Management," U.S. Department of Energy, TID-29442, March.

*Jefferson, R. M., and H. R. Yoshimura, 1977: "Crash Testing of Nuclear Fuel Shipping Containers," Sandia Laboratories, SAND77-1462, December.

Johnson, W. D., Jr., 1977: "The Prospects for Photovoltaic Conversion," *American Scientist,* vol. 65, November–December, pp. 729–736.

*Jones, O. E., 1975: "Advanced Physical Protection Systems for Nuclear Materials," Sandia Laboratories, SAND75-5351, October.

Kammash, T., 1975: *Fusion Reactor Physics: Principles and Technology,* Ann Arbor Science, Ann Arbor, MI.

Kaplan, I. 1963: *Nuclear Physics,* Addison-Wesley, Reading, MA.

Keeny, S. M., et al., 1977: *Nuclear Power Issues and Choices, Report of the Nuclear Energy Policy Study Group,* Ballinger, New York.

Keepin, G. R., 1965: *Physics of Nuclear Kinetics,* Addison-Wesley, Reading, MA.

Keepin, G. R., 1978: "Safeguards Implementation in the Nuclear Fuel Cycle," *Nuclear Materials Management,* vol. 7, no. 3, Fall, pp. 44–58.

Keepin, G. R., (in press): "Classnotes from the International Training Course on Nuclear Materials Accountability and Control for Safeguards Purposes," Los Alamos Scientific Laboratory.

Kelly, B. T., 1966: *Irradiation Damage to Solids,* Pergamon, Elmsford, NY.

Kemeny, J., et al., 1979: "The Need for Change: The Legacy of TMI," Report of the President's Commission on the Accident at Three Mile Island, U.S. Government Printing Office, Washington, D.C., October. (Also in paperback and hard cover by Pergamon, Elmsford, NY, 1979.)

Kendall, H. W., et al., 1977: "The Risks of Nuclear Power Reactors–A Review of the NRC Reactor Safety Study, WASH-1400," Union of Concerned Scientists, Cambridge, MA, August.

Kenton, J., 1977: "Capturing a Star: Controlled Fusion Power," *EPRI Journal,* December, pp. 6–13.

Knief, R. A. (ed.), 1981: "A Basic Textbook on Nuclear Criticality Safety–Lectures from a Short Course Sponsored by the University of New Mexico in May 1979" (tentative title, to be published).

Konstantinov, L., 1976: "Commercial Reactors in the USSR," Argonne Center for Educational Affairs, Argonne, IL, November 2.

*Koplik, C. M., et al., 1979: "Status Report on Risk Assessment for Nuclear Waste Disposal," Electric Power Research Institute, EPRI NP-1197, October.

Kovan, D., 1978: "Safeguarding a Plutonium Industry," *Nuclear Engineering International,* vol. 23, no. 275, August, pp. 41–45.

Krall, N. A., and A. W. Trivelpiece, 1973: *Principles of Plasma Physics,* McGraw-Hill, New York.

Kramer, A. W., (no date): *Nuclear Energy–What It Is–How It Acts, Power Engineering Magazine,* Barrington, IL.

Kratzer, M. B., 1980: "Prospective Trends in International Safeguards," *Nuclear News,* vol. 23, no. 12, October, pp. 56–60.

*Krause, F. R., C. A. Trauth, J. B. Sorensen, and B. D. Zak, 1977: "Potential Environmental Policy Constraints on Nuclear Fuel Development Programs," Sandia Laboratories, SAND77-1337, October.

Kreith, F., and J. F. Kreider, 1978: *Principles of Solar Engineering,* Hemisphere, Washington, DC.

Lamarsh, J. R., 1966: *Nuclear Reactor Theory,* Addison-Wesley, Reading, MA.

Lamarsh, J. R., 1975: *Introduction to Nuclear Engineering,* Addison-Wesley, Reading, MA.

Lapedes, D. N. (ed.), 1976: *McGraw-Hill Encyclopedia of Energy,* McGraw-Hill, New York.

Lapp, R., 1976: *America's Energy,* Reddy Communications, Greenwich, CT.

Lapp, R. E., 1977: *Radioactive Waste: Society's Problem Child,* Reddy Communications, Inc., Greenwich, CT.

Lapp, R. E., 1979: *The Radiation Controversy,* Reddy Communications, Inc., Greenwich, CT.

*LA-79-29, 1979: "Laser Fusion Program at Los Alamos," Los Alamos Scientific Laboratory, LA-79-29.

Lave, L. B., and L. C. Freeburg, 1973: "Health Effects of Electricity Generation from Coal, Oil, and Nuclear Fuel," *Nuclear Safety,* vol. 14, no. 5, September–October, pp. 409–428.

Leach, L., and G. McPherson, 1980: "LOFT Nuclear Tests," *Nuclear Safety,* vol. 21, no. 4, July–August, pp. 461–468.

Leach, L. P., and L. J. Ybarrondo, 1978: "LOFT Emergency Core-Cooling System Experiments: Results from L1-4 Experiment," *Nuclear Safety,* vol. 19, no. 1, January–February, pp. 43–49.

Lederer, C. M., and V. S. Shirley (eds.), 1978: *Table of Isotopes,* 7th ed., Wiley, New York.

LeDoux, J. C., and C. Rehfuss, 1978: "The NRC Program of Inspection and Enforcement," *Nuclear Safety,* vol. 19, no. 6, November–December, pp. 671–680.

*Leverenz, F. L., and R. C. Erdmann, 1975: "Critique of WASH-1400," Electric Power Research Institute, EPRI 217-2-3, June.

*Leverenz, F. L., Jr., and R. C. Erdmann, 1979: "Comparison of the EPRI and Lewis Committee Review of the Reactor Safety Study," Electric Power Research Institute, EPRI NP-1130, July.

Levine, S., 1978: "The Role of Risk Assessment in the Nuclear Regulatory Process," *Nuclear Safety,* vol. 19, no. 5, September–October, pp. 556–564.

Lewin, J., 1977: "The Russian Approach to Nuclear Reactor Safety," *Nuclear Safety,* vol. 18, no. 4, July–August, pp. 438–450.

Lewins, J., 1978: *Nuclear Reactor Kinetics and Control,* Pergamon, New York.

Lewis, E. E., 1977: *Nuclear Power Reactor Safety,* Wiley-Interscience, New York.

*Lewis, H., et al., 1978: "Risk Assessment Review Group Report to the U.S. Nuclear Regulatory Commission," U.S. Nuclear Regulatory Commission, NUREG/CR-0400, September. (Summary, Findings, and Recommendations in "Report of the NRC Risk Assessment Review Group on the Reactor Safety Study," *Nuclear Safety,* vol. 20, no. 1, January–February, 1979).

Lewis, H. W., 1980: "The Safety of Fission Reactors," *Scientific American,* vol. 242, no. 3, March, pp. 53–65.

*Lichtblau, J. H., and H. J. F. Frank, 1978: "Outlook for World Oil Into the 21st Century," Electric Power Research Institute, EPRI EA-745, May.

Linebarger, J. H., D. L. Batt, and V. T. Berta, 1979: "LOFT Isothermal and Nuclear Experiment Results," 14th Intersociety Energy Conversion Engineering Conference, Boston, August 5–10, volume II, pp. 1518–1521.

Lish, K. C., 1972: *Nuclear Power Plant Systems and Equipment,* Industrial Press, Inc., New York.

Litai, D., 1980: "A Risk Comparison Methodology for the Assessment of Acceptable Risk," Ph.D. thesis, Massachusetts Institute of Technology.

*LLL Fusion, 1977: "Laser Fusion Program," Lawrence Livermore Laboratory, UCRL-52000-77-8, August.

*Lomenick, T. F., 1970: "Earthquakes and Nuclear Power Plant Design," Oak Ridge National Laboratory, ORNL-NSIC-28.

Long, J. T., 1978: *Engineering for Nuclear Fuel Reprocessing,* American Nuclear Society, Hinsdale, IL.

*Lotts, A. L., and J. H. Coob, 1976: "HTGR Fuel and Fuel Cycle Technology," Oak Ridge National Laboratory, ORNL/TM-5501, August.

Lovett, J. E., 1974: *Nuclear Materials—Accountability Management Safeguards,* American Nuclear Society, Hinsdale, IL.

Lovins, A. B., 1977: *Soft Energy Paths: Toward a Durable Peace,* Ballinger, Cambridge, MA.

Marshall, W., 1978: "Nuclear Power and the Proliferation Issue," Graham Young Memorial Lecture, University of Glasgow, February 24. (Reprinted in *Combustion,* June 1978.)

*McBride, J. P., R. E. Moore, J. P. Witherspoon, and R. E. Blanco, 1978: "Radiological Impact of Airborne Effluents of Coal-Fired and Nuclear Power Plants," Oak Ridge National Laboratory, ORNL-5315. (Summarized in an article of similar title by the same authors in *Science,* vol. 202, December 8, 1978, pp. 1045–1050.)

*McCloskey, D. J., et al., 1977: "Protection of Nuclear Power Plants Against Sabotage," Sandia Laboratories, SAND77-0116C, October.

*McCormick-Barger, M. L., 1979: "Experiment Data Report for LOFT Power Ascension Test L2-2," EG&G Idaho, NUREG/CR-0492, TREE-1322, February.

*McDonald, J. S., and S. Golan, 1978: "Engineering Aspects of the Pool Type LMFBR, 1000 MWe," Electric Power Research Institute, EPRI NP-645-SY, April.

*McElroy, J. L., and R. E. Burns, 1979: "Nuclear Waste Management Status and Recent Accomplishments," Electric Power Research Institute, EPRI NP-1087, May.

McIntyre, M. C., 1975: "Natural Uranium Heavy-Water Reactors," *Scientific American,* vol. 233, no. 4, October, pp. 17–27.

*McNelly, M. J., and H. E. Williamson, 1977: "Study of the Developmental Status and Operational Features of Heavy Water Reactors," Electric Power Research Institute, EPRI NP-365, February.

Meghreblian, R. V., and D. K. Holmes, 1960: *Reactor Analysis,* McGraw-Hill, New York.

Met. Ed., 1979: "Entry and Decontamination of the Reactor Containment Building at Three Mile Island Unit 2," Metropolitan Edison Company, Report Number Four, Reading, PA, August 27.

Meyer, W., S. K. Loyalka, W. E. Nelson, and R. W. Williams, 1977: "The Homemade Bomb Syndrome," *Nuclear Safety,* vol. 18, no. 4, July–August, pp. 427–438.

*Mihalka, M., 1979: "International Arrangements for Uranium Enrichment," Rand Corporation, R-2427-DOE, September.

Mining Eng., 1974: *Society of Mining Engineers,* August (Uranium Issue).

Moeller, D. W., 1977: "Current Challenges in Air Cleaning at Nuclear Facilities," *Nuclear Safety,* vol. 18, no. 5, September–October, pp. 633–646.

Morgan, K. Z., 1978: "Cancer and Low Level Ionizing Radiation," *The Bulletin of the Atomic Scientists,* vol. 34, no. 7, September, pp. 30–40.

Murakami, M., and H. P. Eubank, 1979: "Recent Progress in Tokamak Experiments," *Physics Today,* vol. 32, no. 5, May, pp. 25–32.

Murray, R. L., 1975: *Nuclear Energy,* Pergamon, Elmsford, NY (2nd ed. available 1980).

NAE, 1975: "Problems of U.S. Coal Resources and Supply to the Year 2000," National Academy of Engineering, Washington, DC.

*NASAP, 1979: "Nuclear Proliferation and Civilian Nuclear Power: Report on the Nonproliferation Alternative Systems Assessment Program," U.S. Department of Energy, DOE-NE-0001, December (draft).

NCRP/39, 1971: "Basic Radiation Protection Criteria," NCRP report number 39, National Council on Radiation Protection and Measurements, Washington.

Negin, C. A., 1979: "Regulatory Turbulence in the Wake of Three Mile Island," *Nuclear Engineering International,* vol. 24, no. 291, October, pp. 41–43.

Nero, A. V., Jr., 1979: *A Guidebook to Nuclear Reactors,* University of California Press, Berkeley, CA.

Nesbit, W., 1979: *World Energy—Will There Be Enough in 2020?,* Edison Electric Institute, Washington.

Nesbit, W., 1980: "Going With the Wind," *EPRI Journal,* vol. 5, no. 2, March, pp. 6–17.

Nevada, 1980: "Nevada Nuclear Waste Storage Investigations," U.S. Department of Energy, Nevada Operations Office, Las Vegas, NV, February.

*Ney, J. F., 1976: "Nuclear Safeguards," Sandia Laboratories, SAND76-5519, June.

*Notz, K. J., 1976: "An Overview of HTGR Fuel Recycle," Oak Ridge National Laboratory, ORNL-TM-4747, January.

NRC, 1974: "Regulatory Guide 1.3: Assumptions Used for Evaluating the Potential Radiological Consequences of a Loss of Coolant Accident for Boiling Water Reactors" and "Regulatory Guide 1.4: Assumptions Used for Evaluating the Potential Radiological Consequences of a Loss of Coolant Accident for Pressurized Water Reactors," U.S. Nuclear Regulatory Commission, June (rev. 2).

NRC, 1979: "NRC Statement on Risk Assessment and the Reactor Safety Study Report (WASH-1400) in Light of the Risk Assessment Review Group Report," U.S. Nuclear Regulatory Commission, January 18.

*NSAC-1, 1979: "Analysis of Three Mile Island–Unit 2 Accident," Nuclear Safety Analysis Center, operated by the Electric Power Research Institute, NSAC-1, July; plus NSAC-1 Supplement, October 1979.

Nucl. Eng. & Des.: *Nuclear Engineering and Design* (see description in Chap. 1 Selected Bibliography); special topic issues referenced in the text are:
January 1974–High-Temperature Gas-Cooled Power Reactors
January 1977–Gas-Cooled Fast Breeder Reactor Engineering and Design
February 1980–Fuel Element Performance Computer Modeling

Nucl. Eng. Int.: *Nuclear Engineering International* (see description in Chapter 1 Selected Bibliography); special topic issues referenced in the text are:

June 1968–Winfrith SGHWR
April 1970–Oconee PWR
June 1970–Pickering CANDU-PHW
July 1971–Phenix LMFBR
August 1971–PFR/Dounreay LMFBR
November 1971–Heysham AGR
August 1972–FFTF LMFBR
November 1972–Gentilly 1 CANDU–BLW
November 1973–Douglas Point BWR
June 1974–Survey of Canada
July 1974–Phenix LMFBR
August 1974–Fulton HTGR
October 1974–Clinch River LMFBR
March 1975–Economics
June/July 1975–LMFBR Prototypes
June/July 1975–BN 600 LMFBR
September 1975–Fessenheim PWR
November 1975–Snupps PWR
July 1976–Kalkar/SNR-300 LMFBR
September 1976–Forsmark 3 BWR
November 1976–Waste Management
December 1976–Survey of France
February 1977–Doublet/Theta-Pinch Experiments
March 1977–JET Fusion Experiment
May 1977–Creys–Malville/Super Phenix LMFBR
May 1977–LMFBR Heat Exchangers
June 1977–Point Lepreau CANDU-PHW
November 1977–Survey of the United States

December 1977–Cherokee–Perkins PWR
January 1978–Final Disposal of Waste
March 1978–High Current Fusion
May 1978–Security and Emergency Services
June 1978–Creys–Malville/Super Phenix LMFBR
August 1978–Reprocessing and Recycling
September 1978–Trillo PWR
November 1978–Joyo LMFBR
November 1978–Nuclear Costs
November 1978–Uranium Resources
December 1978–GCFR Prospects
December 1978–Uranium Reporting
January 1979–LMFBR Cooperation
April 1979–Fuel Management
May 1979–Rasmussen/Lewis
July 1979–LMFBR Progress
August 1979–Fugen HWR
October 1979–Regulations and Licensing
November 1979–Alternative Fuel Cycles
December 1979–Economics
February 1980–Gösgen PWR
March 1980–Operator Training
April 1980–Uranium Supply and Fuel Production
August 1980–Reactor Control and Instrumentation
October 1980–Enrichment
November 1980–Reactor Safety/Reliability
December 1980–Transport of Nuclear Materials

Nucl. Ind.: *Nuclear Industry* (see description in Chap. 1 Selected Bibliography).
Nucl. Mat. Man.: *Nuclear Materials Management* (see description in Chap. 20 Selected Bibliography).
Nucl. News: *Nuclear News* (see description in Chap. 1 Selected Bibliography).
Nucl. News, 1978: "Supreme Court Hands Down Landmark Decision," *Nuclear News*, vol. 21, no. 7, May, pp. 30–32.
Nucl. News, 1979: "The Ordeal at Three Mile Island," *Nuclear News* Special Report, April 6, pp. 1–6. Also several articles in *Nuclear News*, vol. 22, no. 7, May, pp. 32–40.
Nucl. Safety: *Nuclear Safety* (see description in Chap. 1 Selected Bibliography).
Nucl. Sci. Eng.: *Nuclear Science and Engineering* (see description in Chap. 1 Selected Bibliography).
Nucl. Tech.: *Nuclear Technology* (see description in Chap. 1 Selected Bibliography).
Nucl. Tech., 1974: Waste Management Symposium Issue, *Nuclear Technology*, vol. 24, no. 3, December.

*NUREG-0404, 1978: "Generic Environmental Impact Statement on Handling and Storage of Spent Light-Water Reactor Fuel," U.S. Nuclear Regulatory Commission, NUREG-0404. (Summarized in *Nuclear Safety*, vol. 20, no. 1, January–February 1979, pp. 54–62.)

*NUREG-0558, 1979: "Population Dose and Health Impact of the Accident at the Three Mile Island Nuclear Station," U.S. Nuclear Regulatory Commission, NUREG-0558, May.

*NUREG-0585, 1979: "TMI-2 Lessons Learned Task Force Final Report," U.S. Nuclear Regulatory Commission, NUREG-0585, October.

*NUREG-0600, 1979: "Investigation into the March 28, 1979 Three Mile Island Accident by Office of Inspection and Enforcement," Office of Inspection and Enforcement, U.S. Nuclear Regulatory Commission, NUREG-0600, August.

*NVO-210, 1980: "Safety Assessment Document for the Spent Reactor Fuel Geologic Storage Test in the Climax Granite Stock at the Nevada Test Site," U.S. Department of Energy, Nevada Operations Office, NVO-210, January.

Nye, J. S., Jr., 1979: "Balancing Nonproliferation and Energy Security," *Technology Review*, vol. 81, no. 3, December/January, pp. 48–57.

Oceanus, 1977: "High Level Nuclear Wastes in the Seabed?," *Oceanus*, vol. 20, no. 1, Winter.

*O'Dell, R. D. (ed.), 1974: *Nuclear Criticality Safety*, U.S. Atomic Energy Commission, TID-26286.

O'Donnell, E. P., 1979: "The Need for a Cost-Benefit Perspective in Nuclear Regulatory Policy," *Nuclear Engineering International*, vol. 24, no. 291, October, pp. 47–51.

O'Donnell, E. P., and J. J. Mauro, 1979: "A Cost-Benefit Comparison of Nuclear and Nonnuclear Health and Safety Protective Measures and Regulations," *Nuclear Safety*, vol. 20, no. 5, September–October, pp. 525–540.

Okrent, D., 1979: "Risk-Benefit Evaluation for Large Technological Systems," *Nuclear Safety*, vol. 20, no. 2, March–April, pp. 148–164.

Okrent, D., 1980: "Comment on Societal Risk," *Science*, vol. 208, April. pp. 372–375.

*Olander, D. R., 1976: *Fundamental Aspects of Nuclear Reactor Fuel Element*, U.S. Energy Research and Development Administration, TID-26711-P1.

Olander, D. R., 1978: "The Gas Centrifuge," *Scientific American*, vol. 239, no. 2, August, pp. 37–43.

Olds, F. C., 1978: "Fusion Power Developments," *Power Engineering*, vol. 82, no. 11, November, pp. 46–56.

Olds, F. C., 1979*a*: "Outlook for Breeder Reactors," *Power Engineering*, vol. 84, no. 3, March, pp. 58–66.

Olds, F. C., 1979*b*: "The Impressive Super Phenix," *Power Engineering*, vol. 84, no. 8, August, pp. 62–66.

Olds, F. C., 1979*c*: "Outlook for Nuclear Power," *Power Engineering*, vol. 84, no. 10, October, pp. 66–70.

Olds, F. C. 1979*d*: "Coal Resources and Outlook," *Power Engineering*, vol. 84, no. 10, October, pp. 71–78.

Olds, F. C., 1980*a*: "Post TMI Plant Designs," *Power Engineering*, vol. 84, no. 8, August, pp. 54–62.

Olds, F. C., 1980*b*: "Spent Fuel Storage Facility," *Power Engineering*, vol., 84, no. 9, September, pp. 64–67.

Olds, F. C., 1980*c*: "INFCE: The Promise, the Findings, and the Final Frustration," *Power Engineering*, vol. 84, no. 11, November, pp. 52–60.

Onega, R. J., 1975: *An Introduction to Fission Reactor Theory*, University Publications, Blacksburg, VA.

*ORNL-4451, 1970: "Siting of Fuel Reprocessing Plants and Waste Management Facilities," Oak Ridge National Laboratory, ORNL-4451.

*ORNL-4782, 1972: "Molten-Salt Reactor Program," Oak Ridge National Laboratory, ORNL-4782, October.

OTA, 1977: "Nuclear Proliferation and Safeguards," U.S. Office of Technology Assessment, April.

Parkins, W. E., 1978: "Engineering Limitations of Fusion Power Plants," *Science*, vol. 199, March 31, pp. 1403–1408.

Perla, H. F., 1973: "Power Plant Siting Concepts for California," *Nuclear News*, vol. 16, no. 13, October, pp. 47–51.

Phys. Today: Physics Today (see description in Chap. 1 Selected Bibliography).

*PMC-XX-YY, 19XX: "Technical Review: Clinch River Breeder Reactor Plant," Clinch River Breeder Reactor Plant, PMC-XX-YY or CRBRP-PMC-XX-YY (progress reports 19XX, number YY).

Power (see description in Chap. 1 Selected Bibliography).

Power Eng.: Power Engineering (see description in Chap. 1 Selected Bibliography).

*Prassinos, P. G., B. M. Galusha, and D. B. Engleman, 1979: "Experiment Data Report for LOFT Power Ascension Experiment L2-3," EG&G Idaho, NUREG/CR-0792, TREE-1326, July.

Profio, A. E., 1979: *Radiation Shielding and Dosimetry,* Wiley-Interscience, New York.

Pruce, L. M., 1979a: "Let the Sun Shine " *Power,* vol. 123, no. 5, May, pp. 33–37.

Pruce, L. M., 1979b: "Using Geothermal Energy for Power," *Power,* vol. 123, no. 10, October, pp. 37–43.

Pruce, L. M., 1980: "Power From Water," *Power,* vol. 124, no. 4, April, pp. S1–S16.

*Reeder, D. L., 1978: "LOFT System and Test Description (5.5 ft. Nuclear Core I LOCE)," EG&G Idaho, NUREG/CR-0247, TREE-1208, July.

Reeder, D. L., and V. T. Berta, 1979: "The Loss-of-Fluid Test (LOFT) Facility," 14th Intersociety Energy Conversion Engineering Conference, Boston, August 5–10, volume II, pp. 1512–1517.

Reilly, T. D., & M. L. Evans, 1977: "Measurement Reliability for Nuclear Material Assay . . . ," *Nuclear Materials Management,* vol. 6, no. 2, Summer, pp. 41–46.

RESSAR (Generic Safety Analysis Report for Pressurized Water Reactor), Westinghouse Electric Corporation, Pittsburgh, PA.

Rippon, S., 1974: "The Rasmussen Study on Reactor Safety," *Nuclear Engineering International,* vol. 19, no. 223, December, pp. 1001–1071.

Rippon, S., 1980: "France Forging Ahead on Fast Reactor," *Nuclear News,* vol. 23, no. 14, November, pp. 99–104.

Rittenhouse, R. C., 1980: "Solar Power and Conservation: Helping to Carry the Load," *Power Engineering,* vol. 84, no. 3, March, pp. 38–48.

Robertson, J. A. L., 1969: *Irradiation Effects in Nuclear Fuels,* Gordon and Breach Science Publishers, New York.

*Robertson, R. C., 1971: "Conceptual Design Study of a Single Fluid Molten-Salt Breeder Reactor," Oak Ridge National Laboratory, ORNL-4541, June.

Rogers, J. T., 1979: "CANDU Moderator Provides Ultimate Heat Sink in a LOCA," *Nuclear Engineering International,* vol. 24, no. 280, January, pp. 38–41.

*Rogovin, M., and G. T. Frampton, Jr., 1980: "Three Mile Island: A Report to the Commissioners and to the Public," U.S. Nuclear Regulatory Commission, January.

*Rose, R. P., 1979: "Design Study of a Fusion-Driven Tokamak Hybrid Reactor for Fissile Fuel Production," Electric Power Research Institute, EPRI ER-1083, May.

Rose, D. J., and M. Clark, Jr., 1961: *Plasmas and Controlled Fusion,* MIT Press, Cambridge, MA.

Rose, D. J., and M. Feirtag, 1976: "The Prospect for Fusion," *Technology Review,* vol. 79, no. 2, December, pp. 20–43.

Rose, D. J., and R. K. Lester, 1978: "Nuclear Power, Nuclear Weapons and International Stability," *Scientific American,* vol. 238, no. 4, April, pp. 45–57.

Ross, M., 1980: "Efficient Use of Energy Revisited," *Physics Today,* vol. 31, no. 5, February, pp. 24–31.

Rossin, A. D., 1980: "The Soft Energy Path—Where Does It Really Lead?," *The Futurist,* June, pp. 57–63.

Rossin, A. D., and T. A. Rieck, 1978: "Economics of Nuclear Power," *Science,* vol. 201, August 18.

Rotblat, J. 1978: "The Risks for Radiation Workers," *The Bulletin of the Atomic Scientists,* vol. 34, no. 7, September, pp. 41–46.

*Rudisill, C. L., 1977: "Coal and Nuclear Generating Costs," Electric Power Research Institute, EPRI PS-455-SR, April. (Summarized in Rudisill, C. L.: "Coal and Nuclear Cost Comparison," *EPRI Journal,* vol. 2, no. 8, October 1977, pp. 14–17.)

Rudman, R. L., and C. G. Whipple, 1980: "Time Lag of Energy Innovation," *EPRI Journal,* vol. 5, no. 3, April, pp. 14–20.

Rydin, R. A., 1977: *Nuclear Reactor Theory and Design,* University Publications, Blacksburg, VA.

Safety Analysis Reports [SAR]: Preliminary Safety Analysis Reports [PSAR] and Final Safety Analysis Reports [FSAR] required for each reactor licensed by the U.S. Nuclear Regulatory Commission. (Generic SARs for standardized plants include B-SAR-241, CESSAR, GASSAR, GESSAR, and RESSAR, listed elsewhere in this bibliography.)

Sagan, L. G., 1974: *Human and Ecologic Effects of Nuclear Power Plants,* Thomas, Springfield, IL.

*SAND77-0274, 1977: "WIPP Conceptual Design Report," Sandia Laboratories, SAND77-0274, June.

*SAND77-1401, 1978: "Hydrologic Investigations of the Los Medanos Area, Southeastern New Mexico," Sandia Laboratories, SAND77-1401, January.

*SAND79-0182, 1979: "Sandia Irradiator for Dried Sewage Solids," Sandia Laboratories, SAND79-0182, February.

Sandia, 1980: "Particle Beam Fusion," Sandia National Laboratories, Albuquerque, NM, January.

*Schaeffer, N. M., 1973: *Reactor Shielding for Nuclear Engineers,* U.S. Atomic Energy Commission, TID-25951.

Schumacher, E. F., 1973: *Small Is Beautiful,* Harper & Row, New York.

Schwarz, H. J., and H. Hora, 1974: *Laser Interaction and Related Phenomena,* vol. 3, Plenum Press, New York.

Schwarz, H. J., and H. Hora, 1977: *Laser Interaction and Related Phenomena,* vol. 4, Plenum Press, New York.

Sci. Am.: Scientific American (see description in Chap. 1 Selected Bibliography).

Scientific American, 1971: *Energy and Power,* W. H. Freeman, San Francisco, CA.

Science: see description in Chap. 1 Selected Bibliography.

Seaborg, G. T., and J. L. Bloom, 1970: "Fast Breeder Reactors," *Scientific American,* vol. 223, no. 5, November, pp. 13–21.

*Sesonske, A., 1973: *Nuclear Power Plant Design Analysis,* U.S. Atomic Energy Commission, TID-26241.

*Shapiro, N. L., J. R. Rec, and R. A. Matzie, 1977: "Assessment of Thorium Fuel Cycles in Pressurized Water Reactors," Electric Power Research Intitute, EPRI-NP-359, February.

Simnad, M. T., 1971: *Fuel Element Experience in Nuclear Power Reactors,* Gordon and Breach Science Publishers, New York.

*Slade, D. H. (ed.), 1968: "Meteorology and Atomic Energy," U.S. Atomic Energy Commission, TID-24190, July.

Smith, C. B., 1976: *Efficient Electricity Use: A Handbook for an Energy-Constrained World,* Pergamon Press, Elmsford, NY.

Smith, D. R., and A. M. Valentine, 1979: "Ex-Reactor Fuel Cycle Safety," Los Alamos Scientific Laboratory, June.

Smith, D. B., and I. G. Waddoups, 1976: "Safeguarding Nuclear Materials and Plants," *Power Engineering,* vol. 80, no. 11, November.

*Snyder, H. J., J. R. Lindgren, M. Peeks, and W. Jung, 1978: "Status of Development of GCFR Core Assemblies," General Atomic Company, GA-A15032, San Diego, CA, June.

Stacy, W. M., Jr., 1969: *Space-Time Nuclear Kinetics,* Academic Press, New York.

Starr, C., 1969: "Social Benefits vs. Technological Risk," *Science,* vol. 168, September 19, pp. 1232–1238.

Steiner, D., and J. F. Clarke, 1978: "The Tokamak: Model T Fusion Reactor," *Science,* vol. 199, March 31, pp. 1395–1403.

Stickley, C. M., 1978: "Laser Fusion," *Physics Today,* vol. 31, no. 5, May, pp. 50–58.

Stiefel, M., 1979: "Soft and Hard Energy Paths: The Roads Not Taken," *Technology Review,* vol. 82, no. 1, October, pp. 56–66.

Stobaugh, R., and D. Yergin, 1979: "After the Second Shock: Pragmatic Energy Strategies," *Foreign Affairs,* Spring.

Stoller, S. M., and R. B. Richards (eds.), 1961: *Reactor Handbook–Volume II: Reprocessing,* 2nd ed., Interscience, New York.

Strauss, S. D., 1979: "U.S. Nuclear Focus: Closing the Fuel Cycle and Beefing It Up," *Power,* vol. 123, no. 3, March, pp. 37–41.

Symposium, 1978: Proceedings of the Fifth International Symposium on Packaging and Transportation of Radioactive Materials, Las Vegas, Nevada, May 7–12. (Sponsored by Sandia Laboratories, Albuquerque, NM.)

Tech. Rev.: Technology Review (see description in Chap. 1 Selected Bibliography).

Technology Review, 1979: "Nuclear Power: Can We Live with It?," *Technology Review,* vol. 81, no. 7, June/July, pp. 33–47.

10CFR, 1979: *Code of Federal Regulations: 10 Energy, Parts 0 to 199,* Office of the Federal Register, 1979. (Paperbound edition updated annually; loose-leaf version updated with periodic supplements.)

Thompson, T. J., and J. G. Beckerley, 1964: *The Technology of Nuclear Reactor Safety, Volume 1: Reactor Physics and Control,* MIT Press, Cambridge, MA.

Thompson, T. J., and J. G. Beckerley, 1973: *The Technology of Nuclear Reactor Safety, Volume 2: Reactor Materials and Engineering,* MIT Press, Cambridge, MA.

*Tong, L. S., 1972: *Boiling Crisis and Critical Heat Flux,* U.S. Atomic Energy Commission, TID-25887.

*Tonnessen, K. A., and J. J. Cohen, 1977: "Survey of Naturally Occurring Hazardous Materials in Deep Geological Formations: A Perspective on Relative Hazard of Deep Burial of Nuclear Wastes," Lawrence Livermore Laboratory, UCRL-52199, January 14.

Trans. Am. Nucl. Soc.: Transactions of the American Nuclear Society (see description in Chap. 1 Selected Bibliography).

U. Oklahoma, 1975: "Energy Alternatives," University of Oklahoma, Norman, OK.

U.S. Govt., 1977: "Nuclear Proliferation Handbook," Committee on International Relations, U.S. House of Representatives and Committee on Governmental Affairs, U.S. Senate, U.S. Government Printing Office, Washington, DC, September 23.

*van Erp, J. B., 1977: "Preliminary Evaluation of Licensing Issues Associated with U.S.-Sited CANDU–PHW Nuclear Power Plants," Argonne National Laboratory, ANL-77-97, December.

*Varnado, S. G., J. L. Mitchner, and G. Yonas, 1977: "Civilian Applications of Particle-Beam-Initiated Inertial Confinement Fusion Technology," Sandia Laboratories, SAND77-0516, May.

Vendryes, G. A., 1977: "Superphenix: A Full-Scale Breeder Reactor," *Scientific American*, vol. 236, no. 3, March, pp. 26–35.

Veziroglu, T. N., 1978: *Alternative Energy Sources, Volume 5: Nuclear Energy,* Hemisphere, Washington.

Ward, J. E., 1978: "The Need for Licensing Reform: A Technical Perspective," *Nuclear News,* vol. 21, no. 5, April, pp. 53–58.

*WASH-740, 1957: "Theoretical Possibilities and Consequences of Major Accidents in Large Nuclear Power Plants," U.S. Atomic Energy Commission, WASH-740.

*WASH-1101, 1973: "Liquid Metal Fast Breeder Reactor: LMFBR Program Plan," U.S. Atomic Energy Commission, WASH-1101.

*WASH-1174-74, 1974: "The Nuclear Industry, 1974," U.S. Atomic Energy Commission, WASH-1174-74.

*WASH-1222, 1972: "Evaluation of the Molten Salt Breeder Reactor," U.S. Atomic Energy Commission, WASH-1222, September.

*WASH-1248, 1974: "Environmental Survey of the Nuclear Fuel Cycle," U.S. Atomic Energy Commission, WASH-1248.

*WASH-1250, 1973: "The Safety of Nuclear Power Reactors (Light-Water Cooled) and Related Facilities," U.S. Atomic Energy Commission, WASH-1250.

*WASH-1297, 1974: "High Level Radioactive Waste Alternatives," U.S. Atomic Energy Commission, WASH-1297.

*WASH-1400, 1975: "Reactor Safety Study: An Assessment of Accident Risks in U.S. Commercial Nuclear Power Plants," U.S. Nuclear Regulatory Commission, WASH-1400 (NUREG-74/014).

*WASH-1535, 1974: "Proposed Final Environmental Statement, Liquid Metal Fast Breeder Reactor Program," U.S. Atomic Energy Commission, WASH-1535, December.

*Weart, W. D., 1977: "A Radioactive Waste Pilot Plant," *Sandia Technology,* vol. 3, no. 1, SAND77-0701, March.

Weinberg, A. M., and E. P. Wigner, 1958: *The Physical Theory of Neutron Chain Reactors,* The University of Chicago Press, Chicago.

Weisman, J., and L. Eckart, 1978: "Fuel Rod Design," Nuclear Fuel Cycle Education Committee, NFCEC-1, CES-S1, University of Cincinnati, Cincinnati.

Westinghouse, 1975: "PWR Information Course," Westinghouse Electric Corporation, Pittsburgh, PA, April.

Westinghouse, 1979a: "Summary Description of Westinghouse Pressurized Water Reactor Nuclear Steam Supply System," Westinghouse Water Reactor Division, Pittsburgh, PA.

Westinghouse, 1979b: "Spent Fuel Handling and Packaging Program," Westinghouse Advanced Energy Program Report, EM4D Facility, Jackass Flats, NV, January.

*Wett, J. F., 1978: "Large Pool LMFBR Design," Electric Power Research Institute, EPRI NP-883, August.

Whipple, C. 1980: "The Energy Impacts of Solar Heating," *Science,* vol. 208, April 18, pp. 262–266.

Whitaker, R., 1980: "Tapping the Mainstream of Geothermal Energy," *EPRI Journal,* vol. 5, no. 4, May, pp. 6–15.

Wilkinson, W. D., and W. F. Murphey, 1958: *Nuclear Reactor Metallurgy,* D. Van Nostrand, New York.

Williams, J. R., 1974: "Solar Energy: Technology and Applications," Ann Arbor Science Publishers, Ann Arbor, MI.

*Williams, D. C., and B. Rosenstrock, 1978: "A Review of Nuclear Fuel Cycle Alternatives Including Certain Features Pertaining to Weapons Proliferation," Sandia Laboratories, SAND77-1727, January.

Willrich, M., and T. B. Taylor, 1974: *Nuclear Theft: Risks and Safeguards,* Ballinger, Cambridge, MA.

Wilson, C. L., 1977: *Energy: Global Prospects 1985-2000, Report of the Workshop on Alternative Energy Strategies (WAES),* McGraw-Hill, New York. (Wilson, 1977a.)

Wilson, R., 1977: "How to Have Nuclear Power Without Weapons Proliferation," *The Bulletin of the Atomic Scientists,* vol. 33, no. 9, November, pp. 39–44. (Wilson, 1977b).

Yadigoraglo, G., and S. O. Anderson, 1974: "Novel Siting Solutions for Nuclear Power Plants," *Nuclear Safety,* vol. 15, no. 6, November–December, pp. 651–664.

*Yonas, G., J. R. Freeman, A. J. Toepher, T. H. Martin, K. R. Prestwich, and P. A. Miller, 1976: "Particle Beam Fusion," *Sandia Technology,* vol. 2, no. 3, SAND76-0615, October.

Yonas, G., 1978: "Fusion Power with Particle Beams," *Scientific American,* vol. 239, no. 5, November, pp. 50–61.

Zaleski, P., 1976: *Nuclear Energy Maturity, Proceedings of April 1975 Conference: Volume 12: Enrichment and Fusion,* Academic Press, New York.

Zare, R. N., 1977: "Laser Separation of Isotopes," *Scientific American,* vol. 236, no. 2, February, pp. 86–98.

Zebroski, E., and M. Levenson, 1976: "The Nuclear Fuel Cycle," *Annual Review of Energy,* vol. 1, pp. 101–130.

Zweifel, P. F., 1973: *Reactor Physics,* McGraw-Hill, New York.

LATE ENTRIES

*Levenson, M., and F. Rahn, 1980: "Realistic Estimates of the Consequences of Nuclear Reactor Accidents," Electric Power Research Institute, November.

Spiewak, I., and J. N. Barkenbus, 1980: "Nuclear Proliferation and Nuclear Power: A Review of the NASAP and INFCE Studies," *Nuclear Safety,* vol. 21, no. 6, November–December, pp. 691–702.

INDEX

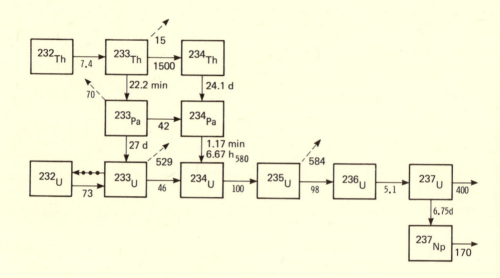

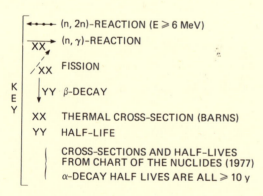

K
E
Y

• • • • → (n, 2n)–REACTION (E ⩾ 6 MeV)

$\overline{XX}$ → (n, γ)–REACTION

╱XX   FISSION

↓YY   β–DECAY

XX   THERMAL CROSS–SECTION (BARNS)

YY   HALF–LIFE

CROSS–SECTIONS AND HALF–LIVES
FROM CHART OF THE NUCLIDES (1977)
α–DECAY HALF LIVES ARE ALL ⩾ 10 y